ANNUAL REVIEW OF NUCLEAR AND PARTICLE SCIENCE

ANNUAL REVIEW OF NUCLEAR AND PARTICLE SCIENCE

J. D. JACKSON, *Editor*
University of California, Berkeley

HARRY E. GOVE, *Associate Editor*
University of Rochester

ROY F. SCHWITTERS, *Associate Editor*
Harvard University

VOLUME 30

1980

ANNUAL REVIEWS INC.   4139 EL CAMINO WAY   PALO ALTO, CALIFORNIA 94306

ANNUAL REVIEWS INC.
Palo Alto, California, USA

REPRINTS   The conspicuous number aligned in the margin with the title of
each article in this volume is a key for use in ordering reprints. Available
reprints are priced at the uniform rate of $1.00 each postpaid. The minimum
acceptable reprint order is 5 reprints and/or $5.00 prepaid. A quantity discount
is available.

International Standard Serial Number: 0163-8998
International Standard Book Number: 0-8243-1530-8
Library of Congress Card Number: 53-995

FILMSET BY TYPESETTING SERVICES, LTD, GLASGOW, SCOTLAND
PRINTED AND BOUND IN THE UNITED STATES OF AMERICA

SOME RELATED ARTICLES IN OTHER
ANNUAL REVIEWS

Annual Review of Nuclear and Particle Science
Volume 30, 1980

CONTENTS

Ann. Rev. Nucl. Part. Sci. 1980. 30 : 1–52

THE PARITY NON-CONSERVING ELECTRON-NUCLEON INTERACTION ✖5613

E. D. Commins and P. H. Bucksbaum·

Department of Physics, University of California, and Materials and Molecular Research Division, Lawrence Berkeley Laboratory, Berkeley, California 94720

CONTENTS

1   INTRODUCTION

It is now known that parity is violated in the electron-nucleon interaction $e+N \rightarrow e+N$. The results of diverse experiments, employing scattering of high energy polarized electrons and low energy atomic spectroscopy, imply that e and N (or in more basic terms e and quark q) engage in a neutral weak coupling, in addition to their much more powerful electromagnetic interaction. This neutral weak (eq) coupling now takes its place (see Table 1) with the observed neutral weak interactions (vq) and (ve), knowledge of which is by now quite extensive. Yet to be observed in their own right are the neutral weak couplings (ee) or (qq). Nevertheless the experimental facts from the vq, ve, and eq sectors are already sufficient to provide a stringent test of theoretical models, and the evidence is strongly in favor of the unified theory of weak and electromagnetic interactions proposed by Glashow,

1

0163-8998/80/1201-0001$01.00

Table 1 Neutral weak interactions

	Neutrino	Electron	Footnote	Quark	Footnote
Neutrino	$\nu + \nu \to \nu + \nu$	$\nu_\mu(\bar{\nu}_\mu) + e \to \nu_\mu(\bar{\nu}_\mu) + e$ $\bar{\nu}_e + e \to \bar{\nu}_e + e$	(a) (b)	$\left.\begin{array}{l}\nu_\mu(\bar{\nu}_\mu) + N \to \nu_\mu(\bar{\nu}_\mu) + X, \\ \to \nu_\mu(\bar{\nu}_\mu) + N + \text{Meson}\end{array}\right\}$ $\bar{\nu}_e + D \to \bar{\nu}_e + n + p$	(c) (d)
Electron		$e + e \to e + e$ $e^+ + e^- \to \mu^+ + \mu^-$	(g) (h)	$e + N \to e + X$ $e + N \to e + N$	(e) (f)
Quark				$N + N \to N + N$ $p + p \to Z + \cdots$ $p + \bar{p} \to Z + \cdots$	(i) (j)

Reactions a–f have been observed. Reactions e and f are the subject of this article.
a Scattering of high energy $\nu_\mu(\bar{\nu}_\mu)$ by e^-.
b Scattering of low energy (reactor) $\bar{\nu}_e$ by e^-. This process also occurs by coupling of charged weak currents.
c High energy neutrino-nucleon scattering (Cline & Fry 1977).
d Low energy reactor $\bar{\nu}_e$-deuteron scattering (Pasierb et al 1979).
e High energy polarized electron scattering (see Section 3).
f Atomic physics (see Section 4).
g This reaction can also cause small parity violation effects in atoms.
h Interference between photon and Z^0 exchange (parity conserving) affects the angular distribution of $\mu^+\mu^-$.
i Parity violation in nuclear forces. The charged weak currents also contribute.
j Z production in high energy pp, $p\bar{p}$ collisions.

Weinberg, and Salam. In the remainder of this section we present general ideas concerning the neutral weak eq interaction. Then we describe the salient features of the Weinberg-Salam model (Section 2), discuss in detail the principles and methods of the SLAC polarized electron scattering experiment (Section 3) and atomic physics experiments (Section 4), and summarize neutral weak interaction results and their implications (Section 5).

Unlike the (vq) and (ve) interactions, the (eq) interaction is described by an amplitude containing an electromagnetic as well as a weak portion: $A = A_{EM} + A_W$, with $A_{EM} \gg A_W$ for all cases of interest in this review. In each (eq) process, the transition probability, proportional to $|A|^2$, thus contains an interference term $\sim 2A_W A_{EM}$ in addition to the dominant term $|A_{EM}|^2$ and the very small term $|A_W|^2$ (which we neglect). If the weak interaction violates parity, the interference term contains a pseudoscalar portion A_{WP}, the sign of which depends on the "handedness" of the system on which the experiment is performed. When the rates for a process $I_\pm$ in the two coordinate frames of opposite handedness are compared, one obtains an asymmetry

$$\Delta = \frac{I_+ - I_-}{I_+ + I_-} \approx \frac{A_{WP}}{A_{EM}}. \qquad 1.$$

In the case of high energy electron scattering at SLAC, the handedness is defined by the helicity of the longitudinally polarized electron beam. We may estimate Δ crudely in this case by employing the Feynman diagrams for the electromagnetic and neutral weak interactions in lowest order (Figure 1).[1] The former proceeds by photon exchange, with $A_{EM} = 4\pi\alpha/q^2$.

The latter is presumed to be mediated by a neutral vector boson Z_0. Even in theories less specific than the Weinberg-Salam model that attempt to establish a connection between weak and electromagnetic interactions, its

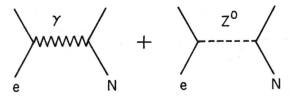

Figure 1 Feynman diagrams for photon (γ) and neutral intermediate boson (Z^0) exchange between e and N.

[1] We adhere to the relativistic conventions employed by Bjorken & Drell (1964); q is the invariant four momentum transfer from electron to nucleon, and $q^2 < 0$ for spacelike q. Also $\alpha = e^2/4\pi\hbar c \simeq 1/137$; we employ units $\hbar = c = 1$ unless otherwise noted, and the Fermi coupling constant is $G = 1.02 \times 10^{-5} \, m_p^{-2}$, where m_p is the proton mass.

mass should be $m_Z \sim (\pi\alpha/G)^{1/2}$. *For $|q^2| \ll m_Z^2$ the weak amplitude should be of order G.* Assuming that parity is violated nearly maximally for the neutral weak interactions as it is for charged weak interactions, we find

$$\Delta = \frac{\sigma_+ - \sigma_-}{\sigma_+ + \sigma_-} \approx A_W/A_{EM} \approx Gq^2/4\pi\alpha \sim q^2/m_Z^2. \qquad 2.$$

For $|q^2| \approx (1 \text{ GeV})^2$ this yields $\Delta \approx 10^{-4}$, an estimate originally given by Zel'dovich (1959). According to a more detailed analysis presented in Section 3, the electron scatters from individual quark-partons in the nucleus (Cahn & Gilman 1978). The weak and electromagnetic contributions from each quark are coherent and interfere, but the sum over all quarks is incoherent. The asymmetry predicted on the basis of the Weinberg-Salam model, and actually obtained in experiment (Prescott et al 1978, 1979), is still of order 10^{-4}.

One employs the same amplitudes (Figure 1) in low energy atomic physics, but they are described in somewhat different language. Photon exchange between atomic electron and nucleus is expressed by the ordinary atomic Hamiltonian H_0, whose eigenstates, the usual atomic states, may be separated into two classes $|\psi_n^0\rangle$, $|\chi_m^0\rangle$ of opposite parity. Exchange of the massive Z^0 is described by an effective zero-range potential H' of order G, which contains scalar and pseudoscalar parts: $H' = H_S + H_P$ (Bouchiat & Bouchiat 1974a,b, 1975). H_S is in principle observable (it leads to energy shifts), but these are so small that they cannot be separated from effects due to H_0 in any known practical experiment because of small uncertainties in H_0. H_P is a significant perturbation on H_0, however, since it causes a state $|\psi^0\rangle$ to be admixed with states $|\chi^0\rangle$ of opposite parity:

$$|\psi^0\rangle \rightarrow |\psi\rangle = |\psi^0\rangle + \sum_n \frac{|\chi_n^0\rangle \langle \chi_n^0 | H_P | \psi^0\rangle}{E(\psi^0) - E(\chi_n^0)}. \qquad 3.$$

Let us consider how the electronic and nucleonic currents contribute to the pseudoscalar portion of the Hamiltonian H_P. A priori these currents may possess scalar (S), vector (V), tensor (T), axial vector (A) and pseudoscalar (P) components. Thus, neglecting momentum-transfer-dependent terms we may write

$$H_P = \sum_{V,A,S,P,T} Gc_k(\bar{\psi}_N \Gamma_k \psi_N)(\bar{\psi}_e \Gamma_k \gamma_5 \psi_e),$$

where the c_k are coefficients to be determined. However, the term corresponding to $\Gamma_k = \gamma_5$ (P term) vanishes in the nonrelativistic nucleon limit. Also it can be shown that the S and T terms are time reversal odd as well and lead to a permanent electric dipole moment of the atom (Hinds et al 1976, Sandars 1968). Experiments to search for linear Stark effect in Xe,

Cs, and Tl place an upper limit on the coefficients c_S and c_T of less than 10^{-3} (Hinds et al 1976, Bouchiat 1975, Gould 1970). Furthermore, it can be shown that the S, T, and P terms yield no interference with electromagnetism in high energy electron scattering where quark and electron masses can be neglected. Thus, we shall assume that the electronic and nucleonic weak neutral currents contain only vector and axial vector components:

$$J_e = V_e + A_e \qquad J_N = V_N + A_N.$$

Then, we may write $H_P = H_P^{(1)} + H_P^{(2)}$, where $H_P^{(1)}$, $H_P^{(2)}$ arise from the combinations $A_e V_N$ and $V_e A_N$, respectively. Ignoring momentum-transfer-dependent terms, we have:

$$H_P^{(1)} = \frac{G}{\sqrt{2}} \sum_i \bar{\psi}_e \gamma_\lambda \gamma_5 \psi_e (C_{1p} \bar{\psi}_{pi} \gamma^\lambda \psi_{pi} + C_{1n} \bar{\psi}_{ni} \gamma^\lambda \psi_{ni}) \qquad 4.$$

$$H_P^{(2)} = \frac{G}{\sqrt{2}} \sum_i \bar{\psi}_e \gamma_\lambda \psi_e (C_{2p} \bar{\psi}_{pi} \gamma^\lambda \gamma_5 \psi_{pi} + C_{2n} \bar{\psi}_{ni} \gamma^\lambda \gamma_5 \psi_{ni}), \qquad 5.$$

where the sum is taken over all protons (p) and neutrons (n) in the nucleus (Feinberg & Chen 1974).

The coupling coefficients $C_{1p}, C_{1n}, C_{2p}, C_{2n}$ are model dependent and must be determined by experiment. As is demonstrated in Section 2, the Weinberg-Salam model (Weinberg 1967, Salam 1968) predicts

$$C_{1p} = \tfrac{1}{2}(1 - 4 \sin^2\theta) \qquad\qquad\qquad 6a.$$

$$C_{1n} = -\tfrac{1}{2} \qquad\qquad\qquad 6b.$$

$$C_{2p} = \frac{g_A}{2}(1 - 4 \sin^2\theta) \qquad\qquad\qquad 6c.$$

$$C_{2n} = -\frac{g_A}{2}(1 - 4 \sin^2\theta), \qquad\qquad\qquad 6d.$$

where $g_A = 1.25$ is the axial vector coupling constant of neutron beta decay, and θ is the Weinberg angle (Weinberg 1972). Experimentally, $\sin^2\theta = 0.228 \pm 0.009$ (Langacker et al. 1979). Assuming Equations 6 and performing a nonrelativistic reduction of the nucleonic currents, we obtain from Equation 4 the effective electronic weak Hamiltonian corresponding to $A_e V_N$:

$$H_W^{(1)} = \frac{G}{2\sqrt{2}} Q_W \rho_N(\mathbf{r}) \gamma_5 \qquad 7.$$

where $Q_W = (1 - 4 \sin^2\theta)Z - N$ and $\rho_N(\mathbf{r})$ is the nuclear density, $\mathbf{r}$ being the electron position. In the limit of a point nucleus and a nonrelativistic electron, Equation 7 yields

$$H_W^{(1)} = \frac{G}{4\sqrt{2}} \frac{1}{m_e c} Q_W[\boldsymbol{\sigma} \cdot \mathbf{p}\delta^3(\mathbf{r}) + \delta^3(\mathbf{r})\boldsymbol{\sigma} \cdot \mathbf{p}]$$ 8.

where $\boldsymbol{\sigma}$ and $\mathbf{p}$ refer to the electron. In the first (second) term on the right-hand side (RHS) of Equation 8, $\mathbf{p} = -i\hbar\nabla$ is understood to apply to the electronic wave function on the left (right) in $\langle\chi|H_W^{(1)}|\psi\rangle$. The factor $Q_W \sim Z$ arises because we sum coherently over each nucleon (since the atomic electron wavelength is much larger than the nuclear diameter).

In the case of $H_P^{(2)}$, which corresponds to $V_e A_N$, a nonrelativistic reduction of the axial nucleonic current yields factors proportional to nucleon spin, and these cancel in pairs in the sum over nucleons, leaving at most two unpaired spins. Thus in the nonrelativistic limit of the electron and for a point nucleus, the Weinberg-Salam model yields an effective Hamiltonian:

$$H_W^{(2)} = \frac{G}{4\sqrt{2}} \frac{1}{m_e c} (1 - 4\sin^2\theta) g_A \boldsymbol{\sigma}_N \cdot \boldsymbol{\sigma}[\boldsymbol{\sigma} \cdot \mathbf{p}\delta^3(\mathbf{r}) + \delta^3(\mathbf{r})\boldsymbol{\sigma} \cdot \mathbf{p}].$$ 9.

$H_W^{(2)}$ contains no enhancement factor Q_W and its effects are therefore of order $Z^{-1}(1 - 4\sin^2\theta)$ relative to $H_W^{(1)}$. $H_W^{(2)}$ is therefore quite negligible compared to $H_W^{(1)}$ for heavy atoms where $Z \gg 1$.

We now consider an electromagnetic transition between two atomic states, $|\psi_1\rangle$ and $|\psi_2\rangle$, of the same nominal parity (i.e. absorption or emission of an optical or microwave magnetic dipole photon). Examples would be the transitions $1^2S_{1/2} \rightarrow 2^2S_{1/2}$ in H, $6^2P_{1/2} \rightarrow 7^2P_{1/2}$ in Tl. From Equation 3, the transition amplitude is, to order G,

$$\langle\psi_2|O_{EM}|\psi_1\rangle \simeq \langle\psi_2^0|O_{EM}|\psi_1^0\rangle + \mathscr{E}_P,$$ 10.

where the first term on the RHS is the zero-order M1 amplitude $\mathscr{M}$, and $\mathscr{E}_P$, given by

$$\mathscr{E}_P = \sum_n \left\{ \frac{\langle\psi_2^0|O_{EM}|\chi_n^0\rangle\langle\chi_n^0|H_P|\psi_1^0\rangle}{E(\psi_1^0) - E(\chi_n^0)} + \frac{\langle\psi_2^0|H_P|\chi_n^0\rangle\langle\chi_n^0|O_{EM}|\psi_1^0\rangle}{E(\psi_2^0) - E(\chi_n^0)} \right\},$$ 11.

is an electric dipole (E1) amplitude caused by parity violation. In a standard phase convention $\mathscr{M}$ is real; then from Equations 8 or 9, $\mathscr{E}_P$ is imaginary, and in general this is required by time reversal (T) invariance.

Experiments have been proposed to detect the existence of $\mathscr{E}_P$ in the isotopes of atomic hydrogen, and in heavy atoms (Cs, Tl, Bi). So far, parity violation has been observed in Tl and Bi. The hydrogenic atom experiments are important because the electronic wave functions are known exactly, so no uncertainty is introduced by atomic theory, and in principle one can measure all four coupling constants C_{1p}, C_{1n}, C_{2p}, C_{2n}. However, the expected effects are small and very difficult to observe. In heavy atoms the effects are larger than in hydrogen but uncertainties in atomic theory make precise calculations of $\mathscr{E}_P$ a difficult task. Moreover, since $H_W^{(1)}$ greatly dominates over $H_W^{(2)}$ for large Z, only the coefficients C_{1p} and C_{1n} can be studied. Indeed, in the Weinberg-Salam model, with $\sin^2 \theta = 0.23$, the contribution of C_{1p} is much less than that of C_{1n}.

2   THE WEINBERG-SALAM MODEL

2.1   Local Gauge Invariance

The unified theory of weak and electromagnetic interactions is based on the principle of local gauge invariance, according to which the Lagrangian density describing the various particles and their interactions must be invariant under a *local* gauge transformation; that is, one which can vary in an arbitrary manner from one space-time point to another. The Dirac Lagrangian density for a free electron,

$$\mathscr{L}_D = \tfrac{1}{2}\bar{\psi}(i\gamma^\mu\partial_\mu - m)\psi - \tfrac{1}{2}[i(\partial_\mu\bar{\psi})\gamma^\mu + m\bar{\psi}]\psi, \qquad\qquad 12.$$

is invariant under the infinitesimal *global* U(1) gauge transformation

$$\delta\psi = \psi' - \psi = i\beta\psi$$

(β is an infinitesimal real constant). However, under the *local* U(1) gauge transformation $\delta\psi = i\alpha(x_\mu)\psi$ (α is a real infinitesimal depending on the x_μ), $\mathscr{L}_D$ is not invariant. In fact,

$$\delta\mathscr{L}_D = \mathscr{L}'_D - \mathscr{L}_D = -(\partial_\mu\alpha)\cdot\bar{\psi}\gamma^\mu\psi.$$

Nevertheless, the invariance is restored by adding to $\mathscr{L}_D$ the term

$$\mathscr{L}_I = -g'\bar{\psi}\gamma^\mu\psi \cdot B_\mu, \qquad\qquad 13.$$

where g' is a constant, and B_μ is a vector field, provided we stipulate that under the gauge transformation

$$\delta B_\mu(x) = B'_\mu(x) - B_\mu(x) = -\frac{1}{g'}\partial_\mu\alpha. \qquad\qquad 14.$$

The addition of $\mathscr{L}_I$ to $\mathscr{L}_D$ is equivalent to the replacement of the ordinary derivative by the "covariant" derivative $\partial_\mu \to D_\mu \equiv \partial_\mu + ig'B_\mu$ in $\mathscr{L}_D$. If we were to put $g' = e$ and identify B_μ as the ordinary vector potential A_μ, Equation 14 would become the familiar gauge transformation condition of electrodynamics. $\mathscr{L}_I$ in Equation 13 would then be the "minimal" interaction Lagrangian, seen here to arise from the assumption of local gauge invariance. Of course, in electrodynamics it is necessary to include an additional gauge invariant term corresponding to the field alone:

$$\mathscr{L}_{EM} = -\tfrac{1}{4}F_{\mu\nu}F^{\mu\nu} \tag{15}$$

where

$$F_{\mu\nu} = \partial_\mu A_\nu - \partial_\nu A_\mu. \tag{16}$$

These ideas are readily extended to include an isospin symmetry. We consider the case of an isodoublet fermion field $\psi = \begin{pmatrix} \psi_1 \\ \psi_2 \end{pmatrix}$ (Yang & Mills 1954). The local SU(2) gauge transformation is given by

$$\delta\psi = \psi' - \psi = i\varepsilon^i\left(\frac{\tau_i}{2}\right)\psi \equiv i\hat{\varepsilon}\psi, \tag{17}$$

where the τ_i ($i = 1, 2, 3$) are 2×2 Pauli matrices, the real infinitesimals ε^i depend on the x_μ, and we adopt the repeated index summation convention for index i. Defining a Lagrangian $\mathscr{L}_D$ for the doublet ψ as in Equation 12, we find

$$\delta\mathscr{L}_D = -\bar{\psi}\gamma^\mu(\partial_\mu\hat{\varepsilon})\psi. \tag{18}$$

Now defining a triplet of vector fields $\mathbf{A} = A_\mu^{1,2,3}$ and $\hat{A}_\mu \equiv A_\mu^i(\tau_i/2)$, we restore the invariance of $\mathscr{L}_D$ by making the replacement

$$\partial_\mu \to D_\mu = \partial_\mu + ig\hat{A}_\mu$$

(g is a constant), provided that

$$\delta\hat{A}_\mu = -\frac{1}{g}\partial_\mu\hat{\varepsilon} + i[\hat{\varepsilon}, \hat{A}_\mu]. \tag{19}$$

A term analogous to that in Equation 15 must be added to describe the "field energy":

$$\mathscr{L}_{YM} = -\tfrac{1}{4}f_{\mu\nu}f^{\mu\nu}. \tag{20}$$

But now, since $\hat{A}_\mu$ and $\hat{A}_\nu$ are 2×2 matrices, in general noncommuting, we require

$$f_{\mu\nu} = \partial_\mu\hat{A}_\nu - \partial_\nu\hat{A}_\mu + ig[\hat{A}_\mu, \hat{A}_\nu] \tag{21}$$

in order for $\mathscr{L}_{YM}$ to be gauge invariant. The term in Equation 21 proportional to g has no analog in electrodynamics. Physically it corresponds to the fact that the fields $A_\mu^{1\pm i2}$ themselves carry "charge", and it implies that nonlinear self-interaction terms of order g ("A^3") and g^2 ("A^4") appear in Equation 20.

When the fields $\mathbf{A}_\mu$ and $\mathbf{B}_\mu$ are quantized,[2] four massless vector quanta

[2] See, for example, Abers & Lee (1973).

appear: two neutral (A_μ^3, B_μ), and two charged $(1\sqrt{2})$ $(A_\mu^{1 \pm i2})$. Attractive possibilities are thus suggested for a unified theory of weak interactions (charged and neutral), and electromagnetic interactions (neutral), in which vector fermion currents are coupled to the aforementioned vector fields. However, at this stage the theory is still unacceptable for several reasons. First, weak currents have axial vector as well as vector components. This requires that $\mathcal{L}_D$ also be invariant under *chiral* gauge transformations $\delta\psi = i\eta^i(\tau_i/2)\gamma^5\psi$, where the η^i are real infinitesimals depending on the x_μ. It can easily be shown that this requires $m_{\text{fermion}} = 0$ in $\mathcal{L}_D$, and an additional means must therefore be found to give mass to the fermions without spoiling gauge invariance. Second, the weak interactions are short ranged, and thus at least some of the Yang-Mills quanta must gain mass, without spoiling the local gauge invariance or renormalizability, another attractive feature of the massless theory.

2.2 Spontaneous Symmetry Breaking

The problem of massive gauge quanta is solved by means of "spontaneous symmetry breaking", a crucial feature added to the SU(2) × U(1) Yang-Mills theory by Weinberg (1967) and Salam (1968). Spontaneous symmetry breaking refers to the situation in which a Lagrangian possesses a symmetry not shared by the ground state of the system. In field theories not of the gauge (Yang-Mills) type, it can be shown (Goldstone 1961, Goldstone et al 1962) that there is massless spin-zero excitation (the so-called Goldstone boson) for each degree of freedom in which the symmetry is spontaneously broken. However, the proof of this statement is based on two assumptions: manifest covariance and a positive metric in Hilbert space. In a gauge theory, one or the other of these conditions is always invalid. The net effect in the case of a Yang-Mills field is that the Goldstone theorem is evaded [the Higgs phenomenon, see Higgs (1964a,b, 1966), Englert & Brout (1964), Guralnik et al (1964)]; each of the unwanted Goldstone bosons disappears, and in its place a corresponding *massive* gauge field appears. In this process, renormalizability is retained, as was first proved by 't Hooft. (See 't Hooft 1971, Lee 1972, Lee & Zinn-Justin 1972, Abers & Lee 1973). In an oft-quoted example, we may consider a complex scalar isodoublet (charged) field

$$\phi = \begin{pmatrix} \dfrac{\phi_1 + i\phi_2}{\sqrt{2}} \\ \dfrac{\phi_3 + i\phi_4}{\sqrt{2}} \end{pmatrix}$$

where $\phi_1, \ldots, \phi_4$ are real scalar fields. We employ the SU(2) × U(1) globally invariant Lagrangian density

$$\mathcal{L} = (\partial_\mu \phi^\dagger)(\partial^\mu \phi) - \mu^2 \phi^\dagger \phi - \lambda(\phi^\dagger \phi)^2, \qquad\qquad 22.$$

where $\mu^2 > 0$ and $\lambda > 0$ correspond to mass (ϕ^2) and "ϕ^4" interaction terms, respectively. We now regard μ^2 as a variable parameter. For $\mu^2 > 0$, Equation 22 describes four real scalar fields with the same mass, and also the vacuum state of the system (classically, the state with lowest energy) is $\phi = 0$. However, for $\mu^2 < 0$ the symmetry of the ground state is broken; the lowest energy state is not $\phi = 0$, and one obtains three scalar fields with zero mass (these correspond to Goldstone bosons) and one scalar field with finite mass $(-2\mu^2)^{1/2}$.

However, let us now require $\mathcal{L}$ in Equation 22 to be rendered invariant under *local* SU(2) × U(1) gauge transformations, by the replacement

$$\partial_\mu \to D_\mu = (\partial_\mu + ig\hat{A}_\mu + \frac{ig'}{2} B_\mu I) \qquad\qquad 23.$$

where I is the 2 × 2 identity matrix. The constants g and g', corresponding to SU(2) and U(1) gauge transformations respectively, are independent. We also add to Equation 22 the terms $-\frac{1}{4}f_{\mu\nu}f^{\mu\nu} - \frac{1}{4}G_{\mu\nu}G^{\mu\nu}$, where $f_{\mu\nu}$ is defined in Equation 21, and

$$G_{\mu\nu} = \partial_\mu B_\nu - \partial_\nu B_\mu. \qquad\qquad 24.$$

We make the substitutions

$$\eta^2 \equiv -\mu^2/2\lambda \quad (\eta^2 > 0 \text{ for } \mu^2 < 0)$$

$$g'/g \equiv \tan\theta$$

and define the new vector fields

$$Z_\mu = \cos\theta\, A_\mu^3 - \sin\theta\, B_\mu \qquad\qquad 25a.$$

$$A_\mu = \sin\theta\, A_\mu^3 + \cos\theta\, B_\mu, \qquad\qquad 25b.$$

where A_μ is a new field, not to be confused with $\mathbf{A}_\mu$. In a straightforward analysis in which higher order terms are dropped, the modified Lagrangian becomes

$$\mathcal{L} = [\tfrac{1}{2}(\partial_\mu \sigma)(\partial^\mu \sigma) - \tfrac{1}{2}(-2\mu^2)\sigma^2]$$

$$-\tfrac{1}{4}Z_{\mu\nu}Z^{\mu\nu} + \tfrac{1}{2}\frac{g^2\eta^2}{2\cos^2\theta}\, Z_\mu Z^\mu$$

$$-\tfrac{1}{4}(A_{\mu\nu}A^{\mu\nu} + A_{\mu\nu}^2 A^{2,\mu\nu})$$

$$+\tfrac{1}{2}\frac{g^2\eta^2}{2}(A_\mu^1 A^{1\mu} + A_\mu^2 A^{2\mu})$$

$$-\tfrac{1}{4}A_{\mu\nu}A^{\mu\nu}. \qquad\qquad 26.$$

Here $Z_{\mu\nu} \equiv \partial_\mu Z_\nu - \partial_\nu Z_\mu$, etc. The first term in square brackets on the RHS of Equation 26 corresponds to the kinetic energy and finite mass of the single scalar boson ("Higgs" boson). Such objects must necessarily arise in the procedure of spontaneous symmetry breaking even if a somewhat different choice is made for $\mathscr{L}$ in Equation 22. The second and third terms correspond to a neutral vector gauge boson Z^0 with finite mass $m_Z^2 = g^2\eta^2/(2\cos^2\theta)$. The fourth and fifth terms correspond to charged vector gauge bosons $W^\pm = (1/\sqrt{2})(A_\mu \mp iA_\mu^2)$ with finite mass $m_W^2 = g^2\eta^2/2$. Finally the last term corresponds to the field energy of a massless vector field A_μ, which we naturally identify as the EM field. Then using Equations 25a and 25b, we can rewrite the term $(g/2)A_\mu^3\tau_3 + (g'/2)B_\mu I$ in Equation 23 as

$$\frac{g}{2}A_\mu^3\tau_3 + \frac{g'}{2}B_\mu I = A_\mu g \sin\theta\left(\frac{I+\tau_3}{2}\right) + Z_\mu g \cos\theta\left(\frac{\tau_3 - \tan^2\theta \cdot I}{2}\right). \qquad 27.$$

The first term on the RHS of Equation 27 contains A_μ and the charge operator $\frac{1}{2}(I+\tau_3)$. Therefore, we can identify $g\sin\theta$ as the electric charge:

$$g\sin\theta = e. \qquad 28.$$

2.3   Coupling of Gauge Fields to Leptons and Quarks

We now introduce the material particles (leptons and quarks) and couple them to the gauge fields. Our choices here are largely constrained by the requirement that the new theory reproduce the known and valid results of electrodynamics and of the old "current-current" charged weak interaction theory (Feynman & Gell-Mann 1958, Sudarshan & Marshak 1958). The charged weak interactions involve only the left-handed fields

$$e_L = \tfrac{1}{2}(1-\gamma_5)e, \ \mu_L = \tfrac{1}{2}(1-\gamma_5)\mu, \ u_L = \tfrac{1}{2}(1-\gamma_5)u,$$

etc. Many experimental facts suggest the arrangement of all fermions in "weak left-handed isodoublets" (Glashow et al 1970):

$$\begin{pmatrix} v_e \\ e \end{pmatrix}_L, \begin{pmatrix} v_\mu \\ \mu \end{pmatrix}_L, \begin{pmatrix} v_\tau \\ \tau \end{pmatrix}_L, \dots (leptons)$$

$$\begin{pmatrix} u \\ d_C \end{pmatrix}_L, \begin{pmatrix} c \\ s_C \end{pmatrix}_L, \dots (quarks)$$

where $d_C = d\cos\theta_C + s\sin\theta_C$, $s_C = -d\sin\theta_C + s\cos\theta_C$, θ_C is the Cabibbo angle, and u, d, s, c are the up, down, strange, and charmed quarks. The right-handed fields $e_R = \tfrac{1}{2}(1+\gamma_5)e$, $\mu_R = \tfrac{1}{2}(1+\gamma_5)\mu$, etc, do not participate in any known way in the charged weak interaction, and this provides a clue as to their weak isomultiplet assignments. In the Weinberg-Salam model one assumes that the components $e_R, \mu_R, \dots, u_R, d_R, \dots$ are all (right-handed) weak *isosinglets* under SU(2). Within SU(2) × U(1) other possibilities may be entertained but as we shall see, these are now ruled out by experiment.

It remains to determine the transformation properties of the fermions under U(1) from the electromagnetic couplings. We introduce the "weak hypercharge" Y, defined by the U(1) gauge transformation $\delta\chi = (i\alpha/2)(Y\chi)$ where α is a real infinitesimal depending on the x_μ, χ is a column matrix of the fermion fields:

$$\chi = \begin{pmatrix} \left.\begin{matrix} v_{eL} \\ e_L \\ v_{eR} \\ e_R \\ \cdots \end{matrix}\right\} \text{leptons} \\ \left.\begin{matrix} u_L \\ d_{cL} \\ u_R \\ d_R \\ \cdots \end{matrix}\right\} \text{quarks} \end{pmatrix}$$

and Y is an $N \times N$ diagonal matrix. When the SU(2) × U(1) covariant derivative is introduced into the fermion Dirac Lagrangian, and if we assume the Weinberg-Salam weak isomultiplet structure, a minimal interaction coupling term arises, of the form

$$\mathscr{L}_{INT} = -\bar{\chi}\gamma^\mu[gA_{\mu 1}L^1 + gA_{\mu 2}L^2 + gA_{\mu 3}L^3 + \tfrac{1}{2}g'B_\mu Y]\chi \qquad 29.$$

where $L^i = [(1-\gamma_5)/2](\tau^i/2)$. From Equations 25 and 28 we find:

$$gA_{\mu 3}L^3 + \tfrac{1}{2}g'B_\mu Y = e\left(L^3 + \frac{Y}{2}\right)A_\mu + g\cos\theta\left(L^3 - \frac{Y}{2}\tan^2\theta\right)Z_\mu. \qquad 30.$$

The first term on the RHS of Equation 30 represents the EM coupling of each fermion; therefore $L^3 + (Y/2)$ is plainly the charge operator Q. The second term on the RHS of Equation 30 may then be expressed in terms of L^3 and Q as $(g/\cos\theta)Z_\mu(L^3 - \sin^2\theta\,Q)$, which yields directly the neutral weak coupling of each fermion in terms of L^3 and Q.

To summarize, we write:

$$\mathscr{L}_{INT} = \mathscr{L}_{EM} + \mathscr{L}_C + \mathscr{L}_N \qquad 31.$$

for EM, charged weak, and neutral weak interactions respectively. Here $\mathscr{L}_{EM} = -eJ^\lambda_{EM}A_\lambda$, where

$$J^\lambda_{EM} = \bar{\chi}\gamma^\lambda Q\chi. \qquad 32.$$

Next,

$$\mathscr{L}_C = -\frac{g}{2\sqrt{2}}(J_C^{\lambda+}W_\lambda^+ + J_C^{\lambda-}W_\lambda^-) \qquad 33.$$

where $W_\lambda^\pm = (1/\sqrt{2})A^{1\pm i2}$,

$$J_C^{\lambda\pm} = \bar{\chi}\gamma^\lambda(1-\gamma_5)\frac{\tau^\pm}{2}\chi, \qquad 34.$$

and $\tau^\pm = \tau_x \pm i\tau_y$. Finally, for the neutral weak interaction,

$$\mathscr{L}_N = -\frac{g}{2\cos\theta}J_N^\lambda Z_\lambda \qquad 35.$$

where

$$J_N^\lambda = \bar{\chi}\gamma^\lambda(1-\gamma_5)\frac{\tau_3}{2}\chi - 2\sin^2\theta\, J_{EM}^\lambda. \qquad 36.$$

We remind the reader that the τ_i refer to *weak* isospin. The constant g may be expressed in terms of the more familiar Fermi coupling constant G by comparing expressions for the muon decay amplitude in the new theory and in the old V-A theory. As is well known, the latter describes muon decay accurately in terms of the amplitude:

$$\mathscr{A} = \frac{G}{\sqrt{2}}u_{\nu_\mu}\gamma_\lambda(1-\gamma_5)u_\mu \cdot \bar{u}_e\gamma^\lambda(1-\gamma_5)v_{\bar{\nu}_e} \qquad 37.$$

where the u's and v are single-particle Dirac spinors. From Equations 33 and 34, the new theory yields:

$$\mathscr{A} = -\frac{g^2}{8}\bar{u}_{\nu_\mu}\gamma_\lambda(1-\gamma_5)u_\mu \cdot \left[\frac{g^{\lambda\sigma}-q^\lambda q^\sigma/m_W^2}{q^2-m_W^2}\right]\bar{u}_e\gamma_\sigma(1-\gamma_5)v_{\bar{\nu}_e} \qquad 38.$$

where the factor in square brackets corresponds to the W propagator. In muon decay, $|q^2| \ll m_W^2$, hence Equation 38 reduces to Equation 37 if we put $G/\sqrt{2} = g^2/8m_W^2$. Thus,

$$m_W^2 = \frac{\pi\alpha}{\sqrt{2}G}\frac{1}{\sin^2\theta} = \frac{(37.5\text{ GeV})^2}{\sin^2\theta} \qquad 39.$$

and

$$m_Z^2 = \frac{m_W^2}{\cos^2\theta}. \qquad 40.$$

From Equations 35 and 36 we may also obtain an effective current-current Lagrangian for neutral weak processes in the limit $q^2 \ll m_Z^2$. This is

$$\mathscr{L}_{\text{eff}} = -\frac{1}{\sqrt{2}} G J_N^\lambda \cdot J_{N\lambda} \qquad\qquad 41.$$

In order to apply Equation 41 to the neutral weak e-N coupling in low energy atomic physics, we must obtain matrix elements of $J_{N\lambda}$ between physical nucleon states in the limit of zero momentum transfer (Weinberg 1972). Application of simple isospin arguments, of the conserved vector current hypothesis, and neglect of all momentum-transfer-dependent terms, results in the following amplitudes:

$$\mathscr{A}(ep \to ep) = -\frac{G}{2\sqrt{2}} \bar{u}_e' \gamma_\lambda (1 - 4 \sin^2 \theta - \gamma_5) u_e \cdot \bar{u}_p' \gamma^\lambda (1 - 4 \sin^2 \theta$$
$$- g_A \gamma_5) u_p \qquad\qquad 42.$$

$$\mathscr{A}(en \to en) = +\frac{G}{2\sqrt{2}} \bar{u}_e' \gamma_\lambda (1 - 4 \sin^2 \theta - \gamma_5) u_e \bar{u}_n' \gamma^\lambda (1 - g_A \gamma_5) u_n. \qquad 43.$$

Thus for a nucleus with Z protons and N neutrons, the $A_e V_n$ component is:

$$\mathscr{A}(A_e V_N) = +\frac{G}{2\sqrt{2}} [(1 - 4 \sin^2 \theta) Z - N] \bar{u}_e' \gamma_\lambda \gamma_5 u_e \cdot \bar{\psi}_N' \gamma^\lambda \psi_N \qquad 44.$$

where ψ_N is a nucleon spinor. In the nonrelativistic limit,

$$\psi_N \to \begin{pmatrix} \chi_N \\ 0 \end{pmatrix}$$

where χ_N is a two-component Pauli spinor. Thus $\bar{\psi}_N' \gamma^\lambda \psi_N = \chi_N'^\dagger \chi_N \, (= 0)$; if $\lambda = 0 \, (= 1, 2, 3)$; and we obtain Equation 7. The nonrelativistic limit of the electron (Equation 8) is easily obtained by noting that

$$\gamma_5 = \begin{pmatrix} 0 & 1 \\ 1 & 0 \end{pmatrix} \text{ and } u_e \simeq \begin{pmatrix} \chi_e \\ \dfrac{\sigma \cdot \mathbf{P}_e}{2 m_e c} \chi_e \end{pmatrix}$$

Similar considerations apply to Equation 9.

3   POLARIZED ELECTRON-NUCLEON SCATTERING

3.1   *Theoretical Analysis*

We now consider the helicity-dependent asymmetry in scattering of high energy polarized electrons by nucleons, measured in the experiment at

SLAC (Prescott et al 1978, 1979). Our discussion here is initially independent of any particular gauge theory, and follows the treatment of
Cahn & Gilman (1978), who employed the quark-parton model. Here the
nucleon is assumed to consist of three "valence" quarks plus a "sea" of
virtual quark-antiquark pairs. This is appropriate in the regime of large
energies and momentum transfers actually used, where electron and quark-
parton masses may be neglected. In this limit $(1 \pm \gamma_5)/2$ are $\pm$ helicity
projection operators, respectively, while vector (γ^μ) and axial vector $(\gamma^\mu \gamma_5)$
interactions preserve helicity. The photon couples to quark or electron
through the vector current γ_μ with a strength given by its charge Q_f^γ. Using
$\gamma_\mu = \gamma_\mu(1+\gamma_5)/2 + \gamma_\mu(1-\gamma_5)/2$ we may define left- and right-handed charges

$$Q_{Lf}^\gamma = Q_{Rf}^\gamma = Q_f^\gamma. \qquad 45.$$

The Z^0 couples to left- and right-handed fermions with strengths Q_{Lf}^Z and
Q_{Rf}^Z, respectively; these are in general different. Thus the Z^0-fermion Dirac
vertex has the form:

$$Q_{Rf}^Z \gamma_\mu \frac{(1+\gamma_5)}{2} + Q_{Lf}^Z \gamma_\mu \frac{(1-\gamma_5)}{2}. \qquad 46.$$

Now consider the scattering of electron and quark in the CM frame. It is
easy to show that if both (massless) fermions have the same helicity, the
scattering is isotropic; if they have opposite helicity, the angular distri-
bution is proportional to $(1 + \cos \hat{\theta})^2$, where $\hat{\theta}$ is the scattering angle. In both
cases the helicity of each fermion remains unchanged in the scattering
process. In the lab frame this translates into a dependence of the differential
cross section on $(1 - y)^2$ ($y = 1 - E'/E \equiv v/E$ where E' and E are the final and
initial lepton energies, respectively). Recalling that the electromagnetic and
weak interactions contribute coherently to the amplitude for each quark,
we easily obtain the following results for the differential cross sections:

RH e^- on RH quark of type i:

$$d\sigma \propto \left| \frac{Q_{Re}^\gamma Q_{Ri}^\gamma}{q^2} + \frac{Q_{Re}^Z Q_{Ri}^Z}{q^2 - m_Z^2} \right|^2; \qquad 47.$$

RH e^- on LH quark of type i:

$$d\sigma \propto \left| \frac{Q_{Re}^\gamma Q_{Li}^\gamma}{q^2} + \frac{Q_{Re}^Z Q_{Li}^Z}{q^2 - m_Z^2} \right|^2 (1-y)^2; \qquad 48.$$

LH e^- on LH quark of type i:

$$d\sigma \propto \left| \frac{Q_{Le}^\gamma Q_{Li}^\gamma}{q^2} + \frac{Q_{Le}^Z Q_{Li}^Z}{q^2 - m_Z^2} \right|^2; \qquad 49.$$

LH e^- on RH quark of type i:

$$d\sigma \propto \left| \frac{Q^\gamma_{Le} Q^\gamma_{Ri}}{q^2} + \frac{Q^Z_{Le} Q^Z_{Ri}}{q^2 - m^2_Z} \right|^2 (1-y)^2. \qquad 50.$$

The asymmetry for longitudinally polarized e^-,

$$\Delta = \frac{d\sigma_R - d\sigma_L}{d\sigma_R + d\sigma_L}, \qquad 51.$$

is now computed simply by multiplying Equations 47–51 by the probability $f_i(x)$ of finding a quark of type i with fractional longitudinal momentum $x = -q^2/2m_p\nu$ in the nucleon and noting that the differential cross sections $d\sigma_{R,L} = d\sigma_{R,L}(x, y)$ involve an incoherent sum over LH, RH quarks in the unpolarized nucleon. Recalling that $Q^\gamma_{Le} = Q^\gamma_{Re} = Q^\gamma_e = -e$, and $Q^\gamma_{Li} = Q^\gamma_{Ri}$, and putting $g_V = (Q^Z_R + Q^Z_L)/2$ and $g_A = (Q^Z_R - Q^Z_L)/2$, we obtain, to order $-q^2/m^2_Z$,

$$\Delta(x,y) = -\frac{(-2q^2)}{m^2_Z} \frac{\sum_i f_i(x)(Q^\gamma_i/e) \left\{ g_{A,e} g_{V,i} + \frac{[1-(1-y)^2]}{[1+(1-y)^2]} g_{V,e} g_{A,i} \right\}}{\sum_i f_i(x)(Q^\gamma_i)^2} \qquad 52.$$

The sum over quark types i in Equation 52 must include both quarks and antiquarks at a given x. However, for $x \gtrsim 0.2$ it is known from various deep-inelastic scattering experiments that antiquarks may safely be neglected, and we confine ourselves to this region henceforth. Also we now restrict ourselves to $SU(2) \times U(1)$ (but not yet to the Weinberg-Salam model). Then the weak charges of the fermions are given by

$$Q^Z_L = \frac{e}{\sin\theta\cos\theta}(L_3 - Q^\gamma \sin^2\theta) \qquad 53.$$

$$Q^Z_R = \frac{e}{\sin\theta\cos\theta}(R_3 - Q^\gamma \sin^2\theta) \qquad 54.$$

where L_3 and R_3 are the third-components of the weak isospin for left- and right-handed fermions, respectively, as in Section 2. In the Weinberg-Salam model, of course, $L_3 = +\frac{1}{2}$ for up-quark, $L_e = -\frac{1}{2}$ for down-quark and electron, and $R_3 = 0$ for all fermions. Also for an isosinglet target (deuterium), $f_u(x) = f_d(x)$. Assuming this, neglecting antiquarks, and employing conventional L_3 assignments, we obtain from Equation 52 the formula:

$$\frac{\Delta_{e,d}(x,y)}{(-q^2)} = a_1 + a_2 \left[\frac{1-(1-y)^2}{1+(1-y)^2} \right], \qquad 55.$$

where

$$a_1 = -\frac{G}{2\pi\sqrt{2}\,\alpha}\frac{9}{10}(1+2R_3^e)(1-\frac{20}{9}\sin^2\theta+\frac{4}{3}R_3^u-\frac{2}{3}R_3^d) \qquad 56.$$

and

$$a_2 = -\frac{G}{2\pi\sqrt{2}\,\alpha}\frac{9}{10}(1-4\sin^2\theta-2R_3^e)(1-\frac{4}{3}R_3^u+\frac{3}{2}R_3^d). \qquad 57.$$

The quantity $\Delta_{e,d}/(-q^2)$ is plotted in Figure 2 as a function of y for various $SU(2) \times U(1)$ models, identified by their assignment of RH fermions. The Weinberg-Salam model has right-handed isosinglets only. The other models place the RH electron in a weak isodoublet with a hypothetical neutral heavy lepton E_0 ("hybrid" model) or assume one or more of the

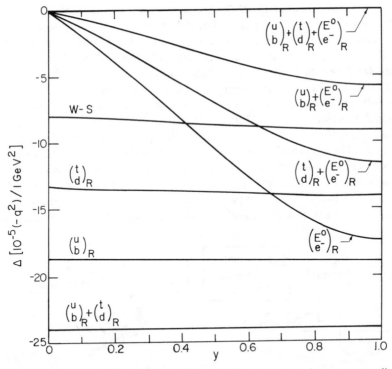

Figure 2   Theoretical asymmetry in scattering of polarized electrons by deuterons, according to various $SU(2) \times U(1)$ gauge models, plotted as a function of y. In all models, the leptons and quarks are assigned to left-handed weak isodoublets. Models differ according to assignment of particles to right-handed weak isomultiplets, as indicated. E_0 is a hypothetical neutral heavy lepton; $\sin^2\theta = 0.23$ is assumed.

quarks to be in RH doublets. In the past, models based on other gauge groups have also been considered. Most of the $SU(2)_L \times SU(2)_R \times U(1)$ models predict no parity violation: $\Delta_{e,d}(x, y) = 0$. Most of the other models $[SU(3) \times U(1), SU(3) \times SU(3),$ etc] are ruled out by the results of neutrino experiments.

3.2   *The SLAC Experiment*

The asymmetry Δ was measured in the scattering of 19.4-GeV polarized electrons on a stationary target (Prescott et al 1979). Since Δ is expected to be small, $(\Delta \simeq 10^{-4})$, the experimenters abandoned traditional single-particle counting techniques. On each 1.5-μs pulse of the accelerator, they detected $\sim 10^3$ scattered electrons in the same detector and integrated the total signal. In this way a statistical precision of 10^{-5} was possible after only 10^7 pulses, or about one day's running.

3.2.1 POLARIZED ELECTRON SOURCE   A block diagram of the SLAC experiment is shown in Figure 3. The polarized electrons were produced in a GaAs crystal by exciting transitions from the $J = \frac{3}{2}$ valence band to the $J = \frac{1}{2}$ conduction band with circularly polarized 7100-Å light from a flashlamp-pumped dye laser. These conduction electrons then left the surface of the crystal and were injected into the Linac. The maximum possible polarization obtainable in this way is $P = 0.5$, but during the actual experiment, $P \simeq 0.37$. P was measured every eight hours by Møller scattering of the electrons on a magnetized iron foil placed behind the target. (The target was removed for the polarization measurement.) A statistical precision of 3% in P was obtained in 20 minutes. The helicity of the beam could be reversed by reversing the polarization of the light. This was accomplished in two ways. The light was first linearly polarized with a Glan-Air prism, and then circularly polarized with a Pockels cell. The sign

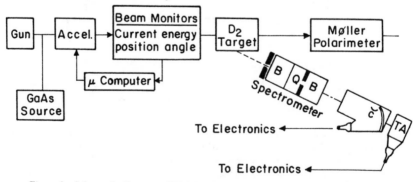

Figure 3   Schematic diagram of SLAC polarized electron scattering experiment.

of the voltage on the Pockels cell was chosen randomly from pulse to pulse: sign reversal changes the polarization of the light. The Glan polarizer could also be rotated by 90°, which reverses the circular polarization as well. In addition, unpolarized electrons could be produced, either by substituting the normal thermionic SLAC source for the GaAs crystal, or by rotating the Glan prism 45°, in which case the Pockels cell has no retardation effect.

Approximately 3×10^{12} electrons per pulse were injected into the Linac, and about 10^{11} electrons remained after acceleration. The beam energy and position were carefully monitored and stabilized by computer-controlled feedback loops. After acceleration, the beam was bent through an angle of 24.5° in the SLAC switchyard to reach the experimental area. The spins of these highly relativistic electrons underwent a $g - 2$ precession such that, for the SLAC bending angle, the spin precessed ahead of the trajectory by

$$\theta = \frac{E\pi}{3.237}$$

where E is in GeV. The beam was longitudinally polarized only when $E/3.237$ was an integer. This provided a source-independent way of reversing the electron helicity or making it zero.

3.2.2 TARGET AND SPECTROMETER After passing through the beam transport system, the electrons struck a 30-cm target containing liquid D_2. Electrons that scattered at 4° in the vertical plane entered the spectrometer, which analyzed the momentum in the horizontal plane and accepted a very broad momentum range, as shown in Figure 4. For most of the experiment, the beam energy was 19.4 GeV, and the momentum acceptance peaked at 14.5 GeV/c. Note that electrons from elastic and resonant scattering also fall within the acceptance range. These contributed a few percent to the cross section. Figure 4 also shows the pion cross section. Since individual particles cannot be discriminated against in a flux counting experiment, the kinematics were chosen to reduce the π, K, and μ background to a few percent.

3.2.3 DETECTORS Two electron detectors were used: a Cerenkov counter and a lead-glass shower counter. They were placed in series, as shown in Figure 3. Since coincidence counting techniques could not be used, the detectors each measured the asymmetry separately. The Cerenkov counter was filled with N_2 at atmospheric pressure. A single spherical mirror focused the light onto a photomultiplier tube. The shower counter consisted of nine radiation lengths of lead glass viewed by four photomultipliers. During the second experimental run the data from the two halves of the counter, each representing half of the momentum acceptance,

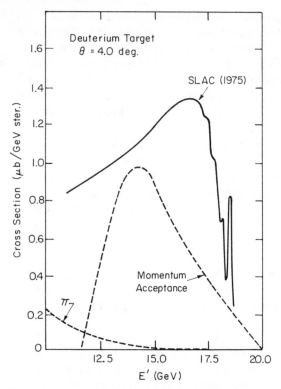

Figure 4 Momentum acceptance from the SLAC polarized electron experiment.

were analyzed separately, thereby providing an asymmetry at two values of y simultaneously.

Behind the shower counter were six inches of lead, which absorbed electrons. The background particles (π, K, μ) penetrated this barrier or formed hadronic showers, and were detected by a shower counter behind it. The background asymmetry observed with this detector gave less than a 1% correction to the final measurement and was consistent with zero.

3.2.4 DATA Data collection was divided into runs of approximately 3.5 hours each. During each run the Pockels cell polarization reversed randomly from pulse to pulse. In between runs, changes were made that reversed the helicity relative to the Pockels cell voltage or eliminated it altogether. These were (*a*) rotation of the Glan prism by 45° (null experiment) or 90° (polarization reversal); and (*b*) changes of the beam energy by an amount causing the electron spins to precess in the switchyard by integral (reversal) or half integral (null) multiples of π. The changes in Δ under these conditions are shown in Figures 5 and 6.

Table 2 Systematic error summary, SLAC polarized electron experiment[a]

Parameter	Measured difference $(+)-(-)$	Correction to $\Delta/-q^2$
Beam energy (%)	$(1.5\pm0.28)\times10^{-4}$	-0.37×10^{-5}
Beam current (mA)	$(2.2\pm0.4)\times10^{-3}$	-0.03
Position parameters (μm)		
x	$(-8.9\pm3.3)\times10^{-2}$	$+0.04$
y	$(-0.65\pm1.8)\times10^{-2}$	-0.02
Beam angle (μrad)		
θ_x	$(-0.37\pm0.7)\times10^{-3}$	0.00
θ_y	$(1.5\pm0.9)\times10^{-3}$	$+0.01$
Total systematic correction		-0.37×10^{-5}

[a] From Prescott et al (1978).

3.3 Results of the SLAC Experiment

Data were obtained with an e^- beam energy of 19.4 GeV, and at values of y corresponding to scattered e^- energies from 10.2 to 16.3 GeV. Corrections for radiative effects were made by assuming that $\Delta_{e,d}$ has kinematic dependence given by Equation 55 and by applying previously measured cross sections and radiative correction formulae (3% correction). Higher order weak processes were ignored. The errors assigned to the data arose mainly from counting statistics with some contribution from uncertainty

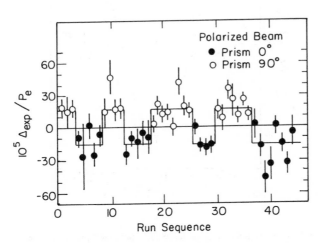

Figure 5 Asymmetry data from the SLAC polarized electron experiment. Beam polarization is reversed by reorienting the polarizing prism from $0°$ to $90°$. (From Prescott et al 1978.)

in beam polarization. However, there were also contributions to the uncertainty from systematic errors: imbalance in beam parameters ($\sim 0.025\Delta_{e,d}$), and uncertainty in beam polarization P ($\sim 2.5\%$). A summary of these systematic errors and corrections is presented in Table 2.

Figure 7 displays the corrected experimental asymmetry as a function of y. Even preliminary data at low values of y ruled out $SU(2)_L \times SU(2)_R \times U(1)$ models and $SU(2) \times U(1)$ models other than the Weinberg-Salam and the "hybrid". The best-fit y dependencies of these latter two are displayed in Figure 7 along with a model-independent parametrization of the data assuming only the form of Equation 55 with a_1 and a_2 as constants to be determined. The results are clearly consistent with the Weinberg-Salam model, and give a best fit of $\sin^2 \theta = 0.224 \pm 0.020$ with a χ^2 probability of 40% (a value in good agreement with that obtained in neutrino experiments). The results are just as clearly inconsistent with the

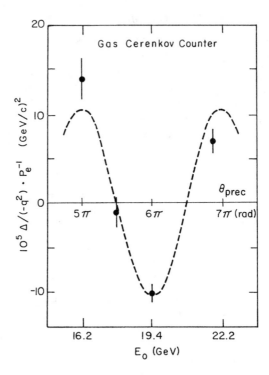

Figure 6 Observed asymmetry from the SLAC polarized electron experiment (discrete points). Dotted curve represents expected energy dependence of asymmetry due to electron spin precession. (From Prescott et al 1978.)

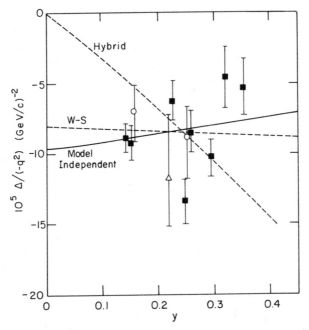

Figure 7 Experimental asymmetry in scattering of polarized electrons on deuterium (SLAC), as a function of y. Solid squares: $E = 19.4$ GeV; open triangle: $E = 16.2$ GeV; open circles: $E = 22.2$ GeV. (From Prescott et al 1979.)

hybrid model; here, a best fit gives $\sin^2 \theta = 0.015$ (widely at variance with neutrino results) and a χ^2 probability of 6×10^{-4}. The best fit for model-independent parameters is

$$a_1 = -(9.7 \pm 2.6) \times 10^{-5} \qquad\qquad 58.$$

$$a_2 = (4.9 \pm 8.1) \times 10^{-5}. \qquad\qquad 59.$$

Further discussion of these results is given in Section 5.

4   PARITY VIOLATION IN ATOMS

4.1   *Hydrogenic Atoms*

4.4.1 ANALYSIS All hydrogenic atom experiments now being pursued utilize the special properties of the $2^2S_{1/2}$ and $2^2P_{1/2}$ states (see Figure 8). At zero magnetic field B, the $2^2S_{1/2}$ and $2^2P_{1/2}$ levels are separated by the Lamb shift $S = 1058$ MHz. The natural lifetime of $2^2P_{1/2}$ is short $[\tau(2P) = 1.6 \times 10^{-9}$ s] since a 2P atom can decay to the ground state by allowed E1 photon emission (Lyman α). The 2P state has natural width $\Gamma_{2P} = 100$ MHz. In the absence of external electric fields (which mix 2P and 2S by Stark effect) the $2^2S_{1/2}$ state is metastable $[\tau(2S) = 1/8$ s] since its only effective mode of decay is by two-photon emission. Thus one can form a beam of 2S atoms that exists over the length of a practical apparatus (meters). The zero B field hyperfine structure (hfs) splittings of the $2^2S_{1/2}$ and $2^2P_{1/2}$ states are precisely calculable, as is the Zeeman effect; and the $2^2S_{1/2}$ hfs splitting has been measured accurately. Parity violation causes a mixing of $2^2S_{1/2}$ and $2^2P_{1/2}$ Zeeman components with the same m_F (e.g. mixing of the levels $\beta_0 e_0$ or $\beta_0 f_0$ in $_1H^1$, see Figure 8). The matrix elements $\langle e_0 | H_P | \beta_0 \rangle$, etc are extremely small, but by way of compensation the effective energy denominator ΔE in Equation 11 is also small. It is in fact, $\Delta E = \Delta E_0 + i\Gamma_{2P}/2$ where ΔE_0 is the real energy separation ($\Delta E_0 = S$ at zero magnetic field and $\Delta E_0 \rightarrow 0$ at a level crossing).

One may carry out a general analysis of parity-violating effects starting from the forms given in Equations 4 and 5 (Dunford et al 1978, Cahn & Kane 1977). We begin with the matrix element of $H_P^{(1)}$ in the nonrelativistic

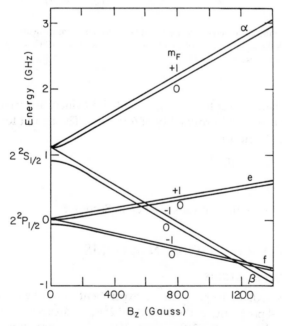

Figure 8   Zeeman effect in the hyperfine structure of $2^2S_{1/2}$ and $2^2S_{1/2}$ states of atomic hydrogen. Parity violation causes mixing of levels β_0, e_0; β_0, f_0; $\beta_{-1}f_{-1}$.

point-nucleon limit, between states $|\psi\rangle$ and $|\chi\rangle$:

$$\langle\chi|H_P^{(1)}|\psi\rangle = \frac{G}{\sqrt{2}}(\chi^\dagger\gamma_5\psi)_{r=0}\cdot C_1 \qquad\qquad 60.$$

where $C_1 = ZC_{1p} + NC_{1n}$. Next we carry out a two-component reduction of $(\chi^\dagger\gamma_5\psi)_{r=0}$. In the special case where χ and ψ correspond to $n^2P_{1/2}$ and $n^2S_{1/2}$ states, respectively, we find

$$\langle n^2P_{1/2}|H_P^{(1)}|n^2S_{1/2}\rangle = iC_1\frac{3G}{\sqrt{2}}\frac{\hbar}{8\pi\,m_ec}R'_{nP}(0)R_{nS}(0)\delta_{m'_jm_j}\delta_{m'_Im_I} \qquad 61.$$

where the R's are Schroedinger radial wave functions, and I and J are nuclear and atomic angular momenta, respectively. Employing explicit values for the R's we obtain:

$$\langle n^2P_{1/2}|H_P^{(1)}|n^2S_{1/2}\rangle = iC_1\bar{V}\delta_{m'_jm_j}\delta_{m'_Im_I} \qquad\qquad 62.$$

where

$$\bar{V} = \frac{G\alpha}{2\pi\sqrt{2}\,a_0^3}\left(\frac{Z}{n}\right)^4(n^2-1)^{1/2}$$

$$= 0.118\left(\frac{Z}{n}\right)^4(n^2-1)^{1/2}\text{ Hz.} \qquad\qquad 63.$$

Similarly, for $H_P^{(2)}$, one obtains:

$$\langle n^2P_{1/2}|H_P^{(2)}|n^2S_{1/2}\rangle = -2i\bar{V}C_2\langle m'_jm'_I|\boldsymbol{\sigma}_e\cdot\mathbf{I}|m_jm_I\rangle \qquad 64.$$

where $C_2 = C_{2p}$ for hydrogen, and $C_2 = \frac{1}{2}(C_{2p}+C_{2n})$ for deuterium (neglecting the small $^3D_1$ component in the nuclear wave function). In terms of individual hyperfine components $|Fm_F\rangle$, where $\mathbf{F} = \mathbf{I}+\mathbf{J}$, the matrix elements are

$$\langle n^2P_{1/2},F'm'_F|H_P^{(1)}|n^2S_{1/2},Fm_F\rangle = iC_1\bar{V}\delta_{F'F}\delta_{m'_Fm_F} \qquad 65.$$

$$\langle n^2P_{1/2},F'm'_F|H_P^{(2)}|n^2S_{1/2},Fm_F\rangle = -2i\bar{V}C_2\delta_{F'F}\delta_{m'_Fm_F}[F(F+1)-I(I+1)$$

$$-(3/4)] \qquad\qquad 66.$$

We next consider the effect of an external magnetic field B (Zeeman effect). For $2^2S_{1/2}$, $2^2P_{1/2}$ states the energy levels of hydrogen in finite B are shown in Figure 8. The states α_{+1}, α_0, β_0, etc may be expressed in terms of states $|2L_{1/2},m_J,m_I\rangle$ as follows:

$$|\alpha_{+1}\rangle = |2S_{1/2}, +\tfrac{1}{2}, +\tfrac{1}{2}\rangle$$

$$|\alpha_0\rangle = \cos\theta_S |2S_{1/2}\tfrac{1}{2}, -\tfrac{1}{2}\rangle + \sin\theta_S |2S_{1/2}, -\tfrac{1}{2}\tfrac{1}{2}\rangle$$

$$|\beta_0\rangle = \cos\theta_S |2S_{1/2}, -\tfrac{1}{2}\tfrac{1}{2}\rangle - \sin\theta_S |2S_{1/2}\tfrac{1}{2}, -\tfrac{1}{2}\rangle$$

$$|\beta_{-1}\rangle = |2S_{1/2}, -\tfrac{1}{2}\tfrac{1}{2}\rangle$$

$$|e_{+1}\rangle = |2P_{1/2}\tfrac{1}{2}\tfrac{1}{2}\rangle \qquad\qquad\qquad 67.$$

$$|e_0\rangle = \cos\theta_P |2P_{1/2}\tfrac{1}{2}, -\tfrac{1}{2}\rangle + \sin\theta_P |2P_{1/2}, -\tfrac{1}{2}\tfrac{1}{2}\rangle$$

$$|f_0\rangle = \cos\theta_P |2P_{1/2}, -\tfrac{1}{2}\tfrac{1}{2}\rangle - \sin\theta_P |2P_{1/2}\tfrac{1}{2}, -\tfrac{1}{2}\rangle$$

$$|f_{-1}\rangle = |2P_{1/2}, -\tfrac{1}{2}, +\tfrac{1}{2}\rangle,$$

where

$$\cos\theta_{S,P} = x_{S,P}[x_{S,P}^2 + (\sqrt{x_{S,P}^2 + 1} - 1)^2]^{-1/2},$$

$$x_{S,P} = \frac{[g_{J,(S,P)} + g_I]\mu_0 B}{a_{S,P}},$$

μ_0 is the electron Bohr magneton, g_J is the Landé g factor, and $a_{S,P}$ = zero field hfs splitting. (In fact, $\theta_S \simeq \theta_P$ and the S,P subscripts may be disregarded.) From Formulae 67 it is easy to show that the only nonzero matrix elements of H_P between states crossing in a magnetic field are:

$$\langle e_0 | H_P | \beta_0 \rangle = -2C_{2p} \cos 2\theta \cdot i\bar{V} \qquad\qquad 68.$$

$$\langle f_0 | H_P | \beta_0 \rangle = [C_{1p} + (1 + 2\sin 2\theta)C_{2p}] i\bar{V} \qquad 69.$$

$$\langle f_{-1} | H_P | \beta_{-1} \rangle = (C_{1p} - C_{2p}) i\bar{V}. \qquad\qquad 70.$$

Note that mixing of e_0 and β_0 involves only C_{2p}; thus according to the Weinberg-Salam model the matrix element is proportional to the small quantity $C_{2p} = 0.62(1 - 4\sin^2\theta) \sim 0.05$ for $\sin^2\theta = 0.23$. Formulae similar to Equations 68–70 are easily obtained for deuterium and tritium.

We now consider the mixing of $2^2S_{1/2}$ and $2^2P_{1/2}$ components, according to the equation:

$$|2^2S_{1/2}\rangle' = |2^2S_{1/2}\rangle + \frac{\langle 2^2P_{1/2} | H_P | 2^2S_{1/2}\rangle}{E_{2^2S_{1/2}} - E_{2^2P_{1/2}}} \cdot |2^2P_{1/2}\rangle. \qquad 71.$$

As noted previously in the energy denominator we must include a term for damping of the 2P state:

$$E_{2S} - E_{2P} = \Delta E_0 + i \frac{\Gamma_{2P}}{2} \qquad\qquad 72.$$

where $\Gamma_{2P} = 100 \text{ MHz}$, and ΔE_0 is the real part of the energy splitting. Also,

ΔE_0 is variable in a magnetic field B because of Zeeman effect; e.g. it becomes zero at the $\beta_0 e_0$ crossing at 545 gauss; or the $\beta_0 f_0$ crossing at 1160 gauss. For the $\beta_0 e_0$ crossing we find from Equations 68 and 63 and 64:

$$\frac{\langle 2^2P_{1/2}, e_0 | H_P | 2^2S_{1/2}, \beta_0 \rangle}{i\Gamma_{2P}/2} = -2C_2 \frac{0.013 \text{ Hz}}{100 \text{ MHz}} = -2.6 \times 10^{-10} C_2. \quad 73.$$

4.1.2 SCALARS AND PSEUDOSCALARS From Equation 73 it is plain that parity-violating effects in hydrogenic atoms are exceedingly small. In order to design effective experiments to detect them it is useful to consider all possible scalar and pseudoscalar forms that can be constructed from the various vectors describing experimental arrangements. (Such an exercise is also helpful for heavy atom experiments.) In general, the relevant vectors are external static electric ($\mathbf{E}$) and magnetic ($\mathbf{B}$) fields, electric vector $\hat{\varepsilon}$ and direction $\hat{k}$ of a light beam, $\hat{\varepsilon}_R$, $\hat{k}_R$ and magnetic vector $\hat{m}_R$ for a microwave field ($\hat{\varepsilon}_R$ and $\hat{m}_R$ are not necessarily orthogonal), initial and final atomic polarization $\mathbf{J}$, and so on. Any term, scalar or pseudoscalar, appearing in the transition probability W, must be expressible in terms of these vectors. However, whether a term actually appears in W, and if so, its magnitude, depends on detailed physical arguments rather than the present symmetry considerations.

For example consider a circular dichroism experiment. Here the only available vectors are those describing a circularly polarized light beam, namely $\hat{k}$ and $\hat{\varepsilon}$ (which is complex). The only pseudoscalar we can form is $i\hat{\varepsilon}^* \times \hat{\varepsilon} \cdot \hat{k}$, which is the photon helicity h.

Certain general considerations restrict our choice of possible forms. First, since reversal of $\hat{\varepsilon}$ produces no physical change, any acceptable term must contain only even powers of $\hat{\varepsilon}$ (as in $h = i\hat{\varepsilon}^* \times \hat{\varepsilon} \cdot \hat{k}$). Next, if $\mathbf{E}$ appears in a pseudoscalar term, it is contained in a Stark amplitude which interferes with $\mathscr{E}_P$; thus $\mathbf{E}$ must appear linearly. Third, under time reversal (T), $\mathbf{E} \to \mathbf{E}$, $\mathbf{B} \to -\mathbf{B}$, $\hat{k} \to -\hat{k}$, $\hat{\varepsilon} \to \hat{\varepsilon}$, $\hat{m} \to -\hat{m}$, $h \to h$, and $\mathbf{J} \to -\mathbf{J}$. If damping is negligible (the case for all heavy atom experiments and hypothetical hydrogenic atom experiments in which 2S-2P level separations are large compared to Γ_{2P}), then scalar and pseudoscalar terms must be even under T. (We assume T-invariance.) For example, $(\hat{\varepsilon} \cdot \mathbf{B})(\hat{\varepsilon} \cdot \mathbf{E} \times \mathbf{B})$ is a T-even pseudoscalar and an acceptable possibility, while $(\hat{\varepsilon} \cdot \hat{\varepsilon})(\mathbf{E} \cdot \mathbf{B})$ is T-odd and thus unacceptable. However, if damping is important (as in hydrogen experiments where 2S-2P level separations are small compared to Γ_{2P}), then it can be shown that formally T-odd terms are acceptable (Bell 1979). Loosely speaking, this occurs because an extra factor of i appears in $\mathscr{E}_P$, due to $\Delta E = \Delta E_0 + (i\Gamma/2) \to i\Gamma/2$. An example of a pseudoscalar term that is acceptable in these circumstances, though formally T-odd, is $(\hat{\varepsilon}_R \cdot \mathbf{E})(\hat{\varepsilon}_R \cdot \mathbf{B})$. Dunford et al (1978) have given a complete list of possible scalar and pseudoscalar terms for hydrogen experiments.

4.1.3 EXPERIMENTS   Experiments are now underway at Seattle (Trainor 1979), Michigan (Dunford et al 1978), and Yale (Hinds 1979). The important parameters in each experiment are summarized in Table 3. Each employs a beam of $2^2S_{1/2}$ atoms traveling along a coaxial B field. The beam is polarized, and then enters one or more resonance regions where transitions are induced to another $2^2S_{1/2}$ state at a 2S-2P level crossing. At present, each experiment is designed to work at the $\beta_0 e_0$ crossing and is thus sensitive to C_{2p}. Future extensions to the $\beta_0 f_0$ and $\beta_{-1} f_{-1}$ crossings and the use of deuterium or tritium as well as hydrogen may permit measurements of all four coupling constants.

Since the parity mixing indicated by Equation 73 is so small, efforts must be made to suppress the parity-conserving amplitude as much as possible, but this results in low counting rates and dilution of signal by background. Asymmetries at the 10^{-7} level are expected if the coupling constants are as described by the Weinberg-Salam model.

In the Michigan experiment (Figure 9) a metastable beam is produced by passing a proton beam from a duoplasmatron source through a cesium charge transfer canal; this yields a flux of $\sim 10^{14}$ atoms/s in the 2S state. A 575-gauss magnetic field is applied in the "β quench" region, and an electric field mixes β and e states, causing decay of all β's. The beam, now pure α, continues to the interaction region, which consists of a $TM_{0,0}$ cylindrical microwave cavity that is tilted at an angle $\phi = 5°$ with respect to the beam axis and $\mathbf{B}$, and tuned to the $\alpha_0\beta_0$ transition. The rf polarization $\hat{\varepsilon}_R$ is along the cavity axis. A dc electric field $\mathbf{E}$ of 1 V/cm Stark-mixes β_0 and e_- states.

Table 3   Experimental parameters for the hydrogen experiments

Experiment	Michigan	Yale
Transition	α_0-β_0 Stark induced	β_0-β_- Stark induced
Magnetic field	545 G ($\beta_0 e_0$ crossing)	500–600 G
PNC mixing	$\beta_0 e_0$	$\beta_0 e_0$
Coupling constant	C_{2p}	C_{2p}
Pseudoscalar	$(\hat{\varepsilon} \cdot \mathbf{E})(\hat{\varepsilon} \cdot \mathbf{B})$	$(\mathbf{E}_1 \cdot \mathbf{B})(\hat{\varepsilon}_{R1} \times \mathbf{B}) \cdot (\hat{\varepsilon}_{R2} \times \mathbf{B})$ or
		$(\mathbf{E}_1 \cdot \mathbf{B})(\hat{\varepsilon}_{R1} \times \hat{\varepsilon}_{R2} \cdot \mathbf{B})$
Size of expected asymmetry for $\sin^2 \theta = 0.23$	3×10^{-7}	1.5×10^{-6}
Expected running time to 1σ	400 h	16 h

E is applied perpendicular to **B** with thin wire electrodes inside the cavity. The $\alpha_0\beta_0$ transition rate is proportional to

$$r_s(\hat{\varepsilon}_R \times \mathbf{B})^2(\mathbf{E} \times \mathbf{B})^2 + r_p(\hat{\varepsilon}_R \cdot \mathbf{E})(\hat{\varepsilon}_R \cdot \mathbf{B}) \qquad\qquad 74.$$

where the term in r_s (r_p) is scalar (pseudoscalar). (As we have noted, the pseudoscalar term is formally T-odd.) It is clear from Expression 74 that the transition rate exhibits an asymmetry with respect to reversal of **E** or **B**, or if $\phi \rightarrow -\phi$. The asymmetry is detected by selective quenching of β_0 atoms in the detection region downstream.

With present running conditions of 10^{13} α_0 atoms per second and $\alpha_0 \rightarrow \beta_0$ conversion fraction of 5×10^{-6}, the expected detection rate is $R = 10^7 (1 \pm 5 \times 10^{-6} C_{2p})$ counts/s, which should yield a statistical precision equal to the asymmetry in 400 hours for $\sin^2 \theta = 0.23$. At present, the $\alpha_0 \rightarrow \beta_0$ transition has been observed, but much work remains, and the apparatus just described may be altered considerably before the experiment is completed. Currently, efforts are being made to reduce diluting backgrounds, which may come from such diverse sources as background gas or wall collisions, scattered Lyman-α radiation from the source, or cascades from higher n states. Future work at Michigan will concentrate on

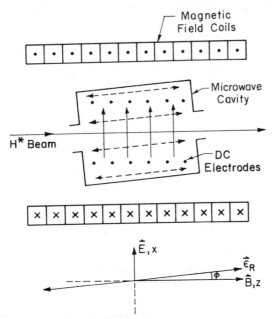

Figure 9 Schematic diagram of the experiment to observe parity violation in hydrogen at Michigan.

detecting and eliminating systematic errors, which have been analyzed by Dunford et al (1978). Reduction of false asymmetries to a level below the signal (for $C_{2p} \simeq 0.1$) requires beam alignments of better than 10^{-3} radians, and electric and magnetic field reversals accurate to 10^{-4} or better. Stray electric fields, and the average motional electric field due to misalignments or divergence of the beam, must be smaller than 5×10^{-6} V/cm!

The Seattle experiment is similar, although details of the interaction region differ slightly. The Seattle group employs an rf cavity with $\hat{\varepsilon}_R$ parallel to the beam and $\mathbf{B}$, and alters the axis of quantization by the small angle ϕ, by applying a small perpendicular $\mathbf{B}$ field. A static electric field is applied to cancel out the resultant motional field in addition to the Stark-mixing $\mathbf{E}$ field. The $\alpha_0 \rightarrow \beta_0$ transition is the same, as are many of the systematic problems, and the experimental parameters and estimated running time are similar. Reduction of stray and motional electric fields may be the chief difficulty in these two experiments. The problem may be simplified by a "separated oscillating fields technique", suggested by Hinds (1979) of Yale, who proposes measuring C_{2p} by driving the transition $\beta_{-1} \rightarrow \beta_0$ near the $\beta_0 e_0$ crossing. Unlike the transition $\alpha_0 \rightarrow \beta_0$, this is a proton spin flip. The experiment is done in two separate regions, 1 and 2, which contain oscillating fields $\mathscr{E}_{R1}$ and $\mathscr{E}_{R2}$ (see Figure 10). Region 1 also contains a static field $\mathbf{E}_1$, and the principal contribution to the $\beta_- \rightarrow \beta_0$ rate is a Stark-

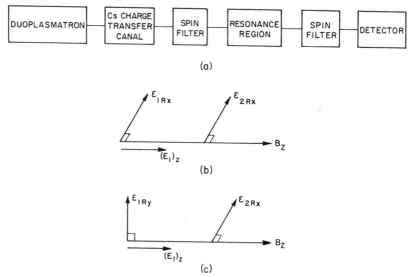

Figure 10   (a) Schematic diagram of the Yale hydrogen experiment. (b) Orientation of vectors in the Yale hydrogen experiment: ε_{R1}, ε_{R2} refer to microwave electric fields in regions 1, 2 respectively; $\mathbf{E}_1$ is static electric field; $\mathbf{B}$ is magnetic field.

induced electric dipole transition in region 1. Region 2 contains no static electric fields, and $\mathscr{E}_{R2}$ drives the principal contribution to the parity nonconserving (PNC) electric dipole transition. As long as phase coherence between $\mathscr{E}_{R1}$ and $\mathscr{E}_{R2}$ is maintained, the total transition rate contains an interference term between these two amplitudes, proportional to the pseudoscalars

$$C_{2p}(\mathbf{E}_1 \cdot \mathbf{B})(\mathscr{E}_{R1} \times \mathbf{B}) \cdot (\mathscr{E}_{R2} \times \mathbf{B}) \qquad 75.$$

or

$$C_{2p}(\mathbf{E}_1 \cdot \mathbf{B})(\mathscr{E}_{R1} \times \mathscr{E}_{R2} \cdot \mathbf{B}), \qquad 76.$$

depending on the relative angle of $\mathscr{E}_{R1}$ and $\mathscr{E}_{R2}$. The scheme has considerable flexibility since the size of the asymmetry now depends not only on $|\mathbf{E}|$, but also on $|\mathscr{E}_{R2}|/|\mathscr{E}_{R1}|$. This, together with the fact that the PNC transition takes place in a field-free region, may decrease the effects of stray electric fields. The major advantage, however, comes from the ability to change the relative phase of the parity-conserving and nonconserving amplitudes by varying the $\mathscr{E}_{R1}$, $\mathscr{E}_{R2}$ phase. During the experiment this will be rapidly varied back and forth by π, which reverses the sign of the interference term. In addition, sign changes occur for $\mathbf{E}_1 \rightarrow -\mathbf{E}_1$ and $\mathbf{B} \rightarrow -\mathbf{B}$. The observed asymmetry can also be distinguished by its magnetic field dependence near the $\beta_0 e_0$ crossing, which differs from the field dependence of stray or motional field asymmetries.

Hinds has carried out an extensive study of possible systematic errors, and has concluded that stray E fields must be kept to the level of $\sim 3 \times 10^{-4}$ V/cm in region 2, and the average beam alignment to 10^{-4} radians. The pseudoscalar contribution to the overall rate is expected to be 1.5×10^{-5} C_{2p}, and a precision of 1 σ after 16 hours of running is expected. Since the $\beta_{-1} \rightarrow \beta_0$ transition is a nuclear spin flip, it is relatively insensitive to backgrounds caused by collisions or stray light. The resonance region can be longer than in the Michigan and Washington experiments because of the lower frequency (100 MHz vs 1.5 GHz) and this results in greater sensitivity.

The advantages of the two-cavity experiment are not restricted to nuclear spin transitions, but may be equally well applied to the $\alpha_0 - \beta_0$ experiments. Such modifications are currently under investigation at Washington and Michigan.

In one version, under study at Washington, the Stark-effect transition occurs in the first cavity, in which $\mathbf{E}_1$ and $\mathscr{E}_{R1}$ are both perpendicular to the beam. The second cavity has $\mathscr{E}_{R2}$ along the beam. The $\mathscr{E}_{R1}$, $\mathscr{E}_{R2}$ phase is adjusted for maximum weak-EM interference. This scheme has the added advantage that the large Stark-allowed $\alpha_+ \beta_0$ and $\alpha_0 \beta_-$ transitions, which cause backgrounds in the one-cavity experiment, cannot occur with $\mathscr{E}_{R1} \parallel \mathbf{E}_1$.

In addition to the atomic hydrogen experiments described so far, experiments involving the decay of muonic hydrogenic atoms have also been suggested (Moskalev 1974a,b, Feinberg & Chen 1974).

4.2   Heavy Atoms

4.2.1   GENERAL CONSIDERATIONS   Two parity-violating effects are of practical importance for heavy atoms. These are "optical rotation," and "circular dichroism," each of which may be understood from consideration of the "M1" transition $\gamma + |\psi_1\rangle \to |\psi_2\rangle$ using circularly polarized photons. If $\hat{k}$ is the direction of photon momentum and $\hat{\varepsilon}_\pm = (\hat{i} \pm i\hat{j})/\sqrt{2}$ describes a circular polarization state of helicity ± 1, then O_{EM} in Equation 11 is $O_{EM} = -\hat{\varepsilon}_\pm \cdot \mathbf{d} - \hat{k} \times \hat{\varepsilon}_\pm \cdot \boldsymbol{\mu}$, where $\mathbf{d}$ and $\boldsymbol{\mu}$ are E1 and M1 transition operators, respectively, for the atom. From this expression we easily obtain $O_{EM} = -\hat{\varepsilon}_\pm \cdot (\mathbf{d} \mp i\boldsymbol{\mu})$, which implies a transition probability

$$W \propto |\mathscr{M}|^2 + |\mathscr{E}_P|^2 \pm 2\,\mathrm{Im}(\mathscr{E}_P\mathscr{M}^*) \qquad 77.$$

and a characteristic asymmetry $\Delta \sim 2\,\mathrm{Im}(\mathscr{E}_P)/\mathscr{M}$. From Equation 11, $\mathscr{E}_P \approx (ea_0/\Delta E)\,\langle\chi|H_W^{(1)}|\psi_1\rangle$, where for purposes of a simple estimate a single state $|\chi\rangle$ is assumed to dominate in the sum, and the electric dipole matrix element $\langle\psi_2|O_{EM}|\chi\rangle$ is assumed to be that of an allowed E1 transition, $\sim 3ea_0$, where a_0 is the Bohr radius. Also, we take ΔE to be a typical spacing between atomic energy levels $\Delta E \approx 0.05e^2/a_0$. Now, matrix elements of $H_P^{(1)}$ are nonzero only for atomic orbitals of opposite parity with nonvanishing value/gradient at the origin ($s_{1/2}$, $p_{1/2}$ orbitals). From Equations 7 or 8, one finds: $\langle\chi|H_W^{(1)}|\psi\rangle \approx 10^{-19}(e^2/a_0)Z^3K$ where K is a relativistic correction factor ($K \sim 10$ for $Z \sim 80$). (This was first demonstrated by Bouchiat & Bouchiat 1974b.) Thus one obtains $|\mathscr{E}_P| \sim 5 \times 10^{-18}\,ea_0Z^3K$. This is, of course, an extremely small electric dipole amplitude in comparison to allowed E1 transition amplitudes (of order ea_0). However, the enhancement factor Z^3K helps considerably for large Z. We find

$$|\mathscr{E}_P| \approx 10^{-10}\,ea_0 \text{ for } Z \sim 80. \qquad 78.$$

It is therefore clear why heavy atoms are chosen for study.

In optical rotation experiments, originally suggested by a number of authors (Zel'dovich 1959, Khriplovich 1974, Sandars 1975, Soreide & Fortson 1975), a beam of linearly polarized light with frequency close to resonance traverses a cell of length L cm containing an atomic vapor of density N cm^{-3}. Optical rotation of the plane of polarization occurs because the linear polarization is a superposition of circular polarization states ($\pm$) that propagate with different indices of refraction $n_\pm$:

$$n_\pm = 1 - \frac{2\pi N}{\hbar}\,\overline{[|\mathcal{M}|^2 \pm 2\,\mathrm{Im}(\mathcal{E}_P\mathcal{M}^*)]}\left\langle \frac{1}{\omega - \omega_0 + (v/c)\omega_0 + i(\Gamma/2)} \right\rangle \qquad 79.$$

Here ω is the photon frequency, ω_0 is the transition frequency, Γ is the natural width of the excited state, v is the thermal atomic velocity in the direction of the light beam, and the angle brackets indicate an average over the Doppler width of the line. Also the bar over the matrix element squared indicates a sum over final, and average of initial, atomic polarizations. Absorption occurs (by slightly differing amounts for the $\pm$ components), and therefore the light emerging from the cell is elliptically polarized. The absorption coefficients are

$$\alpha_\pm = \frac{2\omega}{c}\,\mathrm{Im}(n_\pm). \qquad 80.$$

The optical rotation angle is easily found to be:

$$\phi = \frac{\omega}{2c}l \cdot \mathrm{Re}(n_+ - n_-) \qquad 81.$$

where l is one absorption length. The rotation angle ϕ follows a dispersion-like dependence on $\omega - \omega_0$, and near resonance

$$\phi_{\max} \approx l\frac{\mathrm{Im}(\mathcal{E}_P)}{\mathcal{M}}.$$

For experiments actually performed on the allowed M1 transitions in bismuth, $Z = 83$ (see Figure 11):

$6p^3, J = 3/2$ (ground state) $\rightarrow 6p^3, J = 5/2 \qquad \lambda = 648$ nm

$6p^3, J = 3/2$ (ground state) $\rightarrow 6p^3, J = 3/2 \qquad \lambda = 876$ nm.

$\mathcal{M}$ is approximately one Bohr magneton: $\mathcal{M} \approx ea_0\alpha$. Thus from Equations 78 and 82 we estimate

$$\phi_{\max}/l \sim 10^{-8}\text{--}10^{-7} \text{ radians/absorption length.} \qquad 82.$$

As we shall see, rotations of this order are in fact observed (Barkov & Zolotorev 1978a,b, 1979).

One may also investigate circular dichroism δ in certain forbidden M1 transitions in heavy atoms (Cs, $6^2S_{1/2} \rightarrow 7^2S_{1/2}$), (Tl, $6^2P_{1/2} \rightarrow 7^2P_{1/2}$) (see Figures 12 and 13). By definition, $\delta = (\sigma_+ - \sigma_-)/(\sigma_+ + \sigma_-)$ where $\sigma_\pm$ are the cross sections for resonant absorption of circularly polarized photons with

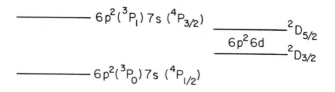

$_{83}Bi^{209}$ $I = 9/2$

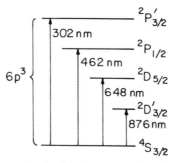

Figure 11   Low-lying energy levels of the bismuth atom. Optical rotation experiments have been carried out using the 648-nm and 876-nm transitions.

$_{55}Cs^{133}$ $I = 7/2$

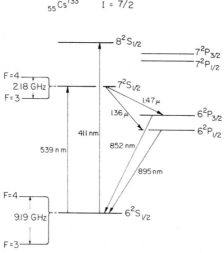

Figure 12   Energy levels of the cesium atom. The forbidden M1 transition $6^2S_{1/2} \rightarrow 7^2S_{1/2}$ (539 nm) is employed by the Paris group to search for parity violation.

helicities ± 1. From Equation 77 we obtain

$$\delta = \frac{\sigma_+ - \sigma_-}{\sigma_+ + \sigma_-} = \frac{2\,\mathrm{Im}(\mathscr{E}_\mathrm{P}\mathscr{M}^*)}{|\mathscr{M}^2| + |\mathscr{E}_\mathrm{P}|^2} \approx \frac{2\,\mathrm{Im}(\mathscr{E}_\mathrm{P})}{\mathscr{M}}. \qquad 83.$$

The M1 amplitudes in these cases are extremely small: $\mathscr{M} \approx (10^{-4}\text{--}10^5)ea_0$, and so $\delta = 2\,\mathrm{Im}(\mathscr{E}_\mathrm{P})/\mathscr{M} \approx 10^{-3}\text{--}10^{-4}$. Unfortunately, full advantage cannot be taken of this relatively large asymmetry because the M1 intensity itself is very small compared to background signals in actual experiments. To overcome this it is necessary to apply an external electric field $\mathbf{E}$ that causes Stark-mixing of $|\psi\rangle$ with the $|\chi_n\rangle$ and introduces an E1 transition amplitude $\mathscr{E}_\mathrm{S}$

$$\mathscr{E}_\mathrm{S} = -e\mathbf{E}\cdot\sum_n \frac{\langle\psi_2^0|O_\mathrm{EM}|\chi_n^0\rangle\langle\chi_n^0|\mathbf{r}|\psi_1^0\rangle}{E(\psi_1^0) - E(\chi_n^0)} + \frac{\langle\psi_2^0|\mathbf{r}|\chi_n^0\rangle\langle\chi_n^0|O_\mathrm{EM}|\psi_1^0\rangle}{E(\psi_2^0) - E(\chi_n^0)}. \qquad 84.$$

For sufficiently large $\mathbf{E}$, we obtain: $|\mathscr{E}_\mathrm{S}|^2 \gg |\mathscr{M}|^2$ and the transition strength becomes greater than background. To detect $\mathscr{E}_\mathrm{P}$ one exploits a pseudoscalar interference term $\sim \mathscr{E}_\mathrm{P}\mathscr{E}_\mathrm{S}$.

4.2.2 CALCULATIONS OF OPTICAL ROTATION For a heavy atom with many (N) electrons, calculations of $\mathscr{E}_\mathrm{P}$ from Equation 11 is a formidable problem in atomic theory, containing a number of subtle difficulties. First,

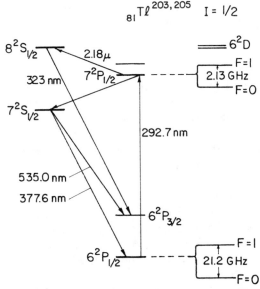

Figure 13 Energy levels of the thallium atom. Parity violation has been observed in the transition $6^2\mathrm{P}_{1/2} \to 7^2\mathrm{P}_{1/2}$ (293 nm) at Berkeley.

evaluation of matrix elements of $H_P^{(1)}$ requires knowledge of wave functions and gradients at the nucleus, where relativistic effects are important. Second, Equation 11 includes a sum over intermediate states, possibly with significant contributions from the continuum and from states of the form $N+1$ electrons, 1 positron (e^+e^- "pair" states). Moreover, $\mathscr{E}_P$ may be seriously affected by shielding corrections, especially in the bismuth optical rotation transitions (Harris et al 1978).

Hiller et al (1979) have discussed a basic theoretical framework that makes possible a consistent treatment due to e^+e^- pairs, from which they have shown there is an appreciable contribution. They have also shown that when the Dirac "velocity" operator α is eliminated in favor of the "length" operator $i\omega r$ in the factors $\langle \psi | O_{EM} | \chi \rangle$ appearing in Equation 11, cancellations result in an accurate formula for $\mathscr{E}_P$, involving only the positive energy N-electron eigenstates of the so-called *no-pair* Hamiltonian H_+. The latter is defined as

$$H_+ = \sum_{i=1}^{N} H_D(r_i) + L_+ \left(\sum_{i<j}^{N} \frac{e^2}{r_{ij}} \right) L_+.$$

85.

Here $L_+ = L_+(1)...L_+(N)$, where $L_+(i)$ is a positive energy projection operator for the i electron, defined by

$$L_+(i)\phi(r_i) = \sum u_n(r_i)\, u_n | \phi \rangle,$$

and the u's are normalized *positive* energy eigenfunctions of the external field one-body Dirac Hamiltonian:

$$H_D = \alpha \cdot \mathbf{p} + \beta m - eA_{nuc}^0(x).$$

Thus specific calculations including only positive energy intermediate states and the length form of the dipole operator are justified. Furthermore, Hiller et al have shown that when the central-field approximation is used, the sum in Equation 11 can be extended over negative energy states with very small error, since these states make only minor contributions. This is important for central-field calculations utilizing a Green's function approach, in which the sum is automatically carried over all intermediate states.

Bismuth has three equivalent p electrons outside of closed shells, and jj coupling is dominant, though not perfect. Thus Bi has a complex structure; it is not surprising that calculations of $\mathscr{E}_P$ are difficult, and that various estimates in the literature disagree by as much as a factor of two. In the calculation of optical rotation in bismuth by the Novosibirsk group (Novikov et al 1976, Khriplovich 1979) the closest levels of opposite parity to $6p^3$, namely the $6p^27s$ levels, are considered, and it is assumed that the 7s electron is added to the $6p^2$ configuration of the Bi-II ion without changing the state of the latter. Effective principal quantum numbers of the 6p and 7s electrons are computed from the known energy levels, and the admixture of $6p^2 7s$ states to $6p^3$ states is thus obtained. Numerical calculations of dipole radial integrals are checked by comparing with lifetimes of Bi excited states. Appreciable contributions also arise from states where a 6s electron is promoted to a 6p orbital: $6s6p^4$. However, the experimental data from Bi II are insufficient to yield numerical values here, so an extrapolation is made from the effective principal quantum number of the 6s electron in Pb and an overall correction factor is applied, since calculated and observed E1 rates in Pb do not agree. Configuration interaction corrections are found to be relatively small from comparison of calculated and observed hfs splittings. The results are in good agreement with other single-particle calculations using the length form of the dipole operator (see Table 4). However, when one goes beyond this approximation, and employs ab initio methods (e.g. the Hartree-Fock method) the situation becomes complicated and some-what unsatisfactory. Sandars (1979) has reviewed the various calculations (see Table 4) and has suggested three possible sources of difficulty:

1. Complications due to the breakdown of jj coupling.
2. Exchange effects, of special importance when the velocity form of the dipole operator is employed. Possibly there exist large corrections due to pair states in this case, as suggested by Hiller et al (1979).

Table 4 Calculations of $\mathscr{E}_P$ in bismuth[a]

| Method | Reference | Im($\mathscr{E}_P$) × 10^{10} au | |
		$J = 3/2 \to J' = 3/2$	$J = 3/2 \to J' = 5/2$
Semiempirical	(Novikov et al 1976)	−3.24	0.94
Parametric potential	(Harris et al 1978)	−4.16	1.31
Parametric potential with shielding	(Harris et al 1978)	−2.78	0.73
Cowan potential (Hartree)	(Henley et al 1977) (Henley & Wilets 1976)	−5.31	
Hartree-Fock length	(Carter & Kelley 1979)	−3.40	0.95
Multiconfiguration Hartree-Fock (MCHF), length[b]	(Grant et al 1979)	−2.82	1.32
MCHF, velocity[b]	(Grant et al 1979)	−2.52	1.53
HF velocity	(Carter & Kelley 1979)	−2.81	0.13

[a] From Sandars (1979). $\sin^2 \theta = 0.23$ is assumed throughout.
[b] 6s and 7s contributions only.

3. Shielding effects by parity-conserving E1 excitations of the remaining electrons. It has been shown that the importance of this effect diminishes as photon frequency increases, and it is expected to enter but play a relatively minor role in the Cs $(6^2S_{1/2} \rightarrow 7^2S_{1/2})$ and Tl $(6^2P_{1/2} \rightarrow 7^2P_{1/2})$ circular dichroism experiments.

Evidently, more work remains to be done on ab initio calculations of $\mathscr{E}_P$ in bismuth. Finally, we note that in optical rotation experiments one requires knowledge of $\mathscr{M}$ as well as $\mathscr{E}_P$. Although the M1 amplitudes have not been measured directly, they can be calculated quite accurately (see Novikov et al 1976).

4.2.3 OPTICAL ROTATION EXPERIMENTS   Three groups have published results of bismuth optical rotation experiments: Seattle (Lewis et al 1977), Oxford (Baird et al 1977), and Novosibirsk (Barkov & Zolotorev 1978a,b, 1979). The first two groups independently reported no parity violations at a level that seemed to exclude predictions on the Weinberg-Salam model; but this was contradicted by the Novosibirsk results, which are in agreement with the predictions of Novikov et al (1976), based on the Weinberg-Salam model. More recently, the Seattle and Oxford groups have also observed parity violation in agreement with predictions of the Weinberg-Salam model.

As noted earlier, the optical rotation per unit absorption length ϕ/l follows a dispersion shape, and the peaks of dispersion correspond to rotations of $1-2 \times 10^{-7}$ radians (see Figure 14). An effect this small cannot be measured with a bismuth cell between crossed polarizers, because the intensity of transmission is proportional to the angle squared. Seattle and Oxford overcome this difficulty by introducing an external optical rotation ϕ_F by means of a Faraday cell. This interferes with the optical rotation ϕ_{PNC} due to parity violation in a known way. The total intensity of light transmitted through the system is $I = I_0[(\phi_{PNC} + \phi_F + \phi_R)^2 + b]$, where I_0 is the incident laser intensity, b is the residual angle-independent transmission through the polarizer (for instance, due to a small birefringence in a window), and ϕ_R is any residual optical rotation not caused by parity nonconservation. By varying the magnetic field in the Faraday cell, one modulates ϕ_F at frequency ω, and lock-in techniques are employed to isolate the ω-dependent part of the intensity:

$$I_\omega = 2I_0\phi_F(\phi_{PNC} + \phi_R).$$

The signal-to-background ratio is

$$\frac{I_\omega}{I} \approx \frac{2\phi_F(\phi_{PNC} + \phi_R)}{\phi_F^2 + b} \simeq \frac{(\phi_{PNC} + \phi_R)}{\phi_F^2} \approx \hat{k}10^{-4} \quad \text{(for Oxford)}$$

where the optimum condition $\phi_F \approx b$ has been assumed. All three groups depend solely on the dispersion shape of ϕ_{PNC} to eliminate all possible sources of ϕ_R, which cannot be distinguished from ϕ_{PNC} in any other way.

The Oxford group is working on the 648-nm $^4S_{3/2} \to {}^2D_{5/2}$ line. This was chosen because at the time it was most accessible to narrow bandwidth continuous wave (CW) dye lasers, which must be used to resolve the details of the lineshape. Unfortunately, molecular bismuth absorption in this band limits the density to that corresponding to about one atomic absorption length. Seattle chose the 876-nm $^4S_{3/2} \to {}^2D_{3/2}$ line, which has no molecular absorption, and this permits much higher densities. Recent advances in tunable diode lasers make available a narrow bandwidth source for 876 nm as well.

In both experiments, the light enters the first polarizer and goes through a Faraday cell. Then it enters the bismuth cell, which is enclosed by an oven and surrounded by magnetic shielding. (Magnetic fields must be kept below 10^{-4} G because of the Faraday effect in bismuth, which is a serious potential source of ϕ_R.) The light then passes out of the cell, through the

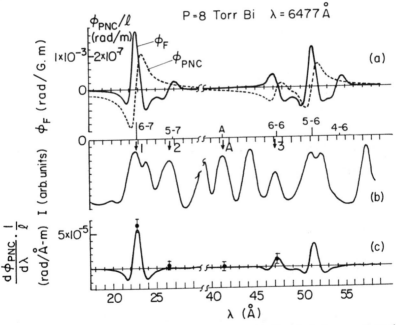

Figure 14 Novosibirsk optical rotation experiment. (*a*) The dashed curve gives the theoretical prediction for parity-violating optical rotation vs wavelength λ. The solid line is calculated Faraday rotation. (*b*) Observed absorption spectrum. (*c*) Calculated curve $(1/l)d\phi_{PNC}/d\lambda$ and results of measurements of the lines 1, 2, 3, *A*. The numbers 6–7, 5–7, etc refer to hyperfine components *F*, *F'*. *A* is a control line.

second polarizer, and is detected. The cell contains a buffer gas that keeps the bismuth vapor from condensing on the cool windows. Data are taken by scanning over the hyperfine structure (Seattle) or by switching between points of maximum dispersion (Oxford). The Oxford group can move its cell in and out of the beam (see Figure 15*a*) to perform null experiments. Other control experiments include the use of a quadrupole resonance or a molecular absorption line, where there is no parity effect.

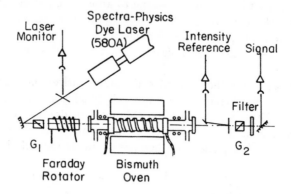

a) OXFORD

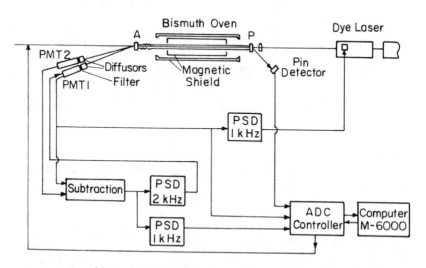

b) NOVOSIBIRSK

Figure 15 Schematic diagrams of optical rotation experiments. (*a*) Oxford. (*b*) Novosibirsk.

The Novosibirsk group also employs the 648-nm transition, but instead of modulating a Faraday cell, they deliberately misalign their polarizers by $\theta = \pm 4$ mrad (see Figure 15b). The detected light has the form $I = I_0 \sin^2 (\theta + \phi_{PNC} + \phi_R) \approx \theta^2 I_0[1 + 2(\phi_{PNC} + \phi_R)/\theta]$. A Spectra Physics 375 CW dye laser, operating single mode, is modulated back and forth by 416 MHz every millisecond. The signal is phase-locked to this frequency, and it is proportional to $d\phi_{PNC}/d\lambda$ (see Figure 14). For technical reasons it is only possible to obtain data with the laser frequency centered on the absorption peaks, but null experiments are still possible by locking to quadrupole or molecular lines in the 648-nm band. The angle θ is reversed every minute or so, and the optics are aligned so that this change does not deviate the beam. The calibration was made by applying a known magnetic field and measuring the Faraday effect. The result of the Novosibirsk experiment may be expressed in terms of $R = Im(\mathscr{E}_P)/\mathscr{M}$:

$$R_{expt} = (-20.6 \pm 3.2) \times 10^{-8}.$$

This result is in agreement with theoretical calculations of Novikov et al (1976): $R_{expt}/R_{theo} = 1.07 \pm 0.14$. It yields the value $Q_W(Bi) = -140 \pm 40$.

4.2.4 CESIUM AND THALLIUM CALCULATIONS   For the transitions Cs $(6^2S_{1/2} \rightarrow 7^2S_{1/2}, 8^2S_{1/2})$ and Tl $(6^2P_{1/2} \rightarrow 7^2P_{1/2}, 7^2P_{3/2})$, relatively straightforward semiempirical calculations of $\mathscr{E}_P$ are expected to be reasonably accurate, since the ground state and low-lying excited states of these atoms are quite well described by a single valence electron outside a spherically symmetric core (one-electron central-field approximation). Core excitation effects on $\mathscr{E}_P$ are believed to be small in each case and susceptible to effective treatment by perturbation theory.

The first central-field calculation was performed by Bouchiat & Bouchiat (1975), who evaluated $\mathscr{E}_P$ for Cs $(6^2S_{1/2} \rightarrow 7^2S_{1/2})$ with a finite sum over $^2P_{1/2}$ terms in Equation 11. Matrix elements of $H_W^{(1)}$ were calculated by means of a modified Fermi-Segre technique, and relativistic corrections were applied. A correction for the contribution of continuum states was also taken into account. Sushkov et al (1976) employed a somewhat similar semiempirical method in calculations of $\mathscr{E}_P$ (Tl, $6^2P_{1/2} \rightarrow 7^2P_{1/2}$), as did Novikov et al (1976) in calculations $\mathscr{E}_P$ (Tl, $6^2P_{1/2} \rightarrow 6^2P_{3/2}$). Neuffer & Commins (1977a) also used the central-field approximation for $\mathscr{E}_P$ (Tl, $6^2P_{1/2} \rightarrow 7^2P_{1/2}, 6^2P_{3/2}, 7^2P_{3/2}$) by fitting a modified Tietz potential (which yields a good approximate solution to the Thomas-Fermi equation) to the $6^2P_{1/2}, 7^2P_{1/2}$ levels of Tl and solving the Dirac equation numerically for the valence electron. $\mathscr{E}_P$ was calculated by a finite sum over nearest $^2S_{1/2}$ states and also by means of a Green's function (sum over all states, including auto-ionizing states in the continuum). A similar calculation was carried out on Cs (Neuffer & Commins 1977b). The wave functions generated in these calculations were used to compute many auxiliary quantities, which could be compared with atomic-beam and spectroscopic data, especially complete for Cs and Tl. These are

1. Allowed E1 transition rates and excited state lifetimes (test of wave functions at large r),
2. Energies and fine structure splittings,
3. Hyperfine structure splittings (test of wave functions near the origin, which is especially important for $\mathscr{E}_P$),
4. Stark-induced E1 amplitudes $\mathscr{E}_S$ (see Equation 84),
5. M1 transition amplitudes, and
6. Anomaly in g_J for ground state.

In general, agreement between calculation and experiment for these auxiliary quantities is very satisfactory, when certain configuration-interaction corrections are taken into account; these are of special importance in $(6^2S_{1/2} \rightarrow 7^2S_{1/2})$ and $g_J (6^2S_{1/2})$ for Cs (see Khriplovich 1975, Flambaum et al 1978). Thus one has confidence that these one-electron central-field approximation calculations are quite adequate for estimating $\mathscr{E}_P$ to $\sim 20\%$ accuracy in each transition. The predicted values of $\mathscr{E}_P$ and δ for Cs and Tl transitions are summarized in Table 5.

Core polarization effects may cause small but systematic deviations from the one-electron central field calculations. Bardsley & Norcross (1980) have shown how to incorporate this into the model for Tl; the results are in better agreement with the available spectroscopic data, especially for

Table 5 Calculations of $\mathscr{E}_P$ and δ in cesium and thallium[a]

Element/ Transition	Reference	$\mathrm{Im}(\mathscr{E}_P) \times 10^{10}$au	$\delta = 2\,\mathrm{Im}(\mathscr{E}_P)/\mathscr{M}$
Cs $6^2S_{1/2} \rightarrow 7^2S_{1/2}$	(Bouchiat & Bouchiat 1974b, 1975)	-0.12	
	(Loving & Sandars 1975, Brimicombe et al 1976)	-0.15	
	(Neuffer & Commins 1977b)	-0.09	1.17×10^{-4}
Tl $6^2P_{1/2} \rightarrow 7^2P_{1/2}$	(Sushkov et al 1976)	-0.76	
	(Neuffer & Commins 1977a)	-0.83	2.2×10^{-3}
$6^2P_{1/2} \rightarrow 6^2P_{3/2}$	(Novikov et al 1976)	-3.3	
	(Neuffer & Commins 1977a)	-3.5	
	(Henley et al 1977, Henley & Wilets 1976)	$-4.04,$ -2.76	
$6^2P_{1/2} \rightarrow 7^2P_{3/2}$	(Neuffer & Commins 1977a)	$+0.76$	

[a] $\mathrm{Sin}^2\,\theta = 0.23$ assumed throughout.

$6P_{1/2} \to nD_{5/2,3/2}$ oscillator strengths. Although the correction to $\mathscr{E}_P$ should be small, the modification is important in calculating Stark effect matrix elements, which are used to calibrate the Tl and Cs experiments.

4.2.5 CIRCULAR DICHROISM EXPERIMENTS IN CESIUM AND THALLIUM The transition Cs $(6^2S_{1/2} \to 7^2S_{1/2})$ is being investigated at Paris (Bouchiat & Pottier 1976b, 1979) while parity violation has been observed at Berkeley (Conti et al 1979) in Tl $(6^2P_{1/2} \to 7^2P_{1/2})$, and work continues to refine the results. One of the original motivations for these experiments is that large circular dichroisms δ are predicted by the Weinberg-Salam model (see Table 5). However, for δ to be observed directly it would be necessary to detect the M1 transition itself with a signal clearly discernible above background. Unfortunately, this has not been possible so far, because in an actual experiment, one must utilize atomic vapor at rather high densities in order to achieve acceptable signals. In this case, random local electric fields due to collisions (and possibly also molecular effects), make weak photon absorption possible over a rather broad band of frequencies, and with a strength much larger than that expected from the extremely feeble M1 amplitude. Thus, $\mathscr{M}$ as well as $\mathscr{E}_P$ must be measured indirectly by interference with a Stark-induced E1 amplitude due to external electric field $\mathbf{E}$ (Equation 84). It is helpful in considering these effects to proceed with a general analysis of possible scalar and pseudo-scalar terms in the transition probability, as was considered for hydrogenic atoms. If we restrict ourselves to *linearly* polarized light and external $\mathbf{E}$ field, and also measure the polarization J of the final atomic state $(7^2S_{1/2}$ in Cs, $7^2P_{1/2}$ in Tl), only the following T-invariant scalars representing Stark-M1 interference can be formed:

$$\hat{\varepsilon} \cdot \mathbf{E} \, \hat{k} \times \hat{\varepsilon} \cdot \mathbf{J} \qquad\qquad 86.$$

$$\hat{\varepsilon} \cdot \mathbf{J} \, \hat{k} \times \hat{\varepsilon} \cdot \mathbf{E}. \qquad\qquad 87.$$

Figure 16 gives the orientation of the various vectors. Choosing $\hat{k}$ along $\hat{x}$, $\mathbf{E}$ along $\hat{y}$, and $\hat{\varepsilon}$ along $\hat{y}$, we find that Expression 86 permits a final polarization along z, which reverses sign with $\hat{k}$ and $\mathbf{E}$. Such a polariz-ation actually occurs in the transitions (Cs, $6^2S_{1/2} \to 7^2S_{1/2}$) and (Tl, $6^2P_{1/2} \to 7^2P_{1/2}$) for $\Delta F = 0$ and $m_F = 0$. An example is the so-called α transition of Tl where $F = 1 \to F' = 1$ and $\Delta m_F = 0$. If $\hat{\varepsilon}$ is along $\hat{z}$ instead, one again has a polarization along z, which reverses sign with $\hat{k}$ and $\mathbf{E}$. This occurs, for example, in the so-called β transition $F = 0 \to F = 1$ in Tl, $(\Delta m_F = \pm 1)$. Detection of such polarizations by observations of the circular polarization of decay fluorescence yielded measurements of $\mathscr{M}$ in Cs (Bouchiat & Pottier 1976a) and in Tl (Chu et al 1977).

If circularly polarized light is employed with helicity $h = \pm 1$, one can form the T-invariant pseudoscalar

$$h\hat{k} \times \mathbf{E} \cdot \mathbf{J}, \qquad\qquad 88.$$

which represents $\mathscr{E}_\mathrm{P}$-Stark interference. With $\hat{k}$ along $\hat{x}$, $\mathbf{E}$ along $\hat{y}$, and $\hat{\varepsilon} = (\hat{y} \pm i\hat{z})/\sqrt{2}$, Expression 88 yields a polarization in the z direction that reverses with h and $\mathbf{E}$, but is independent of the sign of $\hat{k}$. Both the Paris and Berkeley experiments utilize this effect.

At Paris, a CW ring dye laser is employed to excite a single hyperfine component in the 5395-Å transition (Cs, $6^2S_{1/2} \to 7^2S_{1/2}$). The laser light is circularly polarized with a Pockels cell. It then enters a cesium cell, where it is reflected back and forth about 100 times by a pair of curved mirrors inside the cell. This "multipass" design has two advantages: It amplifies the signal, and at the same time essentially eliminates the polarization caused by the Stark-M1 interference, since this reverses with $\hat{k}$.

The static electric field is maintained with plane parallel electrodes inside the cell. The 7S → 6P fluorescence at 1.36 μm is collected and collimated by a lens, passes through an interference filter and circular polarization analyzer, and is then focused onto a germanium detector. The laser

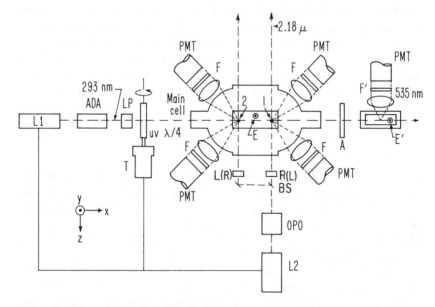

Figure 16 Schematic diagram of the Berkeley thallium experiment. L_1, L_2: Flashlamp-pumped dye lasers. ADA: Nonlinear doubling crystal. LP: linear polarizer. UV $\lambda/4$: 293-nm quarter-wave plate. OPO: Optical parametric oscillator. BS: Beamsplitter. L(R), R(L): reversible 2.18-μm quarter-wave plates. F: Filters for 323-nm radiation.

polarizer and fluorescence analyzer are both modulated, and lock-in detection techniques are used to extract the parity-violating signal. Background is chiefly due to black-body radiation at 1.36 μm from the oven and cell, and a Stark field of ~ 300 V/cm is required to produce a signal large enough for convenient use.

In the Tl $6^2P_{1/2} \to 7^2P_{1/2}$ experiment (see Figure 16), the transition wavelength of 2927 Å is produced by nonlinear second harmonic generation from the output of a flashlamp-pumped pulsed dye laser tuned to 5854 Å. The light is circularly polarized by passing it through a Pockels cell or crystalline quartz quarter-wave plate, which is capable of producing circularly polarized light of great purity (intensity of unwanted polarization $\lesssim 10^{-4} \times$ intensity of wanted polarization). The light then enters the quartz thallium cell, which is in an oven heated to 750–800°C. Oven and cell are inside a rough vacuum, and the cell itself is connected to an ultrahigh vacuum system, which is employed to reduce molecular backgrounds due to contaminants. Technical difficulties have prevented using a multipass system; the laser beam either makes a single pass, or makes one reflection and returns. At best, this only reduces the M1 polarization by $\sim 80\%$. **E** is produced by plane parallel electrodes either inside or outside the cell. The $F = 0 \to F = 1$ transition is used for all parity data. However, the 2.13-GHz hyperfine splitting of the $7^2P_{1/2}$ state is easily resolved, and so it is possible to tune to the $F = 0 \to F' = 0$ line. In this case, there is no final state polarization; the $0 \to 0$ line is thus used for a null experiment.

In thallium the polarization of the $7^2P_{1/2}$ state is not easily analyzed by observation of circular polarization of decay fluorescence. Black-body radiation prevents viewing the 1.3-μm $7P_{1/2} \to 7S_{1/2}$ transition, and the 5350-Å $7S_{1/2} \to 6P_{3/2}$ light has only one-twelfth the original polarization. Instead, a second laser is utilized. It drives an optical parametric oscillator that produces photons circularly polarized along the z axis and tuned to the 2.18-μm $7P_{1/2} \to 8S_{1/2}$ E1 transition. By selective excitation of the $7P_{1/2}$ $F = 1$, $m_F = +1$ or -1 states to $8^2S_{1/2}$, and by observation of the $8^2S_{1/2} \to 6^2P_{3/2}$ fluorescence, the analyzing power of the $7^2P_{1/2}$ polarization is about 70%. The chief source of background in the thallium experiment is UV fluorescence from the cell scattered into the detectors at the wavelength 3230 Å corresponding to the $8^2S_{1/2} \to 6^2P_{3/2}$ transition. Signals appreciably greater than background are achieved for a Stark field of 170 V/cm or more.

In order to eliminate laser fluctuations, two separate interaction regions are used; these have opposite handedness for each laser pulse. This is accomplished by splitting the 2.18-μm beam and polarizing the two halves oppositely in the two regions. Thus the polarization asymmetry can be measured on each pulse. In addition to reversing the UV $\lambda/4$ plate on successive pulses, the IR polarization and the electric field directions are also reversed periodically. All of these reversals, plus the 0-0 null experiment, reduce possible systematic errors to acceptable levels. The most critical reversal is the UV polarization, which is needed to cancel the M1 asymmetry. By contrast, in the cesium experiment the multipass cell eliminates the M1, but the extremely small size of the expected effect makes the electric field reversal critical. For instance, if the laser beam in either experiment is misaligned so that small components of $\mathbf{E}$ exist along $\hat{x}$ (the beam direction) and $\hat{z}$, there will be a Stark-effect polarization along $\hat{z}$ of $P = 2E_x E_z/E^2$ that mimics the parity effect. (This is described by the scalar invariant $h\, \mathbf{J} \cdot \mathbf{E}\, \hat{k} \cdot \mathbf{E}$.) The result of the Berkeley experiment (Conti et al 1979) is $\delta = (5.2 \pm 2.4) \times 10^{-3}$, which yields $Q_W = -280 \pm 140$.

Alternative versions of these experiments with attractive features have been considered independently by Bouchiat et al (1980) and Commins (Bucksbaum 1979) and are being carried out. Here one employs linearly polarized light and selects the m_F components of the final state by using an external magnetic field to split Zeeman components. The pseudoscalar of interest is now proportional to $\hat{\varepsilon} \cdot \mathbf{B}\, \hat{\varepsilon} \cdot \mathbf{E} \times \mathbf{B}$ and $\hat{k}$ is chosen parallel to E. In this version of the experiment, possible systematic errors arising from circularly polarized light are eliminated, and expected signals in the Tl case are much larger than previously attained.

4.2.6 MISCELLANEOUS EFFECTS   There should also exist effects in atoms due to the parity-violating neutral weak electron-electron interaction (Bouchiat & Bouchiat 1974b, Sushkov & Flambaum 1978a), but these are expected to be reduced relative to effects due to $H_W^{(1)}$ by a factor of order Z^{-1} $(1 - 4\sin^2\theta)$ in heavy atoms; that is, comparable to effects due to $H_W^{(2)}$ (Novikov et al 1977). Additional modifications arise from radiative corrections to H_P (Marciano & Sanda 1978).

Parity violation effects in diatomic molecules have been discussed by Sushkov & Flambaum (1978b). Rein et al (1979) have considered parity-violating energy differences between mirror image molecules and have found them to be very small. Other aspects of the latter question have been considered by Harris & Stodolsky (1978). Possible manifestations of parity violation in the Josephson effect have been discussed by Vainshtein & Khriplovich (1974, 1975).

5   SUMMARY AND CONCLUSIONS

The results of electron scattering and atomic physics experiments (eq sector) may be combined with neutrino-nucleon (νq) and neutrino-electron (νe) scattering data to provide a stringent test of neutral weak interaction theories. In carrying out this analysis it is useful to begin with the simplest possible model-independent assumptions (Hung & Sakurai 1979, Sakurai 1979). Thus we start merely by assuming μe universality, that the contributions of heavy quarks c,s,... may be neglected, and that all neutral weak currents possess only vector and axial vector components. We may then write the effective Hamiltonians:

$$H_{vq} = -G\sqrt{2}\,\bar{v}\gamma_\lambda(1-\gamma_5)v\{\tfrac{1}{2}[\bar{u}\gamma_\lambda(\alpha-\beta\gamma_5)u - \bar{d}\gamma_\lambda(\alpha-\beta\gamma_5)d]$$
$$+\tfrac{1}{2}[\bar{u}\gamma_\lambda(\gamma-\delta\gamma_5)u + \bar{d}\gamma_\lambda(\gamma-\delta\gamma_5)d]\}; \qquad 89.$$

$$H_{ve} = -\frac{G}{\sqrt{2}}[\bar{v}_\mu\gamma_\lambda(1-\gamma_5)v_\mu][e(g_V\gamma_\lambda - g_A\gamma_\lambda\gamma_5)e]; \qquad 90.$$

$$H_{eq} = -\frac{G}{\sqrt{2}}\left\{\bar{e}\gamma_\lambda\gamma_5 e\left[\frac{\tilde{\alpha}}{2}(\bar{u}\gamma_\lambda u - \bar{d}\gamma_\lambda d) + \frac{\tilde{\gamma}}{2}(\bar{u}\gamma_\lambda u + \bar{d}\gamma_\lambda d)\right]\right.$$
$$\left. + \bar{e}\gamma_\lambda e\left[\frac{\tilde{\beta}}{2}(\bar{u}\gamma_\lambda\gamma_5 u - \bar{d}\gamma_\lambda\gamma_5 d) + \frac{\tilde{\delta}}{2}(\bar{u}\gamma_\lambda\gamma_5 u + \bar{d}\gamma_\lambda\gamma_5 d)\right]\right\}. \qquad 91.$$

The ten coupling constants $\alpha, \beta, \ldots, \tilde{\beta}, \tilde{\delta}$ must be determined by experiment. Note that Equation 91 reduces to Equations 4 and 5 provided we put

$$C_{1p} = -\tfrac{1}{2}(3\tilde{\gamma}+\tilde{\alpha}), \; C_{1n} = -\tfrac{1}{2}(3\tilde{\gamma}-\tilde{\alpha}),$$

which follows from the assumption that $p = 2u+d, n = 2d+u$.

The neutrino-nucleon scattering data yield values of α, β, γ, and δ that are determined up to an overall sign ambiguity. These are listed in Table 6, column 1. For a review of neutrino-scattering, see Cline & Fry (1977). Note that β is determined from a recent measurement of the cross section for

$$\bar{v}_e + D \rightarrow \bar{v}_e + n + p \qquad 92.$$

in which low energy electron-antineutrinos are employed (Pasierb et al 1979). In $\bar{v}_e + e \rightarrow \bar{v}_e + e$, charged and neutral currents can participate. All other neutrino scattering experiments have so far utilized high energy v_μ or $\bar{v}_\mu$.

The coupling constants g_V, g_A for ve scattering have now been measured to be (Armenize et al 1979)

$$g_A = -0.52\pm0.06 \qquad g_V = 0.06\pm0.08. \qquad 93.$$

Actually there is an ambituity involving $g_A \leftrightarrow g_V$ here; we have written the "axial vector dominant" solution.

Next, the results of the SLAC electron scattering experiment are expressed in terms of the coefficients a_1, a_2 appearing in $\Delta/(-q^2)$ (Equation 52).

Writing

$$a_1 = \frac{G}{\sqrt{2}\,e^2}\frac{9\tilde{\alpha}+3\tilde{\gamma}}{5}$$ 94.

$$a_2 = \frac{G}{\sqrt{2}\,e^2}\frac{9\tilde{\beta}+3\tilde{\delta}}{5}$$ 95.

and recalling the experimental values for $a_{1,2}$ (Equations 58 and 59), we find

$$\tilde{\alpha}+\frac{\tilde{\gamma}}{3} = -0.60\pm0.16$$ 96.

$$\tilde{\beta}+\frac{\tilde{\delta}}{3} = 0.31\pm0.51.$$ 97.

The results of the Novosibirsk and Berkeley Tl and Bi experiments: $Q_{\mathrm{W}}(\mathrm{Bi}) = -140\pm40$ and $Q_{\mathrm{W}}(\mathrm{Tl}) = -280\pm140$ are expressible in terms of $\tilde{\alpha}$ and $\tilde{\gamma}$ by means of the relation:

$$Q_{\mathrm{W}} = -\tilde{\gamma}(Z+N)+\tilde{\alpha}(Z-N).$$ 98.

Table 6 Determination of neutral current coupling parameters (from Sakurai 1979)

Parameter	Model independent[a]	Factorization dependent[b]	Weinberg-Salam model	Weinberg-Salam model, $\sin^2\theta = 0.23$
α	$\pm0.58\pm0.14$	0.58 ± 0.14	$1-2\sin^2\theta$	0.54
β	$\pm0.92\pm0.14$	0.92 ± 0.14	1	1
γ	$\mp0.28\pm0.14$	-0.28 ± 0.14	$-\frac{2}{3}\sin^2\theta$	-0.153
δ	$\pm0.06\pm0.14$	0.06 ± 0.14	0	0.0
g_{V}	0.00 ± 0.18 or -0.52 ± 0.13	0.03 ± 0.12	$-\frac{1}{2}(1-4\sin^2\theta)$	-0.04
g_{A}	-0.56 ± 0.14 or -0.07 ± 0.15	-0.56 ± 0.14	$-\frac{1}{2}$	-0.5
$\tilde{\alpha}$	-0.72 ± 0.25	-0.72 ± 0.25	$-(1-2\sin^2\theta)$	-0.54
$\tilde{\beta}$	—	0.06 ± 0.21	$-(1-4\sin^2\theta)$	-0.08
$\tilde{\gamma}$	0.38 ± 0.28	0.38 ± 0.28	$\frac{2}{3}\sin^2\theta$	0.153
$\tilde{\delta}$	—	0.00 ± 0.02	0	0.0
$\tilde{\alpha}+\frac{1}{3}\tilde{\gamma}$	-0.60 ± 0.16	-0.60 ± 0.16	$-(1-\frac{20}{9}\sin^2\theta)$	-0.489
$\tilde{\beta}+\frac{1}{3}\tilde{\delta}$	0.31 ± 0.51	0.06 ± 0.21	$-(1-4\sin^2\theta)$	0.08

[a] Coupling constants determined without recourse to factorization or gauge theory considerations.
[b] Coupling constants determined with factorization constants included.

We plot the experimental constraints on $\tilde{\alpha}$ and $\tilde{\gamma}$ in Figure 17. Thus we obtain

$$\tilde{\alpha} = -0.72 \pm 0.25 \qquad\qquad\qquad 99.$$

$$\tilde{\gamma} = 0.38 \pm 0.28. \qquad\qquad\qquad 100.$$

Further restrictions on the coupling constants are obtained if one assumes a model with single Z boson exchange (the "factorization" hypothesis). Assuming μe universality, this is characterized by seven independent parameters for coupling Z to v_L, $u_{L,R}$, $d_{L,R}$, and $e_{L,R}$. Since we started with ten parameters $\alpha, \beta, \ldots, \tilde{\delta}$ there must be three independent "factorization" relations connecting the latter. Hung & Sakurai (1977a,b) have shown these to be

$$\tilde{\gamma}/\tilde{\alpha} = \gamma/\alpha \qquad\qquad\qquad 101.$$

$$\tilde{\delta}/\tilde{\beta} = \delta/\beta \qquad\qquad\qquad 102.$$

$$g_V/g_A = \alpha\tilde{\beta}/\tilde{\beta}\alpha. \qquad\qquad\qquad 103.$$

Let us plot the allowed values of γ/α from vq data (Table 6) on Figure 17 and utilize Equation 101. Then we see that the SLAC and Bi, Tl results fall within the region permitted by vq data; this provides model-independent evidence for factorization.

Since $\tilde{\beta}$ and $\tilde{\delta}$ have not yet been separately determined (this would require a measurement of parity violation at the $\beta_0 e_0$ crossing in hydrogen, for

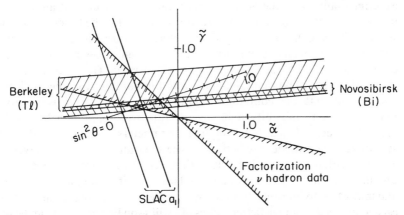

Figure 17 Results of the SLAC polarized electron experiment (parameter a_1) and the Novosibirsk (Bi) and Berkeley (Tl) atomic physics experiments are plotted on the $\tilde{\alpha}$-$\tilde{\gamma}$ plane. The factorization hypothesis together with v-hadron scattering data constrains the allowed region of the $\tilde{\alpha}$-$\tilde{\gamma}$ plane as shown. (From Sakurai 1979.)

example), one cannot test Equation 102 directly. However, combining Equations 101, 102, and 103, one obtains

$$g_V/g_A = \frac{[\alpha+(\gamma/3)][\beta+(\delta/3)]}{[\tilde{\alpha}+(\tilde{\gamma}/3)][\tilde{\beta}+(\tilde{\delta}/3)]}.$$ 104.

From the experimentally determined ratio a_2/a_1 (Equations 96 and 97), one can determine the right-hand side of Equation 104. This determines that only the "axial-vector dominant" solution for g_V,g_A is acceptable in νe scattering, as we have written in Equation 93.

One may use the SLAC experimental result (Equations 96 and 97) and the factorization condition (Equation 102), with δ/β determined from νq data (Table 6), to obtain

$$\tilde{\beta} = 0.29 \, {}^{+0.55}_{-0.51} \qquad \tilde{\delta} = 0.02 \, {}^{+0.17}_{-0.06}.$$ 105.

Another factorization relation,

$$\begin{Bmatrix} \tilde{\beta} \\ \tilde{\delta} \end{Bmatrix} = \begin{Bmatrix} \beta \\ \delta \end{Bmatrix} \frac{g_V}{g_A} \frac{\tilde{\alpha}+(\tilde{\gamma}/3)}{\alpha+(\gamma/3)}$$ 106.

yields more precise limits:

$$\tilde{\beta} = 0.06 \pm 0.21 \qquad \tilde{\delta} = 0.00 \pm 0.02.$$ 107.

Now, the signs of the $\bar{\nu}_e e$ coupling constants g_V,g_A can be determined experimentally, since the neutral current and charged current amplitudes interfere. (We make a standard V-A choice of sign for the latter.) Also, the signs of the eq constants $\tilde{\alpha}$, $\tilde{\beta}$, $\tilde{\gamma}$, $\tilde{\delta}$ are determined because of interference with the electromagnetic amplitude of known sign. One cannot measure directly the absolute sign of νq coupling constants. However, it can be shown (Sakurai 1979) that the factorization hypothesis, together with the assumption that the coupling strength c_ν^2 in $\nu+\nu \rightarrow \nu+\nu$ is positive, removes the sign ambiguity in the νq constants. Table 6, column 2, gives the present values of the ten coupling parameters thus obtained. The Weinberg-Salam model predictions of these parameters are also given in Table 6, column 3, and it is clear that very satisfactory agreement is obtained for $\sin^2 \theta = 0.23$ (column 4). However, it is obviously desirable and important to improve the precision of determination of the parameters, especially γ, δ, $\tilde{\beta}$, $\tilde{\gamma}$, and $\tilde{\delta}$. Better values of $\tilde{\gamma}$ will soon be obtained from heavy atom experiments, and $\tilde{\beta}$, $\tilde{\gamma}$, and $\tilde{\delta}$ may be measured accurately in hydrogen atom experiments within the next few years. Beyond this one may hope that various experiments will eventually shed light on small but important effects such as the electron-electron parity-violating coupling, momentum-transfer-dependent terms and higher order corrections to the Weinberg-Salam model. The latter would be particularly exciting to observe since the ability to predict them is one of the most important features of unified gauge field theories.

We are grateful to the following persons for providing us with materials related to their work before publication: Dr. M. A. Bouchiat and Dr. L. Pottier, Prof. C. Bouchiat, Prof. E. Adelberger, Prof. E. Hinds, Dr. C. Prescott, Prof. P. G. H. Sandars, Prof. J. J. Sakurai, and Prof. W. Williams. We also thank them and Prof. E. N. Fortson, Prof. I. B. Khriplovich, and Prof. V. Telegdi for helpful discussions. This review was written under the auspices of the Chemical Sciences Division, Office of Basic Energy Sciences, US Department of Energy.

Literature Cited

Abers, E. S., Lee, B. W. 1973. *Phys. Rep. Phys. Lett. C* 9C: 1–141

Armenize, N. et al. 1979. *Phys. Lett. B* 86: 225–28

Baird, P. E. G., Brimicombe, M. W., Hunt, R. G., Roberts, G. J., Sandars, P. G. H., Stacey, D. N. 1977. *Phys. Rev. Lett.* 39: 798–801

Bardsley, J. N., Norcross, D. W. 1980. *J. Quant. Spectrosc. Radiat. Transfer.* In press

Barkov, L. M., Zolotorev, M. S. 1978a. *Pis'ma Zh. Eksp. Teor. Fiz.* 27: 379–83 (*JETP Lett.* 27: 357–61)

Barkov, L. M., Zolotorev, M. S. 1978b. *Pis'ma Zh. Eksp. Teor. Fiz.* 28: 544–48 (*JETP Lett.* 28: 503–6)

Barkov, L. M., Zolotorev, M. S. 1979. *Phys. Lett. B* 85: 308–13

Bell, J. S. 1979. *Proc. Workshop Parity Violation in Atoms, Cargese, Corsica, September 1979.* In press

Bjorken, J. D., Drell, S. D. 1964. *Relativistic Quantum Mechanics.* New York: McGraw-Hill. 300 pp.

Bouchiat, C. 1975. *Phys. Lett. B* 57: 284–88

Bouchiat, M. A., Bouchiat, C. 1974a. *Phys. Lett. B* 48: 111–14

Bouchiat, M. A., Bouchiat, C. 1974b. *J. Phys. (Paris)* 35: 899–927

Bouchiat, M. A., Bouchiat, C. 1975. *J. Phys. (Paris)* 36: 493–509

Bouchiat, M. A., Poirier, M., Bouchiat, C. 1980. *J. Phys. (Paris).* In press

Bouchiat, M. A., Pottier, L. 1976a. *J. Phys. Lett. (Paris)* 37: L79–L83

Bouchiat, M. A., Pottier, L. 1976b. *Phys. Lett. B* 62: 327–30

Bouchiat, M. A., Pottier, L. 1979. See Bell 1979. In press

Brimicombe, M. W., Loving, C. E., Sandars, P. G. H. 1976. *J. Phys. B* 9: L237–40

Bucksbaum, P. H. 1979. See Bell 1979. In press

Cahn, R. N., Gilman, F. J. 1978. *Phys. Rev. D* 17: 1313–22

Cahn, R. N., Kane, G. L. 1977. *Phys. Lett. B* 71: 348–52

Carter, S. L., Kelley, H. P. 1979. *Phys. Rev. Lett.* 42: 966–69

Chu, S., Commins, E. D., Conti, R. 1977. *Phys. Lett. A* 60: 96–100

Cline, D., Fry, W. F. 1977. *Ann. Rev. Nucl. Sci.* 27: 209–78

Conti, R., Bucksbaum, P., Chu, S., Commins, E., Hunter, L. 1979. *Phys. Rev. Lett.* 42: 343–46

Dunford, R. W., Lewis, R. R., Williams, W. L. 1978. *Phys. Rev. A* 18: 2421–36

Englert, F., Brout, R. 1964. *Phys. Rev. Lett.* 13: 321–23

Feinberg, G., Chen, M. 1974. *Phys. Rev. D* 10: 190–203, 3789–95

Feynman, R. P., Gell-Mann, M. 1958. *Phys. Rev.* 109: 193–98

Flambaum, V. V., Khriplovich, I. B., Sushkov, O. P. 1978. *Phys. Lett. A* 67: 177–79

Glashow, S. L., Iliopoulos, J., Maiani, L. 1970. *Phys. Rev. D* 2: 1285–92

Goldstone, J. 1961. *Nuovo Cimento* 19: 154–64

Goldstone, J., Salam, A., Weinberg, S. 1962. *Phys. Rev.* 127: 965–70

Gould, H. 1970. *Phys. Rev. Lett.* 24: 1091–93

Grant, I. P., Rose, S. J., Sandars, P. G. H. 1979. *J. Phys. B.* In press

Guralnik, G. S., Hagen, C. R., Kibble, T. W. B. 1964. *Phys. Rev. Lett.* 13: 585–87

Harris, M. J., Loving, C. E., Sandars, P. G. H. 1978. *J. Phys. B* 11 : L749–53

Harris, R. A., Stodolsky, L. 1978. *Phys. Lett.* *B* 78 : 313–16

Henley, E. M., Klapisch, M., Wilets, L. 1977. *Phys. Rev. Lett.* 39 : 994–97

Henley, E. M., Wilets, L. 1976. *Phys. Rev. A* 14 : 1411–17

Higgs, P. W. 1964a. *Phys. Lett.* 12 : 132–33

Higgs, P. W. 1964b. *Phys. Rev. Lett.* 13 : 508–9

Higgs, P. W. 1966. *Phys. Rev.* 145 : 1156–63

Hiller, J., Sucher, J., Feinberg, G., Lynn, B. 1979. Private communication

Hinds, E. A., Loving, C. E., Sandars, P. G. H. 1976. *Phys. Lett B* 62 : 97–99

Hinds, E. A. 1979. See Bell 1979. In press

Hung, P. Q., Sakurai, J. J. 1977a. *Phys. Lett. B* 69 : 323–38

Hung, P. Q., Sakurai, J. J. 1977b. *Phys. Lett. B* 72 : 208–14

Hung, P. Q., Sakurai, J. J. 1979. *UCLA Rep. /79/TEP/9 May 1979*. Los Angeles : Univ. Calif.

Khriplovich, I. B. 1974. *Pis'ma Zh. Eksp. Teor. Fiz.* 20 : 686–88 (*JETP Lett.* 20 : 315–17)

Khriplovich, I. B. 1975. *Yad Fiz.* 21 : 1046 (*Sov. J. Nucl. Phys.* 21 : 538–40)

Khriplovich, I. B. 1979. See Bell 1979. In press

Langacker, P., Kim, J. E., Levine, M., Williams, H. H., Sidhu, D. P. 1979. *Proc. Neutrino 1979 Conf., Bergen, Norway*. In press

Lee, B. W. 1972. *Phys. Rev. D* 5 : 823–35

Lee, B. W., Zinn-Justin, J. 1972. *Phys. Rev. D* 5 : 3121–60

Lewis, L. L., Hollister, J. H., Soreide, D. C., Lindahl, E. G., Fortson, E. N. 1977. *Phys. Rev. Lett.* 39 : 795–98

Loving, C. E., Sandars, P. G. H. 1975. *J. Phys. B* 8 : L336–38

Marciano, W. J., Sanda, A. E. 1978. *Phys. Rev. D* 17 : 3055–64

Moskalev, A. N. 1974a. *Pis'ma Zh. Eksp. Teor. Fiz.* 19 : 229 (*JETP Lett.* 19 : 141–50)

Moskalev, A. N. 1974b. *Pis'ma Zh. Eksp. Teor. Fiz.* 19 : 394 (*JETP Lett.* 19 : 216–18)

Neuffer, D. V., Commins, E. D. 1977a. *Phys. Rev. A* 16 : 844–62

Neuffer, D. V., Commins, E. D. 1977b. *Phys. Rev. A* 16 : 1760–67

Novikov, V. N., Sushkov, O. P., Khriplovich, I. B. 1976. *Sov. Phys. (JETP)* 44 : 872–80

Novikov, V. N., Sushkov, O. P., Flambaum, V. V., Khriplovich, I. B. 1977. *Zh. Eksp. Teor. Fiz.* 46 : 802 (*Sov. Phys. JETP* 46 : 420–22)

Pasierb, E., Gurr, H. S., Lathrop, J., Reines,

F., Sobel, H. W. 1979. *Phys. Rev. Lett.* 43 : 96–99

Prescott, C. Y., Atwood, W. B., Cottrell, R. L., DeStaebler, H., Garwin, E. L., Gonidec, A., Miller, R. H., Rochester, L. S., Sato, T., Sherden, D. J., Sinclair, C. K., Stein, S., Taylor, R. E., Clendenin, J. E., Hughes, V. W., Sasao, N., Schuler, K. P., Borghini, M. G., Lubelsmeyer, K., Jentschke, W. 1978. *Phys. Lett. B* 77 : 347–52

Prescott, C. Y., Atwood, W. B., Cottrell, R. L., DeStaebler, H., Garwin, E. L., Gonidec, A., Miller, R. H., Rochester, L. S., Sato, T., Sherden, D. J., Sinclair, C. K., Stein, S., Taylor, R. E., Clendenin, J. E., Hughes, V. W., Sasao, N., Schuler, K. P., Borghini, M. G., Lubelsmeyer, K., Jentschke, W. 1979. *SLAC Publ. 2319 (1979)*. Palo Alto, Calif. : SLAC

Rein, D. W., Hegstron, R. A., Sandars, P. G. H. 1979. *Phys. Lett. A* 71 : 499–502

Sakurai, J. J. 1979. *UCLA Rep. /79/TEP/1S*. Los Angeles : Univ. Calif.

Salam, A. 1968. In *Relativistic Groups and Analyticity (Nobel Symposium #8)*, ed. N. Svartholm, p. 367. Stockholm : Almqvist & Wiksell

Sandars, P. G. H. 1968. *J. Phys. B* 1 : 499–510

Sandars, P. G. H. 1975. *Atomic Physics IV*, pp. 27–35. New York : Plenum

Sandars, P. G. H. 1979. Private communication

Soreide, D. C., Fortson, E. N. 1975. *Bull Am. Phys. Soc.* 20 : 491 (Abstr.)

Sudarshan, E. C. G., Marshak, R. 1958. *Phys. Rev.* 109 : 1860–61

Sushkov, O. P., Flambaum, V. V. 1978a. *Yad Fiz.* 27 : 1308 (*Sov. J. Nucl. Phys.* 27 : 690–92)

Sushkov, O. P., Flambaum, V. V. 1978b. *Sov. Phys. JETP* 75 : 10

Sushkov, O. P., Flambuam, V. V., Khriplovich, I. B. 1976. *Pis'ma Zh. Eksp. Teor. Fiz.* 24 : 502 (*JETP Lett.* 24 : 461–64)

't Hooft, G. 1971. *Nucl. Phys. B* 35 : 167–88

Trainor, T. 1979. See Bell 1979. In press

Vainshtein, A. E., Khriplovich, I. B. 1974. *Pis'ma Zh. Eksp. Teor. Fiz.* 20 : 80 (*JEPT Lett.* 20 : 34–35)

Vainshtein, A. E., Khriplovich, I. B. 1975. *Zh. Eksp. Teor. Fiz.* 68 : 3 (*Sov. Phys. JETP* 41 : 1–3)

Weinberg, S. 1967. *Phys. Rev. Lett.* 19 : 1264–66

Weinberg, S. 1972. *Phys. Rev. D* 5 : 1412–17

Yang, C. N., Mills, R. N. 1954. *Phys. Rev.* 96 : 191–95

Zel'dovich, Ya. B. 1959. *Zh. Eksp. Teor. Fiz.* 36 : 964–66 (*Sov. Phys. JETP* 9 : 682L–83L)

Ann. Rev. Nucl. Part. Sci. 1980. 30: 53–84
Copyright © 1980 by Annual Reviews Inc. All rights reserved

TRANSIENT MAGNETIC FIELDS ×5614 AT SWIFT IONS TRAVERSING FERROMAGNETIC MEDIA AND APPLICATION TO MEASUREMENTS OF NUCLEAR MOMENTS

N. Benczer-Koller[1]

Department of Physics, Rutgers University, New Brunswick, New Jersey 08903

M. Hass[2]

Department of Nuclear Physics, Weizmann Institute of Science, Rehovot, Israel

J. Sak[1]

Department of Physics, Rutgers University, New Brunswick, New Jersey 08903

CONTENTS

[1] Supported in part by the National Science Foundation NFS-PHY-78-01530.
[2] Supported in part by the Israel-US Binational Science Foundation. Incumbent of the Charles H. Revson Foundation Career Development Chair.

0163-8998/80/1201-0053$01.00

1   INTRODUCTION

The hyperfine interaction between the nucleus and its electronic environment has been studied extensively since it was first observed in a measurement of a nuclear parameter. The present work focuses on the magnetic hyperfine interaction between the magnetic moment of a very short-lived excited nuclear level and the transient magnetic field acting on fast ions traversing a polarized ferromagnet. Such an interaction is manifested by a precession of the angular distribution of decay γ rays; the sense and magnitude of the precession determine the sign and absolute value of the magnetic moment of the corresponding level. For levels with lifetimes of the order of 10^{-12} s, magnetic fields of more than 10^7 G must be present in order to produce a measurable precession. Since such fields cannot be produced on a *macroscopic* scale, one has to look for special atomic or solid-state electronic environments that would produce megagauss fields on a *microscopic* scale. The traditional interplay between nuclear physics and atomic or solid-state physics in hyperfine interaction studies is well demonstrated in these investigations. On the one hand, understanding the nature and origin of such megagauss fields involves detailed and new information regarding ion-solid interactions and related subjects in the atomic physics of highly stripped ions. On the other hand, as demonstrated below, the transient hyperfine field is a "universal" phenomenon, and thus can facilitate measurements of nuclear magnetic moments hitherto unobtainable. Since magnetic moments probe aspects of nuclear structure different from other nuclear properties like energies or transition probabilities, such measurements can significantly contribute to our understanding of nuclear structure.

The nature of the "static" hyperfine field acting on impurities embedded substitutionally or interstitially in a ferromagnet is fairly well understood. Consequently the static hyperfine interaction has long been used as a tool for measuring magnetic moments of nuclear states with lifetimes longer than nanoseconds. It was not, however, until the discovery that very much larger effective magnetic fields can be produced by the interaction between very fast ions and magnetic media that the measurements of nuclear magnetic moments of very short-lived nuclei took on a new preeminence. The existence of these strong magnetic fields acting on *moving* ions inside a ferromagnet, was first discovered by the MIT-Wisconsin group in 1966 (Borchers et al 1966). They noticed a systematic difference between values of the angular precession for the same nucleus depending on whether the probe was embedded in a ferromagnet and experienced the *static* hyperfine field, or whether it was in the process of being implanted into the solid following an energetic collision with another nuclear ion. The magnitudes of the observed magnetic fields obtained by implantation were always more positive than the values obtained for the same nuclear probe bound to the ferromagnet. It was soon realized that this difference is due to a very strong magnetic field which acts on the implanted ions only during their slowing down time (typically 1–2 ps). This transient field is always positive (parallel to the external polarizing field) while the static hyperfine field may be either positive or negative.

Soon afterwards, Lindhard & Winther (1971) proposed a theoretical framework to describe the transient field. The Lindhard-Winther (LW) theory suggests that the transient field originates from the Coulomb scattering of free electrons in the ferromagnet off the bare charge of the moving ion. For low velocities, $v < 2\pi Z v_0$, the electron density at the nucleus is enhanced by a factor $\approx 2\pi Z v_0/v$, where Z is the charge of the nucleus, v its velocity, and $v_0 = e^2/\hbar$ is the Bohr velocity. Since there is a net spin imbalance in the ferromagnet, the charge enhancement gives rise to a net magnetic field at the nucleus. The LW model has essentially only one free parameter, v_p, the Fermi velocity of the polarized electron of the host ferromagnet. Since the field in the LW model is proportional to v_0/v down to $v = v_p$, below which it is assumed to be constant, lower values of v_p would result in larger transient fields. The original Lindhard-Winther paper suggests a value of $v_p/v_0 = 0.5$, which produced a good fit to the available data. Subsequent measurements indicated somewhat larger fields that required an adjustment of the value of v_p to $v_p/v_0 = 0.3$. This correction was incorporated into the "adjusted" Lindhard-Winther model (ALW) (Hubler et al 1974)[3].

In these experiments ions recoiled and subsequently stopped in the ferromagnet. The measured angular precession was the sum of the precession due to the transient field and that resulting from the interaction with the static field. The lifetime of the nuclear level, the time history of the recoiling ion, the stopping powers of the recoiling ion, and the possible radiation damage effects must all be known in order to isolate the information regarding the specific contribution of the transient field to the total measured effects, and to ultimately deduce values of nuclear moments. Although a very valuable insight was obtained from the realization that the transient field is a "universal" phenomenon, and from the first systematic measurements of nuclear magnetic moments of short-lived states, the difficulties mentioned above tended to hinder further progress in this field.

---

[3] Hindsight now reveals that the observed values of the transient field were larger than predicted by the LW theory because experiments were performed at *higher* velocity with the field increasing with v while the parametrization of the ALW model concentrated on the *low* velocities.

In 1975 the Bonn-Strasbourg group found that the hyperfine interaction experienced by oxygen ions that start with an initial velocity $v/c = 0.02$ and that ultimately stop in an iron foil was considerably larger than predicted by even the ALW model (Forterre et al 1975a). Since most data used in obtaining the ALW parametrization were taken at relatively low velocities ($v/c \lesssim 0.01$), the discrepancy was attributed to a different mechanism, which manifested itself at higher velocities and thus did not come into play in earlier experiments. This work was soon followed by other more detailed experiments of the Bonn-Strasbourg (Forterre et al 1975b) and the Utrecht (Eberhardt et al 1975, Eberhardt et al 1977) groups. The new data strongly suggested that the magnetic field *increases* with the velocity of the ion through the ferromagnet rather than decreases, as predicted by the LW model. An experiment on the ^{12}C 2_1^+ excited state with $v/c \simeq 0.03$ shows unambiguously that the magnetic interaction is indeed very large for fast ions (Goldberg 1976a). The low velocity behavior of the field could not be tested in this experiment because, since the 2_1^+ state has a very short lifetime, the excited nucleus decays before it slows down appreciably.

The Rutgers group confirmed the existence of this large magnetic field and its approximate proportionality to the ion velocity by examining the interaction of fast ions traversing "thin" ferromagnetic foils and recoiling out of these foils at high velocity (Hass et al 1976, Brennan et al 1977). Again, only the high velocity part of the magnetic interaction is probed by this technique. The data clearly demonstrated that the velocity dependence of the hyperfine interaction follows not the v_0/v prediction of the LW model but rather a $\sim v/v_0$ parametrization first suggested by the Utrecht group.

Clearly, new models and parametrizations were needed to understand the high velocity data, and the relevant atomic physics phenomena to be taken into account might be very different from the Coulomb scattering processes underlying the LW picture. Actually, an alternative model of the hyperfine field in terms of selective production and decay of s holes in the inner electronic shells of the moving ions was originally presented by Borchers et al (1968) prior to the LW work, but it was not followed up at the time.

New directions in the utilization of these large magnetic fields for measurements of nuclear magnetic moments emerged. The possibility of selecting only the high velocity part of the interaction by carrying out experiments with thin magnetized foils, for example, was particularly attractive because it eliminated many of the difficulties of the early experiments.

A vast array of data has been published for nuclei light and heavy, fast and slow, and moving in a variety of ferromagnets. This paper aims not to provide a complete understanding of the observed phenomena, because no overall picture yet exists, but rather to expose and order the main features of this subject to provide a basis for further investigations.

The following sections therefore describe the general experimental procedure, characterize the current data within a coherent pattern amenable to theoretical scrutiny, discuss parametrizations of the hyperfine field, highlight the most striking magnetic moment measurements, and point out future directions.

2 EXPERIMENTAL PROCEDURE

A variety of experimental arrangements have been used in many laboratories. Here we present the main features of a typical experimental setup and discuss in more detail points of particular interest.

A particle beam is allowed to impinge on a target deposited on a ferromagnetic foil. Ions excited to the nuclear level of interest recoil into the ferromagnet and either stop in it or emerge out into the vacuum or into a nonmagnetic backing. The nuclear level can be excited by various processes such as a particular nuclear reaction, Coulomb excitation, or (HI, xn) reaction, depending on the specific level and the recoil velocity desired. De-excitation γ rays are detected by NaI(Tl) scintillation counters or by Ge(Li) detectors. The γ-ray angular distribution $W(\theta)$ is measured either in coincidence with scattered particles at a particular angle or with respect to a chosen quantization axis without the coincidence requirement. A weak external magnetic field H polarizes the ferromagnetic foil and thus establishes a direction along which the hyperfine field may act. A new angular correlation $W'(\theta) = W(\theta - \Delta\theta)$ is then measured; the sign and magnitude of the angular precession $\Delta\theta$ yield the information on the magnetic interaction between the nuclear moment and the transient field. Actually the equality $W'(\theta) = W(\theta - \Delta\theta)$ is valid only for small values of $\Delta\theta$; it indeed holds for all cases discussed here (Niv et al 1980).

As an example, to demonstrate a typical experiment, we discuss the Coulomb excitation of the 2^+ first excited state in even Sn nuclei (Hass et al 1980). The experimental arrangement is shown schematically in Figure 1. A 108-MeV ^{35}Cl beam impinges on a 500 μg/cm^2 Sn target deposited on a 1.5 mg/cm^2 iron foil backed by a 10 mg/cm^2 copper foil. This thin-iron triple-foil arrangement is discussed below in more detail. The delayed γ radiation from the Coulomb excited 2_1^+ level of Sn isotopes is detected in four 12.7 × 12.7 cm NaI(Tl) crystals in coincidence with backscattered ^{35}Cl ions. The angular correlation, with respect to a quantization axis chosen along the beam direction, is given by

$$W(\theta) = 1 + 0.66\, P_2(\cos\theta) - 1.26\, P_4(\cos\theta)$$

(a)

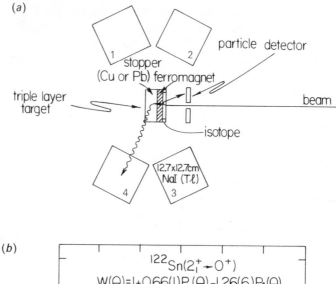

(b)

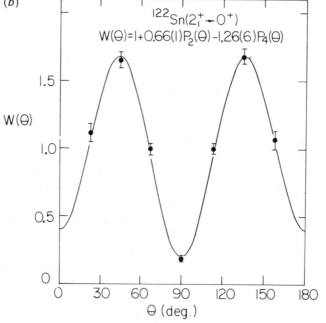

Figure 1 (a) Schematic of a typical experimental setup (not to scale) displaying the triple layer target and the particle and the four γ-ray detectors numbered as described in text. (b) Typical particle γ-ray angular correlation $W(\theta)$ for a $2^+ \xrightarrow{\Delta m = 0} 0^+$ transition following Coulomb excitation by heavy ions. In a typical experiment the γ-ray detectors are placed at or near the angle corresponding to the maximum slope $[1/W]\, dW/d\theta$ (Hass et al 1980).

in agreement with a pure $\Delta m = 0,2 \rightarrow 0$ correlation corrected for the finite size of the detectors. Precession measurements were carried out with the γ detectors at $\pm 67.5°$ and $\pm 112.5°$, where the slope of the correlation is large. The magnetization of the iron layer is reversed periodically to form double ratios, defined as

$$\rho_{ij} = \left[\frac{N(\theta_i)\uparrow \cdot N(\theta_j)\downarrow}{N(\theta_i)\downarrow \cdot N(\theta_j)\uparrow} \right]^{1/2}, \qquad \qquad 1.$$

where $N(\theta_i)\uparrow$ represents, for example, the photo peak intensity detected by the ith detector as the external magnetic field points up with respect to the detectors' plane. The average double ratio $\rho = (\rho_{23}/\rho_{14})^{1/2}$ is free of systematic errors and is related to the angular precession $\Delta\theta$ by

$$\Delta\theta = S^{-1} (\rho - 1)/(\rho + 1) \qquad \qquad 2.$$
$$S = [W(\theta)]^{-1} \, dW(\theta)/d\theta.$$

The double ratios ρ_{13} and ρ_{24} should be equal to unity and serve as a consistency check of the data. It should be noted that such measurements can be performed with two detectors and even with only one. Nevertheless, this multidetector arrangement is desirable in order to eliminate systematic errors.

The measured angular shift $\Delta\theta$, corrected for the effects due to the beam-bending in the external stray magnetic field, is then compared to the expression

$$\Delta\theta = (g\mu_N/\hbar) \int_0^T B[v(t)] \, e^{-t/\tau} \, dt, \qquad \qquad 3.$$

where g is the g-factor of the nuclear level, τ is its mean-life, μ_N is the nuclear magneton, and $B(v)$ is the strength of the transient field at ion velocity v. T is the flight time of the ion through the ferromagnet. The duality of hyperfine interaction studies is again illustrated; if g is known from other independent experiments, one can obtain information on the function $B(v)$, and, conversely, knowing the magnetic field $B(v)$, one can measure nuclear magnetic moments.

Of particular interest to nuclear physics applications is the thin-iron triple-foil technique developed by the Rutgers group (Hass et al 1976, 1978). This particular target assembly was originally suggested in 1969 by Grodzins (1970) but was not explored further because the hyperfine field was then believed to be small at high velocities. In this method the recoiling ions traverse the thin magnetized foil but stop and decay in a nonmagnetic foil (copper or lead). The results are a significant improvement over the results obtained with the conventional thick-iron foils in which the recoiling ions actually stop:

(a) The magnetic field acts only during the transit time through the magnetic foil. Consequently, the measured precession is insensitive to the excited states mean-life, provided it is long compared to the transit time (typically 0.1–0.3 ps).

(b) Poorly known stopping powers at low velocities do not enter into the analysis of the results since by the time the active ion slows down it is no longer in the ferromagnet but finds itself in the nonmagnetic environment of the copper or lead backings.

(c) There is no contribution to the measured precession from the hyperfine static field of the ion in the host ferromagnet. The static field is not always well known and may further be affected by radiation damage effects.

The thin-foil technique is also essential to the investigation of the velocity dependence of the hyperfine interaction. By using thin magnetic foils of various thicknesses and ions with different initial velocities, one can differentially scan the function $B(v)$ (Brennan et al 1978a, Shu et al 1980). In limited situations such differential measurements can also be carried out either by using a very short-lived nuclear level (Goldberg et al 1976a) or by analyzing the Doppler-broadened shape of a γ-ray line (de Raedt et al 1980). Results of such investigations are presented in Section 3.

A very important feature of these experiments concerns the measurement of the saturation magnetization of the ferromagnetic foil. Since the foils may be prepared by a variety of techniques (rolling, electroplating, gluing, sputtering), they will generally not exhibit the same magnetization characteristics as the bulk ferromagnetic material. Therefore the magnetization of the prepared target assembly must be measured for each sample. Practical magnetometers for this purpose are being used in several laboratories (Brennan 1978, Zalm et al 1979).

To reduce beam-bending effects, the polarizing external fields applied must be as low as possible. Magnetization measurements indicate that, for a typical iron foil a few mg/cm^2 thick, a very small polarizing field of less than 300 G is sufficient to saturate the foil. Another ingenious experimental arrangement, employed by the Utrecht group, uses single-crystal iron frames for achieving saturation magnetization at extremely low external fields, but the ferromagnetic layer in these targets is by necessity very thick (Zalm et al 1976, 1979).

3 CHARACTERIZATION AND ORIGIN OF THE TRANSIENT FIELD

3.1 Observations

The transient field, denoted by $B(v,Z)$, corresponds to the difference between the total magnetic field acting on an ion moving with velocity v and containing a nucleus of charge Z and the average field in the ferromagnet. The key to the description of the transient field lies most probably in the state of ionization and excitation of the moving ion, which in turn is governed by the ion velocity. It is likely that not any one mechanism is responsible for the observed interactions. The whole velocity spectrum can be naturally divided into three regions according to the state of ionization of the ion. These regions are characterized by different strengths **a** of the observed magnetic interactions $B \sim \mathbf{a}f(v)$ and the observations are summarized in Table 1.

It will be useful, in view of the incomplete status of the experiments, to further subdivide the observations according to the atomic weight of the ions. This selection arises from the fact that, for experimental reasons, light ions were measured at relatively higher velocities than heavy ions and therefore traverse the solid in a relatively higher state of ionization for which a more quantitative treatment is possible.

3.1.1 FAST IONS: $v \gg Zv_0$ The interaction between a magnetized medium and a charge moving at very high velocities, i.e. velocities at which the ions are completely ionized, is qualitatively easy to understand and calculate. Two sets of experiments have been carried out for this configuration. The first measured very fast ^{12}C (Goldberg et al 1976a,b, Kumbartzki et al 1979) and ^{13}C (Dybdal et al 1976, 1979a) ions stopping in iron foils. The second inspected the rotation upon traversal of magnetized iron plates of the spin

Table 1   Velocity range of characteristic ionization states[a]

Ion velocity	Ionization state of ion	Field strength constant **a**	Approximate velocity dependence $f(v)$
$v < v_0$	small	not known	not known
$v_0 < v < Zv_0$	partial to considerable	~ 100	$\sim v$
$v \gg Zv_0$	complete	~ 10	$1/v$

[a] $v_0 = e^2/\hbar$ is the Bohr velocity, which estimates the velocities of both conduction electrons and outer electrons of the ion, and Zv_0 is the velocity at which the ion approaches a completely stripped charge state.

direction of fast polarized μ^+ (Brennan et al 1978b). In both cases a very small field was observed. For ^{12}C, the field has been mapped below and above Zv_0 and the importance of the bound electrons (Figure 2) at lower ion velocities is clear. In the case of positive muons moving with $v/v_0 \approx 100$, an upper limit for the transient field was obtained: $B \leq 2.6$ kG. At these high velocities the ion is unable to capture electrons into bound states and no atomic structure considerations need to be taken into account. The electrons in the ferromagnet are relatively free to respond to the presence of the ions and are responsible for both the electric and magnetic interactions acting on the ions. These fields are produced from the electric polarization of electrons in the ion's vicinity. This polarization of the medium then results in an enhanced density of electrons near the ion, and since in a spin-polarized ferromagnet the electrons have a net polarization along the direction of the external applied field, they produce a net magnetic field at the site of the ion in the same direction as that of the external field. The magnitude and velocity dependence of the resulting hyperfine interaction have been calculated (Bruno & Sak 1978, 1979, Sak & Bruno 1978).

It is interesting to note the connection between the mechanism that produces the dynamic field in this regime to the interaction of ions with solids and the phenomena associated with wake effects (Vager & Gemmel 1976). These points are discussed in Section 3.2.

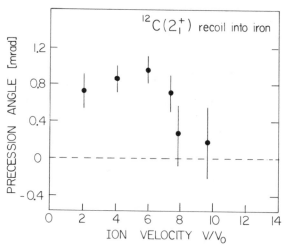

Figure 2   Experimental nuclear precessions as a function of ion velocity for ^{12}C(2_1^+) ions in iron. Because of the very short lifetime of the state ($\tau = 65$ fs), the experimental precession samples the magnitude of the transient field at the indicated velocity. The decrease of the transient field at $v > Zv_0$ is apparent, while the high value of the precession angle at lower velocities reflects the importance of bound electrons (Kumbartzki et al 1979).

3.1.2 INTERMEDIATE VELOCITY IONS: $v_0 < v < Zv_0$ The transient field has been measured over a range of ions covering most of the periodic table. For the purpose of the discussion of the hyperfine field strength and its dependence on the velocity and atomic number of the moving ion, we consider only experiments carried out (*a*) on nuclei whose excited-state magnetic moment has been measured by an unambiguous technique (such as the static fields acting on radioactive sources implanted into a known site by a controlled procedure), and (*b*) for nuclear species where long-lived isotopes allow independent calibration of either the hyperfine field or the magnetic moment. Furthermore, to avoid the problems that arise in the interpretation of experiments on ions *stopping* in the magnetic foil, we concentrate on the results obtained by the thin foil experimental procedure. However, one may add some of the results from experiments carried out with thick-iron foils. Measurements on the light nuclei in particular were carried out for ions entering the magnetic foil at various initial velocities. The resulting data can be reanalyzed by subtracting the contribution obtained for the slowest ions from the effect measured for the faster ones. Thus one can extract out of data on ions that stopped in the magnetic foil the part of the effect that occurs as the ion moves at high speed through the foil.

The early observations suggested that the hyperfine interaction increases with the velocity of the ion and its atomic number. A field that varied linearly with v and Z was introduced phenomenologically by Eberhardt et al (1977):

$$B(v,Z) = \mathbf{a}\ (v/v_0)\ Z\ \mu_B N_p, \qquad\qquad 4.$$

where μ_B is the Bohr magneton, N_p the volume density of polarized electrons such that $\mu_B N_p$ is the magnetization of the foil; for example, $\mu_B N_p = 1752$ G for a saturated iron foil. The experimental results were in agreement with the hypothesis and yielded $\mathbf{a} \simeq 60$. A direct test of this hypothesis was carried out by the Rutgers group who measured the strength of the interaction as a function of magnetic foil thickness L (Hass et al 1976, Brennan et al 1977, 1978a). Under the assumption of linearity of B with v, the observed precession of the gamma-ray angular correlation, $\Delta\theta$, which is proportional to $B\ dt \cong v\ dt$, is also proportional to the thickness of the foil L since $\int v\ dt = L$. These experiments were carried out on Se ions traversing thin-iron foils. The results plotted in Figure 3 generally agree with Equation 4. However, since there is no a priori physical reason for choosing the particular v and Z dependence of Equation 4, a more general expression

$$B(v,Z) = \mathbf{a}'(v/v_0)^{P_v}\ Z^{P_z}\ \mu_B N_p \qquad\qquad 5.$$

was proposed; the best fit to the data at that time gave $P_v = 0.5 \pm 1.0$ and $P_z = 1.5 \pm 0.5$. The hyperfine field should somehow be related to the effective charge of the ion in the solid. Therefore, yet another parametrization for the field was proposed,

$$B(v,Z) = \mathbf{a}'' Z_{\text{eff}}^{P_z} \mu_B N_p, \qquad\qquad 6.$$

which followed the form for the effective charge for ions leaving foils, since the effective charge inside solids is not known. This expression fitted the early data as well as the form of Equation 5.

Clearly data with higher statistical accuracy and covering a broader range of velocities ought to be collected. The effect of the various parametrizations on the extraction of nuclear magnetic moments from the data is discussed in Section 4. Recent experiments on heavier ions, Pd, Nd, Sm, and Pt, traversing very thin magnetic foils were indeed carried out over velocities extending to lower and higher values. In order to avoid confusion over the possible parametrizations of the transient field, the linear velocity expression (Equation 4) was chosen as a framework for presenting the data; any deviation from this form will be manifested in different values of the parameter $\mathbf{a}$ derived from each experimental angular precession. Again, only results for $v \gtrsim v_0$ are included in this discussion.

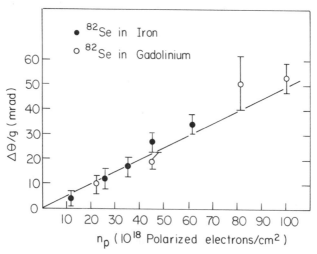

Figure 3 Precession angles $\Delta\theta/g$ as a function of the number of polarized electrons per cm^2 encountered by ^{82}Se ions traversing iron and gadolinium foils. The linearity with respect to the foil thickness indicates that the transient field is approximately proportional to the velocity of the ion, while the fact that the same curves fit both the data for iron and gadolinium foils reflects the direct proportionality of the transient field to the magnetization of the foil (Brennan et al 1978a).

Figure 4 shows the value of **a** as a function of the mean velocity of the ion in the magnetic foil (Shu et al 1980). The main feature of the data in Figure 4 is that the Z and v dependence of Equation 4 satisfies the observations within a factor of two for ions whose atomic number and velocity vary by large factors. This remarkable consistency with the simple phenomenological expression of Equation 4 suggests a common origin of the transient field for all these ions at intermediate velocities. The approximate agreement with the linear Z and v dependence is most probably also due to the fact that, even for a thin foil experiment, the transient field is probed over a broad range of velocities thus yielding only some average value. On the other hand, when one inspects the data more closely, the deviations from Equation 4 are apparent: **a** is not a constant as a function of the ion velocity, it decreases with increasing velocity. An inclusion of a LW, $1/v$ term could explain the observed trend, but, since the applicability of the LW model for low velocity measurements is questionable, it is better not to invoke any additional interactions in the present analysis until a better theoretical justification can be obtained.

The deviations from the linear dependence on Z and v are not surprising since the light ion experiments described in the next section suggest strongly that the large magnitude of the transient field may be accounted for in terms of polarized holes in the inner electronic configurations produced during the scattering of the moving ion by the lattice atoms of the ferromagnetic foil. Structure effects and related discontinuities should appear as the ion

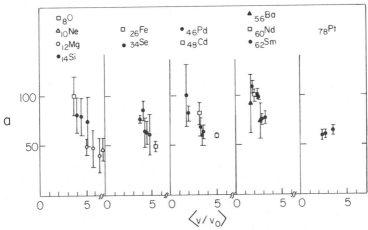

Figure 4   Plot of the constant **a** in the expression $B(v, Z) = \mathbf{a}(v/v_0)Z\,\mu_B N_p$, as a function of ion velocity and atomic number. While the gross features of the data are consistent with this expression, it is apparent that the assumption of a strict linear velocity dependence is not justified.

electronic configurations and their interactions with the polarized electron of the ferromagnet vary with velocity and atomic number. Nevertheless a fairly good fit of most of the data presented in Figure 4 may be obtained with parameters $\mathbf{a}' = 96.7 \pm 1.6$, $p_v = 0.45 \pm 0.18$, and $p_z = 1.1 \pm 0.2$ in Equation 5 (Shu et al 1980). Nevertheless, any parametrization of this kind is only a simplification demonstrating the average properties of a more complex microscopic picture. The trends observed so far dictate that more data be obtained as a function of the ion velocity and with a much smaller velocity spread than obtained heretofore in order to ascertain the relative importance of atomic structure effects and the consequences of an average behavior of ions interacting with polarized ferromagnets. It is clear that both the rough "universality" and the "fine structure" aspects of the hyperfine interaction will be studied extensively in the near future.

3.1.3 LIGHT IONS: $0 \leq v/v_0 \leq 8$ In many experiments on very light ions with $8 \leq Z \leq 14$, the ions entered the magnetic foil at a high initial velocity ($v/c \gg 0.02$) and subsequently stopped in it. In general, the complications arising from having to consider the effects of the static field on the nuclear precession are not too severe for the light nuclei, because, while the static field at these nuclei is not well known, it is nevertheless believed to be small. Furthermore, the nuclei have very short-lived excited states and decay before or shortly after they stop.

The hyperfine interactions observed for these light nuclei were first analyzed by assuming a transient field linear in the ion velocity over the whole velocity range and proportional to the hyperfine field at the nuclear site produced by unpaired s electrons, B_{ns} (Van Middelkoop 1978):

$$B(v,Z) = C(Z)\,(v/v_0),$$

where $C(Z) = \gamma B_{ns}.$ 7.

The parameter $C(Z)$, which relates to the strength of the interaction for a particular ion of atomic number Z, shows sharp discontinuities as a function of Z (Figure 5a). Van Middelkoop explained these results in terms of a molecular interaction between the incoming ion and the magnetic environment by which inner s electrons from the moving ion can be promoted into empty polarized d orbitals of the magnetic ion. This promotion mechanism is very sensitive to the relative energies of the atomic orbitals of the colliding ions and therefore will exhibit structure effects. These will be more pronounced in light nuclei where fewer electronic shells participate in the interaction than in heavier nuclei where averaging over many shells will smear the discontinuities.

These conclusions depend crucially on the expectations that the dynamic field is linear in velocity down to zero velocity, and, in addition, that no unexpected effects occur in the stopping power dE/dx at very low velocities. The first of these assumptions is clearly contradicted by the evidence presented in Section 3.1.2 where the constant representing the strength of the interaction was shown to be a function of the ion velocity. The second of these assumptions should also be regarded with caution, since strong oscillations have in fact been observed in the stopping powers of various ions ($6 \leq Z \leq 36$) stopping in amorphous carbon (Bhalla & Bradford 1968). A similar oscillatory behavior of the ion-solid interaction has affected the measurement by the Doppler-shift attenuation method of nuclear lifetimes of a particular ion stopping in a complete range of stopping materials from $Z = 6$ to $Z = 83$ (Broude et al 1972). These oscillations have been generally attributed to atomic shell structure effects either in the

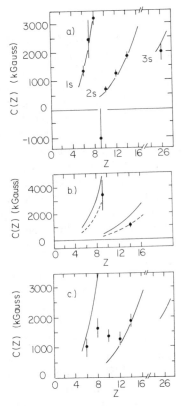

Figure 5   The Z dependence of the transient field obtained for light ions stopping in (*a*) iron, (*b*) cobalt, and (*c*) gadolinium foils. The solid (dashed) lines correspond to a one-parameter fit to the data based on a transient field in iron (cobalt) given by $B = C(Z)(v/v_0)$ (Van Middelkoop 1978, 1979).

slowing ion or in the stopping material, and are not understood well enough to be taken into account quantitatively in the study of the dynamic field at very low velocity.

An experiment was carried out on ^{16}O ions recoiling through thin and thick magnetized iron foils; the resulting strength parameter $a = 96 \pm 20$ obtained from the high velocity experiment with the thin foil can be compared to that obtained for the thick-foil experiment $a = 160 \pm 16$ (Shu et al 1980). This result indicates that the shell effects described above may be primarily a manifestation of interactions at very low velocities. In this respect, it is interesting to note the connection between the linear parametrization and the results from a recent experiment on K-vacancy production on light ions stopping in iron and nickel foils (Dybdal et al 1979b,c,d). This experiment and other x-ray work is discussed in Section 3.2.3.

3.1.4 THE TRANSIENT FIELD IN DIFFERENT FERROMAGNETS   Explicit in the expression of Equation 4 is the proportionality of the transient field to the net magnetization of the foil, $\mu_B N_p$. This hypothesis has been tested both with light and heavy ions. The transient field effect on Se ions traversing both iron and gadolinium foils scales as the magnetization of the foil, namely as $\mu_B N_p = 1752$ and 1541 G respectively for iron and gadolinium foils (Figure 3) (Brennan et al 1978a). Similar experiments have been carried out for Si, Pd, and Pt ions traversing iron and nickel foils. With $\mu_B N_p = 495$ G for nickel, again a transient field proportional to the foil magnetization was observed (Kalish et al 1980). The Stockholm and Aarhus groups have carried out experiments for various ions recoiling and stopping in gadolinium; these also result in an interaction proportional to the magnetization (Fahlander et al 1978). In thick-target experiments on ions with the same initial velocity, the measured angular precession is larger for gadolinium than for iron foils; this is because, for a typical case, the ions' range is larger in gadolinium than in iron, allowing the ion to probe the field for a longer time.

In light nuclei the ion-solid interaction is again more complex. The effect on ^{19}F might be indicative of a mismatch of the molecular orbitals occurring at $Z = 9$ (Van Middelkoop et al 1978). While the field at ^{19}F stopping in cobalt corresponds to promotion of 1s electrons from the F to the cobalt 3d orbit, the field of ^{19}F in iron is considerably smaller (Figure 5b), which suggests that the molecular orbitals for F in iron enhance the promotion of a 2s electron from the F to the iron orbital rather than the promotion of a 1s electron. Salm & Klepper (1976) found very large fields on ^{19}F in nickel and attributed them to a similar molecular promotion mechanism, although a later experiment on ^{18}F did not corroborate the ^{19}F

data (Goldberg et al 1978). A recent experiment on ^{12}C ions in iron, cobalt, and nickel demonstrates that the transient field scales with magnetization at $v/v_0 = 4.1$, but at $v/v_0 = 2.1$ the fields in iron and nickel were comparable, again supporting the molecular orbital promotion mechanism for light ions at low velocity (Speidel et al 1980).

The sharp atomic structure effects appearing in light ions stopping in iron or cobalt (Figure 5) are noticeably modified when the ions stop in gadolinium foils (Figure 5c). The results for gadolinium clearly show less pronounced discontinuities and in fact are not inconsistent with a smooth dependence of $C(Z)$ on the atomic number (Van Middelkoop 1979). It is not surprising that the structure effects are different in iron and gadolinium as the coupling between the s electrons of the moving ion and either the 3d or 4f electrons of the magnetic host must be very different. In addition, since these experiments were carried out for *stopped* ions, the differences between the iron and gadolinium results may be caused by differences in the low energy dependence of the respective stopping powers. Again some of the effects may be correlated with K-hole production and are discussed below.

3.2   Theory and Models

It would obviously be desirable to explain all of those manifestations of the transient field described in the previous section in terms of a unique universal interaction: as of yet, however, this cannot be done by any single model that describes the field's magnitude, velocity, and charge dependence. When the ion moves in the ferromagnet at $v \lesssim Zv_0$, the atomically bound electrons play a dominant role in the magnetic interaction and any realistic model must consider the complex interaction between the ion and the magnetic medium. Only in the region where the moving ion is totally stripped is it possible to carry out rigorous calculations. Within this context, the following discussion proceeds from a description of precise calculations to a pattern of qualitative models.

3.2.1   FAST IONS: $v \gg Zv_0$ The theoretical framework explaining the magnetic interaction between a bare charge moving at high velocity through a magnetic environment and the medium is derived from the original treatment of Lindhard & Winther (1971), even though their work was aimed at explaining the large transient field acting at slow ions. The transient field is described in terms of the Coulomb interaction between the moving charge and the polarized valence electrons of the medium. This interaction leads to an enhancement Q of the electron wave function at the position of the moving ion. Lindhard & Winther obtained

$$B_{LW} = (8\pi/3)QR(1+\xi)\,\mu_B N_p, \qquad\qquad 8.$$

where $\mu_B N_p$ is the magnetization of the medium, $R = (1+Z/84)^{2.5}$ is a relativistic correction factor, and ξ is an asymmetry parameter that vanishes for slow ions but is important for fast ions. Neglecting band structure, Q can be estimated from the wave function of an electron in the Coulomb field of a charge Z:

$$Q = 2\pi\eta/(1-e^{-2\pi\eta}),$$

where $\eta = Zv_0/v_r$ and v_r is the relative velocity of the electron and the ion. In the limit of low velocity, the exponential term may be neglected; this leads (a) to the familiar $1/v$ dependence of the transient field at very low velocities that has been sharply contradicted by experiment, and (b) to the Z dependence that is in general agreement with observation.

As discussed in the introduction, several refinements have been incorporated into the theory, none of which changes these qualitative results. The theory is applicable, however, to the case where $v \gg Zv_0$ where the ion is completely stripped as long as the asymmetry parameter ξ is not neglected. Bruno & Sak (1978) have extended the Lindhard-Winther model to the region of high velocity where $Q \approx 1 + \pi\eta$.

In this case, the transient field contribution appears as a small correction in Q. An asymmetry correction $\xi = -1/4$ was separately calculated. The transient field, more specifically the part of the field due to the presence of the ion, becomes

$$B = 2\pi^2 MZ(v_0/v), \qquad\qquad 9.$$

where $M = \mu_B N_p$ is the magnetization. This result is similar, apart from a numerical factor, to the LW field, but there is a fundamental difference: the LW formula (Equation 8) applies only to low velocities, $\eta > 1$, whereas Equation 9 is correct only in the opposite limit $\eta \ll 1$. Moreover, the field B_{LW} represents the *total* field while B is the *deviation* from the average local field in the target, $(8\pi/3)M$. In the muon experiment, for example, $v/v_0 \approx 100$, and Equation 9 predicts a small value $B = 0.25$ kG, which is consistent with the experimental upper limit of 2.6 kG (Brennan et al 1978b).

Another approach to the calculation of the transient field at very high velocities is based on the observation that the condition $\eta = Zv_0/v \ll 1$ allows one to treat the Coulomb field of the completely stripped ion as a perturbation. Thus the linear response theory is applicable and the transient field can be expressed by means of dielectric function of the target material as a function of the wave vector k and frequency ω:

$$\varepsilon(k,\omega) = 1 + 4\pi[\alpha_\uparrow(k,\omega) + \alpha_\downarrow(k,\omega)]. \qquad\qquad 10.$$

Here $\alpha_\uparrow (\alpha_\downarrow)$ is the dynamic polarizability of the valence electrons with spin up (spin down). In a ferromagnet $\alpha_\uparrow \neq \alpha_\downarrow$ and the transient field can be expressed as:

$$B(Z,v) = (2\mu_B Z/3\pi) \int d^3k \left[\frac{\alpha_\uparrow(\mathbf{k},\mathbf{k} \cdot \mathbf{v}) - \alpha_\downarrow(\mathbf{k},\mathbf{k} \cdot \mathbf{v})}{\varepsilon(\mathbf{k},\mathbf{k} \cdot \mathbf{v})} \right]$$
$$\times [2 + (5/3)P_2(\cos \theta)],$$
11.

where θ is the angle between the vectors $\mathbf{k}$ and $\mathbf{v}$. This expression is exact for high velocities but approximations must be used to calculate $\alpha_{\uparrow(\downarrow)}$. The simplest possibility is offered by the classical expression of Drude, which was used with success for the study of Coulomb explosions (Vager & Gemmel 1976), and also yields the Bethe-Bloch formula for the stopping power. Sak & Bruno (1978) used

$$\alpha_{\uparrow\downarrow}(k,\omega) = -\omega_{p\uparrow\downarrow}^2/[4\pi(\omega + i0)]^2$$
12.

where $\omega_{p\uparrow\downarrow}^2 = 4\pi e^2 n_{\uparrow\downarrow}/m$ are the square of the plasma frequencies for the spin-up and spin-down electrons, $n_{\uparrow\downarrow}$ are the respective concentrations, e and m are the electronic charge and mass, and $i0$ stands for an infinitesimal imaginary part to define $\alpha_{\uparrow\downarrow}$ for $\omega = 0$. The polarizability (α) is purely classical, which is reflected in its independence of the wave vector $\mathbf{k}$. This fact leads in turn to a divergent integral; such divergencies are well known from the derivation of the Bethe-Bloch formula for the stopping power (Vager & Gemmel 1976). The difficulty is removed by realizing that quantum mechanical effects become important at short distances from the ion, of the order of $k_{max}^{-1} = h/mv$. Thus the k integral in Equation 11 may be cut off when k reaches k_{max}. One then obtains a result similar to that of Equation 9;

$$B = 4\pi MZ (v_0/v).$$
13.

The coefficient 4π in this formula cannot be taken too seriously in view of the ambiguity in the cutoff at small distances, but Equation 13 certainly gives the right order of magnitude for the transient field at very high velocity.

A more satisfactory calculation can be done by using a quantum mechanical theory of dielectric response. One such theory is based on the random phase approximation (Lindhard 1954). The integral in Equation 11 must be evaluated numerically with the result (Bruno & Sak 1979)

$$B = (0.86 \pm 0.01) 4\pi MZ(v_0/v),$$
14.

which is quite close to the classical estimate (Equation 13). These results are consistent with the experimental observations of Section 3.1.1.

3.2.2 INTERMEDIATE VELOCITY IONS: $v_0 < v < Zv_0$ In the region of intermediate velocities, the situation is much more complicated. The LW theory is insufficient because it neglects the bound states. Brandt (1977) suggested that Z in Equation 4 be replaced by an effective charge Z_{eff} corresponding to the degree of stripping of the ion. A simple estimate based on Thomas-Fermi model gives (Betz 1972)

$$Z_{eff} = Z[1 - \exp(-v/v_0 Z^{2/3})] \qquad\qquad 15.$$

so that for $v_0 < V < v_0 Z^{2/3}$, $Z_{eff} \approx Z^{1/3}(v/v_0)$ and $B \approx$ a constant times $Z^{1/3}$, independent of the velocity; this is not supported by evidence. However, as stressed by Brandt, the density enhancement may be underestimated by this approach and effects nonlinear in Z could be important. Nonlinear contributions, however, would not change the conclusion that the magnetic field is independent of velocity in the range $v_0 < v < v_0 Z^{3/2}$. Clearly the bound states must play a dominant role. There is strong evidence that the source of the hyperfine field is the partial polarization of deep shells in the moving ions that takes place during collisions between the ion and the atoms of the ferromagnet. During an encounter of the two atoms their atomic orbitals couple to form quasi-molecular states, and vacancies in s shells of the ion can be created by promotion of electrons into the vacant orbitals of the host. If these holes couple to the polarized states of the host, a strong magnetic field will result at the nucleus of the ion. Such phenomena are best analyzed microscopically for light ions.

3.2.3 LIGHT IONS: $0 \le v/v_0 \le 8$ Eberhardt et al (1975, 1977) proposed a mechanism to explain the observed hyperfine interactions in light ions stopping in magnetic materials. It is based on the assumptions that (a) deep holes in s shells are created when electrons are promoted from the moving probe into empty orbitals of the combined molecular state of the ion and host atoms, and (b) these holes are polarized. For $Z \le 8$ two mechanisms for polarization have been discussed (Dybdal et al 1976). For $Z > 8$ 2s electrons are promoted into polarized states of the host atom, are captured into higher excited states of the ion, and consequently cascade down into the bound half-filled s shells with preservation of the polarization.

This analysis leads to the following expression for the transient field

$$B = \sum_n (F_{1^+}^{ns} - F_{1^-}^{ns})B_{ns}, \qquad\qquad 16.$$

where $F_{1^+}^{ns}$ and $F_{1^-}^{ns}$ are fractions of ions having a single electron in the ns orbit with the spin up and down, respectively. B_{ns} is the field of a single electron in the ns orbit and is given by (Kopferman 1956)

$$B_{ns} = 16.7 \, Z(Z-q)/n_{\text{eff}}^3 \, \text{(Tesla)}, \hspace{3cm} 17.$$

where n_{eff} is the effective principle quantum number [$n_{\text{eff}} = 1$ for $n = 1$, $n_{\text{eff}} = 2.67$ for $n = 3$, and $n_{\text{eff}} = 2 - 0.8/(Z-1)$ for $n = 2$]; q is the screening charge and equals 0, 2, and 10 for $n = 1$, 2, and 3 respectively.

It is much more difficult to estimate the fractions $F_{1\pm}^{ns}$. The largest contribution in the sum of Equation 16 probably comes from the lowest ns state that couples directly to the polarized shell of the ferromagnet. For iron the vacancy is in the 3d state and couples to the 1s state of the ion for $Z \leq 8$, whereas for $10 \leq Z \leq 25$ it couples to the 2s orbit; $Z = 9$ lies on the borderline. Van Middelkoop et al (1978) were able to explain the Z dependence of B entirely from the Z dependence of B_{ns} as given by Equation 17. Their analysis requires, to agree with experiment, that $P^{ns} = F_{1+}^{ns} - F_{1-}^{ns}$ be Z independent and approximately linear in velocity.

A simple phenomenological model for P^{ns} was advanced by de Raedt et al (1980). It is based on the kinetics of electron loss and capture in a deep shell. Consider a particular ns orbit and let σ^{ns} be the cross section for the production of the vacancy due to the coupling to the 3d orbit in iron. Let τ^{ns} be the lifetime of the vacancy. Then the vacancy on a given ion will be produced $\sigma^{ns}Nv$ times in one second assuming that the lifetime τ^{ns} is much shorter than the time elapsed between two successive productions: $\tau^{ns}\sigma^{ns}Nv \ll 1$, where N is the concentration of iron atoms. The number $\sigma^{ns}\tau^{ns}Nv$ represents the fraction of the time during which a deep vacancy exists on the ion. This expression must be further multiplied by the degree of polarization. Assuming that the coupling to the 3d orbit dominates and that the polarization of the 3d orbit (0.44) is fully transferred to the deep vacancy, one obtains

$$B(v,Z) = 0.44 \, Nv \sum_n \sigma^{ns}\tau^{ns} \, B_{ns}. \hspace{3cm} 18.$$

The linear dependence of B on velocity is of purely kinematic origin: the number of deep vacancies is proportional to the number of encounters per second of the ion with iron atoms. Since B_{ns} can account for the Z dependence of B, one could obtain a good description of the experimental data if one further assumed that the product $\tau^{ns}\sigma^{ns}$ is only weakly dependent on Z or v. The weak v dependence is not understood but is not unreasonable. However, one would expect that both σ^{ns} and τ^{ns} would decrease with increasing Z. The cross section σ^{ns} decreases because the radius of the ns orbit gets smaller for higher Z and the transfer of the vacancy takes place at smaller values of the internuclear distance. Also the lifetime of the vacancy is a decreasing function of Z (McGuire 1969).

This model is not entirely satisfactory because, on the one hand, the promotion mechanism based on molecular orbit velocity matching is only applicable to few light ion partners and in certain velocity regions, and, on the other hand the polarization mechanism is poorly understood. Nevertheless it does yield reasonable agreement with the observed magnitude of the transient field. Dybdal et al (1979d) calculated the magnitude of the resulting field on the basis of the two polarization mechanisms described earlier, both of which yield about 13% polarization. Their results agree fairly well with the experimental data. They also obtained novel evidence of the general correctness of the ideas incorporated in this model from experiments on K-hole production in O, F, and Si ions stopping in iron, cobalt, and nickel foils (Dybdal et al 1979b,c).

Figure 6 shows that at relatively high velocity, the vacancy production increases with velocity and the general trends are the same for O and F ions in any of the three hosts. However, at low velocities, the K-vacancy cross section remains constant for O in iron and for F in nickel and cobalt, while it decreases with velocity for F ions in iron. Whereas these data give no information on the degree of polarization of the vacancies, they qualitat-

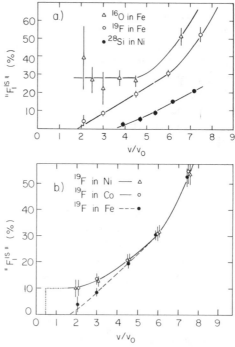

Figure 6 Measured equilibrium K-vacancy fractions, F_1^{1s}, (*a*) for O and F in iron and Si in nickel and (*b*) for F in iron, cobalt, and nickel (the curves serve only to guide the eye) (Dybdal et al 1979c).

ively explain the results obtained with O ions at high and low velocities (Shu et al 1980) and allow for some of the apparent structure observed as a function of Z for light ions (Figure 5). Furthermore, they expose the very subtle atomic interactions at low velocities that play a major role in the interpretation of experiments where the ions actually stop in the host. At somewhat higher velocities, these atomic effects are averaged out, and the more "universal" trends described for the intermediate velocity regime appear.

4 NUCLEAR MAGNETIC MOMENTS

The importance of nuclear magnetic moment measurements to the understanding of nuclear structure has been amply demonstrated. The recent developments in the characterization of the transient field resulted in measurements of many magnetic moments of picosecond nuclear levels previously unobtainable.

It is impossible, within the scope of the present paper, to compile all measurements of nuclear magnetic moments using the transient field technique. We limit ourselves to measurements made since the realization that the hyperfine field increases rather than decreases with the ion's velocity, and we only skim over the wealth of nuclear information obtained by the various groups working on this subject.

In evaluating the nuclear magnetic moments obtained by the transient field technique, one must note that relative values of magnetic moments in series of isotopes are generally well determined; absolute values, however, can be subject to uncertainties due to inadequate calibration or parametrization of the field. Nevertheless, as discussed in Section 3, various parametrizations yield very similar results when applied to isotopes with atomic number and velocity close to calibration points. For example, the magnetic moments of the 2^+ states of even-even Ba isotopes were measured by the thin-foil technique (Brennan et al 1980). The nearest isotopes that can be used to calibrate the field are the Pd-Cd isotopes on one side and Sm isotopes on the other. Various widely different parametrizations of the transient field were chosen to fit the measurements on these calibrating isotopes and were subsequently used to obtain magnetic moments for the Ba isotopes. The same moments were obtained to better than 1%, which suggests that, barring dramatic shell effects, the interpolating procedure yields accurate results irrespective of the functional form chosen for the transient field. However, any parametrization in a case for which either Z or v fall well outside the body of information used to obtain this parametrization should be applied with caution.

Most of the results given below have been obtained with the thin-foil technique, whose advantages are discussed above. While in some medium-heavy nuclei such as the $^{64-70}$Zn isotopes, comparable g factors were measured with both thin iron (Benczer-Koller et al 1978) and thick gadolinium (Fahlander et al 1979) foils, it is still recommended that the thin-foil technique be used in nuclear moment measurements in order to avoid the uncertainties arising from the unknown stopping mechanism.

There are three main regions of interest where measurement of nuclear magnetic moments can contribute to our understanding of nuclear structure. (a) Nuclei around closed shells: the g-factors of excited states in these nuclei can vary appreciably depending on the configuration of the valence nucleons; for many cases even a determination of the *sign* of the magnetic moment can be significant. (b) Low-lying levels of a collective nature where not only absolute measurements of specific states but even relative measurements in series of isotopes help to establish the systematic trends that may differentiate among different theoretical predictions. (c) High spin states such as the higher levels in a rotational band where again a systematic measurement on a series of levels in the same nucleus can shed light on the structure of the band to which they belong.

4.1   Single-Particle Shell Structure Around Magic Numbers

Magnetic moments of low-lying levels in near-magic nuclei are a direct measure of the purity of their configurations. For example, negative g-factors of 2_1^+ levels are rare and can, in fact, be expected only in quite pure shell model configurations of neutrons in stretched angular momentum states ($p_{3/2}$, $d_{5/2}$, $f_{7/2}$, etc). These g-factors are very sensitive to small admixtures of other configurations that tend, in general, to increase the algebraic value of the g-factor. Conversely, g-factors of pure proton configurations are very large and positive.

4.1.1   NEGATIVE MAGNETIC MOMENTS IN THE sd SHELL: O ISOTOPES   The signs of the magnetic moments of the 2_1^+ levels in ^{18}O and ^{20}O have been determined in recoil-into-iron experiments and were found to be negative. These results may be combined with other measurements of the magnitude to yield $g(^{18}$O, $2_1^+) = -0.31 \pm 0.02$ (Forterre et al 1975a), $g(^{20}$O, $2_1^+) = -0.39 \pm 0.04$ (Gerber et al 1976). The sign of these g-factors confirms the $d_{5/2}$ nature of the wave functions of the oxygen isotopes, and the magnitude helps estimate the amount of admixture of $d_{3/2}$ and $s_{1/2}$ configurations. The ^{18}O case was found, in fact, to be the first 2_1^+ level to have a negative magnetic moment; the ^{18}O experiment was also the first to indicate a pronounced deviation from the ALW model.

4.1.2   PARTICLES AND HOLES AROUND THE $1f_{7/2}$ SHELL: Fe AND Ni ISOTOPES   The magnetic moments of the 2_1^+ levels in the $^{54-58}$Fe isotopes were measured using the thin-iron method (Brennan et al 1977). The moment for ^{54}Fe was previously determined in an ion implantation experiment in thick iron, which was analyzed with the ALW parametrization (Hubler et al 1974). A subsequent measurement was carried out with a thick gadolinium foil (Fahlander et al 1977). The three experiments are in agreement and yield a weighted average $g(^{54}\text{Fe}, 2^+) = +1.52 \pm 0.22$. Figure 7a shows the magnetic moments of the even-even Fe isotopes; one can note here the transition from the semi-magic nucleus ^{54}Fe, where the g-factor corresponds to that of the configuration of two proton holes in the $f_{7/2}$ shell and a closed neutron shell, to ^{58}Fe, which is described by collective parameters.

In the $^{58-64}$Ni isotopes, it is the proton $f_{7/2}$ shell that is closed with 2-8 valence neutrons added to a closed ^{56}Ni core. Figure 8a shows the results obtained for the g-factors of the 2_1^+ levels of even nickel isotopes obtained by the thin-foil method (Hass et al 1978). A systematic increase in the magnitude of the g-factors as neutrons are added is evident in these results. This trend cannot be fully explained by including only $p_{3/2}$-$f_{5/2}$-$p_{1/2}$ neutron configurations; better agreement with experiment is obtained when configurations involving the excitation of a single nucleon from the closed $f_{7/2}$ shell are included. This core excitation is seen to be important for ^{58}Ni and ^{60}Ni but strongly blocked for ^{62}Ni and ^{64}Ni.

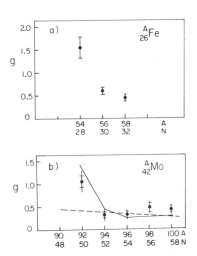

Figure 7   The g-factors of the 2_1^+ states of the even (a) Fe isotopes (Brennan et al 1977) and (b) Mo isotopes (Häusser 1978). The dashed line corresponds to the collective model prediction (Greiner 1966) while the solid line represents a calculation performed in the framework of the BCS theory (Lombard 1968). The large error bars refer to the absolute g-factors, while the small error bars apply to relative g-factors.

4.1.3   THE $Z = 50$ AND $N = 50$ REGION: Mo AND Sn ISOTOPES   The results for the g-factors of 2_1^+ levels in even $^{92-100}_{42}$Mo isotopes are presented in Figure 7b (Häusser 1978). The $^{94-100}$Mo nuclei exhibit vibration-like properties and the g-factor results agree with such pictures. ^{92}Mo, on the other hand, consists of two $g_{9/2}$ protons outside a doubly closed $^{90}_{40}$Zr$_{50}$ core, and indeed the magnetic moment of the 2_1^+ level is large. The Mo isotopes resemble the Fe isotopes in the sense that two protons (or two proton holes in the case of Fe) outside a closed core exhibit single-particle properties; the addition of neutron pairs increases the collective motions and brings the g-factors back to the collective value (Lombard 1968).

The even $^{112-124}$Sn isotopes exhibit the general features of a closed $Z = 50$ proton shell with neutrons occupying the various neutron configurations in this mass region. Because of the very short mean-lives of the 2_1^+ levels of the even Sn isotopes ($0.5 < \tau < 1$ ps), the thin-foil method with a typical interaction time of $T \approx 0.25$ ps is well-suited for g-factor measurements (Hass et al 1980). The results are presented in Figure 8b. The small absolute values of the g-factors and the negative signs for the heavier Sn

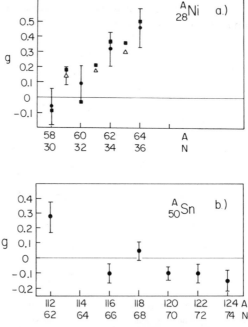

Figure 8   The g-factors of the 2_1^+ states of the (*a*) Ni isotopes (Hass et al 1978) and (*b*) Sn isotopes (Hass et al 1980). The squares represent the theoretical shell model calculations; the dots and triangles denote the experimental results for even and odd nuclei respectively.

isotopes indicate that proton excitations from the $Z = 50$ shell are indeed small, while the $h_{11/2}$ neutron configurations are important. The closure of the $Z = 50$ proton shell is very well demonstrated by comparing the Sn results with the collective values of the neighboring Pd and Cd isotopes. It is also interesting to note the transition from negative g-factors for the heavier Sn isotopes to the positive g-values for the lighter ones. This trend is reproduced by calculations based on the BCS approach (Lombard 1968).

4.1.4   HOLES IN THE ^{206}Pb CORE: Tl ISOTOPES   The g-factors of the $3/2^+$ and $5/2^+$ levels in ^{203}Tl and ^{205}Tl have been measured and were compared to shell model predictions based on the coupling of $3s_{1/2}$, $2d_{3/2}$, and $2d_{5/2}$ protons to the ^{206}Pb core states (Häusser et al 1979). The g-factors of the $5/2^+$ levels had not been measured prior to this experiment. The current data yield $g(^{203}\text{Tl}, 5/2^+) = 1.03 \pm 0.43$ and $g(^{205}\text{Tl}, 5/2^+) = 0.89 \pm 0.26$ and support the calculations stressing the importance of the core-excited 2^+ level in ^{206}Pb.

4.2   Collective Vibrations in Medium Weight Nuclei

The magnetic moments of the 2_1^+ levels in the vibrational region of the Pd-Cd-Ba isotopes is presented in Figure 9 (Brennan et al 1980). The most obvious feature of the sequence of magnetic moments is that they are nearly constant and nearly equal to Z/A, a clear manifestation of the collective nature of the states. The large variations from one isotope to another found for Fe, Ni, and Mo are not present. The measured values, however, fall consistently below the Z/A limit, in rough agreement with the prediction by Nilsson & Prior (1961) and Greiner (1966) (Figure 9), who make a more realistic estimate of the magnetic moments based on a greater pairing force between proton than between neutron pairs, thus allowing for a larger neutron deformation. Nevertheless, the data are still below the calculated value based on a 1.2 ratio between the neutron and proton deformations. Arima & Iachello (1976) developed an alternative framework to describe collective excitations, the interacting boson approximation. In first approximation the g-factors may be expressed in terms of the number of neutron (N_ν) and proton (N_π) bosons outside closed shells:

$$g = A + B_\nu N_\nu + B_\pi N_\pi$$

where B_ν and B_π are constants within the same major shell. The best fit to the Pd-Cd data yields values for the constants A, B_ν, B_π, which, however, have not yet been obtained from a microscopic calculation and have not been compared to the constants needed to fit the magnetic moments of other nuclei within the same shell. Better accuracy is needed for careful comparison with theory, especially in a test of the second-order term in the

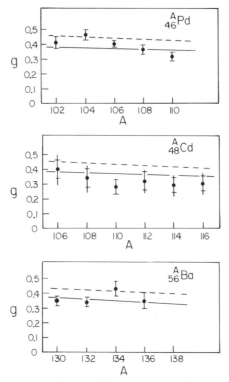

Figure 9 The *g*-factors of the 2^+_1 states of the Pd, Cd, and Ba isotopes. The error bar represents the absolute error resulting from the combination of the experimental statistical errors and the error in the calibration point. The smaller error bars shown on the Cd isotopes indicate just the statistical errors in the experiments (Brennan et al 1980). The dashed line corresponds to the $g = Z/A$ prediction of collective models while the solid line includes the contribution arising from different pairing between protons and neutrons (Greiner 1966).

interacting boson approximation which will relate the *g*-factors to the quadrupole moments. Extension of the systematics in the neighboring isotopes in the shells above and below $Z = 50$ is in progress in several laboratories.

4.3   *High Spin States*

The physics of high spin states has been of major interest in recent nuclear physics studies. Of particular interest is the measurement of magnetic moments of such states since the magnetic moment can be sensitive to dramatic changes observed in the level structure and associated with higher spin values.

One of the most interesting cases for a g-factor measurement is the region of high spin states where the back-bending phenomenon occurs. It is beyond the scope of the present review to go into the details of the well-known back-bending behavior of rotational bands in various nuclei. It is sufficient to mention that because of the single-particle nature of the magnetic moment operator, a g-factor measurement of levels below and above a back-bending region can be very sensitive to the detailed behavior of the nucleons with increasing rotational frequency. Such measurement can then be compared to predictions of models such as the rotational-alignment model and others. Levels in this region have typical lifetimes of the order of 1 ps and the transient field method is well suited for such measurements. In a typical experiment, a high spin state will be populated in a nucleus recoiling at high velocity. The present realization that the transient field *increases* with velocity to values of about 20 MG for a rare-earth nucleus at $v/c \approx 0.04$ has facilitated g-factor experiments for high spin, short-lived levels in heavy nuclei. Since the transient field acts only during this short flight time of the ion in the ferromagnet, it is necessary to ensure that the levels under investigation are indeed populated during this time. This condition is met in Coulomb excitation because the excitation process is prompt and lacks significant side-feeding channels. Reactions leading to the high spin state require very high energy heavy ion beams. Two such projects are now in progress; the Chalk River–Berkeley collaboration has measured magnetic moments in [170–174]Yb isotopes with a Kr beam from the HILAC (Ward et al 1979). The GSI-Bonn-Heidelberg-Weizmann collaboration is using a Pb beam from the UNILAC to Coulomb-excite [158]Dy (M. Hass, private communication). The preliminary Chalk River–Berkeley results are consistent with a constant g-factor of about 0.3 for levels up to $I = 12$. The GSI experiment is now being analyzed. Further results from these two experiments are expected in the near future.

5   SUMMARY AND FUTURE EXPECTATIONS

The interest in the study of the transient hyperfine field is stimulated by the interest in microscopic investigations of ion-solid interactions as well as by its instrumental role in measurements of magnetic moments of short-lived nuclear states. This interest is well demonstrated by the wealth of data acquired in the last several years and reviewed in this paper. It is now well established that a large magnetic field acts on the nuclei within fast ions traversing a ferromagnetic foil and that this field increases with velocity for $v < Zv_0$ and then decreases for $v > Zv_0$. This behavior seems to hold for ions throughout the periodic table and the general features of the transient field can be reproduced by simple parametrizations in terms of the ions' velocity and atomic number. Such parametrizations can be utilized to measure nuclear magnetic moments and thus provide new and significant nuclear structure information. It is also clear that the simple and general parametrizations describing the transient field serve only as a framework within which interesting effects associated with detailed atomic structure can be discussed. While no rigorous quantitative description of the transient field yet exists, several models, supported by experimental information on electron vacancy production in ions inside matter, reproduce its qualitative aspects.

At the risk of being repetitive, we emphasize again the need for differential velocity measurements of the transient field. In order to separate the complex mechanisms responsible for the observed interactions it is essential to probe the various atomic structure effects as functions of velocity without the complication of averaging over a broad velocity profile. Such measurements should then serve as a basis for more theoretical work regarding the various processes taking place when the moving ion interacts with a magnetized medium. For example, different velocities bring different atomic levels into matching conditions with the host polarized electrons and could thus provide information regarding the relative contributions of the different atomic shells to the transient field. Recent work suggests that the potential associated with a moving ion inside matter (Z. Vager, private communication) can give rise to a distortion as well as a polarization of the inner orbitals of the moving ion. Such a model may illuminate the general characteristics of the transient field at complex ions, serving as a complementary theoretical approach to the microscopic atomic calculations and bringing into focus the possible connection between the transient field phenomenon and effects associated with passage of molecular clusters through matter. It may also be interesting to consider the connection between the polarization of inner electronic shells produced in swift ions traversing a magnetized medium and the polarization produced in fast ions recoiling out of a tilted foil (Niv et al 1979).

Even with the restricted understanding of the transient field available to date, considerable progress has been possible in the determination of nuclear moments of short-lived states over the whole periodic table. The systematic measurement of magnetic moments in a series of isotopes has been especially fruitful in critical comparisons with existing nuclear structural models. More magnetic moments of interest in nuclear physics can be measured; in particular, the subject of high spin states has only recently been addressed and will undoubtedly draw much attention in the future. The observations at intermediate velocities suggest that the transient field increases with a power of velocity smaller than one. Since high spin states can generally be populated by using reactions that result in high velocities of the excited nuclei, the velocity dependence of the transient field at velocities higher than heretofore explored should be carefully mapped.

The excitement in further explorations of the transient field stems from the flexibility inherent in the investigation of these phenomena. By choosing the velocity regime, the nucleus, and the magnetic medium, one can address the various atomic and nuclear physics aspects of a subject that will remain an intensive and fruitful area of research.

ACKNOWLEDGMENTS

We would like to thank all our colleagues who have communicated and discussed their results with us. We also wish to thank Professor G. Goldring for helpful comments regarding the manuscript.

Literature Cited

Arima, A., Iachello, F. 1976. *Ann. Phys.* 99 : 253
Benczer-Koller, N., Hass, M., Brennan, J. M., King, H. T. 1978. *J. Phys. Soc. Jpn.* 44 : 341–44
Betz, H. D. 1972. *Rev. Mod. Phys.* 44 : 465
Bhalla, C. P., Bradford, N. J. 1968. *Phys. Lett. A* 27 : 318
Borchers, R. R., Bronson, J. D., Murnick, D. E., Grodzins, L. 1966. *Phys. Rev. Lett.* 17 : 1099–1102
Borchers, R. R., Herskind, B., Bronson, J. D., Grodzins, L., Kalish, R., Murnick, D. E. 1968. *Phys. Rev. Lett.* 20 : 424
Brandt, W. 1977. Penetration phenomena with heavy projectiles. In *Atomic Physics in Nuclear Experiments*, ed. B. Rosner, R. Kalish. pp. 441–75. Bristol: Hilger
Brennan, J. M. 1978. PhD thesis, Rutgers Univ. New Brunswick, N.J.
Brennan, J. M., Benczer-Koller, N., Hass, M., King, H. T. 1977. *Phys. Rev. C* 16 : 899–901
Brennan, J. M., Benczer-Koller, N., Hass, M., King, H. T. 1978a. *Hyperfine Interactions* 4 : 268–77
Brennan, J. M., Benczer-Koller, N., Hass, M., Kossler, W. J., Lindemuth, J., Fiory, A. T., Murnick, D. E., Minnich, R. P., Lankford, W. F., Stronach, C. E. 1978b. *Phys. Rev. B* 18 : 3430–36
Brennan, J. M., Hass, M., Shu, N. K. B., Benczer-Koller, N. 1980. *Phys. Rev. C.* 21 : 574–78
Broude, C., Engelstein, P., Popp, M., Tandon, P. N. 1972. *Phys. Lett. B* 39 : 185–87

Bruno, J., Sak, J. 1978. *Phys. Lett. A* 68 : 463–65
Bruno, J., Sak, J. 1979. *Phys. Rev. B* 19 : 3427
De Raedt, J. A. G., Holthuizen, A., Rutten, A. J., Sterrenburg, W. A., Van Middelkoop, G. 1980. *Hyperfine Interactions* 7 : 455–64
Dybdal, K., Eberhardt, J. L., Rud, N. 1976. *Phys. Lett. B* 64 : 414–416
Dybal, K., Eberhardt, J. L., Rud, N. 1979a. *Hyperfine Interactions* 7 : 29–43
Dybdal, K., Eberhardt, J. L., Rud, N. 1979b. *Phys. Rev. Lett.* 42 : 592–95
Dybdal, K., Forster, J. S., Rud, N. 1979c. *Phys. Rev. Lett.* 43 : 1711–15
Dybdal, K., Forster, J. S., Rud, N. 1979d. In *8th Int. Conf. on Atomic Collisions in Solids*, Hamilton, Canada
Eberhardt, J. L., Van Middelkoop, G., Horstman, R. E., Doubt, H. A. 1975. *Phys. Lett. B* 56 : 329–31
Eberhardt, J. L., Horstman, R. E., Zalm, P. C., Doubt, H. A., Van Middelkoop, G. 1977. *Hyperfine Interactions* 3 : 195–212
Fahlander, C., Johansson, K., Karlsson, E., Possnert, G. 1977. *Nucl. Phys. A* 291 : 241–48
Fahlander, C., Johansson, K., Karlsson, E., Norlin, L. O., Possnert, G 1978. *Phys. Scr.* 17 : 31–38
Fahlander, C., Johansson, K., Possnert, G. 1979. *Z. Phys. A* 291 : 93–96
Forterre, M., Gerber, J., Vivien, J. P., Goldberg, M. B., Speidel, K.-H. 1975a. *Phys. Lett. B* 55 : 56–58
Forterre, M., Gerber, J., Vivien, J. P.,

Goldberg, M. B., Speidel, K.-H., Tandon, P. N. 1975b. *Phys. Rev. C* 11 : 1976–82

Gerber, J., Goldberg, M. B., Speidel, K.-H. 1976. *Phys. Lett. B* 60 : 338–40

Goldberg, M. B., Konejung, E., Knauer, W., Kumbartzki, G. J., Meyer, P., Speidel, K.-H., Gerber, J. 1976a. *Phys. Lett. A* 58 : 269–71

Goldberg, M. B., Kumbartzki, G. J., Speidel, K.-H., Forterre, M., Gerber, J. 1976b. *Hyperfine Interactions* 1 : 429–34

Goldberg, M. B., Knauer, W., Kumbartzki, G. J., Speidel, K.-H., Adloff, J. C., Gerber, J. 1978. *Hyperfine Interactions* 4 : 262–67

Greiner, W. 1966. *Nucl. Phys.* 80 : 417–433

Grodzins, L. 1970. Nuclear moment measurements using heavy ion recoil techniques. In *Nuclear Reactions Induced by Heavy Ions*, ed. R. Brock, W. R. Hering, pp. 367–91. Amsterdam : North-Holland

Hass, M., Brennan, J. M., King, H. T., Saylor, T. K. 1976. *Phys. Rev. C* 14 : 2119–25

Hass, M., Benczer-Koller, N., Brennan, J. M., King, H. T., Good, P. 1978. *Phys. Rev. C* 17 : 997–1000

Hass, M., Broude, C., Niv, Y., Zemel, Y. 1980. *Phys. Rev. C* 22 : 97–100

Häusser, O. 1978. In *Nuclear Interactions, Proc. Canberra Conf. 1978*, ed. B. A. Robson. Berlin : Springer

Häusser, O., Haas, B., Ward, D., Andrews, H. R. 1979. *Nucl. Phys. A* 314 : 161–70

Hubler, G. K., Kugel, H. W., Murnick, D. E. 1974. *Phys. Rev. C* 9 : 1954–64

Kalish, R., Shu, N. K. B., Benczer-Koller, N., Holthuizen, A., Rutten, A. J., Van Middelkoop, G. 1980. *Hyperfine Interactions*. In press

Kopferman, H. 1956. *Kernmomente*. Frankfurt : Akademischer

Kumbartzki, G. J., Hagemeyer, K., Knauer, W., Krösing, G., Kuhnen, R., Mertens, V., Speidel, K.-H., Gerber, J., Nagel, W. 1979. *Hyperfine Interactions* 7 : 253–64

Lindhard, J. 1954. *Mat. Fys. Medd. Dan. Vid. Selsk.* 28 : 8

Lindhard, J., Winther, A. 1971. *Nucl. Phys. A* 166 : 413–35

Lombard, R. J. 1968. *Nucl. Phys. A* 114 : 449–62

McGuire, E. 1969. *Phys. Rev.* 185 : 1–6

Nilsson, S. G., Prior, O. 1961. *Mat. Fys. Medd. Dan. Vid. Selsk.* 32 : 16

Niv, Y., Hass, M., Zemel, A. 1980. *Hyperfine Interactions* 8 : 19–27

Niv, Y., Hass, M., Zemel, A., Goldring, G. 1979. *Phys. Rev. Lett.* 43 : 326–30

Sak, J., Bruno, J. 1978. *Phys. Rev. B* 18 : 3437–39

Salm, W., Klepper, O. 1976. *Phys. Rev. Lett.* 37 : 88–90

Shu, N. K. B., Melnik, D., Brennan, J. M., Semmler, W., Benczer-Koller, N. 1980. *Phys. Rev. C* 21 : 1828–37

Speidel, K.-H., Kumbartzki, G. J., Knauer, W., Krösing, G., Mertens, V., Tandon, P. N., Gerber, J., Freeman, R. M., Goldberg, M. B. 1980. *Hyperfine Interactions*. In press

Vager, Z., Gemmel, D. S. 1976. *Phys. Rev. Lett.* 37 : 1352–54

Van Middelkoop, G. 1978. *Hyperfine Interactions* 4 : 238–56

Van Middelkoop, G. 1979. In *4th Gen. Conf. Eur. Phys. Soc. Trends in Physics, 1978, York, Engl.*, pp. 438–44. Bristol : Hilger

Van Middelkoop, G., De Raedt, J. A. G., Holthuizen, A., Sterrenburg, W. A., Kalish, R. 1978. *Phys. Rev. Lett.* 40 : 24–37

Ward, D., Häusser, O., Andrews, H. R., Taras, P., Skensved, P., Rud, N., Broude, C. 1979. *Nucl. Phys. A* 330 : 225–42

Zalm, P. C., Eberhardt, J. L., Horstman, R. E., Van Middelkoop, G., de Waard, H. 1976. *Phys. Lett. B* 60 : 258–60

Zalm, P. C., Van der Laan, J., Van Middelkoop, G. 1979. *Nucl. Instrum. Methods* 161 : 265–72

Ann. Rev. Nucl. Part. Sci. 1980. 30: 85–157

NUCLEI AT HIGH   ✕ 5615
ANGULAR MOMENTUM [1]

R. M. Diamond and F. S. Stephens

Nuclear Science Division, Lawrence Berkeley Laboratory, University of California, Berkeley, California 94720

CONTENTS

1   INTRODUCTION

One probably thinks of angular momentum in terms of rotation of classical bodies. The galaxies, the earth, or perhaps tops (gyroscopes) come to mind. From these objects we have an intuitive feeling for certain aspects of rotation: the flow pattern can be rigid as in a top, or more complicated (with strong irrotational components) as for water in an oval container, depending on the internal structure. The moment of inertia depends on this flow pattern as well as on the shape. Nuclei can rotate, and the above concepts have a direct applicability, but nuclei are much more complex and

---

[1] This article was written by contractors of the US Government under contract No. W-7405-ENG-48. The US Government has the right to retain a nonexclusive, royalty-free license in and to any copyright covering this paper.

interesting than classical rotors. They have important quantal aspects and further, are "finite" systems, being composed of a rather small number of nucleons. The first of these properties implies some restrictions on the rotation and the second means that there are also important single-particle or noncollective effects in nuclei, as well as a continuous variation between collective and noncollective properties. At one limit, the nucleons act coherently and collective bands develop that follow the $I(I + 1)$ rotational pattern to within a percent or two and have transition probabilities 200 times larger than a single particle would have. At the other limit, a few individual nucleons may carry all the angular momentum of a high spin state. Between these limits we find sometimes a complex behavior and other times a coexistence of these simpler limiting behaviors. The study of nuclei at high angular momenta can be cast into the form of understanding first these two limits themselves, and then the interplay between them as the spin increases.

It is not difficult to trace the development of these two limiting situations. Single-particle angular momentum was implied immediately by the shell model of nuclei, which was conceived in the 1930s (Bethe & Bacher 1936), but bloomed only after 1949 (Mayer) with the recognition of the importance of spin-orbit splitting. Isomers with spins 7/2, 9/2, or even 11/2, based on particles in high j orbitals, were the first high spin states and also some of the early direct evidence in support of the shell model (Goldhaber & Sunyar 1951). It is, in principle, straightforward to align the angular momenta of many particles to make a high spin state, but reasonably pure shell-model (noncollective) states with more than three or four aligned particles (and spins higher than 10 or $12\hbar$) are not so common. The double-closed-shell region of lead is exceptional in that states with spins up to 20 or $30\hbar$ can be explained to high accuracy with only the shell model and modest residual interactions (Blomqvist 1979).

Rotation of nuclei was proposed by Bohr (1951, 1952), based on the understanding that the existence of strongly deformed shapes required such a mode. There was early evidence that such shapes existed (Casimir 1936) and there were ideas as to how they might arise from the appropriate number of particles in a shell-model potential (Rainwater 1950). The art of calculating nuclear potential energy surfaces as a function of spin and shape parameters has been refined many times, using modified harmonic oscillator, Woods-Saxon, and, most recently, Hartree-Fock shell-model potentials, all of which generally give reasonable agreement with the observed shapes. The moments of inertia were a problem for a few years. The way to connect the shell-model structure to a collective moment of inertia, by cranking the potential, was envisioned in 1954 (Inglis) and later shown to lead to the rigid-body value for the moment of inertia (Bohr &

Mottelson 1955), two to three times larger than those observed. This discrepancy was attributed to the neglect of "residual interactions," but a quantitative explanation had to await the deeper understanding of nuclear pairing that came in 1958 (Bohr et al). The pairing correlations introduce strong irrotational components into the flow pattern of nuclei, and thus reduce the moment of inertia (Griffin & Rich 1960, Nilsson & Prior 1961). Recently it has been possible to identify the reduction of the pairing correlations with increasing spin, and the attendant rise of the moment of inertia toward the originally predicted rigid-body value.

It was, of course, recognized that there was likely to be a full range of behavior between these limits (Bohr 1952). For low spins the relevant region of "vibrational-like" nuclei has resisted a simple and satisfying treatment and is still an area of intense study. In the higher spin range this problem came into sharp focus in 1971 (Johnson et al) when a discontinuity (called backbending) was observed in several rotational bands at spins around $20\hbar$. This behavior is due to the breaking of a pair of high j particles, each aligning its angular momentum directly with the collective one generated by the remaining particles (Stephens & Simon 1972). These rotational nuclei are clearly moving toward the noncollective limit. At this point in 1975, Bohr & Mottelson perceived that the noncollective shell-model states could be viewed as states due to the "effective" rotation of a nucleus about its symmetry axis. This has the great advantage of giving the average energy of these states from the moment of inertia for the corresponding rigid rotation. (It has the disadvantage of being frequently mistaken for a true collective rotation.) From this viewpoint, one would conclude that the lowest-lying states over a rather broad range of high spins should, on the average, be noncollective, corresponding to effective rotation about an oblate symmetry axis (the largest moment of inertia for a given deformation in this range). This led to the idea that the high spin states of some nuclei (the maximum spin that can survive fission is $60\text{--}70\hbar$) might be completely noncollective and thus have an inherent irregularity that would give rise to some isomers (Bohr & Mottelson 1974). Such isomers would be very convenient for study, and were indeed found at spins up to $\sim 30\hbar$. However, they do not seem to occur at the very highest spins, where nuclei appear to prefer some kind of compromise between the collective and noncollective behavior. It is important to understand the nature of this compromise. Are there still rotational bands, and, if so, how good are they and what is their moment of inertia? Can one still hope to identify aligned particles, and, if so, how many and which ones? What is the shape of nuclei under these conditions? These are today's questions, and this review aims to define what we know about them and how we are trying to find out more. There have been several reviews covering these aspects of high spin states (Bohr &

Mottelson 1974, Newton 1974, Lieder & Ryde 1978). Reviews covering other, or more limited, aspects are mentioned where appropriate.

The experimental approach to high spin states has relied on only a few methods for production. Radioactive decay was the first. Early evidence for shell-model isomers came mostly from β-decay studies (Goldhaber & Sunyar 1951), and α-decay provided systematic evidence for the occurrence of rotational states up to spins of 6 or $8\hbar$ in the actinide region of nuclei (Asaro & Perlman 1953), just at the time the rotational model was proposed. Coulomb excitation refers to the purely electromagnetic excitation of nuclear states (usually in a collision where there is insufficient energy to penetrate close enough to involve the nuclear forces). This process was essentially born with the rotational model, and immediately established the large electric-quadrupole transition probabilities implied by deformed nuclei (Huus & Zupancic 1953). Projectiles with high charge (heavy ions) were found to excite successively a number of rotational transitions ($\Delta I = 2$) in a single collision; in 1977 Fuchs et al observed the excitation of a state having $30\hbar$ in ^{238}U using ^{208}Pb projectiles. Coulomb excitation has the advantage of giving transition moments as well as high spins, but it can reach only about halfway up to the highest nuclear spins, and thus eventually gave way to a third method. When two nuclei collide at energies above the Coulomb barrier, one of the main processes can be fusion, in which all the angular momentum of the initial system is retained. The amount of angular momentum depends on the projectile, bombarding energy, and impact parameter. Values up to about $25\hbar$ were brought in by ^{4}He projectiles during the first experiments in 1963 (Morinaga & Gugelot); by 1968 (Stephens et al) ^{40}Ar projectiles were used and they can easily bring more than $100\hbar$ into the compound system. This method, then, brings in all the angular momentum the nucleus can hold; the problem is to identify what happens in those events involving the highest angular momentum. It is not so difficult to isolate events from a particular channel leading to a certain final product nucleus, but even then, above about $20\hbar$, one finds essentially no resolved lines because the population is spread over too many states. According to taste, some experimenters have responded by following the resolved lines further up in spin (and down in intensity) or by hunting for isomers that would provide much greater sensitivity for measuring resolved lines. The current spin record for states based on resolved lines is $38\hbar$ (Khoo et al 1978) and an average, with hard work, might be $30\hbar$. Other experimenters have developed techniques to extract information from unresolved spectra. These include multiplicity filters, sum spectrometers, correlation methods, and, projecting slightly into the future, 4π multiplicity detectors. Both approaches are reviewed here, together with some prognosis for the future.

2 PHYSICAL CONCEPTS

This section discusses the concepts underlying the study of high angular momentum states. The factors that limit angular momentum in nuclei are considered in the first part and the areas covered in this review defined. The second part of this section conveys the overview that is currently emerging of the physics of very high spin states. Finally, in the last part the detailed calculations currently made for high spin states are described. We do not concentrate on the calculations themselves, but on the physical input to them and the kind of results that come out.

2.1 Angular Momentum Limits

The phrase "high angular momentum" as applied to nuclei can mean several things. In the very light nuclei it refers generally to the maximum angular momentum that can be generated by valence nucleons; for example, it would be around $4\hbar$ for the p-shell nucleus ^{8}Be, and $8\hbar$ for the sd-shell nucleus ^{20}Ne. This limit exceeds $100\hbar$ for nuclei around mass 170, and is effectively replaced by lower limits, most generally the instability against fission. Between these regions, for $40 \gtrsim A \gtrsim 100$, the highest angular momentum that can be conveniently studied is limited by what can survive the particle evaporation cascade that follows production of the compound nucleus. The present review is limited to the highest of these mass regions, $A \gtrsim 100$. Even with this limitation, "high angular momentum" is now used to refer to two spin ranges: the highest spins accessible ($\sim 60\hbar$) or the highest spins that have been well studied ($\sim 20\hbar$). The latter of these ranges has been exciting, with the discovery of backbending and its subsequent interpretation in terms of band crossing (Johnson & Szymanski 1973, Sorensen 1973). Indeed, a spectroscopy of band crossings is currently developing. However, although the phenomena and concepts from this region are important here, the present review aims at the highest spins it is possible to study.

It is impossible to specify a maximum angular momentum for nuclei without specifying a time scale. For example, a target and projectile nucleus may be in contact momentarily with as much as $500\hbar$ in the system. The way angular momentum is transferred into the internal degrees of freedom of the two nuclei is one of the fascinating aspects of deep inelastic collisions, but such a system cannot fuse and the angular momentum is largely returned to the external degrees of freedom as the system separates. In order to live longer than $\sim 10^{-20}$ s, a system must have a nonvanishing barrier against fission, and the point at which this occurs can be readily estimated using the liquid-drop model (Bohr & Kalckar 1937). This model considers the nucleus as an incompressible fluid with volume, surface, Coulomb, and rotational (based on the equilibrium shape) energies. Its success in giving average nuclear potential energies under various conditions has been impressive from the earliest days of nuclear physics. The angular momentum for which the fission barrier vanishes has been calculated by Cohen et al (1974) and is shown as a function of mass number (along the line of β stability) by the solid curve in Figure 1. This represents the maximum angular momentum that a "cold" idealized nucleus could contain, and is about $100\hbar$ for $A \approx 130$. It is lower both for higher mass numbers (because of the increasing Coulomb repulsion) and for lower mass numbers (because of the lower surface energy and the higher rotational frequencies required by the smaller moments of inertia). If cold nuclei could be produced at these angular momenta, spectroscopic studies might be possible up to near this limit. But the heavy ion fusion reactions that bring in this much angular momentum also bring in of order 40 MeV of excitation energy (or more), greatly increasing the fission probability. In order to prevent fission (where most of the angular momentum goes into the relative motion of the fragments), another process must de-excite the nucleus, and at such

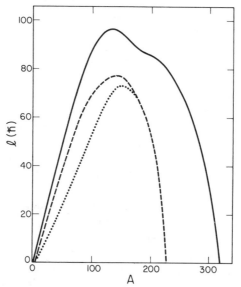

Figure 1   The solid line indicates the angular momentum value for a given mass number A where the fission barrier just vanishes according to the liquid-drop model. The dashed line shows where it is about 8 MeV (Cohen et al 1974). The dotted line is an estimate of the boundary between particle evaporation (*above*) and γ-ray emission (*below*) estimated from data in Newton et al (1977).

excitation energies this can only be particle evaporation. The time scale for particle evaporation (at these excitation energies and for normal binding energies) is 10^{-17}–10^{-18} s, and in order to slow the fission down to such times, a fission barrier of the order of the neutron binding energy (~ 8 MeV) is required. The dashed line in Figure 1 corresponds to an 8-MeV fission barrier, below which particle evaporation should dominate. Between these lines, or in the time range 10^{-17}–10^{-21} s, essentially only fission of the system can be studied, and there is little information about this region. When the decreasing level density slows the particle evaporation sufficiently ($\sim 10^{-15}$ s), or the excitation energy is below the neutron binding energy, γ-ray emission takes over and de-excites the nucleus to its ground state. The angular momentum removed by the particle evaporation is small if neutrons or protons are evaporated ($\sim 1\hbar$ per particle) but can be quite large (up to $20\hbar$) for α particles (Section 3). An estimate of the maximum angular momentum surviving the particle evaporation has been made based on γ-ray and α-particle emission probabilities, and is shown as the dotted line in Figure 1. Between the dashed and dotted lines (10^{-17}–10^{-15} s) the states can be studied by means of the evaporated particles— a method more sensitive to the nuclear structure than fission, but much less sensitive than γ-ray decay. The present review covers only the γ-ray studies of high angular momentum states, which have access to the region below the dotted line in Figure 1. It is easy to understand why these studies have centered first around the region $A \sim 140$, where spins up to $\sim 70\hbar$ are accessible.

2.2 Nuclear Rotation

Our understanding of high spin states has recently improved, making possible a reasonably simple and unified description of this subject. Various aspects of nuclear behavior have been recognized for some time. The most obvious perhaps is the occurrence of collective and noncollective (or single-particle) excitation, and almost as familiar are classical, quantum mechanical, and microscopic features. The present section discusses the relationship of these concepts to one another and describes the interplay among them with increasing angular momentum. The pairing correlations introduce strong irrotational components into the nuclear flow pattern, and make the subject more complex at the lower spin values. However, almost all estimates indicate that these correlations will be quenched as a result of Coriolis effects by angular momenta $\sim 30\hbar$, so this aspect will not be emphasized.

The collective limit is one we must understand, and the basic nuclear system here has been found to be an axially symmetric rotor with quadrupole deformation. For a more complete discussion of this subject see

Bohr & Mottelson (1975). Such a system with two equivalent axes 1,2 about which it can rotate has a Hamiltonian of the form:

$$\mathcal{H}_{rot} = \frac{\hbar^2}{2\mathbf{F}} (I_1^2 + I_2^2),$$
1.

where $\mathbf{F}_1 = \mathbf{F}_2 \equiv \mathbf{F}$ is a moment of inertia[2] that is taken here to be constant, but whose value depends on the internal structure of the system. The energy of a rotor of this type can be expressed either in terms of the angular momentum

$$E_{rot}(I) = \frac{\hbar^2}{2\mathbf{F}} I(I + 1),$$
2.

or in terms of the rotational frequency, ω,

$$E_{rot}(\omega) = \frac{1}{2} \mathbf{F}\omega^2.$$
3.

Deviation from rigid-rotor behavior can be expressed as an expansion either in $I(I + 1)$ or in ω^2; the latter generally converges better. The relationship between these is just

$$\mathbf{F}\omega = \hbar[I(I + 1)]^{1/2}.$$
4.

To relate ω to rotational transition energy, E_γ, we have:

$$\hbar\omega = 2[I(I + 1)]^{1/2} \frac{\partial \mathcal{H}_{rot}}{\partial(I_1^2 + I_2^2)} \approx \frac{\partial E_{rot}}{\partial I}$$

$$\approx \frac{1}{2} [E_{rot}(I + 1) - E_{rot}(I - 1)] = \frac{1}{2} E_\gamma,$$
5.

where the right side is specifically for the collective stretched $(I + 1 \rightarrow I - 1)$ quadrupole transitions. This last relationship, that the rotational frequency is about half the γ-ray transition energy, is often used because the transition energy is a readily measured quantity. It can also be related to the angular momentum:

$$E_\gamma = \frac{\hbar^2}{2\mathbf{F}} (4I - 2).$$
6.

These relationships are for perfect rotors; in real nuclei the moment of inertia is not completely constant, which complicates all these expressions.

The other limit we are considering is a single-particle one, where the motion of each particle is nearly independent of the others. The region between these limits is of considerable importance for high spin studies and involves coupling various numbers of single particles to the collective

[2] The symbol $\mathcal{F}$ is also used on various figures to represent moment of inertia.

core—in the above case, to the rotor. The generalization to include single particles coupled to the symmetry axis of the rotor ($K \neq 0$) was made in the original paper by Bohr (1952), and this does not modify Equations 5 or 6. However, an important coupling for high spin states is of the single particles to the rotation axis, and in such a case Equations 5 and 6 must either be modified, or F must be considered an "effective" value. This idea and its implications are discussed at several points in this review, especially in Section 6.3.

The moment of inertia of a classical rotor depends on both the shape and the flow pattern, of which the latter is expected to be rigid in nuclei at high spins. The shape of a rigid ellipsoid can be expressed in terms of the parameters σ and γ, defined so that the semi-axes r_i are related to the mean radius R by $r_i = a_i R$, where

$$a_1 = \exp\left[\sigma \cos\left(\gamma - \frac{2\pi}{3}\right)\right] \qquad a_2 = \exp\left[\sigma \cos\left(\gamma + \frac{2\pi}{3}\right)\right]$$
$$a_3 = \exp\left[\sigma \cos\gamma\right]. \tag{7}$$

For small deformation this gives $\Delta R/R \approx \varepsilon \approx 1.5\sigma$ and $\beta \approx 1.6\sigma$. Such an ellipsoid has moments of inertia

$$\mathbf{F}_1 = \frac{1}{5} M(r_2^2 + r_3^2) = \frac{1}{2}(a_2^2 + a_3^2)\mathbf{F}_0 \tag{8}$$

where $\mathbf{F}_0$ is the rigid-sphere value, and the axes may be permuted cyclically. Values of $\mathbf{F}_0$ can be obtained from the expression for a rigid sphere given by Myers (1973). From the equivalent sharp radius for the matter distribution (Blocki et al 1977)

$$R_s = 1.28A^{1/3} - 0.76 + 0.8A^{-1/3} \text{ fm}; \tag{9}$$

the value for a sphere is

$$\frac{2\mathbf{F}_0}{\hbar^2} = \frac{4MR_s^2}{5\hbar^2} = 0.01913AR_s^2 \text{ (fm) MeV}^{-1}. \tag{10}$$

The effect of a diffuse surface can be added simply by

$$\mathbf{F}_{\text{diff}} = \mathbf{F}_{\text{sharp}} + 2Mb^2, \tag{11}$$

where b is the width of the diffuse region, normally around 1 fm. For orientation one can use the simpler expression:

$$R = 1.16A^{1/3} \text{ fm}, \tag{12}$$

which leads to

$$\frac{\hbar^2}{2\mathbf{F}_0} = 36A^{-5/3} \text{ MeV}. \tag{13}$$

For an oblate ($\gamma = 60°$) or a prolate ($\gamma = 0°$) shape, rotating parallel to the symmetry axis, $\parallel$, or perpendicular to it, $\perp$, the lowest order expansions of Equation 8 are illustrated in Figure 2. Triaxial shapes will fall between these limits, and for this reason are not considered explicitly here. These expansions begin to deviate significantly from the exact expressions around $\beta = 0.3$, and for β values near 0.8 the exact $P - \perp$ moment of inertia becomes larger than the $O - \parallel$ one, in contrast to the situation shown in Figure 2. It is interesting that the liquid-drop shapes of minimum energy as a function of spin reflect this behavior, being moderately deformed and oblate through most of the spin range and becoming prolate in the $A = 150$ region when β reaches values around 0.8, just prior to fission. The energy trajectories based on these four cases of classical rigid rotation and Equation 2 are shown in the right part of Figure 2 for $\beta = 0.3$. The lowest energies are for an oblate shape rotating about the symmetry axis, corresponding to its largest moment of inertia. The earth is oblate for precisely this reason; however, real rotating nuclei are generally not oblate due to the shell effects, as discussed below.

For systems where the quantal aspects are important, the preceding discussion has to be clarified, since these systems cannot rotate collectively about a symmetry axis—there is no way to orient them with respect to such an axis. It was understood for some time that this meant these degrees of freedom were contained in the single-particle motion. However, when Bohr & Mottelson (1975) considered aligning particle angular momenta along a symmetry axis, they realized that *on the average* the energy was the same as

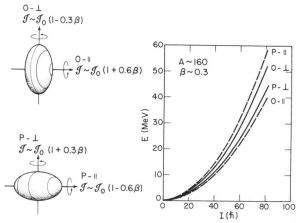

Figure 2   The left side shows the lowest order estimate for the rigid-body moments of inertia of prolate (P) and oblate (O) shapes rotating parallel ($\parallel$) or perpendicular ($\perp$) to the symmetry axis. The right side shows the corresponding energy trajectories for $\beta = 0.3$ and mass number 160.

rotating the system classically about that axis. They strictly showed this only in the Fermi gas approximation, but it is generally believed to be true, or approximately so, for realistic nuclear systems. Particle angular momenta aligned along a nuclear symmetry axis are then viewed as an *effective* rotation of the system about that axis. Thus the trajectories sketched in the right part of Figure 2 all have meaning for nuclei; the solid ones are true collective rotations, having smooth energies and strongly enhanced E2 transition probabilities, whereas the dashed lines are the *average* location of irregularly spaced states having single-particle character. Both features of the latter type states suggest that isomers should be probable, and these expectations have led to a number of searches for them, as discussed in Section 5.

To this picture the microscopic aspects of nuclear structure must be added. Nuclear levels in a potential well are grouped together into shells in very much the same way electrons are in an atom. Certain nucleon numbers ("magic numbers") complete shells and have extra stability in analogy to the noble gas electronic structures. However, when nuclei deform, the shells change, so that the number to complete a shell is different. Thus, in general, a given nucleon number will prefer the shape that makes it look most nearly like a closed shell. These "shell effects" can be as large as 10–12 MeV (the double-closed spherical shell at ^{208}Pb), but on the average might be 3–4 MeV. Compared with the right side of Figure 2, it is apparent that 3–4 MeV is larger than the full spread of liquid-drop shapes up to $I \sim 30$. Thus below this spin (for $A \sim 160$) the nuclear shape is determined mainly by shell effects. Around $I = 60$, however, the spread in Figure 2 is ~ 10 MeV, considerably larger than the normal shell effects, so that here one expects only oblate shapes rotating around the symmetry axis (noncollective behavior with isomers) or prolate shapes rotating collectively (smooth bands and no isomers). Since these behaviors appear quite different experimentally, there is hope to learn about the shapes and dynamics of nuclei at these spin values. Neither the exact evaluation of Equation 8 nor the consideration of larger deformations changes this pattern. The $P - \perp$ and $O - \parallel$ shapes become closer in energy with increasing deformation, and even cross, but they remain well separated from the other types.

Shell effects are the larger features of the microscopic structure of nuclei. There are also smaller features that depend on specific nucleon orbitals. For high angular momentum states the so-called intruder orbitals are especially important. These are the high j orbitals from the next higher shell that are brought down by the spin-orbit interaction ($h_{11/2}$ into the $N = 4$ shell or $i_{13/2}$ into the $N = 5$ shell, for example). When nuclei rotate, there is a Coriolis force on the individual nucleons, which tends to align the particle angular momentum with that of the core rotation (Stephens 1975), and which is in general proportional to j (it is an $\mathbf{I} \cdot \mathbf{j}$ term in the Hamiltonian). The intruder orbitals thus feel the strongest aligning force, and the phenomenon known as backbending corresponds in the $A = 160$ region to breaking a pair of $i_{13/2}$ neutrons and aligning the angular momentum of each one ($\sim 10\hbar$ total) with the core rotational angular momentum, which is also of order $10\hbar$. This alignment is interesting with respect to the picture we have been developing of nuclei at high spins. It is a step toward the noncollective limit. From the ground state, where all the angular momentum is collective, the nucleus has gone over to a state where only half is collective and the other half is carried by two individual neutrons that do not follow the general collective motion at all. From a slightly different point of view one can say that these two neutrons follow rather circular orbitals in the plane in which the prolate nucleus is rotating. They therefore constitute a triaxial bulge whose build-up will lead toward an oblate nucleus with angular momentum along its symmetry axis. This subject is considered again later in this section and in Section 6.3 where detailed behavior at very high angular momentum is discussed.

At the other limit, where the angular momentum is noncollective, the individual particle orbits can in principle, and sometimes in fact, be identified. Here, collective features come in as residual interactions between the particles, and when enough particles are involved, these build up coherently and lead to deformations. This subject is discussed in Section 5. It now appears that at high spins the nucleus strongly prefers to use both collective and noncollective modes for carrying angular momentum, and there is some hope that this will be largely understandable as a superposition of several nearly pure single particles coupled to a collective core.

2.3   Calculations

It is not our purpose to discuss the detailed calculations of nuclear properties at high angular momentum. However, we give a brief summary to orient those unfamiliar with this subject, to provide some references (Ragnarsson et al 1978) for easy access to more details, and to give some background for results that are used later. The whole shell-model concept involves calculating the individual nucleon orbits in the average potential generated by the rest of the nucleons. It was the introduction in 1949 (Mayer) of a large spin-orbit $(\mathbf{l} \cdot \mathbf{s})$ term to a harmonic oscillator or square-well potential that made the shell model work so well for spherical nuclei. By 1955 Nilsson had introduced an $r^2 Y_{20}$ deformation term to the harmonic oscillator potential (and an l^2 term to flatten it) and calculated the levels of deformed nuclei. Since this "ad hoc" potential did not have the proper behavior for large deformations, Strutinsky (1966) devised a procedure to normalize its *average* behavior to that of the liquid-drop model. Pairing could be added in a reasonably simple (BCS) approximation (Bohr et al 1958). This modified harmonic oscillator (MHO) model has been enormously successful for deformed nuclei. It was reasonably straightforward to "crank" this potential around the x axis by adding a term, $-\omega j_x$, and calculate the orbits in a potential rotating with frequency ω (Bengtsson et al 1975, Neergaard & Pashkevitch 1975, Neergaard et al 1976, Andersson et al 1976). Some results of this procedure are shown below. A variation is to use a Woods-Saxon (WS) potential (Neergaard et al 1977). This solves problems in the average moment of inertia having to do with the l^2 term in the MHO potential, but on the whole seems to give very similar results. A basic improvement has been the introduction of the Hartree-Fock (HF) potential, where self-consistency between the potential and calculated orbits is required. Pairing can be included in a more fundamental way in the Hartree-Fock-Bogulubov (HFB) method, which makes this approach very promising (Mang 1975, Faessler et al 1976a, Goodman 1979). Results are now becoming available for the high spin region and again they seem not to differ much from the MHO or WS potentials. At present, the greater development of, and experience with, the MHO potential makes it still quite competitive with the others.

The moment of inertia is a concept used throughout this review, and it seems worthwhile to note how it is obtained from such microscopic calculations. The cranking formula was developed in 1954 by Inglis and is given by

$$\mathbf{F}_x = 2\hbar^2 \sum_{i \neq f} \frac{\langle f | j_x | i \rangle^2}{\varepsilon_f - \varepsilon_i}, \qquad\qquad 14.$$

where j_x is the operator for infinitesimal rotations about the x axis and ε_i is the eigenvalue for the state i. Bohr & Mottelson showed in 1955 that an evaluation of Equation 14 for an anisotropic harmonic oscillator leads to the rigid-body moment of inertia, in spite of the fact that only partly filled j shells contribute and individual terms can be as large as 10–20% of the sum even when there are 200 nucleons. Pairing decreases $\mathbf{F}$ by reducing the matrix elements in the numerator of Equation 14, and also by increasing the energy difference in the denominator (Nilsson & Prior 1961); however, the details of that reduction are not important here. An essential point is that to obtain the moment of inertia for collective rotation the sum in Equation 14 runs only over the nondiagonal matrix elements. The diagonal matrix elements give the aligned angular momentum. For the case where the rotation axis (x) is the symmetry axis, j_x is exactly diagonal, resulting in no nondiagonal matrix elements in Equation 14, and a vanishing collective moment of inertia. This result corresponds to the inhibition of collective rotations about a symmetry axis for a quantal system, as discussed earlier. If one chooses to include part or all of the aligned angular momentum into the definition of a moment of inertia, it must be an *effective* moment of inertia. This point is discussed further in Section 6.

The first result from any of the calculations mentioned above is the behavior of individual orbitals with increasing rotational frequency. In the spherical shell model there is a $(2j + 1)$ degeneracy of levels in a j shell, which is largely lifted by deforming the potential, as is indicated in Figure 3. This is a portion of a Nilsson diagram and shows the energy of levels as a function of deformation, δ, where $\delta = 0$ is the spherical shape. The energy is

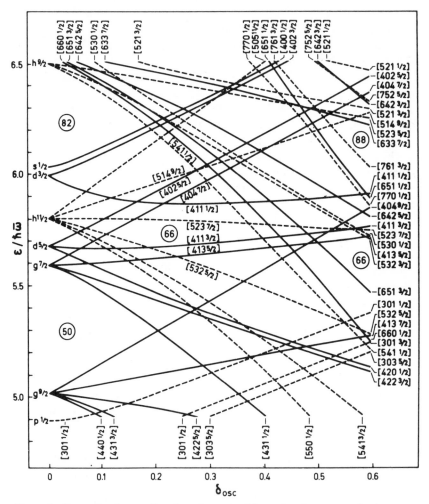

Figure 3 Energy levels as a function of prolate deformation for protons in the range $50 \leqq Z \leqq 82$, calculated by Gustafson et al (1967) using the MHO potential.

given in units of the oscillator energy:

$$\hbar\omega_0 \approx 41A^{-1/3} \text{ MeV.}$$

15.

The resulting levels are twofold degenerate corresponding to time reversal symmetry of the nucleon motion, and are characterized by their projection on the symmetry axis, Ω. If a given deformation, say $\varepsilon = 0.2$, is chosen, the resulting system can be cranked, and Figure 4 shows a portion of the levels

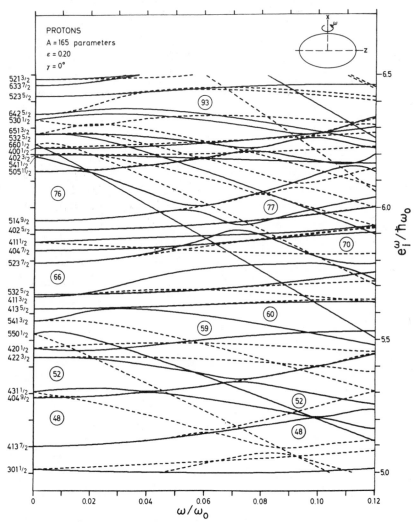

Figure 4   Energy levels as a function of rotational frequency (in units of oscillator frequency ω_0) for protons in a prolate nucleus having $\varepsilon = 0.2$ and mass number 165 as calculated by Andersson et al (1976) using the MHO potential. The solid and dashed lines correspond to states having different symmetry with respect to $\exp(i\pi j_x)$ (different "signatures").

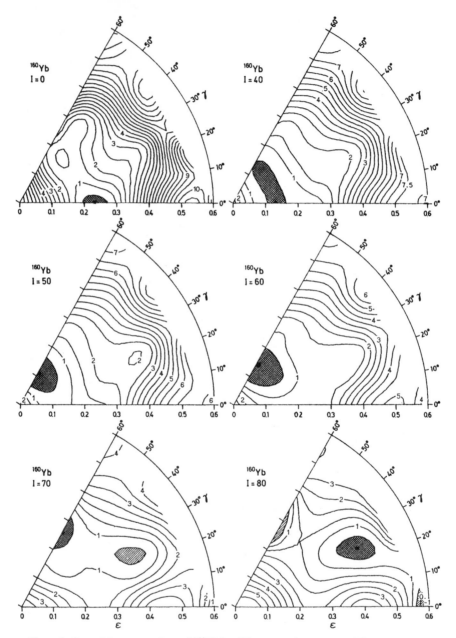

Figure 5   Potential energy surfaces of ^{160}Yb for different angular momenta. The contours are 0.5 MeV apart and the shaded regions are those of lowest energy. Taken from Andersson et al (1976).

in Figure 3 at $\varepsilon = 0.2$ as a function of cranking frequency ω, given in units of the oscillator frequency, ω_0. The energy $e'(\omega)$ is in the rotating frame and in units of $\hbar\omega_0$. The time reversal degeneracy of the levels is lifted, and the slope of these levels $de'(\omega)/d\omega$ is proportional to the aligned angular momentum. The strongly aligned high j orbits are the steeply down-sloping ones, with a limiting slope corresponding to their maximal alignment. From such information one can sum up energies and calculate potential energy surfaces as a function of rotational frequency.

The potential energy surfaces calculated using the MHO potential for ^{160}Yb at several angular momenta are shown in Figure 5 (Andersson et al 1976). The coordinate system plots deformation, ε, radially outward, and γ (Equation 7) as an angle from the horizontal; $\gamma = 60°$ corresponds to an oblate shape rotating about its symmetry axis. The nucleus ^{160}Yb is a prolate nucleus at low spin values, in agreement with the potential energy minimum calculated for $I = 0$ in Figure 5. At higher spins the calculations suggest it becomes triaxial ($I = 40$), then oblate ($I = 50, 60$), before finally fissioning around $I = 80$. The information on the energy minima can be more compactly presented by a line which is their locus. Such curves are shown in Figure 6 for several Yb isotopes. For the lighter isotopes like ^{158}Yb oblate shapes are predicted for spins up to $60\hbar$, and one could hope to see noncollective behavior, and perhaps isomers, in this range.

One can also calculate where the shell effects will favor particular shapes. For example the "superdeformed" shape with axis ratio $2:1$ has been of interest for some time. In Figure 7 the shell effects that occur with this shape are shown for neutrons (and are similar for protons) (Leander et al 1979). The fissioning isomers in the $^{240}_{94}$Pu region have this shape, and arise from the minima in Figure 7 at ~ 140 neutrons (and a corresponding one around 90 protons). Searches are being made for such shapes in some lower mass regions at high spins. Extensive calculations now exist for all three types of potential, and are discussed where relevant in other sections.

3   PRODUCTION OF HIGH SPIN STATES

The method used for producing high spin states depends on what kind of high spin state is meant. In the light elements one would probably choose a transfer reaction with an appropriately high l window. In the spin 20–30$\hbar$ region of the heavier elements considered here ($A > 100$), Coulomb excitation would be an extremely important method. However, the present review is centered on the very highest spins, 60–70$\hbar$, and only the heavy ion fusion reactions have been used to study such states. Other methods may eventually be found. The deep inelastic collisions can populate such spins, but so far the information flow has been the other way—our knowledge of

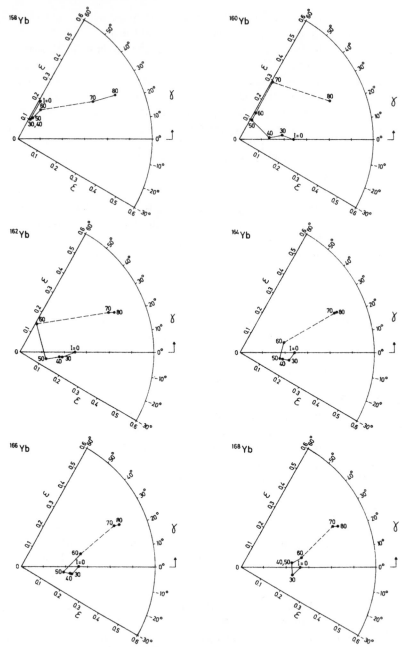

Figure 6   Trajectories in the ε, γ plane of the calculated equilibrium shape for isotopes of Yb as a function of total angular momentum, *I*. The dashed lines indicate a discontinuous change of shape. From Andersson et al (1976).

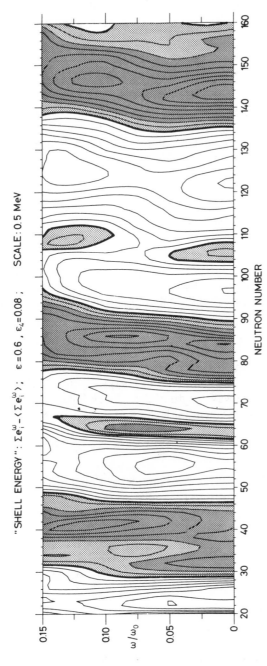

Figure 7 The neutron shell energy for a prolate rotating system with axes ratio 2 : 1 ($\varepsilon = 0.6$) and hexadecapole deformation, $\varepsilon_4 = 0.08$. The contour interval is 0.5 MeV, and the negative energy regions are shaded. The calculations were made with the MHO potential as reported by Leander et al (1979).

high spin states has been used to estimate the angular momentum transfers in these collisions. The present section, therefore, deals only with heavy ion fusion reactions.

3.1 Heavy Ion Fusion Reactions

The idea that a target and projectile nucleus fuse to form a compound system, whose subsequent decay is independent of its formation, goes back to Niels Bohr (1936). Independent decay means that the system remembers nothing of the entrance channel, except that required by conservation laws, notably here, angular momentum. No evidence contrary to this idea has been found, though we now know that "composite" systems can be formed which, for various reasons, live for a much shorter time than the compound systems ($\sim 10^{-17}$–10^{-18} s), and consequently remember more about the entrance channel. This section discusses the formation and decay of the compound system not in depth, but rather to understand how angular momentum gets into and out of such a system.

An essential feature of heavy ion collisions, for the present purposes, is that they are quite classical. The heavy ion wavelengths are considerably smaller than the dimensions of the colliding systems. Thus, considering the nuclei as charged black spheres,

$$\sigma_R = \pi(R_1 + R_2)^2(1 - E_{CB}/E_{CM}) = \pi \bar{\lambda}^2 l_{max}(l_{max} + 1), \qquad 16.$$

where σ_R is the reaction cross section, R_1 and R_2 the target and projectile radii, E_{CM} the center-of-mass bombarding energy, E_{CB} the Coulomb barrier height, $\bar{\lambda}$ the de Broglie wavelength in the center-of-mass system, and l_{max} the maximum angular momentum leading to a reaction. From Equation 16, l_{max} can be evaluated:

$$l_{max} = 0.219(R_1 + R_2)[\mu(E_{CM} - E_{CB})]^{1/2}, \qquad 17.$$

with R_1 and R_2 in fm and given, for example, by Equations 9 or 12; μ is the reduced mass in mass units; and E_{CM} and E_{CB} are in MeV. The more distant collisions, involving the highest l waves, lead to transfer reactions and deep inelastic collisions, so that the evaporation residue cross section, σ_{er}, is given by the right side of Equation 16, but with a lower l value, l_{er}, replacing l_{max}. Thus a measured fusion cross section implies an l_{er} given as

$$l_{er}^2 = 1.5\sigma_{er}\mu E_{CM}, \qquad 18.$$

where σ_{er} is in barns and the other units are as above. These equations are approximate and fail for high projectile velocities ($E_{CM}/E_{CB} > 2$) and, of course, for angular momenta in excess of that which the compound system can hold. Nevertheless, they are excellent for orientation, and Equation 17 shows, for example, that 200-MeV ^{40}Ar projectiles on a ^{124}Sn target involve angular momenta up to $96\hbar$, more than the ^{164}Er* compound nucleus can hold. The distribution of angular momenta is given by

$$\sigma(l) = \pi \lambdabar^2 (2l+1) T_l, \qquad\qquad 19.$$

where T_l is a transmission coefficient; that for the black sphere (or sharp cut-off) model is unity up to l_{max} and zero above that. This gives the familiar "triangular" distribution of cross section with l where, fortunately, the larger angular momenta have the greater cross section.

The decay of the compound nucleus is described by the statistical model of nuclear reactions introduced by Weisskopf (Blatt & Weisskopf 1952) and Ericson (1960). More details can be found in Grover & Gilat (1967), Thomas (1968), and Fleury & Alexander (1974). This model assumes that each state decays independently of its formation into one of the open channels, according to the width of that channel. The open channels in the present case are mainly fission and the evaporation of neutrons, protons, and α particles. For a given channel, the width of a state of energy E_i and spin J_i is just an emission rate, R, to a given final state E_f, J_f, summed over final states. This emission rate is determined by the level density, ρ, and a transmission coefficient according to

$$R(E_i J_i, E_f J_f) = \frac{1}{\hbar} \frac{\rho(E_f J_f)}{\rho(E_i J_i)} \sum_l T_l, \qquad\qquad 20.$$

where the sum runs over all possible l values consistent with J_i and J_f. The T_l sum can be related to a product of the inverse cross section σ_c times the particle kinetic energy, ε, where the former of these is expected to be relatively constant over reasonable ranges of E_f. For energies, E, well above the yrast[1] energy, E_y, the level density is approximately

$$\rho(E) \propto (E - E_y)^{-2} \exp 2[a(E - E_y)]^{1/2}, \qquad\qquad 21.$$

where a is a level density parameter, roughly $A/8$ MeV^{-1}. This leads finally to relative values for R given by

$$R(E_f) \propto (E_f - E_y)^{-2} \varepsilon \sigma_c \exp 2[a(E_f - E_y)]^{1/2}. \qquad\qquad 22.$$

This equation gives the spectrum of evaporated particles and also, through the level widths, the competition among channels. One should realize this equation is rather approximate, and applicable only at high temperatures, but it will serve to illustrate the points to be made here.

The angular momentum enters Equation 22 mainly through the level densities and particularly through $E_f - E_y$. The yrast energy is given roughly by $E_y = I(I+1)\hbar^2/2\text{F}$, where F is approximately a rigid-body value. Thus for any given excitation energy E, the level density will decrease

---

[1] The yrast line traces out the lowest energy level in the nucleus for each spin value.

as I increases until at some point $(E = E_y)$, it goes to zero. This suggests that an evaporated particle will tend to carry away angular momentum, since it will find a higher final level density by doing so; and this tendency will become stronger for higher spin values since the slope of the yrast line becomes steeper with spin. However, particles evaporated with angular momentum l have to overcome a centrifugal barrier given by

$$E_l = l(l+1)\hbar^2/2\mu R^2,\qquad\qquad 23.$$

where μ is the reduced mass of the emitted particle and R is the interaction radius. For protons or neutrons from a nucleus around mass 150, E_l is about 2 MeV for $l = 2$. The slope of the yrast line in this mass region is given by the observed γ rays, and is about 1.5 MeV for $\Delta I = 2$ at $I \sim 60$. Thus protons or neutrons cannot gain by carrying out angular momentum even at these highest spins. However, α particles have a centrifugal barrier about four times lower than protons or neutrons, and thus can remove large amounts of angular momentum at high spins. Fortunately, α particles have a large Coulomb barrier in this mass region, and it is possible to find circumstances in which they are not emitted (discussed below). However, for $A < 100$, the Coulomb barrier is much lower _and_ the yrast line has a steeper slope because of the smaller moment of inertia, so that α particles inevitably remove angular momentum, and thus set a limit on the maximum spins observable through γ decay in this mass range.

The competition among channels also enters Equation 22 mainly through the level densities, and therefore E_f. Neglecting the temperature-related kinetic energy and assuming a final kinetic energy equal to the Coulomb barrier E_{CB}, we can relate the final energy to the initial energy by

$$E_f \approx E_i - B - E_{CB},\qquad\qquad 24.$$

where B is the binding energy of the emitted particle. Consider proton and neutron evaporation from nuclei with $A \sim 150$ around the region of β stability where the binding energies are about the same. Then for similar final product nuclei, the E_f would be ~ 8 MeV lower for proton evaporation due to its Coulomb barrier. This is sufficiently large so that only neutrons are evaporated. On the other hand, as one moves off the line of β stability toward the neutron-deficient side, the neutron binding energy goes up and the proton (and α) binding energy goes down until, with about 15% fewer neutrons, the neutron binding energy is equal to the proton binding energy _plus_ its Coulomb barrier. At this point neutrons and protons are evaporated with equal probability, and if one goes further neutron deficient, protons are preferentially evaporated, driving the system back toward the equal probability line. Since it is not possible to make neutron-excess nuclei in heavy ion fusion reactions (no sufficiently neutron-rich targets and/or projectiles exist), the accessible region is from the β-stability line to a line having about 15% fewer neutrons. Alpha binding energies vary with neutron number like those for protons, so that by producing nuclei as neutron rich as possible, evaporation of both can be essentially completely avoided, conserving the angular momentum and reducing the number of open channels. These are the systems most attractive for very high spin studies, where the (HI, xn) reactions dominate.

The results from a more realistic, statistical model calculation are shown in Figure 8 (Hillis et al 1979). The system is 147-MeV ^{40}Ar projectiles and a

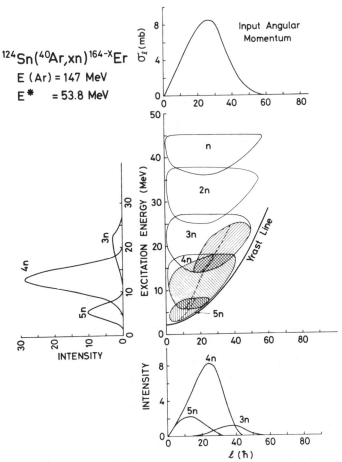

Figure 8   Statistical model calculations for the reaction ^{124}Sn(^{40}Ar, xn)$^{164-x}$Er at 147 MeV as given by Hillis et al (1979). The upper plot shows the angular momentum input and excitation energy (53.8 MeV), and those below it show the populations after emission of 1–5 neutrons. The shaded regions do not emit more particles and de-excite by γ-ray emission. The dashed line is the entry line. The side and bottom plots show the correlation of reaction channel with excitation energy and angular momentum, respectively.

^{124}Sn target to give ^{164}Er* as the compound nucleus. The initial excitation energy and angular momentum distribution are shown at the top of Figure 8 and the contour lines on the main excitation energy vs angular momentum plot show the region populated following each neutron evaporation. After the third neutron evaporation, part of the population is left to γ decay (shaded) and part goes on to emit a fourth neutron (white). This repeats for the fourth neutron, producing a small cross section for emitting five neutrons. There is a reasonably clear line about one neutron binding energy above the yrast line, separating the γ-emitting region from the particle evaporation region. Two types of correlation are shown in the side and bottom plots. The side one shows that the total γ-ray energy is correlated with the reaction channel, higher total energies being associated with fewer neutrons evaporated. The bottom plot shows that higher angular momenta are also associated with fewer neutrons evaporated, often referred to as a "fractionation" of angular momentum among the products. It is apparent from these plots, or the main E vs I plot, that higher total γ-ray energies are associated with higher spin values. These correlations all come about because the energy that is tied up carrying the angular momentum of the system is not available for evaporating neutrons, but later appears as γ-ray energy. All these correlations are used in studies of very high angular momentum in such nuclei.

3.2 Gamma-Ray De-Excitation

The remainder of this review is devoted to studies of γ rays de-exciting the evaporation residues from heavy ion reactions. The present section introduces the general pattern of this de-excitation. The previous section has established that the γ decay occurs in the region between the yrast line and a line roughly one neutron binding energy above it. This region is illustrated in Figure 9, and the decay path is sketched for four different spin inputs. Two types of γ rays occur: those that cool the nucleus to or toward the yrast line, called "statistical"; and those that are more or less parallel to the yrast line and remove the angular momentum, called "yrast-like." At each point there is a competition between these two types. The statistical γ rays (dipole transitions) depend only on average matrix elements and level densities, and are the topic of Section 4. The yrast-like γ rays can be collective or not, and are discussed in Sections 5 and 6. The competition between these types depends on the excitation energy, which affects the level densities, and also on the degree of collectivity, which produces faster (enhanced) yrast-like transitions.

There is another division of the γ rays: those that are resolved in the spectrum and those that are not. In order to be resolved, a transition must have a sufficiently large population (as a rough estimate, 1%), and this is not

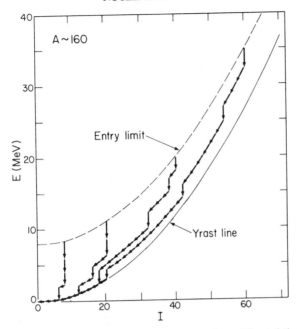

Figure 9 Typical γ-ray de-excitation pathways to the ground state. The statistical transitions, indicated by vertical arrows, lower the temperature of the system, whereas the yrast-like transitions are roughly parallel to the yrast line and remove the angular momentum of the system.

generally the case for the higher excitation energies since there are thousands of possible pathways. Only up to spins 20 or $30\hbar$ (heavy arrows along the yrast line in Figure 9) does enough population collect in the yrast or near-yrast states, to produce resolved lines. In just two or three cases this has gone as high as 36 or $37\hbar$, and these are discussed in Section 5. For the higher spin transitions, and for all the statistical transitions, one must work with unresolved, or "continuum," spectra. The many techniques being developed for this purpose are discussed in Section 6.

4   STATISTICAL GAMMA RAYS

It is not difficult to see the features of the γ-ray spectrum described in the previous section. Figure 10 shows examples of the de-excitation cascades from the ^{174}W product nuclei following the reactions, ^{162}Dy(^{16}O, 4n) and ^{163}Dy(^{16}O, 5n), as measured with a NaI detector (Newton & Sie 1980). One can distinguish three components. Below about 0.6 MeV there are resolved peaks due to the last few transitions in the lowest rotational bands.

Underneath these peaks and extending up to about 1 MeV is the "yrast bump," the major portion of the unresolved yrast-like transitions. Above the yrast bump edge at ~ 1 MeV is a high energy tail that falls roughly exponentially with increasing γ-ray energy. This is composed of statistical γ rays, which are the subject of the present section.

4.1   Theoretical Expectations

The exponential behavior at high energies gives a first indication that these transitions are statistical in nature, as this dependence can be ascribed to the exponential factor in the level-density expression (Equation 21) used in statistical model calculations. The statistical transition rate from state i to state f per unit energy interval of final states and for radiation of multipole l is

$$T(i \rightarrow f) \propto M_{if}^2 E_\gamma^{2l+2} \frac{\rho(E_f - E_y)}{\rho(E_i - E_y)},$$

25.

where $\rho(E)$ is given by Equation 21. This expression assumes a much higher level density at E_i than at E_f, and M_{if} is the transition matrix element. For small γ-ray energies, $E_y = E_i - E_f$, the expression is dominated by the factor E_γ^{2l+2}, but for large γ-ray energies it is dominated by the decreasing

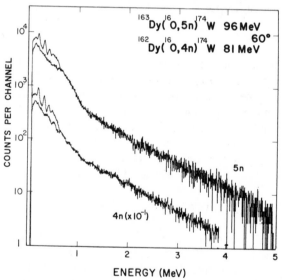

Figure 10   Pulse height spectra from a NaI detector, taken in coincidence with neutrons, for reactions leading to ^{174}W, before and after subtraction of the discrete lines (Newton & Sie 1980).

exponential. It should be noted that E_f cannot become smaller than E_y so that the γ-ray energy cannot exceed $E_i - E_y$. Thus large γ-ray energies are effectively cut off at about the neutron binding energy.

The same sort of general behavior occurs in radiative neutron capture, where a small number of statistical γ rays of moderate energy cool the nucleus to the ground state (Bartholomew et al 1973). In the capture of monoenergetic neutrons all the product nuclei start at the same excitation energy (neutron binding energy plus kinetic energy). However, after emission of the last neutron in a (HI, xn) reaction the resulting nuclei have a distribution of energies above the yrast line ranging from zero to roughly a neutron binding energy, so that the average statistical transition energy should be smaller.

Several computer programs have been developed that produce results in general agreement with experimental spectra, such as shown in Figure 10, and with cross-section data, feeding times, etc. Grover (1967) and Grover & Gilat (1967) calculated the particle and γ-ray de-excitation of a rare earth nucleus produced by a heavy ion reaction and were the first to discuss the nature of the yrast line as the ground state for thermal excitation energy at that spin. Monte Carlo techniques were used by Sarantites & Hoffman (1972) to calculate excitation functions to particular spin states in medium-light product nuclei; more recently Wakai & Faessler (1978) refined and extended such computations, paying special attention to the competition between statistical and collective transitions. Somewhat earlier, Liotta & Sorensen (1978) performed calculations also emphasizing the competition between the (cooling) statistical transitions and collective bands of a specific nature, the wobbling triaxial rotor.

Since all the calculations contain a number of assumptions and depend upon values chosen for γ-ray hindrance or enhancement factors, level-density formulas, etc, only a few common features required to fit the experimental data are mentioned here. These are (a) E1 transitions, even though assumed to be hindered on average by a factor of 10^{-2}–10^{-3} over the single-particle value, dominate over M1 and E2 statistical transitions (at least up to spins of 30–40$\hbar$ considered in the calculations); (b) the collective E2 transitions must be enhanced by 100–200 times more than the single-particle unit in order to compete successfully with the E1 statistical transitions and be consistent with the observed (short) average lifetimes; and (c) in good rotors (the cases chosen for calculations), the competition may occur over a considerable region of the excitation energy-spin space. This last result comes about because the decay rates for the collective rotational transitions do not depend upon the excitation energy above the yrast line, and the rates for the statistical E1 decays do not increase very rapidly with excitation energy. Even so, it is clear that one would expect the statistical E1 transitions to compete best against the collective E2 decays at the highest excitation energies above the yrast line, and also at the lowest spin, where the rotational E2 transitions are of lowest energy and hence slowest.

4.2   Experimental Evidence

It is of interest to consider an experimental spectrum in greater detail. Figure 11 (Simon et al 1977) shows the de-excitation spectra of ^{162}Yb from the reaction (181 MeV) ^{40}Ar + ^{126}Te → ^{166}Yb*, taken with three NaI detectors. The events were selected by requiring a coincidence with a known ^{162}Yb transition identified in a Ge(Li) detector. The yrast bump and statistical regions can be distinguished even in the pulse-height spectrum

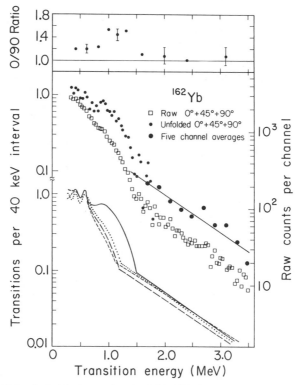

Figure 11   NaI pulse height (*squares*) and unfolded (*dots*) γ-ray spectra (0° + 45° + 90°) from the reaction ^{126}Te(^{40}Ar, 4n)^{162}Yb at 181 MeV. The larger dots are averaged over five channels. At the top is the 0°/90° ratio for the unfolded spectra. At the bottom are schematic unfolded spectra for the same case (*solid line*) and for ^{80}Se(^{86}Kr, 4n)^{162}Yb at 331 MeV (*dotted line*), ^{126}Te(^{40}Ar, 4n)^{162}Yb at 157 MeV (*long-dashed line*), and ^{150}Sm(^{16}O, 4n)^{162}Yb at 87 MeV (*short-dashed line*) (Simon et al 1977).

(open squares). However, if this is "unfolded" (Mollenauer 1961), that is, corrected for the NaI response function to give the primary γ-ray spectrum (filled circles), the components are considerably clearer. The latter curve has been corrected for counter efficiency to give the absolute number of γ-ray transitions per event per transition energy interval (40 keV). Summing the whole curve gives the average γ-ray multiplicity [minus the gate transition in the Ge(Li) detector], in this example, $\langle M \rangle = 24$. Summing the curve above the edge of the yrast bump (~ 1.5 MeV) gives $1\frac{1}{2}$ to 2 transitions for the average number of high energy statistical transitions per event.

To determine the total number of statistical transitions, those that lie below 1.5 MeV must be estimated. The shape of the statistical component may be approximated by an expression of the form:

$$N_\gamma \propto E_\gamma^n \exp[-(E_\gamma/T)], \qquad\qquad 26.$$

where n and T are constants to be fixed by the spectrum above the edge of the yrast bump. In this particular example, another $1\frac{1}{2}$ to 2 transitions can be estimated to lie below 1.5 MeV, giving a total of 3–4 statistical transitions. This value is in agreement with other methods for estimating this number in this mass region (Hillis et al 1979, Sarantites et al 1978), and does not preclude some variation with total angular momentum.

At the top of Figure 11 is shown the ratio of intensities of the unfolded spectrum from the 0° NaI detector to the one at 90°. It can be seen that in the high energy region the γ-ray transitions are essentially isotropic (ratio slightly greater than unity), in contrast to the yrast bump region. This has been found in several other cases (Simon et al 1979, Newton et al 1978, Ockels 1978a, Deleplanque et al 1978a) although there is at least one example (Trautmann et al 1975) that seems to differ. The isotropy may be explained if the high energy region consists of a mixture of stretched dipole and stretched quadrupole transitions, or stretched and unstretched dipole transitions (or some other less likely possibilities). The angular distribution measurements alone cannot distinguish among these possibilities.

However, the multipolarity of the continuum region has also been studied by conversion coefficient measurements. The first such study (Feenstra et al 1977) was on the continuum produced in the reaction ^{160}Gd(α, 4n)^{160}Dy. The γ-ray and electron spectra were measured simultaneously with a Ge(Li) (or NaI) detector and a mini-orange spectrometer, and both were in coincidence with another Ge(Li) detector to select events from the 4n reaction. Rather extensive tests were made with the mini-orange to determine the background under the spectral region of interest. The highest angular momentum at the start of a γ-ray cascade following ~ 47-MeV α-particle bombardment is about $20\hbar$, and the edge of the yrast bump should be below 1 MeV. Above this energy the experimental points followed the E1 curve, a reasonably clear indication of E1 character since no other multipole type has a smaller conversion coefficient. Two further studies have been performed with Ne beams, which unfortunately give somewhat conflicting results (Westerberg et al 1978, Feenstra et al 1979). We discuss only the work of Feenstra et al, which claims to have explained the difference. This group studied the systems ^{146}Nd(^{20}Ne, 4n or 5n)^{162}Yb or ^{161}Yb, again by simultaneous measurements with a NaI detector and a mini-orange spectrometer. Both were in coincidence with a Ge(Li) detector in order to provide gates on the discrete transitions from either ^{162}Yb or ^{161}Yb. These results are shown in Figure 12. Again at energies above 1.5 MeV the conversion coefficients fit those of E1 radiation, while in the yrast bump region the results are consistent with predominantly E2 radiation.

The current picture that has emerged from the study of the high energy statistical tail region is (a) the high energy transitions are probably E1; (b) there are ~2 such transitions per event above the edge of the yrast-like transition and presumably a similar number under the yrast bump; and (c) the angular distribution is essentially isotropic. If (a) and (c) both hold, they imply that the transitions are a mixture of approximately 2/3 stretched and 1/3 unstretched electric dipoles. It is clear that more work is needed. No measurements involving the highest angular momenta have been made, and this is where the slope of the yrast line applies the strongest pressure toward stretched dipole and especially quadrupole transitions. These

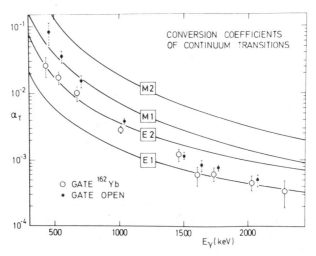

Figure 12   Conversion-electron coefficients (*circles*) determined with a mini-orange spectrometer and a NaI detector, both in coincidence with a Ge(Li) detector in order to pick out the ^{146}Nd(^{20}Ne, 4n)^{162}Yb channel. The points are coefficients obtained with an open gate on the Ge(Li) detector, and the curves are the theoretical E1, E2, and M1 values (Feenstra et al 1979).

conclusions indicate that the statistical γ rays do not give much information about the structure of very high spin states, but rather represent a background against which we must look for that information.

5   NONCOLLECTIVE HIGH SPIN STATES

Noncollective states occur when the nuclear angular momentum is carried by high j particles with large projections on the symmetry axis. This behavior leads (as described in Section 2) to an irregular yrast line, and to no, or weak, collective transitions. Therefore the statistical transitions compete better against the yrast-like ones and cool more of the population intensity to the yrast line at relatively high spin. A result is the possibility of isomers, the "yrast traps" (Bohr & Mottelson 1974). Conversely, the observation of high spin isomers is taken to be an indication of this noncollective rotation.

5.1   Hafnium Region

In the discussion of classical rotating shapes in Section 2, the rotation of an oblate nucleus around its symmetry axis was the lowest energy mode of rotation for a given deformation at moderate spins, while that of a prolate nucleus around its symmetry axis was the highest energy mode and so not expected to occur very often. Nevertheless it does take place in the region of the stable Hf nuclei, in particular in ^{176,178}Hf. These are strongly prolate nuclei with well-formed rotational bands. For such prolate deformation in this region of N and Z, the $g_{7/2}$ and $h_{11/2}$ proton orbitals and the $h_{9/2}$ and $i_{13/2}$ neutron orbitals are nearly full, so the Fermi level comes nearest to their high Ω projections. These projections, then, lie lowest in excitation energy and can form high K states (where angular momentum is aligned along the symmetry axis) without as large an expenditure of energy as required by the collective rotation. An example is the 16^+ four-quasiparticle bandhead in ^{178}Hf, which lies below the 16^+ level of the ground band. In addition, rotational bands built on these bandheads may have lower rotational energies than the corresponding members of the ground band, because the unpairing of particles has decreased the pairing correlations and hence increased the moment of inertia. Thus, in ^{178}Hf (Khoo & Løvhøiden 1977) the higher spin members of the first 8^- two-quasiparticle isomeric state become the yrast states (before the occurrence of the 16^+ four-quasiparticle state) just because their larger moment of inertia more than compensates (by spin $12\hbar$) for the difference between the 1.147-MeV 8^- bandhead energy and the ground-band 8^+ energy of 1.059 MeV.

The partial level scheme for ^{176}Hf is shown in Figure 13 (Khoo et al

1976). The four-quasiparticle (14⁻, 15⁺, 16⁺) and six-quasiparticle (19⁺, 20⁻, 22⁻) bandheads are well described by the Nilsson model, and their projections on the symmetry axis, K, appear to be reasonably good quantum numbers, as is evidenced by the K-forbidden transitions from the 14⁻ and 19⁺ bandheads. The fact that the 16⁺ isomer drops below all other $I = 16$ states has been reproduced in a calculation using the cranked modified oscillator potential at a fixed deformation and taking into account pairing as well as the hexadecapole degree of freedom (Åberg 1978). This is simply a shell effect. The calculation further suggests that around spin $40\hbar$ collective rotation again becomes lowest in energy. Qualitatively similar behavior is indicated for all the even-even Hf nuclei from mass 172 to 180. The interpretation of these high K states as effective rotations about a symmetry axis is in no way contrary to the conventional interpretation based on Nilsson assignments for them. However, it adds a perspective both on their competition with the purely collective states and on their relationship to states in spherical (and oblate) nuclei, which is described in the remainder of this section.

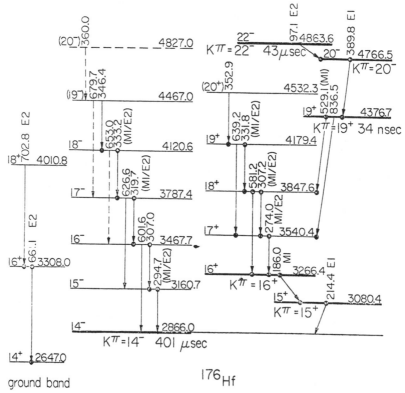

Figure 13   Partial level scheme for ¹⁷⁶Hf showing high spin bandheads (Khoo et al 1976).

5.2   *Lead Region*

The region above and below doubly magic ^{208}Pb has good examples of noncollective high spin states, but here the nuclei are nearly spherical. One example is ^{212}Rn, with four protons beyond closed shells. The levels up to $I = 19$ at 5.7 MeV appear to be rather pure proton shell-model configurations, as shown in Figure 14 (Horn et al 1977). Above this spin, the neutron

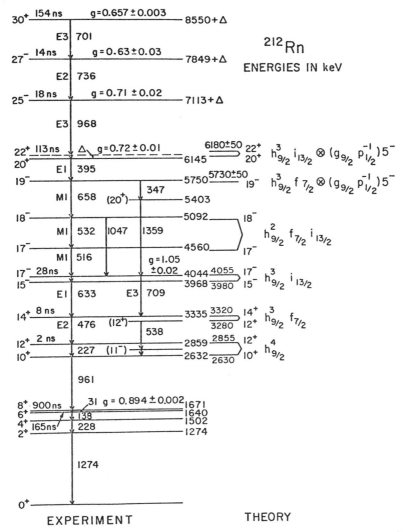

Figure 14 Level scheme for ^{212}Rn [including the unobserved transition $\Delta = E(22^{+}) - E(20^{+})$]. Energies are in keV (Horn et al 1977).

core is excited and the higher spin states involve both neutrons and protons, as indicated by the g factors of the isomeric states. Even such a near-spherical (slightly oblate) nucleus may be considered to have an effective moment of inertia, whose value can be obtained from a plot of excitation energy vs $I(I+1)$, that is,

$$2F/\hbar^2 = d[I(I+1)]/dE. \qquad\qquad 27.$$

Such a plot for ^{212}Rn is shown in Figure 15, where two regions of differing slope are apparent; one below ~ 6 MeV where only protons are excited to make the states, and one above 6 MeV where both protons and neutrons are involved. If collections of independent particles under rotation do behave on average like a rigid rotor, as has been suggested, the high energy region should have approximately the rigid moment of inertia. The rigid-sphere value has been drawn in as the dashed line (arbitrarily normalized vertically), and it can be seen that the slopes are similar. This is taken to indicate a general validity to the idea of effective rotation about a symmetry axis; however, as discussed in the next part of this section, shell effects are expected to be large, and it is not clear to what extent the close agreement with the rigid-body value in Figure 15 is accidental.

5.3 $N = 82$ Region

Most of the presently known examples of effective (noncollective) rotation at high spins come from the region just above 82 neutrons. The various

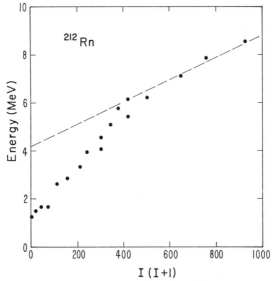

Figure 15   Plot of energy vs $I(I+1)$ for the yrast levels of ^{212}Rn. The dashed line is drawn through points above $20\hbar$.

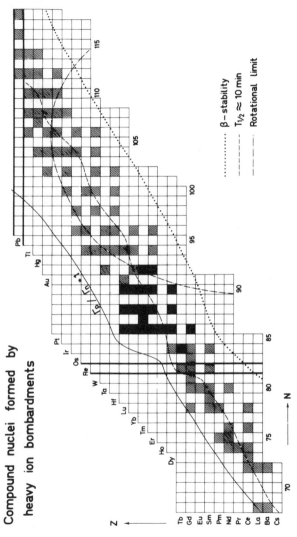

Figure 16 Region of periodic table searched for high spin isomeric states. Compound systems formed via ^{40}Ar, ^{50}Ti, and ^{65}Cu beams are indicated as shaded squares. Systems with positive results have filled squares. The final nuclei probably have 3–5 fewer neutrons (Pedersen et al 1977).

theoretical calculations mentioned in Section 2 predict slightly oblate nuclei for this region, consistent with this behavior. The calculations also predict that a few nuclei in this region, even some that may be prolate initially, will become rather strongly oblate at very high spin ($> 40\hbar$), with the same noncollective mode for carrying angular momentum. This implies the existence of isomeric traps along the yrast line at high spin, offering a challenge to experimentalists.

The first systematic search for high spin isomers was performed by Pedersen et al (1977); targets from Ba ($Z = 56$) to Po ($Z = 84$) were irradiated by pulsed beams of ^{40}Ar, ^{50}Ti, and ^{65}Cu. The recoiling evaporation residues were caught on a thin lead foil, which was viewed by 16 NaI(Tl) counters operated in coincidence. An event in the Pb foil that triggered several detectors would indicate with quite high sensitivity a high spin isomer. Isomers below $35\hbar$ were found to occur systematically in a fairly small region of nonrotational nuclei, as shown in Figure 16. A second study of this region with a total energy γ-ray spectrometer in coincidence with Ge(Li) detectors (Borggreen et al 1980) determined the excitation energies of the isomers as well as their halflives and γ-ray multiplicities. Twenty-two isomers were found or confirmed, and the island of isomers was shown to be limited to the region $82 \leq N \leq 86$ and $Z \leq 68$. The isomeric states have excitation energies ranging from 3 to 12.2 MeV and spins (determined by an empirical relationship) up to $(33 \pm 2)\hbar$. The structure of the isomers can be ascribed to the alignment of a small number (2–8) of shell-model particles in a spherical or slightly oblate potential (Døssing et al 1977, Andersson et al 1978, Cerkaski et al 1977, Faessler et al 1976b). There was no indication of isomeric states at very high spins within the limitations set by the experimental arrangement, $20 < T_{1/2} < 500$ ns, indicating no strongly oblate shapes there.

Two of the best studied nuclei in this group of noncollective ones are ^{152}Dy (Khoo et al 1978, Merdinger et al 1979, Haas et al 1979, Aguer et al 1979) and ^{154}Er (Baktash et al 1979); the level scheme of the former is shown in Figure 17. Both nuclei show the irregular level spacing and the existence of isomeric states characteristic of this noncollective mode of excitation. But the transition energies are not completely random; with very few exceptions they fall within the limits (700 ± 200) keV. Also, there are no E3 transitions, as might be expected from a pure spherical Nilsson potential, and as indeed are seen in the nuclei above ^{208}Pb (^{212}Rn for example, Figure 14). Another aspect, especially apparent in ^{154}Er, is that there are sequences of γ rays of almost constant intensity that are of the same multipolarity and connect levels of the same parity. These two- or three-step sequences have transitions of (600 ± 100) keV with lower energy lines connecting the different sequences. So there are certainly elements of

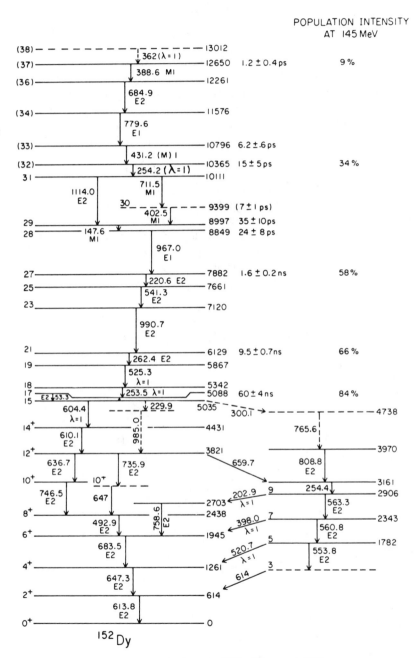

Figure 17   Level scheme for ^{152}Dy (Haas et al 1979).

weak collectivity present, waiting for a more detailed understanding. Also at higher spins ($>40\hbar$) there is evidence for still greater collectivity as described in Section 6.

A striking feature of these level schemes is illustrated in the plot of excitation energy vs $I(I+1)$ for ^{152}Dy in Figure 18. Initially the experimental data rise rapidly because pairing decreases, and then at about spin 16 they form an approximately straight line up to the highest spin, 36. The slope of this long straight portion of the curve leads to the determination of an effective amount of inertia through Equation 27. For ^{154}Er this yields $2F/\hbar^2 = \sim 140$ MeV^{-1} and for ^{152}Dy ~ 142 MeV^{-1}. These values are 10–15% greater than those of a rigid diffuse sphere of the appropriate mass. This increase could be considered an indication of deformation and, if so taken, would correspond to $\beta \approx 0.2$ for a deformed oblate shape, a rather large deformation. Certainly in the next section where collective rotation is considered, the slope of the curve, and hence the effective moment of inertia, is thought to bear a connection to the deformation of the nucleus. But here, near a magic number where the angular momentum is being carried largely by a few particles aligned along

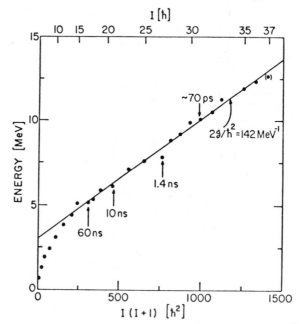

Figure 18 Plot of energy vs $I(I+1)$ for yrast levels of ^{152}Dy assuming $I = 17$ for the 60-ns isomer. For $I > 14$, the data lie close to a straight line with slope corresponding to $2F/\hbar^2 = 142$ (MeV)$^{-1}$ (Khoo et al 1978).

the rotation axis, it seems more likely that shell effects will be large and may affect this slope. Results of several recent calculations support this view. For example, Figure 19 shows curves calculated for several nuclei with neutron numbers that are around $N = 82$ using a *spherical* Nilsson potential and BCS pairing (L. Moretto, unpublished work, 1979), and it can be seen that shell effects create even larger differences in slopes than are seen in experiment. Similar results were found by Leander et al (1979); these authors point out the importance of pairing in calculating these moments of inertia and the configurations of states along the yrast line. Thus, it now seems likely that slopes such as that in Figure 18 cannot be used to indicate deformation. Values for these nuclei must be determined by another method.

The measurement of the static quadrupole moment of a state gives clear information about deformation, and such measurements can in principle be made on nanosecond isomers using perturbed angular distribution techniques. The first such determination at high spin for nuclei in this region was performed on the $(I \sim 49/2^+)$ 500-ns isomer in ^{147}Gd (Häusser et al 1980). The quadrupole moment obtained was $|3.14| \pm 0.17$ b, from which a deformation β of -0.18 was derived (the negative sign was assumed). This is a reasonably large deformation, two or three times that implied by the

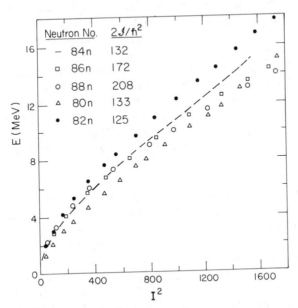

Figure 19   Plot of energy vs I^2 for the yrast levels of nuclei with the indicated neutron number as calculated with a spherical Nilsson potential (L.G. Moretto, unpublished work, 1979).

aligned particles alone. Thus it is the first real indication that sizeable oblate deformations might exist in this region. Nevertheless, these are very difficult experiments and more such data are badly needed.

The study of nuclei in this region of the periodic table is currently an active research area, and there is hope that a systematics will emerge that can shed more light on the structure of these high spin states. However, it no longer seems so likely that isomers will be found at spins significantly above these examples ($\sim 30\hbar$). This is partly because of the steepness of the slope of the yrast line at the higher spin values, but it also might be an indication that the strongly oblate shapes are not quite as stable as the present calculations indicate.

6   COLLECTIVE HIGH SPIN STATES

The present section is concerned with collective rotation, in simplest form the motion of a deformed nucleus around an axis perpendicular to its symmetry axis. Evidence for such a collective mode at high spins comes from the observation of a large proportion of stretched electric quadrupole E2 transitions that are strongly enhanced compared with the single-particle lifetime. Although individual transitions in the cascades cannot be observed because too many pathways are involved, it might be expected that the moments of inertia of these bands would not differ greatly at a given spin. Then the $I \rightarrow I - 2$ transitions would all have a similar energy, and one that is related to the value of I. Section 6.1 presents evidence for such rotational behavior. If the spin and transition energy can be determined, Equation 6 gives the effective moment of inertia, the average value at that spin for all bands involved, and Section 6.2 contains these evaluations. Second-generation coincidence experiments now being performed make possible the determination of the collective moment of inertia based on differences of transition energies; Section 6.3 is concerned with this technique.

6.1   *Evidence for Collective Rotation*

The well-known rotational transitions at low spin, whether in the ground bands of even-even nuclei or the decoupled bands of odd-mass nuclei, are strongly enhanced stretched ($I \rightarrow I - 2$) E2 transitions. For $K \neq 0$ bands, there are also M1 or mixed M1-E2 transitions with $\Delta I = 1$. These properties have been found in the continuum spectra of some nuclei (Sections 6.1.1 and 6.1.2), which leads to the conclusion that they very probably come from rotational bands. Furthermore, rotational transitions have a definite correlation between their energy E_γ and the spin of the de-exciting state I (Equation 6). This is in contrast to the situation with statistical transitions or with other nonrotational, e.g. vibrational or single-particle, ones. Observation of this correlation provides additional evidence for the rotational nature of some of the continuum γ rays (Section 6.1.3).

6.1.1 GAMMA-RAY MULTIPOLARITY   The top of Figure 11 shows the ratio of events at $0°$ to those at $90°$ in the unfolded spectra from the reaction $^{126}Te(^{40}Ar, 4n)^{162}Yb$ at 181 MeV. Although for $E_\gamma > 2$ MeV this ratio is near unity, in the region of the yrast bump between 0.8 and 1.5 MeV it is considerably larger, with ratios as high as 1.5 ± 0.15. This is the value expected in such a beam-gamma correlation for nearly pure stretched E2 transitions. There is some ambiguity; similar, or larger, ratios could be obtained from $\Delta I = 0$ dipole transitions. But these can be ruled out as a major component of the cascades because the number of γ rays measured per event, the γ-ray multiplicity, would otherwise be too small to carry off the angular momentum in the nucleus (as determined, for example, from the evaporation residue cross sections, Equation 18). The majority of the transitions must be "stretched," either dipoles or electric quadrupoles. (Magnetic quadrupole and higher multipolarities are ruled out by lifetime considerations, as discussed in the next section.) At transition energies less than 0.8 MeV the $0°/90°$ intensity ratio falls, and below 0.4–0.5 MeV Figure 11 shows that it goes below one, suggesting mainly stretched dipole transitions. Although different techniques have been employed, results are similar for the continuum cascades in other strongly deformed nuclei (Simon et al 1979, Hagemann 1979, Trautmann et al 1979, Vivien et al 1979). For example, angular distribution coefficients A_2, from the expression

$$W(\theta) = 1 + A_2 P_2(\cos \theta) + A_4 P_4(\cos \theta) + \dots \qquad 28.$$

were derived by Newton & Sie (1980) from unfolded spectra taken with a NaI spectrometer in coincidence with a neutron counter system (the accuracy of the data did not justify using higher terms). Values near 0.35 were found for spectra from the reactions $^{148,149}Sm(^{16}O, xn)^{161}Yb$, $^{163,164}Dy(^{16}O, xn)^{176}W$, and $^{162,163}Dy(^{16}O, xn)^{174}W$ for transitions between 0.5 and 1.2 MeV. This coefficient was nearly zero at 1.5 MeV and above as discussed in Section 4, and fell to -0.5 at γ-ray energies of ~ 0.3 MeV, indicating mainly stretched dipole transitions there. The experimenters suggested that these dipoles were from mixed M1/E2 cascade transitions from $K \neq 0$ rotational bands in the continuum.

The next development in this direction involved beam-gamma-gamma angular correlation measurements. One such measurement used an array of six NaI counters to select the observed nuclei so that the sensitivity to different types of radiation was enhanced (Deleplanque et al 1978a). The angular distribution of the selected events was measured with additional detectors, and depended on both the number and angle of the array detectors that fired. The percentage of different multipolarities for each transition energy (channel) in the γ-ray spectrum was deduced by comparing the experimental and calculated fold spectra for a given set of conditions. The results for several compound systems are given in Figure 20. Five types of transitions were considered in the calculations; quadrupoles with $\Delta I = 0, 1, 2$ and dipoles with $\Delta I = 0, 1$. The results, however, are presented only in terms of plots of stretched quadrupole and stretched dipole transitions. This is because the experimental results showed there was rather little ($< 10\%$) $\Delta I = 0, 1$ quadrupole radiation in the spectrum. The $\Delta I = 0$ dipole type cannot be ruled out this way, since it resembles too closely the stretched quadrupoles, but the argument has already been made that the experimental γ-ray multiplicity and the estimated average spin of the nuclei do not allow many $\Delta I = 0$ transitions. One or two $\Delta I = 0$ transitions are not excluded, and in the high energy statistical region there is

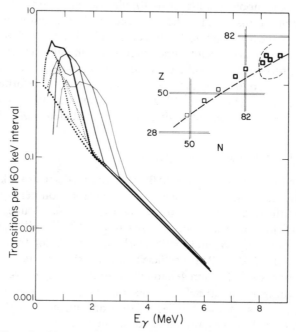

Figure 20   Plots of multipole spectra vs γ-ray transition energy for stretched quadrupole (*full lines*) and stretched dipole (*dotted lines*) components for the principal product nuclei shown by symbols of the same intensity in the section of the periodic table given as an insert. These spectra were determined from the unfolded spectra of 7.6×7.6-cm NaI detectors placed at $0°$, $45°, 90°$ to the beam. A multiplicity filter of six additional NaI counters arranged in a vertical plane helped select nuclei aligned along a particular axis (Deleplanque et al 1978a).

some evidence for them from the conversion coefficient data discussed below and in Section 4. Thus the stretched quadrupole component for E_γ > 2 MeV, where there are very few transitions, could be partly or entirely $\Delta I = 0$ dipole instead.

The spectra in Figure 20 are composites of up to four very similar individual spectra. The location of the most probable product is given by the appropriate symbol on the section of the nuclide chart shown in the insert; the mass region from $A \sim 90$ to $A \sim 170$ is covered. (One region of nuclei, just above the 82 neutron shell, is omitted from this plot and discussed separately in Section 6.1.3.) The results show that the stretched dipole component of the yrast-like transitions occurs at the lower energies. It is a relatively weak component in the good rotational nuclei and occurs there only at energies <0.5 MeV. In the nonrotational nuclei the dipole component runs up to 1 MeV and is much stronger ($\sim 50\%$ of all transitions); it seems clear that these are associated with noncollective behavior. The higher energy part of the yrast-like transitions is invariably seen to be composed of stretched quadrupole transitions. In the region of deformed nuclei these are, no doubt, rotational E2 transitions, and the systematic behavior of these transitions in Figure 20 suggests this may generally be the case. The regular decrease in height of these spectra, together with the edge that moves to higher energies is just what would be expected for maximum spin values like those of Figure 1 and roughly rigid-body moments of inertia.

Linear polarization measurements can be a very useful supplement to the angular distribution experiments, as they can determine the magnetic or electric character of the radiation while the distribution can yield the multipolarity. The polarization measurements are usually made with a Ge(Li) polarimeter and make use of the lower probability of Compton scattering in the direction of polarization of the γ ray. These measurements are most reliable in the low energy γ-ray region because the polarization sensitivity falls with increasing γ-ray energy. A study of the system ^{148}Nd(^{16}O, 4n)^{160}Er began with angular distribution measurements; for unfolded spectra these showed A_2 values near 0.3 for γ-ray transition energies between 0.5 and 1.5 MeV, falling below zero at 0.3 MeV (Vivien et al 1979). Again only stretched dipoles and stretched quadrupoles were considered in the analysis. The degree of linear polarization was measured,

$$P = \frac{1}{Q} \frac{I_\mathrm{V}/I_\mathrm{H} - 1}{I_\mathrm{V}/I_\mathrm{H} + 1}, \qquad\qquad 29.$$

where the ratio $I_\mathrm{V}/I_\mathrm{H}$ is the ratio of measured intensities in the vertical and horizontal analyzing crystals of the polarimeter, and Q is the linear polarization sensitivity of the instrument as determined with known transitions. Values of P near 0.4 were found for transition energies in the region 0.5–0.75 MeV. Combined with the angular distribution data, this result corresponds to about 75% stretched E2 transitions in this region. Below 0.5 MeV, where the dipole component increases, the polarization becomes negative, which indicates M1 transitions there, as has been suggested.

Similar studies have been performed using (S, xn) and (Ar, xn) reactions to produce $^{156-158}$Er (Trautmann et al 1979) and $^{159-161}$Er (Hübel et al, unpublished work, 1978) at much higher spin values than possible with ^{16}O beams, again with a three-Ge(Li) Compton polarimeter. Values for the linear polarization P of 0.3–0.4 were obtained in the transition energy region of 0.5–1.2 MeV. In combination with angular distribution measurements, this suggests that the predominant stretched quadrupole radiation is electric, which agrees with its collective interpretation. Some stretched M1 transitions would be necessary as the remaining dipole component to explain the polarization values, especially at energies of 0.5 MeV and just below.

Measurement of the conversion-electron coefficients also provides information about the nature of the yrast-like transitions, but such experiments, discussed in Section 4 in connection with the statistical γ rays, give more ambiguous results in this energy region. The measured coefficients shown in Figure 12 are consistent with a predominance of E2 multipolarity in the yrast-like regions (with some M1 likely at the lower energies). Mixtures of E1 and M1 transitions cannot be eliminated by these results alone, but, taken together with the angular distribution and polarization results already discussed, it is clear that the higher energy yrast-like transitions are quite generally of stretched E2 character. At 0.5 MeV the results of the two groups are in serious disagreement; those of Feenstra et al (1979) are very close to the theoretical E2 values, while those of Westerberg et al (1978) approach the M1 line. The cleaner coincidence technique of the former group is likely more accurate, but additional work is needed to resolve this problem.

6.1.2 LIFETIMES   Another type of information that shows more directly the collectivity of the continuum cascades is the average transition lifetime. An upper limit can be set by measuring the feeding time to a particular discrete transition near the bottom of the cascade. Dividing this time by the average number of transitions in the cascade gives the average transition lifetime. In early recoil distance Doppler-shift studies to measure the lifetimes of the discrete transitions, these feeding times were determined as a byproduct. The upper limits on the feeding times for a dozen nuclei ranging from ^{120,122}Xe through some rare earth nuclei to ^{184}Hg were all determined to be < 15 ps, and generally ~ 5 ps (Diamond et al 1969, Kutschera et al 1972, Newton et al 1973, Ward et al 1973, Rud et al 1973, Bochev et al 1975). Since there are ~ 20 transitions per cascade, an individual transition, on the average, takes a fraction of a picosecond. The average γ-ray energy is ~ 1 MeV, so that only a dipole transition or a strongly enhanced electric quadrupole transition is fast enough to satisfy this requirement. Magnetic quadrupole and higher multipole radiation are ruled out. This conclusion applies to the bulk of the transitions; slower ones of low intensity (a few percent) cannot be excluded.

A more detailed measurement can be made by comparing the spectrum from a thin target with that from a similar target backed with lead or gold (Hübel et al 1978). States whose lifetimes are short compared to the stopping time in the backing will show Doppler-shifted transitions, while those levels whose lifetimes are longer than the stopping time will not be shifted. To increase the magnitude of the effect, inverse reactions were used. That is, Xe projectiles bombarded targets of ^{27}Al and ^{28}Si to make compound nuclei ^{163}Ho* and ^{164}Er* with velocities 8–9% of light. These spectra are shown at the top of Figure 21. The differences appear quite small, but are magnified when the thin-target spectrum is divided by the backed-target one, as in Figure 21b for the raw data and in Figure 21c for the unfolded spectra. The scheme of analysis assumed a rotational cascade decay along the yrast region and depended upon a single parameter, the intrinsic quadrupole moment, Q_0, of the deformed nucleus, or alternatively, the corresponding enhancement factor over single-particle decay. The value of this parameter was obtained from a fit to the ratio of the unfolded spectra, which is shown in Figure 21c as a solid curve. The enhancement factors found in several experiments are of the order of 200 ± 75, showing that in the spin range 30–50$\hbar$ these nuclei are strongly deformed rotors.

6.1.3 TRANSITION ENERGY-SPIN CORRELATIONS   For rotational nuclei, Equation 6 expresses a very important correlation between a γ-ray transition energy and the initial state spin, namely, they are proportional (neglecting 2 compared with 4I). If the maximum spin given to a compound nucleus at the start of its de-excitation cascades is increased, the maximum γ-ray energy observed in rotational cascades should increase, and the amount of the increase for a given spin change is proportional to the reciprocal of the effective moment of inertia of the nucleus.

At the bottom of Figure 11, a full line has been drawn to repeat schematically the unfolded continuum spectrum (given above) from the 181-MeV ^{40}Ar irradiation of ^{126}Te to make ^{162}Yb by a 4n reaction. The long-dashed line is a similar representation of the unfolded spectrum from a 157-MeV ^{40}Ar bombardment. The former reaction brings, on average, about 40ℏ to the γ-ray cascade, while the latter (lower energy) irradiation brings only 25–30ℏ. It is clear that the higher spin causes additional γ rays (the difference in area of the curves is about six transitions) with higher transition energies, and so moves the edge of the yrast-like transitions to higher energy. This movement was one of the earliest indications that

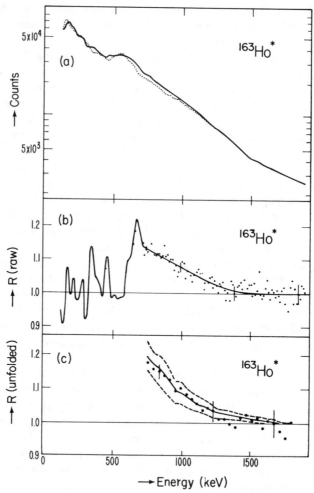

Figure 21 (*a*) Pulse height spectra at 0° for a self-supporting (*solid curve*) and for a gold-backed (*dotted curve*) ^{27}Al target irradiated by ^{136}Xe to give ^{163}Ho*. (*b*) Ratio of the two spectra in (*a*). (*c*) Ratios of the unfolded spectra from (*a*) are given as dots. The solid curve represents the best fit of the calculated ratios (see text) to the data, giving an average B(E2) value of 270 ± 100 single-particle units. The dashed lines show these error limits (Hübel et al 1978).

rotation was involved in many of these de-excitation cascades. The dotted and short-dashed lines in the same figure are representations of the unfolded spectra from the reactions (331 MeV) ^{86}Kr + ^{80}Se and (87 MeV) ^{16}O + ^{150}Sm, respectively, to yield the same products ^{162}Yb + 4n, and both reactions involve, on average, about the same angular momentum (25–30$\hbar$) as the 157-MeV ^{40}Ar case. It can be seen that these three spectra are almost identical, hence the angular momentum is the important variable rather than some other property of the system such as the target-projectile combination.

Another way to demonstrate the effect of differing amounts of initial angular momentum on the continuum cascades is to compare the spectra for different reaction channels. For, as mentioned earlier, there is a fractionation of angular momentum by reaction channel from a given compound nucleus; the fewer neutrons emitted, the higher the average angular momentum remaining. An example is shown in Figure 22 (Simon et

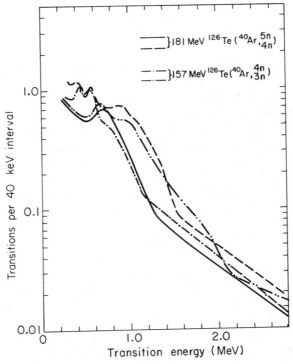

Figure 22 Schematic representations of unfolded Yb continuum spectra following ^{40}Ar reactions. The 5n and 4n channels at 181 MeV correspond to about the same angular momenta as the 4n and 3n channels at 157 MeV (Simon et al 1977).

al 1977). Here the (^{40}Ar, 4n) spectra of Figure 11 are compared with unfolded spectra from the adjacent odd-n reaction channels. Above 0.6 MeV the 5n channel from the 181-MeV ^{40}Ar reaction looks very much like the 4n channel of the 157-MeV ^{40}Ar reaction, because almost the same average angular momentum is involved. Similarly, the 3n reaction at 157 MeV resembles the 4n reaction at 181 MeV. An important point is that the shape of the yrast bump above the discrete lines (>0.8 MeV) seems to depend mostly on angular momentum input and not much on the odd or even nature of the product nucleus.

A quite different but powerful technique to study the spin-E_γ correlation is the measurement of average γ-ray multiplicitly as a function of transition energy. Gamma-ray multiplicities can be measured in a number of ways, some of which have become increasingly sophisticated in recent years. But the early technique is quite straightforward, requiring only a trigger counter in coincidence with the detector in which the spectrum is being measured. Then

$$\frac{N_c}{N_s} = 1 - \left[1 - \frac{\omega}{1-\Omega} \right]^{\langle M-1 \rangle} \xrightarrow[\substack{\omega \to 0 \\ \Omega \to 0}]{} \langle M-1 \rangle \omega, \qquad 30.$$

where N_c and N_s are the coincidence and singles counting rates in the trigger counter, ω and Ω are the efficiencies (including solid angle, γ efficiency, angular correlation, conversion coefficient, etc) of the detector and trigger counters, respectively, and $\langle M \rangle$ is the desired average γ-ray multiplicitly (Tjøm et al 1974, Hagemann et al 1975, der Mateosian et al 1974, Newton et al 1975, Trautmann et al 1975, Fenzl & Schult 1975).

If instead of the single trigger-counter, a number of counters are used, and the number of these counters that are coincident in each event is recorded, one has a "multiplicity filter." These have been given a variety of names depending upon their geometrical appearance: porcupine, halo, hedgehog, urchin, etc (Hagemann et al 1975, Andersen et al 1978, Sarantites et al 1976, Westerberg et al 1977). From the ratios of the intensities of the various folds (events with the same number of trigger counters), not only the average multiplicity can be determined, but also the second moment, $\langle M^2 \rangle - \langle M \rangle^2$, the square of the width of the distribution in M, and, in principle, still higher moments. A number of algorithms have been developed to obtain these moments from the experimental fold distributions (Andersen et al 1978, Westerberg et al 1977, van der Werf 1978, Ockels 1978b, Kohl et al 1978, Kerek et al 1978). Because the γ-ray multiplicity is related to the average angular momentum dissipated in the de-excitation cascades, measurement of the multiplicity and of its second moment has become a useful tool in compound nuclear and deep inelastic reaction studies. For the present discussion, multiplicities are important because they indicate the spin, and because spin-E_γ correlations cause differences in the multiplicity for different transition energies (Newton et al 1977).

It is useful to measure the multiplicity of many individual intervals, or channels, of a γ-ray spectrum and plot these against the γ-ray energy—giving a "multiplicity spectrum." Such spectra for the system ^{40}Ar $+ ^{124}$Sn $\rightarrow$ ^{164}Er* at several bombarding energies are shown in Figure 23 (Deleplanque et al 1978b). There is a pronounced peak at all bombarding energies, and it comes at the upper edge of the yrast-like transitions in the γ-ray spectrum shown above. This multiplicity peak shows clearly that the highest energy yrast-like transitions are associated with the highest spins populated. The right side of Figure 23 shows γ-ray and multiplicity spectra calculated from a very simple model; a purely rotational yrast-like cascade is assumed. All the significant observed features are reproduced. This provides a rather direct confirmation of the collective rotor picture for this nucleus, all the way from the discrete transitions near the ground state to the highest spins observed, $\sim 60\hbar$. The moment of inertia used in the calculation was roughly the rigid-body value and was constant over the whole spin range. This suggests that the deformation of ^{160}Er remains prolate, $\beta \sim 0.3$, up to the highest spins.

Away from the deformed rotors, there is a more complex behavior. The

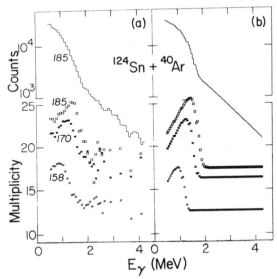

Figure 23 Observed (*left*) and calculated (*right*) multiplicity spectra vs E_γ for the ^{124}Sn + ^{40}Ar system at the indicated bombarding energies. One NaI γ-ray spectrum is also shown (Deleplanque et al 1978b).

^{40}Ar + ^{82}Se → ^{122}Te* system (Figure 24) does not show a peak in the multiplicity spectrum at 121 or 131 MeV, but develops one at still higher bombarding energies. The major reaction product at these energies, ^{118}Te, has a ground band with roughly equidistant level spacing up to spin 14$\hbar$ (quasi-vibrational). There are stretched E2 transitions between these levels that are weakly collective, but certainly not rotational. Above 150-MeV bombarding energy (spin 30–35$\hbar$), a peak appears in the multiplicity spectrum and moves to higher transition energy with increasing ^{40}Ar energy. This is an indication of the onset of rotational behavior. A calculation can reproduce this feature if a rotational spectrum is assumed for $I > 35\hbar$, and an uncorrelated spectrum below this spin. The ^{48}Ca + ^{100}Mo → ^{148}Sm* system also shown in the figure is a striking example; the nearly semi-magic product nuclei do not show formation of a rotational peak until about spin 50$\hbar$.

All of these nuclei appear to become rotational at some spin. The behavior of ^{152}Dy and ^{154}Er would be very interesting in this respect since they are outstanding examples of nonrotational decay from weakly oblate nuclei up to nearly 40$\hbar$. A group at Orsay has recently studied ^{154}Er in comparison with ^{160}Er by bombarding targets of ^{119}Sn and ^{124}Sn with ^{40}Ar beams of appropriate energy (Deleplanque et al 1979). They employed

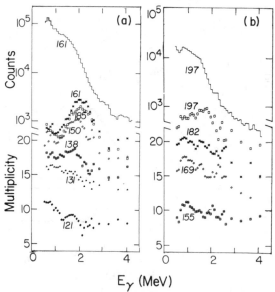

Figure 24 Plots of multiplicity vs γ-ray transition energy for the systems (a) ^{82}Se + ^{40}Ar and (b) ^{100}Mo + ^{48}Ca for the indicated bombarding energies. One NaI γ-ray spectrum is also shown for each system (Deleplanque et al 1978b).

a multiplicity filter, NaI detectors at 0° and 90°, and a Ge(Li) detector to provide a gate on transitions from 4n and 5n products. The ^{164}Er compound nucleus has a γ-ray spectrum (Figure 25) showing the characteristic yrast bump below 1.6 MeV with a flat region below 1 MeV. The ^{159}Er compound nucleus has a very different spectrum showing two bumps below 1.6 MeV separated by a valley at about 1 MeV. The low energy bump has a high intensity and contains all the known discrete transitions around 0.7-MeV connecting states with spins up to 36$\hbar$. The higher bump starts at ∼ 1 MeV, and above its maximum at 1.3 MeV seems to resemble the upper part of the rotational bump in the ^{159,160}Er product nuclei. This second bump develops very strongly with increasing fold number; i.e. it comes from high spin states. The multiplicity spectrum for the ^{154,155}Er products also shows two peaks. The multiplicity estimated for the upper peak, 31, indicates again that it comes from the highest spin states in the cascade, and there is a remarkable similarity to the upper part of the ^{159,160}Er multiplicity spectrum. The multipole composition of the γ-ray spectra, deduced from the 0° and 90° intensities with the assumption of only stretched dipole and quadrupole transitions, shows that the upper bump in ^{154,155}Er consists mainly of stretched quadrupole transitions. This is not true for the lower peak, which has a large fraction of dipoles, in agreement

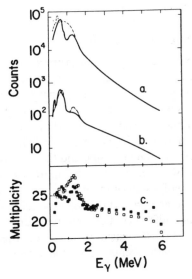

Figure 25   (a) Spectra at 0° for ^{124}Sn + ^{40}Ar system (*dashed line*) and ^{119}Sn + ^{40}Ar system (*solid line*) in fourfold coincidence with a multiplicity filter. (b) Spectra for ^{154}Er (*thick line*) and ^{155}Er (*dot-dashed line*). (c) Multiplicity spectra for the ^{124}Sn + ^{40}Ar system (*open squares*) and the ^{119}Sn + ^{40}Ar system (*filled squares*) (Deleplanque et al 1979).

with the known discrete lines in ^{154,155}Er. The calculated spin at the top of the cascade is ~60$\hbar$ and the moment of inertia is ~150 MeV^{-1}. The evidence seems quite good that, at about spin 40$\hbar$, the products ^{154,155}Er switch from weakly oblate nuclei, to rather strongly deformed (most likely rotational) nuclei. The experimental results do not determine whether these shapes are oblate or prolate, but the similarities with the heavier Er nuclei suggest the latter.

Although the number of cases studied is still quite small, the experimental evidence indicates rather strongly that nuclei in the mass region 100–170 deform (if not deformed already in the ground state) and show rotational behavior above some spin that depends on the detailed structure of the particular nucleus. There appears to be a rather systematic behavior. The $N = 82$ and nearby nuclei have mostly dipole transitions at low energies, and only begin to rotate at very high spins (50–60$\hbar$). The quasi-vibrational nuclei have more stretched E2 radiation at low energies, which is weakly collective but not rotational. They deform and become rotational at lower spins (35–45$\hbar$). Finally, there are well-deformed prolate nuclei that show a high proportion of collective stretched E2 transitions at all spins. There are indications that some lower mass nuclei also show these behavior patterns, but that is not as clear yet. It should be noted that the term rotational, as used here, covers a range of behavior, including some with large deviations from that of a perfect rotor.

6.2 Effective Moments of Inertia

In a rotational nucleus, the energy and spin are related through the moment of inertia. As discussed (Section 2), the yrast line follows, on average, a rotational trajectory in plots of excitation energy vs spin, independent of whether the motion is truly collective or is due to particle alignment. Thus the slope of the yrast line, or of a line parallel to it, provides a measure of the effective moment of inertia via Equation 27. The entry line is assumed to be approximately parallel to the yrast line and thus can be used to determine moments of inertia (Section 6.2.1). The more usual measurements of moments of inertia involve determining the slope of a rotational trajectory from a γ-ray transition energy and the corresponding spin using Equation 6, or measuring the difference between rotational transition energies (Section 6.2.2). None of these measurements is very precise, since it has not yet been possible to isolate events associated with a narrow slice of very high spins.

6.2.1 ENTRY-LINE METHOD   The entry line may be defined as a line tracing the average excitation energy as a function of spin for the first states that will decay by γ emission. Statistical model calculations indicate that the entry line is almost parallel to the yrast line over a certain region, and so it has been suggested that determination of the entry line in this region might provide a measurement for the effective moment of inertia (Simpson et al 1977). An experiment was performed, involving mainly the reaction $^{58}\text{Ni}(^{16}\text{O}, 2p\gamma)^{72}\text{Se}$, to test this idea. The position of the entry line was determined by measuring the average energy of the protons in coincidence with known γ rays of the desired reaction product. The average proton energy was also determined from the summed spectra of two separate proton detectors in coincidence with each other and with γ rays. An excitation for the residual nucleus could be calculated from this average proton energy, the incident bombarding energy, and the Q value of the reaction. The average spin corresponding to this excitation energy was determined from the average γ-ray multiplicity. In two cases, the excitation energy was also determined as the product of the average γ-ray energy and the γ-ray multiplicity, and these values agreed with those from the proton energies. The experimental entry line is in reasonable agreement with the calculated one except at the highest spins, where it does not increase in energy as fast as expected. The experimental points lie along a curve whose slope gives a moment of inertia of $2F/\hbar^2 = 41$ (MeV)$^{-1}$, which can be compared with the average values from known ground-band transitions of 37 MeV^{-1} or with the calculated value for a deformed nucleus with $\beta = 0.3$ of 38 MeV^{-1}.

The use of a sum crystal (Section 6.2.2) permits many points to be determined along the entry line in a single bombardment. With this technique, an experiment, again on $^{58}\text{Ni}(^{16}\text{O}, 2p\gamma)^{72}\text{Se}$, at three bombarding energies gives the entry line points shown in Figure 26 (Tjøm et al 1978). Also shown are the known yrast states to spin $14\hbar$, and an extrapolation to higher spin values. The entry line slope corresponds to a value of $2F/\hbar^2 = 40$ (MeV)$^{-1}$, in agreement with the earlier experiment.

Determination of moments of inertia from the entry line assumes that it is parallel to the yrast line (to a rotational trajectory). This cannot be true in detail. If the actual populations resemble the statistical calculations shown in Figure 8, the emission of each additional neutron makes a step in the entry line, or at least an irregularity that has nothing to do with the moment of inertia. Also, the emission of α particles occurs preferentially at high spin (and high energies above the yrast line) and may drive the entry line down toward the yrast line in this region, giving incorrect moments of inertia there. Nevertheless, this method depends on different quantities than the other one that will be described, and the similarity of the results gives confidence that they are both generally correct.

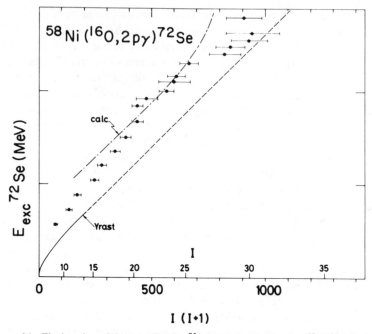

Figure 26 The location of the entry line for ^{72}Se made by the reaction ^{58}Ni(^{16}O, 2pγ), and averaged for 55, 60 and 65-MeV bombarding energies. Values of the excitation energy are obtained from the product $\langle M_\gamma \rangle \langle E_\gamma \rangle$, and the corresponding values of I are derived from $\langle M_\gamma \rangle$. The yrast line at low energies comes from known states, and the extrapolated dashed line is calculated for a rigid rotor with $2F/\hbar^2 = 40$ (MeV)$^{-1}$. The entry line predicted from a statistical model calculation is also shown (Tjøm et al 1978).

6.2.2 TRANSITION ENERGY METHODS   An early method for determining the effective moment of inertia was based on the energy of the yrast bump edge (the highest energy transition in the cascades) and an estimate of the corresponding maximum spin (Banaschik et al 1975, Simon et al 1976). The latter could be made in one or both of two ways: (*a*) from a measurement of the γ-ray multiplicity, or (*b*) from a measurement of the fusion cross section.

Measurement of γ-ray multiplicities was discussed in Section 6.1.3. However, the maximum spin at the top of the γ-ray cascade is needed for calculating the moment of inertia by the above method. To obtain this one must transform average multiplicity into average spin, and then go from average to the maximum. Multiplicity and angular momentum are clearly related in that they must increase together, but the exact relationship differs depending upon the type of nucleus. For the simplest case, a good rotor, all the cascade transitions except the few statistical ones are stretched E2 transitions, so that

$$I \approx 2(M_\gamma - \delta), \qquad\qquad 31.$$

where δ is the effective number of statistical γ rays that remove no angular momentum. The value of δ is quite commonly taken between 2 and 4, and there are indications that this is approximately correct in the case of good rotors (Simon et al 1977, Hagemann et al 1975, Hillis et al 1979, Sarantites et al 1978). However, for nonrotational nuclei that include dipoles in their yrast-like cascades, either the factor 2 (the angular momentum carried off per γ ray) should be reduced or δ should be increased. Thus, considerable knowledge of the system is required in order to choose the relationship between spin and multiplicity. After that, one must still make an estimate of the shape of the angular momentum distribution (triangular, Gaussian etc), in order to be able to extrapolate from the average to the maximum spin. Another way to estimate the spin at the top of the γ-ray cascade is based on the sum of the fusion cross sections up to and including the reaction channel of interest, using Equation 18. In this sum, one must be careful not to omit reaction channels having angular momenta equal to or less than the channel of interest. Fortunately, these two methods give results that are in general agreement.

The spin and transition energy at the top of the γ-ray cascades can be used in Equation 6 to obtain a value of the effective moment of inertia for that spin. To obtain additional values at lower spins, the bombarding energy can be lowered. Or, in the lower spin region of a spectrum where the population is constant (i.e. every cascade goes through that region; for example, in a 3n channel at angular momenta where the corresponding l waves lead mainly to the 4n and 5n channels), a spin value can be obtained for any transition energy by summing the number of transitions up to that point and using Equation 31 (where δ now refers only to the statistical transitions under the region summed). Some estimates for the effective moment of inertia obtained by these methods are shown in Figure 27 for the product nucleus ^{162}Yb (Simon et al 1977). The small dots give the values for the known discrete ground-band transitions, and the open circles are for the isotone ^{160}Er which is known through the backbend. The four points with error bars are calculated by the "edge method," and the large dots connected by a solid line are from the method just described.

Both of these methods depend upon associating a given γ-ray energy with a particular spin value. There also exists a "differential" method that utilizes an expression derived from Equation 6 for stretched E2 transitions:

$$\Delta E_\gamma = 8\hbar^2/2\mathbf{F} - 2E_\gamma \mathrm{dln}\mathbf{F}/\mathrm{d}I. \qquad\qquad 32.$$

For a reasonably constant moment of inertia, this becomes

$$\Delta E_\gamma = 8\hbar^2/2\mathbf{F}. \qquad\qquad 33.$$

The energy difference, ΔE_γ, can be obtained simply from the height of the γ-ray spectrum. For example, on the unfolded spectrum of Figure 11, the height of the spectrum gives the number of rotational transitions (n_γ) per energy interval (0.040 MeV in Figure 11) per event. The reciprocal, $0.040/n_\gamma$, is the average difference between adjacent transition energies, ΔE_γ, and can be related to the moment of inertia by Equation 33. The values so calculated are also shown in Figure 27 as diamonds connected by a dashed line. This method also requires a constant population in the region of application, and so has a limited range of use (between 0.7 and 1.0 MeV in Figure 11). The values from the different methods (shown in Figure 27) are in reasonably good agreement, and above spin $20\hbar$ $[(\hbar\omega)^2 \approx 0.12$ MeV$]$ are within 10% of the value for a rigid diffuse sphere of mass 162. The uncertainties in these values are at least that large, so the results agree with this simple estimate as well as can be expected using this approach.

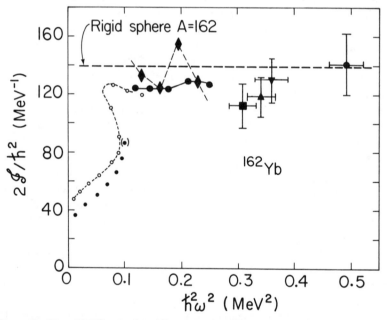

Figure 27   Plot of $2F/h^2$ vs $(\hbar\omega)^2$ for ^{162}Yb. The small dots are the known discrete transitions, and the open circles are the known transitions in the isotone ^{160}Er. The large dots connected by a solid line are values in ^{162}Yb derived by the integral method from the 181-MeV ^{40}Ar spectrum for E_γ between 0.7 and 1 MeV, and the diamonds come from the differential method applied to the same data. The square, triangles, and circles represent values for the moment of inertia derived from the yrast bump edge for 87-MeV ^{16}O, 157-MeV ^{40}Ar, 331-MeV ^{86}Kr, and 181-MeV ^{40}Ar reactions, respectively. The horizontal dashed line corresponds to a rigid diffuse sphere with $A = 162$ (Simon et al 1977).

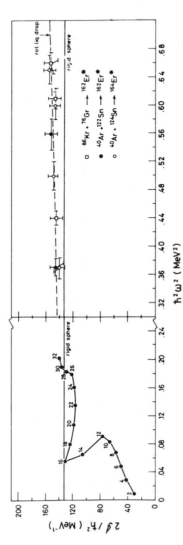

Figure 28 Plot of $2F/\hbar^2$ vs $(\hbar\omega)^2$ for the compound system ^{162}Er*. The data on the left are obtained from the known transitions in ^{158}Er. The right-hand values are obtained by determining the average multiplicity (with a multiplicity filter) and the yrast bump edge from the multiplicity peak in $\langle M_\gamma \rangle$ vs E_γ plots. The dashed line is the moment of inertia predicted from a rotating liquid drop with $A = 158$ (Hillis et al 1979).

A newer study using these edge methods gives the results shown in Figure 28 (Hillis et al 1979). The product nuclei are $^{158-160}$Er, and the edge of the yrast bump is taken from multiplicity curves like those in Figure 23. The spins come from average multiplicity data using Equation 31 (with $\delta = 4$). The measured moments of inertia are seen to fall on the rotating liquid-drop estimate, about 10–15% above the rigid-sphere value. This is in general accord with Figure 27 for Yb nuclei.

Recently there has appeared a somewhat more objective method for

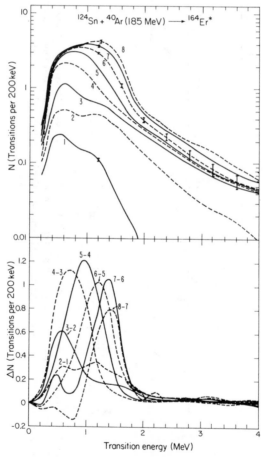

Figure 29   (*Top*) Spectra from a 7.6 × 7.6-cm NaI detector (number of transitions per 200 keV per event) for consecutive ∼4-MeV wide slices of the coincident total γ-ray energy spectrum from 185-MeV ^{40}Ar + ^{124}Sn → ^{164}Er* observed with a 33 × 20-cm sum crystal. (*Bottom*) The difference in spectra from neighboring slices as indicated in the figure (Körner et al 1979).

determining the edge of the yrast-like transitions as a function of angular momentum input. This involves measuring the total γ-ray de-excitation energy by placing the target inside a large NaI crystal with only narrow channels to admit the beam and to let some external counters view the target. The coincident spectrum in the external counters can then be measured for different energy "slices" in the sum spectrum. The top of Figure 29 shows the spectra taken by an external NaI detector in coincidence with $\sim$4-MeV wide slices of summed energy from such a crystal (Körner et al 1979). In the first few spectra the increase in average angular momentum associated with higher sum energy causes an increased yield of essentially all γ rays. After about slice 5 the yield of the lower energy transitions ($E_\gamma < 0.6$ MeV, $I < 20$) is saturated because the feeding occurs at higher spin. The movement of the yrast bump edge to higher energy then becomes noticeable, and by subtracting the spectrum of one slice from the next one, we obtain the curves in the bottom of the figure. Their centroids can be considered to be the average edge, $\bar{E}_\gamma$, for the spectra of the two consecutive slices. The corresponding spin, $\bar{I}$, can be obtained from the average multiplicity for the two spectra determined in the same measurement. Use of $\bar{I}$ and $\bar{E}_\gamma$ in Equation 6 gives an effective moment of inertia for that γ-ray energy, and values so determined for the compound system ^{164}Er* are shown in Figure 30. An important advantage of this method is that the edge and the multiplicity are both determined by the input distribution of angular momentum, and no extrapolation from average to maximum is required.

A type of differential method can also be applied to the spectra obtained in coincidence with slices of the total-energy spectrum. The area of each difference peak (bottom of Figure 29) gives the incremental number of rotational transitions (half the angular momentum increase ΔI) and the difference in the centroids of two consecutive peaks gives the average increase in transition energy, ΔE_γ. Use of Equation 33 then gives effective moments of inertia that depend essentially on the difference in multiplicities rather than on the multiplicities themselves, and so provides a partially independent evaluation of F. As can be seen in Figure 30, the integral and differential methods are in good agreement.

This sum crystal method probably provides the best values presently available for moments of inertia at very high spin. They are in rather close agreement with the liquid-drop model—about 10% or so above the rigid-sphere values. A few measurements have been made for comparable spin ranges outside the rare earth region. They also seem to be near the liquid-drop value. Very careful studies are currently underway to look for the large increase in moment of inertia at the very highest spins predicted by the liquid-drop model, and reinforced in some regions by shell effects (Section

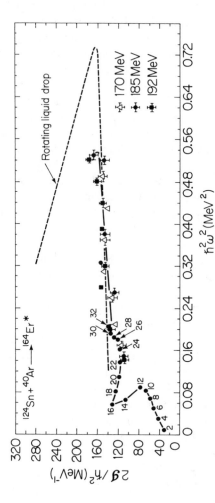

Figure 30 Plot of $2\mathbf{F}/\hbar^2$ vs $(\hbar\omega)^2$ for the compound system ^{164}Er* made by ^{40}Ar + ^{124}Sn. The solid circles to spin $32\hbar$ are the known transitions in ^{158}Er. The symbols without error bars are calculated from the expression $2\mathbf{F}/\hbar^2 = 8/\Delta E_\gamma$ (Equation 41). The pure liquid-drop prediction is indicated by the dashed line.

2). There is some evidence, both in the $A \sim 90$ and the $A \sim 160$ regions (possible upbend at highest energy points in Figure 30), for such a "giant backbend," or at least for moments of inertia well beyond the rigid-sphere value. This is one of the most exciting areas of high spin studies at present.

6.3   The Collective Moment of Inertia

The γ-ray spectrum from a rotational nucleus is highly correlated in time, spatial distribution, and energy. Part of a rotational energy level diagram is shown on the left side of Figure 31, and the corresponding γ-ray spectrum on the right side. This spectrum is seen to be composed of equally spaced lines, up to some maximum energy corresponding to the decay of the state with highest angular momentum, I_{max}. There are at least two kinds of correlation in this spectrum. One is of the maximum γ-ray energy with I_{max}; the previous section discussed how this led to the evaluation of effective moments of inertia in rotational nuclei. This section discusses another correlation, that between γ-ray energies. One illustration of this correlation, which is immediately obvious from Figure 31, is that no two γ rays have the same energy.

A schematic two-dimensional spectrum of one γ-ray energy against another for a rotational nucleus is shown in Figure 32. The dots represent the location of coincidences between γ rays de-exciting states up to $I = 14$ in a perfect rotational band with moment of inertia $\mathbf{F}$. However, different bands might have somewhat different moments of inertia, and the lines through the dots represent bands having moments of inertia differing by $\pm 10\%$ from $\mathbf{F}$. It is also well known that nuclei align angular momentum at

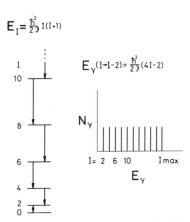

Figure 31   Schematic energy level spectrum of a rotor on the left, and its γ-ray spectrum on the right. Each vertical line on this spectrum represents a transition between adjacent rotational levels.

relatively low spin values, and the crosses in Figure 32 are coincidences between transitions from spins 16–26ℏ in a band having 11ℏ aligned, also with a moment of inertia **F**. The light lines through the crosses again represent aligned bands differing in **F** by $\pm 10\%$. It is clear that even if bands are populated that differ considerably in moment of inertia and aligned angular momentum, a strong pattern remains in the two-dimensional γ–γ coincidence spectrum. The valley along the diagonal, representing the absence of transitions of the same energy, is not at all filled, and the first ridge adjacent to it is rather clear. (Notice, however, that a single band with a *changing* moment of inertia or alignment has not been considered.) It is then of considerable interest to know whether these correlations are present in real spectra at very high spins.

To perform such correlation experiments requires good statistics and a method to reduce the number of uncorrelated events. The first successful experiment was made by Andersen et al (1979). They used four detectors, which resulted in six independent pairs, and, after equalizing the gains, plotted every coincidence on a single two-dimensional spectrum. A background of uncorrelated events, $\tilde{N}_{ij}$, was subtracted from the γ–γ matrix. Basically this background is calculated for the point ij from the projections of the row, $\sum_k N_{ik}$, and the column, $\sum_l N_{lj}$, on the assumption

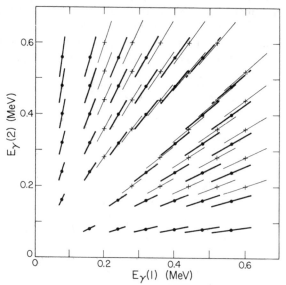

Figure 32 Schematic correlation plot for a rotational nucleus. The dots locate the coincidences for a band with spins up to 14ℏ and moment of inertia **F**, and the heavy lines show the effect of bands where **F** differs by $\pm 10\%$. The crosses show the location of coincidences in a band with spins 16–26ℏ, 11ℏ of aligned angular momentum, and moment of inertia **F**. The light lines again show the effect of bands differing in **F** by 10%.

that every observed γ ray has equal probability of being in coincidence with every other observed γ ray. The correlated two-dimensional spectrum is then

$$\Delta N_{ij} = N_{ij} - \tilde{N}_{ij} = N_{ij} - \sum_{k} N_{ik} \sum_{l} N_{lj} \bigg/ \sum_{l,k} N_{lk}.$$ 34.

There are some problems with this subtraction method, and others are being explored, but it seems clear that the results discussed here are not much affected by these problems. The original experiments were limited by beam energies to rather low angular momenta. We only discuss a later study (Deleplanque et al 1980), which applied this method to a system where the maximum angular momentum the nucleus can hold was brought in.

The correlated spectrum for the system $^{124}\text{Sn} + ^{40}\text{Ar} \rightarrow ^{164}\text{Er}^*$ at 185-MeV ^{40}Ar energy is shown in Figure 33. Three features, believed to be

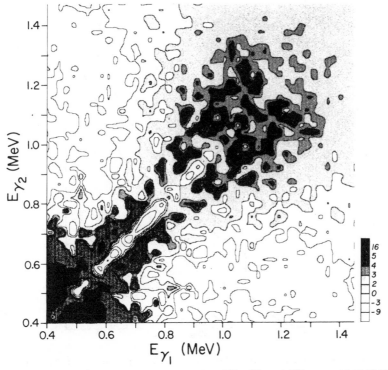

Figure 33 Correlation spectrum from the reaction $^{124}\text{Sn}(^{40}\text{Ar}, \text{xn})^{164-x}\text{Er}$ at 185 MeV. The data were taken on Ge(Li) detectors and treated according to Equation 34. The plot shows contours of equal numbers of correlated events, where the darker regions have more counts according to the scale at the right edge. From Deleplanque et al (1980).

general, have been pointed out. First, there is a distinct valley along the diagonal up to about 1 MeV (spin 40$\hbar$) having a measurable width, and there is some possibility this valley also exists in the region above 1.1 MeV. Second, there are a few bridges across this valley beginning as low as 0.55–0.60 MeV and continuing as far up as the valley persists. Also there are irregularities in the ridges alongside this valley. Finally, there is a general filling of the valley above ~ 1 MeV, which is rather complete around 1.1 MeV. There are many other features in Figure 33 that one would hope to understand; however, thus far only these three have been carefully considered.

The interpretation of these data is somewhat speculative, but seems sufficiently interesting to describe here. The type of decay pathway envisioned is illustrated in Figure 34. The scalloped patterns represent successions of rotational bands through which the population flows to the

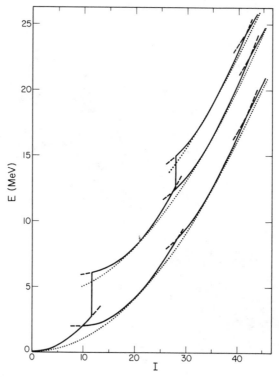

Figure 34 Schematic illustration of the bands in a decay pathway (*solid lines*) and their envelopes (*dotted lines*). The plot parameters are somewhat arbitrary but were taken to be $2F_{coll}/\hbar^2 = 50 \text{ MeV}^{-1}$, $2F_{eff}/\hbar^2 = 100 \text{ MeV}^{-1}$; $j_a = 0, 10, 18,$ and $24\hbar$ for the bands, in order of increasing energy.

ground state. The transition between bands might be a simple band crossing as shown, or might occur via a statistical transition from a higher band (change of temperature); we assume that the band character is not a strong function of temperature. (This is not known, but represents the simplest starting assumption; there is no evidence to the contrary.) The mathematical description of such a band structure is reasonably straightforward. The energy in each band is given by

$$E(I) = \frac{\hbar^2}{2\mathbf{F}_{coll}} R^2 + E(j_a) = \frac{\hbar^2}{2\mathbf{F}_{coll}} (I - j_a)^2 + E(j_a), \qquad 35.$$

where $\mathbf{F}_{coll}$ is the moment of inertia for collective rotation, j_a is the aligned angular momentum, $E(j_a)$ is the bandhead energy for the aligned bands, and one is neglected compared to $I - j_a$. One possible assumption is that $\mathbf{F}_{coll}$ and j_a change smoothly with spin in a band; however, that is not what happens at low spins in the backbending region and it is not suggested by the irregularity of the observed correlation spectrum at high spins. The other limiting assumption is that taken in Figure 34: that $\mathbf{F}_{coll}$ and j_a are approximately constant in a band and change sharply between bands. In fact, $\mathbf{F}_{coll}$ is also taken to be constant in Figure 34, and only j_a and $E(j_a)$ change between bands. It is apparent that Equation 35 describes a parabola, displaced from the origin horizontally by j_a and vertically by $E(j_a)$, and the bands in Figure 34 are sections of these parabolas. The γ-ray transition energy within one of these bands, holding $\mathbf{F}_{coll}$ and j_a constant, is

$$E_\gamma = 2\left(\frac{dE}{dI}\right)_{j_a, \mathbf{F}_{coll}} = \frac{\hbar^2}{2\mathbf{F}_{coll}} 4(I - j_a). \qquad 36.$$

Equation 6, the usual form for a rotor, can be approximated:

$$E_\gamma \approx \frac{\hbar^2}{2\mathbf{F}_{eff}} 4I, \qquad 37.$$

where $j_a = 0$, corresponding to no horizontal displacement. This just gives the envelope of the bands and is the dotted curve in Figure 34. The moment of inertia must be "effective" since the spin I is composed of both collective rotations, R, and aligned angular momentum, j_a. The relationship between the moments of inertia is then

$$\mathbf{F}_{eff} = \frac{I}{I - j_a} \mathbf{F}_{coll}. \qquad 38.$$

Since j_a is expected to be large—around $\frac{1}{2}I$ in the backbend region, and probably not relatively smaller at higher spins—$\mathbf{F}_{coll}$ must be much smaller than $\mathbf{F}_{eff}$, reflecting the fact that a single particle cannot contribute fully both to the aligned angular momentum and to the collective moment of inertia. Only I and E_γ are usually measured, so it is F_{eff} that is normally determined and found to be roughly equal to the rigid-body value.

The width of the valley in a correlation spectrum is related to the *difference* between successive transition energies, or to the curvature of the energy expression (Equation 35). Assuming again that these occur within a band of constant F_{coll} and j_a,

$$\Delta E_\gamma = 4\left(\frac{d^2 E}{dI^2}\right)_{j_a, F_{coll}} = 8\,\frac{\hbar^2}{2F_{coll}}.$$

39.

Thus, the valley width measures F_{coll} and should give values considerably smaller than the F_{eff} determined from γ-ray energies (Section 6.2). The first results on this point (Deleplanque et al 1980), suggest that this is indeed the case, with values of F_{coll} as low as $\sim 0.6 F_{eff}$. However, these results are tentative and more data are needed. If F_{coll}, F_{eff}, and I can be reliably measured, then one can obtain j_a, which would be a very interesting quantity to know for these high spin states.

The bridges are easy to understand, but require a short digression to review some features of the behavior in the backbending region at spins

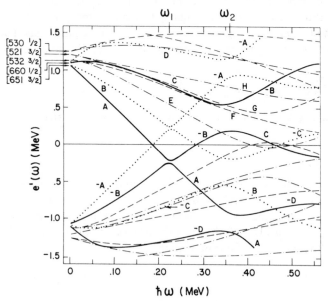

Figure 35 Cranking model calculations of energy in the rotating frame, $e'(\omega)$ vs ω for neutrons around $N = 90$. The levels marked A, B, C, and D are components of the $i_{13/2}$ orbital, while those labeled E, F, G, and H have negative parity ($N = 5$ shell). The different types of lines refer to different signatures of the levels. From Riedinger et al (1980).

$\gtrsim 20\hbar$. Figure 35 is a diagram of the type shown in Figure 4 except that pairing is included, and it covers only a very restricted region, the five states nearest the Fermi level. The $e'(\omega) = 0$ line is the Fermi level, and each state is reflected around this line, so that it appears twice. A rough way to think about this is that in one case $[e'(\omega) < 0$ for low $\omega]$ the level is included in the pairing correlations, and in the other $[e'(\omega) > 0$ for low $\omega]$ it is not included, so that it is a quasiparticle. If particles are placed in the levels labeled A and B (thereby necessarily emptying those labeled $-$A and $-$B), a two-quasiparticle excited band is formed, which corresponds to the most aligned $i_{13/2}$ components. As the frequency increases the energy of this band drops relative to the fully paired vacuum ($e' = 0$) until at the point labeled ω_1 it crosses the vacuum and becomes lowest in energy. This crossing has been observed in many nuclei and corresponds to the first backbend. If one follows only the lowest band, then this process may be expressed as AB unpairing (or aligning) at frequency ω_1. The important point is that AB will unpair in essentially every band at this frequency. As another example, the two-quasiparticle band EF will be crossed by the four-quasiparticle band ABEF, again at ω_1, and for EF one could substitute CD, or CDEF, etc. The frequency ω is a measurable quantity ($\sim E_\gamma/2$), and the above argument says that at certain transition energies many bands at widely different excitation energies will experience a backbending (band crossing). However, a complication occurs if state A has a quasiparticle in it (either a single one in an odd-mass nucleus or one of a pair such as AE in an even-even nucleus). Then AB cannot unpair—A is blocked—and this backbending will not occur. The frequency then increases normally with spin to about $\omega = 0.36$ MeV, the point marked ω_2, where BC will unpair and align. Again this band crossing can happen in many bands. It is worth noting that at about this same frequency, 0.36 MeV, AD unpairs, so that if B rather than A is blocked then AD can unpair and cause a band crossing.

The experimental data that suggest such a picture are shown in Figure 36 (Riedinger et al 1980). Here the aligned angular momentum i is plotted vs the frequency ω. The aligned angular momentum is obtained from the experimental data by subtracting the spin of the ground band (no quasiparticles) from the spin of the band of interest at the same rotational frequency (transition energy). The upbends or backbends in Figure 36 correspond to band crossings, and the sharp backbend in the ground band at frequency 0.27 MeV is in reasonable agreement with the AB unpairing frequency of 0.23 MeV in Figure 35. The remarkable feature of Figure 36 is the coincident upbending of six bands at frequencies around ω_2, 0.36 MeV. These include three bands where A is blocked (AE and AF in ^{160}Yb and A itself in ^{161}Yb), two where B is blocked (BF in ^{160}Yb and B itself in ^{161}Yb), and the pairing vacuum as an excited band above the point where AB crossed. The yrast band has a second band crossing at a frequency ω_3 around 0.42 MeV that is probably due to a pair of protons and thus is not shown on Figure 35. There are other fascinating aspects to this picture, but it is not really in the spin range we want to discuss in detail. Important for the very high spin region are the concepts: (a) alignments occur and keep occurring as the frequency increases, (b) these occur in many bands at a few characteristic frequencies where orbitals unpair or cross, and (c) they cause major disruptions in the rotational bands when they occur.

The lowest energy bridges in Figure 33 are due to known backbends in the nuclei produced. The large one at 0.55 MeV ($\omega \sim 0.27$ MeV) corresponds to the first backbends in ^{158}Er and ^{160}Er (the major even-even products). Backbending or upbending behavior implies several γ rays of similar energies in the band and thus tends to fill the valley. The arguments just developed in connection with Figure 35 demonstrate that a given level crossing will appear in many bands at the same rotational frequency (the same γ-ray energy). This must be true for the second large bridge at ~ 0.8 MeV ($\omega \sim 0.42$ MeV), at the location of the second backbend in ^{158}Er, since the population of that backbend in the yrast sequence is quite weak but the bridge is a prominent feature. It is known that the first backbend in this region of nuclei involves the alignment of two $i_{13/2}$ neutrons, and the second probably the alignment of two $h_{11/2}$ protons. There are at least two more higher energy bridges in Figure 33, each of which must involve many bands

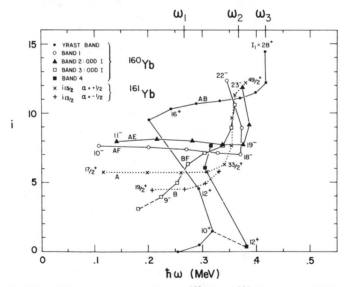

Figure 36 Plot of alignment i vs ω for bands in ^{160}Yb and ^{161}Yb. The proposed labeling of the bands is according to the nomenclature on Figure 35. From Riedinger et al (1980).

since they have not been observed in any discrete-line studies. It is not yet clear which orbitals are involved in these higher bridges. However, the general behavior up to about 1 MeV γ-ray energy seems reasonably clear—a deep valley reflecting good rotational behavior, and a few large irregularities in both the valley and ridge structure resulting from alignment of specific high j orbitals.

Above 1 MeV in Figure 33 the valley is largely filled, and completely so in places. It is not really clear what causes this, but a reasonable extension of the features described above seems able to do it. The levels calculated for protons by rotating a prolate potential was shown in Figure 4. The Er nuclei produced here have 68 protons, so the Fermi level is in a basically $N = 4$ shell into which an $N = 5$ ($h_{11/2}$) orbital has intruded. If the 68-proton region (around the level 523, 7/2) in Figure 4 is followed to higher frequency, it intersects some strongly down-sloping levels at $\omega/\omega_0 \approx 0.07$ ($E_\gamma \approx 1.1$ MeV for $A = 160$). These levels come from the shell above ($N = 5$ with $N = 6$ intruder), and in fact correspond to the highest aligned $i_{13/2}$ ($N = 6$) components together with the most aligned $h_{9/2}$ level. An $N = 6$ level coming down into the basically $N = 4$ shell ($\Delta N = 2$) is particularly interesting since it has the same parity and thus can mix into these levels. The neutrons in this region do the same thing [$N = 7$ ($j_{15/2}$) coming into the $N = 5$ shell] at about the same frequency. Thus one might expect many level crossings and very disturbed rotational bands. Perhaps this can account for the general filling of the valley in this region. If true, it could be a reasonably general phenomenon implying tendencies for the nucleus to become more triaxial and more deformed. It will be interesting to try to verify such behavior.

These E_γ-E_γ correlation experiments certainly represent a frontier in the continuum work. There are detailed features in the low energy region of Figure 33 that are not yet understood, and the cause for the filling of the valley at high spins represents a major effect to be studied. This kind of experiment could also be done in coincidence with an energy sum spectrometer, to concentrate the observed population more on the highest spins. These correlation studies seem to offer real hope for a detailed understanding of higher spin states without resolving the spectrum.

7 CONCLUSIONS

Most nuclei apparently compromise between purely collective and purely noncollective behavior at very high spins. Noncollective behavior in nuclei has been seen only as high as $38\hbar$, at which point a more collective structure seems to develop (Section 5). These angular momenta are the very highest observed for discrete states, and even in these cases there is some weak collectivity. It would be interesting to understand better how the collectivity comes into these states and just what kind of collective motion is involved. On the other hand, with increasing spin in deformed nuclei, particles begin to decouple from the symmetry axis and align their angular momenta directly with that resulting from the collective rotation. The alignment of the first two particles (backbending) is well studied in many nuclei, and that of the second pair, in a few nuclei. In Section 6.3 it became apparent that this process continues up to 6 or 8 aligned particles, at which point a qualitative change occurs. It was suggested that this change might be the occupation of aligned $\Delta N = 2$ orbitals, but certainly other possibilities exist. Regardless of their initial nature, nuclei seem to move toward a structure where both collective and noncollective features exist at very high spins. The deformations involved at these highest spins seem from the moment of inertia values in Section 6.2 to be moderate, not much larger than those in deformed nuclei at low spins, though there are intriguing hints that it might sometimes become much larger at the very highest spins. The shapes are not known, though they seem likely to be triaxial—a compromise between oblate, favored by the noncollective motion, and prolate, favored by the collective motion. Overall it is an exciting time; we have learned a lot, but we can see that there is much more to understand.

On the experimental side, several new techniques have been invented and are being exploited to attack the unresolved γ-ray spectra from the highest spin states. These include the multiplicity filter and total-energy spectrometer discussed in Sections 6.1 and 6.2 and the correlation methods described in Section 6.3. One of the most important needs is to isolate narrower populations of very high spin states. The best possible so far is about 70% full width at half maximum (FWHM). Currently under construction are 4π NaI detectors with 100–200 elements, which will be able to select multiplicity (and therefore, in most cases, spin) distributions with a 20–25% FWHM. This will provide much sharper views of all high spin properties. Furthermore, present detectors must always be chosen either for good resolution, Ge(Li), or for good response function, NaI. Development is underway on detectors, liquid xenon, that combine both properties. Thus, in this field that has already yielded many new ideas and new techniques, there are certainly more to come.

ACKNOWLEDGMENTS

We wish to thank Dr. W. J. Swiatecki for many helpful comments. Discussions with a number of colleagues and visitors have been greatly appreciated. This work was supported by the Nuclear Science Division of the US Department of Energy under Contract No. W-7405-ENG-48.

Literature Cited

Åberg, S. 1978. *Nucl. Phys. A* 306: 89

Aguer, P., Bastin, G., Charmant, A., El Masri, Y., Hubert, P., Janssens, R., Michel, C., Thibaud, J. P., Vervier, J. 1979. *Phys. Lett B* 82: 55

Andersen, O.; Bauer, R., Hagemann, G. B., Halbert, M. L., Herskind, B., Neiman, M., Oeschler, H., Ryde, H. 1978. *Nucl. Phys. A* 295: 163

Andersen, O., Garrett, J. D., Hagemann, G. B., Herskind, B., Hillis, D. L., Riedinger, L. L. 1979. *Phys. Rev. Lett.* 43: 687

Andersson, C. G., Hellström, G., Leander, G., Ragnarsson, I., Åberg, S., Krumlinde, J., Nilsson, S. G., Szymanski, Z. 1978. *Nucl. Phys. A* 309: 141

Andersson, C. G., Larsson, S. E., Leander, G., Moller, P., Nilsson, S. G., Ragnarsson, I., Åberg, S., Bengtsson, R., Dudek, J., Nerlo-Pomorska, B., Pomorski, K., Szymanski, Z. 1976. *Nucl. Phys. A* 268: 205

Asaro, F., Perlman, I. 1953. *Phys. Rev.* 91: 763

Baktash, C., der Mateosian, E., Kistner, O. C., Sunyar, A. W. 1979. *Phys. Rev. Lett.* 42: 637

Banaschik, M. V., Simon, R. S., Colombani, P., Soroka, D. P., Stephens, F. S., Diamond, R. M. 1975. *Phys. Rev. Lett.* 34: 292

Bartholomew, G. A., Earle, E. D., Ferguson, A. J., Knowles, J. W., Lone, M. A. 1973. *Adv. Nucl. Phys.* 7: 229

Bengtsson, R., Larsson, S. E., Leander, G., Moller, P., Nilsson, S. G., Åberg, S., Szymanski, Z. 1975. *Phys. Lett. B* 57: 301

Bethe, H. A., Bacher, R. F. 1936. *Rev. Mod. Phys.* 8: 82

Blatt, J. M., Weisskopf, V. F. 1952. *Theoretical Nuclear Physics.* New York: Wiley

Blocki, J., Randrup, J., Swiatecki, W. J., Tsang, C. F. 1977. *Ann. Phys.* 105: 427

Blomqvist, J. 1979. *Proc. Symp. on High-Spin Phenomena in Nuclei, ANL/PHY-79-4.* Argonne Natl. Lab., Ill.

Bochev, B., Karamyan, S. A., Kutsarova, T., Subbotin, V. G. 1975. *Yad. Fiz.* 22: 665; Transl. *Sov. J. Nucl. Phys.* 22: 343

Bohr, A. 1951. *Phys. Rev.* 81: 134

Bohr, A. 1952. *Mat. Fys. Medd. Dan. Vid. Selsk.* 26: No. 14

Bohr, A., Mottelson, B. R. 1955. *Mat. Fys. Medd. Dan. Vid. Selsk.* 30: No. 1

Bohr, A., Mottelson, B. R., Pines, D. 1958. *Phys. Rev.* 110: 936

Bohr, A., Mottelson, B. R. 1974. *Phys. Scr.* 10A: 13

Bohr, A., Mottelson, B. R. 1975. *Nuclear Structure,* Vol. 2. Reading, Mass: Benjamin

Bohr, N. 1936. *Nature* 137: 344

Bohr, N., Kalckar, F. 1937. *Mat. Fys. Medd. Dan. Vid. Selsk.* 14: No. 10

Borggreen, J., Bjørnholm, S., Christensen, O., Del Zoppo, A., Herskind, B., Pedersen, J., Sletten, G., Folkmann, F., Simon, R. S. 1980. *Z. Phys.* In press

Casimir, H. B. G. 1936. *On the Interaction Between Atomic Nuclei and Electrons.* Prize Essay. Tweede, Haarlem: Teyler's

Cerkaski, M., Dudek, J., Szymanski, Z., Andersson, C. G., Leander, G., Åberg, S., Nilsson, S. G., Ragnarsson, I. 1977. *Phys. Lett. B* 70: 9

Cohen, S., Plasil, F., Swiatecki, W. J. 1974. *Ann. Phys.* 82: 557

Deleplanque, M. A., Byrski, Th., Diamond, R. M. Hübel, H., Stephens, F. S., Herskind, B., Bauer, R. 1978a. *Phys. Rev. Lett.* 41: 1105

Deleplanque, M. A., Husson, J. P., Perrin, N., Stephens, F. S., Bastin, G., Schück, C., Thibaud, J. P., Hildingsson, L., Hjorth, S., Johnson, A., Lindblad, Th. 1979. *Phys. Rev. Lett.* 43: 1001

Deleplanque, M. A., Lee, I. Y., Stephens, F. S., Diamond, R. M., Aleonard, M. M. 1978b. *Phys. Rev. Lett.* 40: 629

Deleplanque, M. A., Stephens, F. S., Andersen, O., Ellegaard, C., Garrett, J. D., Herskind, B., Fossan, D., Neiman, M., Roulet, C., Hillis, D. L., Kluge, H., Diamond, R. M., Simon, R. S. 1980. *Phys. Rev. Lett.* In press

der Mateosian, E., Kistner, O. C., Sunyar, A. W. 1974. *Phys. Rev. Lett.* 33: 596

Diamond, R. M., Stephens, F. S., Kelly, W. H., Ward, D. 1969. *Phys. Rev. Lett.* 22: 546

Døssing, T., Neergaard, K., Matsuyanagi, K., Chang, H. C. 1977. *Phys. Rev. Lett.* 39: 1395

Ericson, T. 1960. *Adv. Phys.* 63: 479

Faessler, A., Devi, K. R. S., Grummer, F., Schmid, K. W., Hilton, R. R. 1976a. *Nucl. Phys. A* 256: 106

Faessler, A., Ploszajczak, M., Devi, K. R. S. 1976b. *Phys. Rev. Lett.* 36: 1028

Feenstra, S. J., Ockels, W. J., van Klinken, J., deVoigt, M. J. A., Sujkowski, Z. 1977. *Phys. Rev. Lett. B* 69: 403

Feenstra, S. J., van Klinken, J., Pijn, J. P., Janssens, R., Michel, C., Steyaert, J., Vervier, J., Cornelis, K., Huyse, M., Lhersonneau, G. 1979. *Phys. Lett. B* 80: 183

Fenzl, M., Schult, O. W. B. 1975. *Z. Phys.* 272: 207

Fleury, A., Alexander, J. M. 1974. *Ann. Rev. Nucl. Sci.* 24:279
Fuchs, P., Emling, H., Folkman, F., Grosse, E., Schwalm, D., Simon, R., Wollersheim, H. J., Idzko, J., Pelte, D. 1977. *Jahresbericht.* Darmstadt, Germany: GSI
Goldhaber, M., Sunyar, A. W. 1951. *Phys. Rev.* 83:906
Goodman, A. L. 1979. *Adv. Nucl. Phys.* 11:263
Griffin, J. J., Rich, M. 1960. *Phys. Rev.* 118:850
Grover, J. R. 1967. *Phys. Rev.* 157:832
Grover, J. R., Gilat, J. 1967. *Phys. Rev.* 157:802, 814, 823
Gustafson, C., Lamm, I. L., Nilsson, B., Nilsson, S. G. 1967. *Ark. Fys.* 36:613
Haas, B., Andrews, H. R., Häusser, O., Horn, D., Sharpey-Schafer, J. F., Taras, P., Trautmann, W., Ward, D., Khoo, T. L., Smither, R. K. 1979. *Phys. Lett. B.* 84:178
Hagemann, G. B. 1979. See Blomqvist 1979, p. 55
Hagemann, G. B., Broda, R., Herskind, B., Ishihara, M., Ogaza, S., Ryde, H. 1975. *Nucl. Phys. A* 245:166
Häusser, O., Mahnke, H.-E., Sharpey-Schafer, J. F., Swanson, M. L., Taras, P., Ward, D., Andrews, H. R., Alexander, T. K. 1980. *Phys. Rev. Lett.* 44:132
Hillis, D. L., Garrett, J. D., Christensen, O., Fernandez, B., Hagemann, G. B., Herskind, B., Back, B. B., Folkmann, F. 1979. *Nucl. Phys. A* 325:216
Horn, D., Häusser, O., Faestermann, T., McDonald, A. B., Alexander, T. K., Beene, J. R., Herrlander, C. J. 1977. *Phys. Rev. Lett.* 39:389
Hübel, H., Smilansky, U., Diamond, R. M., Stephens, F. S., Herskind, B. 1978. *Phys. Rev. Lett.* 41:791
Huss, T., Zupancic, C. 1953. *Mat. Fys. Medd. Dan. Vid. Selsk.* 28: No. 1
Inglis, D. R. 1954. *Phys. Rev.* 96:701
Johnson, A., Ryde, H., Sztarkier, J. 1971. *Phys. Lett. B* 34:605
Johnson, A., Szymanski, Z. 1973. *Phys. Rep. C.* 7:181
Kerek, A., Kihlgren, J., Lindblad, Th., Pomar, C., Sztarkier, J., Walus, W., Skeppstedt, Ö., Bialkowski, J., Kownacki, J., Sujkowski, Z., Zglinski, A. 1978. *Nucl. Instrum. Methods* 150:483
Khoo, T. L., Bernthal, F. M., Robertson, R. G. H., Warner, R. A. 1976. *Phys. Rev. Lett.* 37:823
Khoo, T. L., Løvhøiden, G. 1977. *Phys. Lett. B* 67:271
Khoo, T. L., Smither, R. K., Haas, B., Häusser, O., Andrews, H. R., Horn, D., Ward, D. 1978. *Phys. Rev. Lett.* 41:1027 and private communication, 1980

Kohl, W., Kolb, D., Giese, I. 1978. *Z. Phys. A* 285:17
Körner, H. J., Hillis, D. L., Roulet, C. P., Aguer, P., Ellegaard, C., Fossan, D. B., Habs, D., Neiman, M., Stephens, F. S., Diamond, R. M. 1979. *Phys. Rev. Lett.* 43:490
Kutschera, W., Dehnhardt, D., Kistner, O. C., Kump, P., Povh, B., Sann, H. J. 1972. *Phys. Rev. C* 5:1658
Leander, G., Andersson, C. G., Nilsson, S. G., Ragnarsson, I., Åberg, S., Almberger, J., Døssing, T., Neergaard, K. 1979. See Blomqvist 1979, p. 197
Lieder, R. M., Ryde, H. 1978. *Adv. Nucl. Phys.* 10:1
Liotta, R. J., Sorensen, R. A. 1978. *Nucl. Phys. A* 297:136
Mang, H. J. 1975. *Phys. Rep.* 18:325
Mayer, M. G. 1949. *Phys. Rev.* 75:1969
Merdinger, J. C., Beck, F. A., Byrski, T., Gehringer, C., Vivien, J. P., Bozek, E., Styczen, J. 1979. *Phys. Rev. Lett.* 42:23
Mollenauer, J. F. 1961. *Lawrence Radiat. Lab. Rep. UCRL-9748*
Morinaga, H., Gugelot, P. C. 1963. *Nucl. Phys.* 46:210
Myers, W. D. 1973. *Nucl. Phys. A* 204:465
Neergaard, K., Pashkevich, V. V. 1975. *Nucl. Phys. A* 268:205
Neergaard, K., Pashkevich, V. V., Frauendorf, S. 1976. *Nucl. Phys. A* 262:61
Neergaard, K., Toki, H., Ploszajczak, M., Faessler, A. 1977. *Nucl. Phys. A* 287:48
Newton, J. O. 1974. In *Nuclear Spectroscopy and Reactions.* New York: Academic
Newton, J. O., Lee, I. Y., Simon, R. S., Aleonard, M. M., El Masri, Y., Stephens, F. S., Diamond, R. M. 1977. *Phys. Rev. Lett.* 38:810
Newton, J. O., Lisle, J. C., Dracoulis, G. D., Leigh, J. R., Weisser, D. C. 1975. *Phys. Rev. Lett.* 34:99
Newton, J. O., Sie, S. H. 1980. *Nucl. Phys. A* 334:499
Newton, J. O., Sie, S. H., Dracoulis, G. D. 1978. *Phys. Rev. Lett.* 40:625
Newton, J. O., Stephens, F. S., Diamond, R. M. 1973. *Nucl. Phys. A* 210:19
Nilsson, S. G. 1955. *Mat. Fys. Medd. Dan. Vid. Selsk* 29: No. 16
Nilsson, S. G., Prior, O. 1961. *Mat. Fys. Medd. Dan. Vid. Selsk.* 32 No. 16
Ockels, W. J., 1978a. Thesis, Univ. Groningen, The Netherlands
Ockels, W. J. 1978b. *Z. Phys. A* 286:181
Pedersen, J., Back, B. B., Bernthal, F. M., Bjørnholm, S., Borggreen, J., Christensen, O., Folkmann, F., Herskind, B., Khoo, T. L., Neiman, M., Pühlhofer, F., Sletten, G. 1977. *Phys. Rev. Lett.* 39:990
Ragnarsson, I., Nilsson, S. G., Sheline, R. K.

1978. *Phys. Rep.* 45:1
Rainwater, J. 1950. *Phys. Rev.* 79:432
Riedinger, L. L., Andersen, O., Frauendorf, S., Garrett, J. D., Gaardhøje, J., Hagemann, G. B., Herskind, B., Makovetsky, Y. V., Waddington, J. C., Guttormsen, M., Tjøm, P. O. 1980. *Phys. Rev. Lett.* 44:568
Rud, N., Ward, D., Andrews, H. R., Graham, R. L., Geiger, J. S. 1973. *Phys. Rev. Lett.* 31:1421
Sarantites, D. G., Barker, J. H., Halbert, M. L., Hensley, D. C., Dayras, R. A., Eichler, E., Johnson, N. R., Gronemeyer, S. A. 1976. *Phys. Rev. C* 14:2138
Sarantites, D. G., Hoffman, E. 1972. *Nucl. Phys. A* 180:177
Sarantites, D. G., Westerberg, L., Dayras, R. A., Halbert, M. L., Hensley, D. C., Barker, J. H. 1978. *Phys. Rev. C* 17:601
Simon, R. S., Banaschik, M. V., Colombani, P., Soroka, D. P., Stephens, F. S., Diamond, R. M. 1976. *Phys. Rev. Lett.* 36:359
Simon, R. S., Banaschik, M. V., Diamond, R. M., Newton, J. O., Stephens, F. S. 1977. *Nucl. Phys. A* 290:253
Simon, R. S., Diamond, R. M., El Masri, Y., Newton, J. O., Sawa, P., Stephens, F. S. 1979. *Nucl. Phys. A* 313:209
Simpson, J. J., Tjøm, O., Espe, I., Hagemann, G. B., Herskind, B., Neiman, M. 1977. *Nucl. Phys. A* 287:362
Sorensen, R. A. 1973. *Rev. Mod. Phys.* 45:353
Stephens, F. S. 1975. *Rev. Mod. Phys.* 47:43
Stephens, F. S., Simon, R. S. 1972. *Nucl. Phys. A* 183:257

Stephens, F. S., Ward, D., Newton, J. O. 1968. *Jpn. J. Phys. Suppl.* 24:160
Strutinsky, V. M. 1966. *Yad. Fiz.* 3:614; Transl. *Sov. J. Nucl. Phys.* 3:449
Thomas, T. D. 1968. *Ann. Rev. Nucl. Sci.* 18:343
Tjøm, P. O., Espe, I., Hagemann, G. B., Herskind, B., Hillis, D. L. 1978. *Phys. Lett. B* 72:439
Tjøm, P. O., Stephens, F. S., Diamond, R. M., de Boer, J., Meyerhof, W. E. 1974. *Phys. Rev. Lett.* 33:596
Trautmann, W., Proetel, D., Häusser, O., Hering, W., Riess, G. 1975. *Phys. Rev. Lett.* 35:1694
Trautmann, W., Sharpey-Schafer, J. F., Andrews, H. R., Haas, B., Häusser, O., Taras, P., Ward, D. 1979. *Phys. Rev. Lett.* 43:991
van der Werf, S. Y. 1978. *Nucl. Instrum. Methods* 153:221
Vivien, J. P., Schutz., Y., Beck, F. A., Bozek, E., Byrski, T., Gehringer, C., Merdinger, J. C. 1979. *Phys. Lett. B* 85:325
Wakai, M., Faessler, A. 1978. *Nucl. Phys. A* 307:349
Ward, D., Andrews, H. R., Geiger, J. S., Graham, R. L. 1973. *Phys. Rev. Lett.* 30:493
Westerberg, L., Sarantites, D. G., Geoffrey, K., Dayras, R. A., Beene, J. R., Halbert, M. L., Hensley, D. C., Barker, J. H. 1978. *Phys. Rev. Lett.* 41:99
Westerberg, L., Sarantites, D. G., Lovett, R., Hood, J. T., Barker, J. H., Currie, C. M., Mullani, N. 1977. *Nucl. Instrum. Methods* 145:295

Ann. Rev. Nucl. Part. Sci. 1980. 30:159–210
Copyright © 1980 by *Annual Reviews Inc. All rights reserved*

LARGE TRANSVERSE ⚹5616
MOMENTUM HADRONIC
PROCESSES

P. Darriulat
CERN, 1211 Geneva 23, Switzerland

CONTENTS

1 A HISTORICAL SURVEY

The decade just elapsed has witnessed drastic changes in our understanding of deep hadronic structure and of strong interaction dynamics. The study of the production mechanism of large transverse momentum hadrons in hadron-hadron collisions, hereafter abbreviated as LPTH (for large p_T hadrons), has been a major—if perhaps not spectacular—contributor to this evolution.

I introduce the subject with a historical survey of the progress achieved over the last decade. Being an experimentalist I discuss, among the theoretical ideas that blossomed in this field, only those that were

159

0163-8998/80/1201-0159$01.00

particularly efficient—owing to their strong predictive power—at motivating and emulating experimental efforts.

1.1 The Sources

In the mid-1960s, despite the many successes of the quark model (Dalitz 1965) in describing the spectroscopy of hadrons, their electromagnetic properties, and the symmetries of their weak decays, serious attempts to understand strong interactions in terms of quark dynamics were still discouraged by the apparent complexity of available experimental data. Possible connections between LPTH and the electromagnetic proton form factors had been discussed by Wu & Yang (1965) but they remained of a rather speculative nature.

It was not until 1969 that a spectacular breakthrough took place. Stimulated by recent electroproduction measurements in the deep inelastic region (Panofsky 1968, Breidenbach et al 1969), Bjorken (1969) formulated the scaling behavior of the nucleon structure functions, and Feynman (1969) proposed the parton model according to which a hadron, when viewed in a frame in which it has infinite momentum, is composed of independent point-like constituents (partons). An explicit parton model of the nucleon with "valence" quarks and "sea" quark-antiquark pairs was presented by Bjorken & Paschos (1969) and later supported by the results of neutrino experiments (Eichten et al 1973).

At the same time the CERN Intersecting Storage Rings (ISR) were coming in operation, offering a new possibility to probe short-range hadron interactions and to compare the results with those of deep inelastic leptoproduction (Berman & Jacob 1970).

The parton picture provided a framework in which to formulate efficiently such comparisons. The main properties of LPTH could be outlined independently from the exact nature of the partons and of their interactions: a systematic investigation along these lines was presented by Berman, Bjorken & Kogut in 1971.

1.2 LPTH and the Parton Model

In parton models (Berman, Bjorken & Kogut 1971, Ellis & Kislinger 1974) LPTH result from the binary collision of two incident partons which carry significant fractions of the colliding hadron momenta (Figure 1). The remaining partons act as spectators and final state interactions play a minor role, even when—as is the case for quark partons—their presence is necessary to redistribute quantum numbers among the collision products. Three ingredients are therefore necessary for a complete description of the collision process: the parton wave functions within the colliding hadrons, the scattering matrix describing the binary parton collision and the fragmentation functions of the scattered partons. The contribution of the hard scattering, which takes place over a very short time scale, factors out from that of the other, slower, processes.

The parton wave functions are assumed to obey Bjorken scaling: their dependence upon the incident hadron momentum P_{inc} and upon the longitudinal parton momentum k_L is reduced to a dependence upon the single scaling variable $x = k_L/P_{inc}$. Deep inelastic leptoproduction data

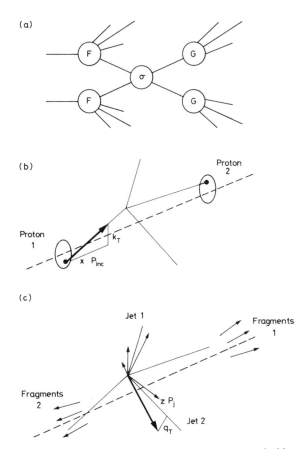

Figure 1   The naive parton model. (*a*) The hard scattering (σ) between two incident partons factors out from the structure functions (F) and from the fragmentation functions (G). (*b*) Initial state: the partons taking part in the hard scattering have momenta decomposed as xP_{inc} along the beam and k_T transversally. Scaling x-distributions and strongly damped k_T's are characteristic of the naive parton model. (*c*) Final state: the struck partons fragment into hadrons with momenta zP_j along the parton momentum and q_T transversally. Scaling z-distributions and strongly damped q_T's are characteristic of the naive parton model. The debris of the incident hadrons travel near the beam directions.

suggest that x distributions take the form $F(x)/x$ where the structure function $F(x) \to 0$ when $x \to 1$. The presence of x in the denominator corresponds to a smooth dependence upon rapidity. This variable, whose differential $dy = dp_L/E$ is a relativistic invariant against longitudinal Lorentz transformations, appears naturally whenever the problem under study does not impose the choice of a specific frame of reference (Bjorken 1971). Dependence upon k_T, the transverse parton momentum, is assumed to be strongly damped, with $\langle k_T \rangle \sim 300$ MeV/c in accordance with the observed properties of typical hadronic final states.

The parton fragmentation functions depend, of course, very much on the nature of the partons. In the case of quark partons, Berman, Bjorken & Kogut (1971) conjectured a behavior similar to that of the incident wave functions. The hadron fragments have limited transverse momenta with respect to their parent parton and carry a fraction z of its momentum; z distributions are independent of the actual value of the parton momentum and take a form $G(z)/z$, with $G(z)$ similar to $F(x)$.

As a result of these simple assumptions, fixed-angle inclusive cross sections exhibit scaling properties when expressed as functions of $x_T = 2p_T \cdot s^{-1/2}$ where p_T is the transverse momentum of the produced hadron and $s^{1/2}$ the total energy ($\simeq 2P_{\text{inc}}$ since all variables are referred to the c.m. system of incident hadrons). In particular, when the parton-parton interaction is mediated via the exchange of a photon or of a vector gluon, p_T and $s^{1/2}$ are the only momentum scales of the problem and the invariant cross section for production of a hadron h with 4-momentum (E, p) at angle θ reads

$$E \frac{d^3}{dp^3} \sigma(h_1 + h_2 \to h + \ldots) = s^{-2} f(x_T, \cos \theta) = p_T^{-4} g(x_T, \cos \theta).$$

Many outstanding properties of the final state structure proceed directly from the parton picture. In particular the large transverse momentum products are expected to appear along two directions coplanar with the incident beams (the momenta of the scattered partons), resulting in a "double-core" or "double-jet" pattern. While the jets are predicted to have opposite azimuths, their polar angles may differ widely because it is possible for the parton-parton system to have a large longitudinal momentum.

The spectator partons are expected to produce a set of hadrons with limited transverse momenta with respect to the incident beams and with smooth rapidity distributions.

1.3   The ISR Discovery: LPTH Within Reach

It was in such a context of ideas that three ISR groups (Alper et al 1973a, Banner et al 1973, Büsser et al 1973), while measuring single-particle inclusive cross sections in the large transverse momentum region, observed much larger yields than commonly expected from a naive extrapolation of the low p_T exponential behavior. This result came as a surprise: in spite of the success of the quark-parton model in describing deep inelastic leptoproduction results, hadronic processes were still considered by most physicists as belonging to a different chapter of physics with its own specific laws. Instrumentation at the CERN ISR and at the newly operational Fermilab synchrotron was indeed quite inadequate to tackle LPTH experiments. This lack of experimental preparation has been a severe handicap and it took several years until adequately instrumented detectors became available. The ISR discovery nonetheless demonstrated the feasibility of exploring the large p_T region with existing machines and immediately triggered a large experimental effort.

The data of Büsser et al (1973) are shown in Figure 2. Invariant cross

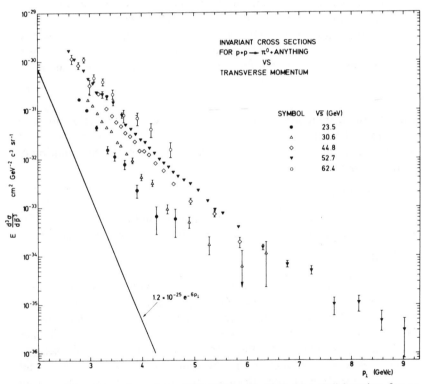

Figure 2 The pioneer data of the CERN–Columbia–Rockefeller Collaboration, first to extend the explored p_T scale up to 9 GeV/c, were presented at the 16th Int. Conf. on High Energy Physics (Chicago) in 1972. Invariant π^0 production cross sections are displayed vs p_T for different values of $s^{1/2}$ in the ISR energy range. The measured points fall orders of magnitude above the extrapolation of low p_T data (*solid line*). The increase with $s^{1/2}$ is well described by a scaling form in $p_T^{-8.24}$.

sections for the process $p + p \rightarrow \pi^0 + \dots$ are displayed versus p_T, the π^0 transverse momentum, for several values of the total c.m. energy $s^{1/2}$. The measured spectra are much less steep than exponential and are well described by a form (units are GeV and cm^2)

$$E \frac{d^3}{dp^3} \sigma = (1.54 \pm 0.10) \times 10^{-26} p_T^{-(8.24 \pm 0.05)} \exp \left[-(26.1 \pm 0.5)x_T \right].$$

While the observation of x_T scaling suggested the validity of the parton model in describing LPTH, the value $-(8.24 \pm 0.05)$ obtained for the exponent of p_T seemed to discourage simple interpretations in terms of point-like parton interactions.

1.4 Mesons as Partons

For several years the development of new models has been dominated by the impact of this important result. An immediate effect was to bring into fashion the constituent interchange model of Blankenbecler, Brodsky & Gunion (1972a,b, 1973a,b), which predicted scaling $\propto p_T^{-8}$. This prediction was based on the remark that the contribution of subprocesses such as

$$m + q \rightarrow m + q,$$

where m and q stand for meson and quark partons respectively, might be dominant in LPTH. The reason is that mesonic partons, say pions, would not be revealed in leptoproduction experiments where the contribution of subprocesses such as

$$m + e \rightarrow m + e$$

is strongly depressed by the presence of the meson form factor. In LPTH instead the interaction can take place via quark exchange, for example

$$\pi^+ + d \rightarrow \pi^0 + u.$$

A convenient framework in which to discuss such processes was provided by the dimensional counting rules (Brodsky & Farrar 1973, 1975, Matveev, Muradyan & Tavkhelidze 1973), which relate the inclusive cross section for the reaction $h_1 + h_2 \rightarrow h + \dots$ to the number n_a of "active" quark lines taking part in the hard scattering process and the number n_p of "passive" quark lines wasting momentum in the transitions between hadrons and quarks (Figure 3). According to these rules, the fixed-angle cross section takes the form

$$E \frac{d^3 \sigma}{dp^3} (h_1 + h_2 \rightarrow h + \dots) \propto (p_T^2)^{-(n_a - 2)} f(x_T)$$

with

$$\lim_{x_T \to 1} f(x_T) = (1 - x_T)^{(2n_p - 1)}.$$

A variety of other subprocesses have been considered, as in the quark-fusion model of Landshoff & Polkinghorne (1973a,b,c). The many successes of these approaches in describing the scaling behavior of LPTH cross sections and the hadron form factors have been reviewed by Sivers, Blankenbecler & Brodsky (1976). For a number of years one was led to believe that several subprocesses with $n_a \geq 6$ could be at play in LPTH, while the simple quark-quark subprocess ($n_a = 4$) would have played no role in the explored range of p_T and $s^{1/2}$.

1.5   The Structure of LPTH Final States

Experiments aimed at an analysis of the structure of LPTH final states were conducted at the CERN ISR as soon as it was realized that the relatively large LPTH cross sections made them feasible. It took, however, four years until a clear picture could emerge. Apart for the lack of adequately instrumented detectors, such experiments had to face two additional difficulties: the necessity to use very large transverse momentum triggers to differentiate between the spectators and the collision products directly involved in the hard scattering process, and the distortion of the event

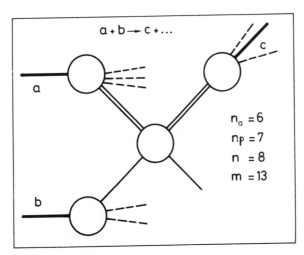

Figure 3   Illustration of the dimensional counting rules for the inclusive cross section $a + b \to c + \ldots$. Active quark lines (taking part in the hard scattering) are indicated as solid lines. Here, the hard-scattering term is the constituent interchange term $q + m \to q + m$ with $n_a = 6$. Passive quark lines are indicated as dashed lines; $n_p = 7$ in the example shown (5 from the initial state and 2 from fragmentation).

structure resulting from the use of single-particle triggers rather than jet triggers.

The first of these difficulties is inherent to LPTH and contrasts with the much cleaner situation in e^+e^- annihilation or leptoproduction experiments. For a produced particle to be unambiguously assigned to a large transverse momentum jet, its transverse momentum must exceed 0.5 GeV/c, or even better 1 GeV/c. This implies that soft jet fragments will never be uniquely identified as such and that very large transverse momentum jets are necessary to permit the observation of a reasonable fraction of their fragments. Experimenters have devoted much effort to constructing detectors providing larger and larger transverse momentum triggers: this has been a strong element of progress in the LPTH history. For the same reason the highest energy accelerators are best suited to LPTH studies, and I restrict the present review to Fermilab and ISR data.

The second difficulty can be overcome by using jet triggers rather than single-particle triggers. This is indeed what has been done at Fermilab in recent years using hadron calorimeters with much success. At the ISR, however, the implementation of a jet trigger is a difficult experimental problem. Hadron calorimeters have lower performance at smaller energies. Systems having the same transverse momenta have much larger laboratory energies at Fermilab than at the ISR. In addition none of the existing ISR detectors could conveniently accommodate such a calorimeter [it was not until 1979 that a detector (R807) optimized for LPTH studies could be installed at the ISR; it is expected to operate in 1980]. As a result ISR data suffer from the so-called trigger-bias effect (Bjorken 1974, Jacob & Landshoff 1976). Since the jet-production cross section has a much steeper p_T dependence than the fragmentation process, it is very likely for a particle of given p_T to be a leading fragment of a relatively soft jet. The resulting artificial asymmetry between the trigger jet and its partner, the "away-side jet," makes it difficult to perceive their similarities.

It is instructive at this stage to briefly review the main properties of the LPTH event structure as they were known to us toward the end of 1976, excellent reviews of which are available in the literature (Ellis & Stroynowski 1977, Jacob & Landshoff 1978).

Figure 4 shows the particle density phase-space distribution averaged over several LPTH interactions.

1. A large fraction of the particles produced are unaffected by the large transverse momentum process. They retain the main properties of "normal" events in which no large p_T particle is produced. However, their longitudinal momentum distribution does not exhibit the "diffraction peak" associated with some low multiplicity interactions. They are associated with the spectator fragments.

2. The large transverse momentum trigger particle is often accompanied by another particle, at small angle to it and having a transverse momentum distribution somewhat broader than in single-particle inclusive spectra. When expressed in terms of a correlation coefficient at fixed p_T, this corresponds to very large values (one or two orders of magnitude) because of the steep fall of the production spectra. The presence of strong $\pi^0 - \pi^0$ correlations indicates that resonances do not play a dominant role. The rapidity gap between the trigger and its companion decreases when the trigger transverse momentum increases, as though both particles were members of the same jet.

3. An increased particle density is observed in an azimuthal wedge opposite to the trigger particle over a wide range of polar angles. Their coplanarity with the incident protons and the trigger improves when the trigger transverse momentum increases, and they exhibit strong transverse momentum correlations with the trigger. Their jet structure is only

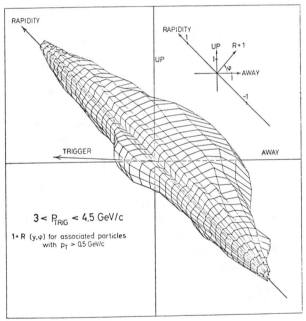

Figure 4   Phase-space density of LPTH final states obtained from actual data of the British–French–Scandanavian (BFS) Collaboration. Particle densities are shown in the rapidity-azimuth space; they are averaged over a large number of LPTH events and compared to "normal" events on which no special trigger requirement is imposed. The trigger jet is visible, despite the trigger-bias effect, in the vicinity of the trigger particle. The away-side jet is responsible for the broad enhancement at opposite azimuth (marked "away"). Owing to the averaging over several events, it spans a wide range of rapidity.

apparent when each event is considered separately, and for this reason was not immediately recognized. The jet axis spans a wide rapidity range, which broadens the particle density distribution when several events are considered together.

These features supported a qualitative interpretation in terms of a parton model: both jets had been identified and the remaining collision products could be assigned to the spectator partons. At the same time, progress had been reported in the study of hadronic final states of e^+e^- annihilations, revealing their jet structure (Hanson et al 1975, 1978). This encouraged many semiquantitative analyses of the LPTH results which strongly confirmed the validity of the parton picture (Bjorken 1975, Ellis, Jacob & Landshoff 1976, Ranft & Ranft 1976, Baier et al 1977, Kripfganz & Ranft 1977, and many others). Moreover quark-quark scattering was usually favored over quark-meson scattering. The dissimilarity observed between the trigger jet and the away-side jet was found consistent with the trigger-bias effect but could not be accounted for by constituent interchange models having a meson and a jet rather than two jets in the final state.

1.6   The Advent of Quantum Chromodynamics

In order to reconcile (a) the successes of the constituent interchange model in accounting for the scaling properties of inclusive cross sections and (b) the successes of the quark-parton model in describing the final state structure, several authors suggested that quark-quark scattering could indeed be the dominant subprocess if instead of being mediated via vector gluon exchange it were not scale free. In particular Hwa et al (1976, 1978) noted that the same effective quark-form-factor would account for the scaling violation measured in deep inelastic leptoproduction and for the p_T^{-8} dependence of LPTH cross sections (see also Contogouris & Gaskell 1977). At the same time the Caltech group (Feynman, Field & Fox 1977, Field & Feynman 1977, 1978) undertook a major phenomenological analysis of a vast amount of data (LPTH, leptoproduction, e^+e^- annihilation) following a similar approach. They postulated for the quark-quark scattering cross section an ad hoc form:

$$\frac{d\sigma}{dt} = \frac{2300\,\text{mb}\,\text{GeV}^6}{st^3},$$

where $s^{1/2}$ and $(-t)^{1/2}$ are the total energy and momentum transfer in the c.m. system of the two quarks. Such studies commonly reveal the need to allow for large internal transverse momenta in the quark wave function, up to about 1 GeV/c (Levin & Ryskin 1976, Della Negra et al 1977).

Since then, progress has been very fast, both on the experimental and on the theoretical fronts. A new generation of experiments has come to fruition at Fermilab and at the ISR. The results of two Fermilab experiments using hadron calorimeters have been extensively studied. It was noted very early (Bjorken 1973, Selove 1972) that the geometry of fixed-target machines is well suited for efficiently using hadron calorimeters to provide a bias-free trigger on a whole jet. The technique has now proven successful and has allowed for direct measurements of jet-production cross sections.

At the CERN ISR (Figure 5), experiments have been conducted with triggers requiring much larger transverse momentum than before, up to 16 GeV/c. As a result, the p_T^{-8} scaling law has been found unable to describe inclusive cross sections in the larger p_T region where lower powers (typically

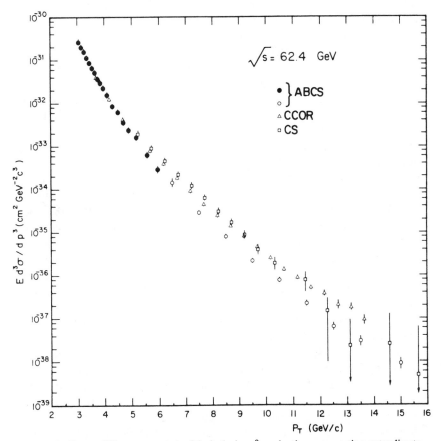

Figure 5   Recent ISR measurements of the inclusive π^0 production cross section, extending to $p_T = 16$ GeV/c, have been performed by three collaborations. The discrepancy observed between the three measurements may be explained by different calibrations of the p_T scale; however, the three experiments agree on the change of scaling behavior toward larger p_T (see Figure 7).

p_T^{-6}) are measured. One could then hope to start sensing the gluon exchange quark-quark diagram (Clark et al 1978a, Angelis et al 1978, Kourkoumelis et al 1979b). In addition adequately instrumented detectors have become available, opening the door to detailed measurements of the jet properties.

At the same time hadronic final states have been the subject of careful studies in neutrino interactions, deep inelastic leptroproduction, and e^+e^- annihilation data and have revealed the role played by gluons.

On the theoretical front the formulation of QCD as a gauge theory of strong interactions has given an unprecedented impetus to our understanding of their dynamics. The suggestion that gluons could play a dominant role in strong interaction dynamics, a suggestion supported by the observation in deep inelastic neutrino interactions of a missing longitudinal momentum to the quark partons (Eichten et al 1973, Deden et al 1975), was taken very seriously in 1973 when Politzer (1973) and Gross & Wilczek (1973) showed the existence of a class of Yang-Mills gauge theories that behave, up to calculable logarithmic corrections, as a free-field theory. It was then attractive to imbed strong interactions in such a theory based on colored quarks and colored octet-gluons (Fritzsch, Gell-Mann & Leutwyler 1973). It took some years for QCD calculations to show their ability to describe deep inelastic leptroproduction and e^+e^- annihilation data, during which LPTH experiments had become mature and could be considered in a more fundamental framework than that provided by the ad hoc approaches mentioned at the beginning of this section. Cutler & Sivers (1977) and Combridge, Kripfganz & Ranft (1977) realized that the one gluon exchange quark-quark term, once renormalized for QCD corrections, could at least partially account for the p_T^{-8} scaling law measured at the ISR below $p_T = 8$ GeV/c. Many subsequent QCD calculations have tackled seriously the problem of describing LPTH data: their successes make it reasonable to hope for a unified description of the various processes probing the deep hadronic structure. I comment briefly on these in Section 3 after having presented the main experimental results in Section 2. For now the following superficial remarks should be sufficient.

Inclusive cross sections are now understood over the whole explored range of p_T. If vector gluon exchange is restored as the basic quark-quark interaction, quark-gluon and gluon-gluon interactions play a dominant role over much of the p_T range. Gluon bremsstrahlung is an important ingredient and its relevance to the description of LPTH final states is essential. While the crudeness of perturbative QCD calculations and the lack of a precise knowledge of the gluon structure functions and fragmentation functions preclude very accurate predictions, it is commonly believed that all observed features of LPTH will soon be understood in the QCD framework. Constituent interchange processes, which are associated with higher-twist effects in QCD, can still play a significant role at moderate p_T.

2 EXPERIMENTAL RESULTS: A SUMMARY

It is outside the scope of this review to give a fair and complete account of all LPTH data. I have tried in the Appendix to provide the reader with a detailed list of published experimental results[1] including brief descriptions of the corresponding detectors. In this section I select a few, usually recent, experiments, that illustrate the present state of our knowledge. This section owes very much to the recent and excellent reviews of Di Lella (1979), Jacob (1979), and Selove (1979).

2.1 Single-Particle Inclusive Cross Sections

Single-particle inclusive cross sections have been the subject of numerous measurements both at the ISR and at Fermilab. ISR measurements benefit from the large available energy range, well adapted to scaling studies, but are restricted to p-p collisions. Fermilab experiments provide flexibility in the choice of the initial state and can be performed with relatively smaller solid-angle detectors, but they cannot reach such large energy values.

Very high values of p_T, up to 16 GeV/c, have been reached in three recent ISR measurements (CS, CCOR, ABCS) of the pp $\rightarrow \pi^0 + \ldots$ cross section in the central region. This channel has always been favored by ISR experimenters owing to the high performance of electromagnetic calorimeters (lead glass, and lead liquid argon sandwiches) in detecting π_0's over large solid angles with a good energy resolution and a high rejection power against other hadrons.

Earlier measurements of the π inclusive production cross section had been analyzed in terms of parton models in which ad hoc scaling violations, inspired from the results of deep inelastic leptoproduction data, had been introduced (Hwa, Spiessbach & Teper 1976, 1978, Feynman, Field & Fox 1977, Field & Feynman 1977, Contogouris, Gaskell & Nicolaidis 1978a, Fishbach & Look 1977a,b,c, Duke 1977).

The success of these approaches led Cutler & Sivers (1977) and Combridge, Kripfganz & Ranft (1977) to point out that a QCD calculation including all gluon corrections to the quark-quark Born term could yield the same result. The new ISR measurements allow us to test the validity of such QCD approaches in a much larger range of p_T. The calculations (Owens, Reya & Glück 1978, Contogouris, Gaskell & Papadopoulos 1978, Field 1978, Halzen, Ringland & Roberts 1978) include some or all of the following factors: quark-quark, quark-gluon, and gluon-gluon terms; parton transverse momentum k_T distribution; Bjorken scaling violation in

---

[1] In the course of the present section, reference is often made to experimental collaborations by means of their acronyms: the reader will find the corresponding references in the Appendix.

the fragmentation and structure functions. The structure functions and their scaling violations are taken from QCD analyses of leptoproduction data, but gluon distributions and parton transverse momentum distributions are educated guesses. The results (Figures 6a and 6b) show the role played by gluons in part of the p_T range. They are not very sensitive to the k_T distribution, in particular at larger p_T and $s^{1/2}$. The reasonable agreement achieved is very encouraging indeed.

It is now clear, as had been suspected long ago, that a scaling form of the type

$$I(x_T, p_T) \equiv E \frac{d^3\sigma}{dp^3} = p_T^{-n} f(x_T)$$

or

$$J(x_T, s) \equiv E \frac{d^3\sigma}{dp^3} = s^{-n/2} g(x_T)$$

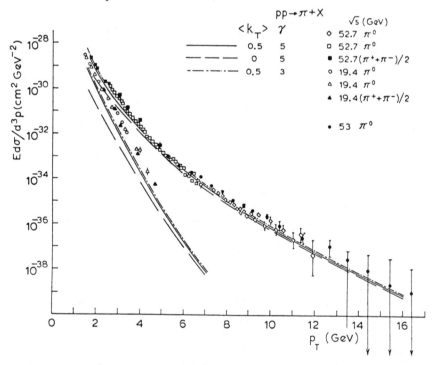

Figure 6a The smooth curves show a typical QCD calculation of inclusive π production (Contogouris, Gaskell & Papadopoulos 1978). The constituent internal transverse momentum, k_T, is illustrated for $\langle k_T \rangle = 0$ and 0.5 GeV/c. Gluon distributions are taken of the form $(1-x)^\gamma$.

is inadequate to describe the data over the whole explored range. In fact, the function

$$n(x_T, p_T) = -\frac{p_T}{I} \frac{\partial}{\partial p_T} I(x_T, p_T)$$

or

$$n(x_T, s) = -2\frac{s}{J} \frac{\partial}{\partial s} J(x_T, s)$$

decreases when p_T, x_T, and/or s increase (CCOR). The data suggest that it may approach 4, the perturbative value, as the collision gets harder (Figure 7).

Recent Fermilab experiments (CP, CLFCI) using an incident π^- beam illustrate well how informative can be the availability of a different initial state. The measurement is of the ratio of the inclusive cross sections for π^- and π^+ production in π^-p collisions. It is considered a crucial test of the constituent interchange model, which obviously predicts very large values

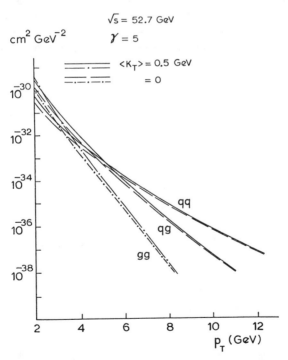

Figure 6b   The effects of the q-q, q-g, and g-g terms on inclusive π production (Contogouris, Gaskell & Papadopoulos 1978).

for this ratio. The data show instead a value close to unity in good agreement with QCD predictions (Figure 8). The π^+/π^- ratio has also been measured in p-p and p-n collisions (CP). As expected from most hard scattering models, it increases with x_T in p-p collisions and remains equal to unity in p-n collisions (Owens, Reya & Glück 1978).

The production of particles other than pions has been the subject of many measurements (CP, BS). These include stable particles ($K^\pm$, $p^\pm$, η) as well as resonances (ρ, ω, K^*). In particular the η/π^0 ratio has been measured up to $p_T = 11$ GeV/c and found equal to 0.55 over the whole range (ABCS).

Unfortunately most available data on particle ratios are restricted to a region of moderate p_T. It may well be that reliable QCD perturbative calculations cannot be performed for moderate values of p_T. In particular the trigger-bias effect would enhance minor contributions from higher-twist subprocesses (Blankenbecler, Brodsky & Gunion 1978, Jones & Gunion 1979), the QCD analogue of constituent interchange, with respect to quark fragmentation. In this context Gustafson & Månsson (1979) studied the scaling behavior of the cross-section difference between K^+ and K^- production (BS, CP) and found a $p_T^{-5.4}$ law. Since QCD scale-breaking effects are expected to partially compensate in the cross-section difference, they consider this support for quark-quark scattering.

Proton production data are especially disturbing: they are an order of

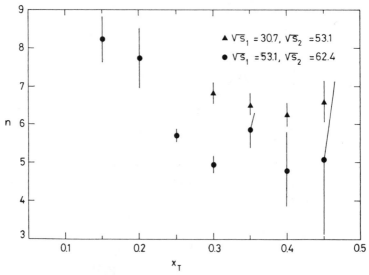

Figure 7 The effective scaling parameter $n(x_T, s)$ ha been evaluated by the CCOR Collaboration from their π^0 production data. It decreases when x_T and/or s increases and it approaches 4, the quark-quark term value, at large x_T and s.

magnitude above QCD estimates using fragmentation functions deduced from $e^+e^- \to p$... (Owens 1979a) while their scaling behavior is well described by the constituent interchange model. Even if this disagreement is nothing but an illustration of our inability to perform accurate QCD calculations, it deserves serious attention. Other modes of proton formation have been considered by Escobar (1979) and Chih Kwan Chen (1978). In general, baryon production is not well under control. Bjorken (1975) had already noted the relation

$$(p/K^+) \simeq 4(\bar{p}/K^-)$$

measured in p-p collisions at $p_T = 3$ GeV/c instead of the expected (p/K^+) $\simeq (\bar{p}/K^-)$.

A few measurements of single-particle inclusive production have been

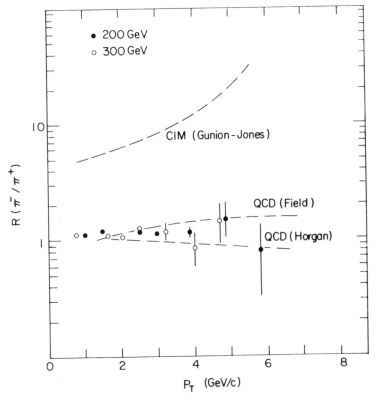

Figure 8   Chicago–Princeton data on the ratio between π^- and π^+ production in $\pi^- p$ collisions. The data rule out any significant constituent interchange contribution but support QCD expectations [model predictions as quoted by Jacob (1979) in his recent review].

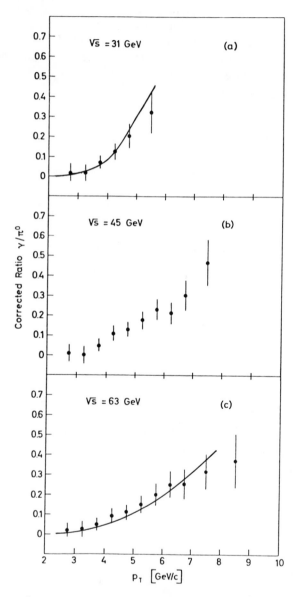

Figure 9   The γ/π^0 ratio as measured by the ABCS Collaboration at three ISR energies: (*a*) $s^{1/2} = 30.6$ GeV, (*b*) $s^{1/2} = 53.1$ GeV, (*c*) $s^{1/2} = 62.4$ GeV. The curves are QCD predictions (Halzen & Scott 1978a,b).

performed at forward angles (NN, CCHK, BCB). Their main conclusion is the validity of "radial" scaling: the angular dependence is accounted for by simply replacing x_T by $x_R = (x_T^2 + x_L^2)^{1/2}$ in the 90° data.

I conclude this brief review of inclusive cross-section data with direct photon production. This channel is of particular interest, and its study, which is just starting, should develop into a very rich source of information. Direct photon production is expected to be rather copious on very general grounds. Photons are electromagnetically coupled to quarks and their weaker coupling is compensated by the fact that they do not fragment. The γ/π ratio is expected to increase with x_T, following the decrease of the fragmentation function of quarks and gluons into a pion (Farrar & Frautschi 1976, Farrar 1977, Escobar 1975, 1977, Fritzch & Minkowski 1977).

In perturbative QCD the dominant source in p-p collisions is quark + gluon → quark + photon and reliable predictions can be made (Halzen & Scott 1978a,b, Llewellyn Smith 1978a, Rückl, Brodsky & Gunion 1978, Contogouris, Papadopoulos & Hongoh 1979, Frazer & Gunion 1979, Jones & Rückl 1979).

Three measurements of the γ/π^0 ratio (ABCS, CRB, FJH) have been published recently. It may be premature to formulate quantitative comments on these new results, but the important point is that a clear and unambiguous signal has been observed (Figure 9), a great advance over earlier inconclusive attempts (the identification of direct photons over a copious background of decay photons is not a trivial experimental problem). The large measured values (γ/π reaches 30% at $p_T = 7\,\text{GeV}/c, s^{1/2} = 62$ GeV) encourage undertaking dedicated experiments in which final states containing a large transverse momentum direct photon are selected; the angular dependence, the scaling behavior, and the correlation with other large transverse momentum products should be measured for different initial states. The absence of fragmentation and the impossibility of confusion with the spectator debris should make direct photons a most powerful probe of the deep hadronic structure and greatly simplify the interpretation of experimental results. This contrasts with inclusive hadronic production, which implicitly contains the whole complication of the LPTH mechanism.

2.2   The Final State Structure: Some General Features

An ideal bias-free LPTH experiment would use a trigger on the total transverse energy E_T generated during the collision. What is often done is to trigger instead on a single large transverse momentum particle, which results in a double bias: a distortion of the mode of fragmentation of the trigger jet and a transverse momentum imbalance between the trigger jet and its partner. In particular one may wonder whether the positive correlation (CCRS) measured between two azimuthally back-to-back particles is not a pure effect of energy momentum conservation. The ABCS Collaboration overcome this objection by presenting an elegant analysis of $\pi^0\pi^0$ azimuthal correlations that demonstrates the dynamical nature of the two-jet production mechanism. E_T is estimated from the missing transverse momentum to the π^0 pair; it must be noted that this is in fact an overestimate to the extent that some transverse energy is transferred to the spectators, and therefore irrelevant. The azimuthal opening $\Delta\phi$ of the pair is studied in fixed intervals of the estimated E_T. Its distribution exhibits two back-to-back peaks as soon as E_T exceeds $\sim 10\,\text{GeV}$ (Figure 10). Kripfganz & Schiller (1978) have shown that a rough QCD calculation using quark-quark scattering only and ignoring fragmentation gives a fair account of this result.

Another demonstration of the coplanarity of the large transverse momentum products is given by the CS Collaboration (Figure 11). They measured the density and momentum distributions of charged particles produced over a 30° azimuthal wedge under three different trigger

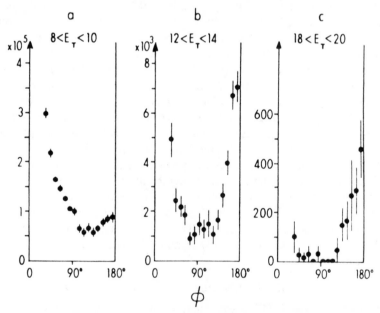

Figure 10   Azimuthal correlation between two large transverse momentum π^0's produced at the ISR, as measured by the ABCS Collaboration. Distributions in ϕ, the azimuthal opening of the pair, are shown for three intervals of the total transverse energy E_T (evaluated from the missing transverse momentum to the π^0 pair). A clear back-to-back structure emerges as soon as E_T exceeds ~ 10 GeV.

conditions, a π^0 with $p_T > 5$ GeV/c being detected either within the same wedge or at azimuthal differences of 90° or 180°. The data are compared with "normal" collisions, i.e. without any specific trigger requirement. At 90° out of the trigger plane no sign of the hard collision is visible.

We note that both experiments show no indication of 3-jet final states as could be expected from the qq → qqg term. Combridge (1978) has pointed out that the observation of such configurations may be very difficult in the geometry of existing ISR experiments with limited azimuthal coverage and specific trigger configurations. In particular a more favorable topology should be that of three coplanar jets, one on one side of the beams, the others at opposite azimuth and separated in rapidity.

The above examples provide good evidence for the coplanarity of the large transverse momentum products but are not sufficient to demonstrate the existence of a jet structure that implies, in addition, collimation in polar angle. Data of the CCOR Collaboration obtained with a solenoid spectrometer triggered on a large $p_T(>7$ GeV/c) π^0 provide a good illustration of the polar angle collimation first observed in other experiments (C, CCHK, BFS). The data (Figure 12) are for charged particles in the azimuthal hemisphere opposite to the trigger (away-side particles). No

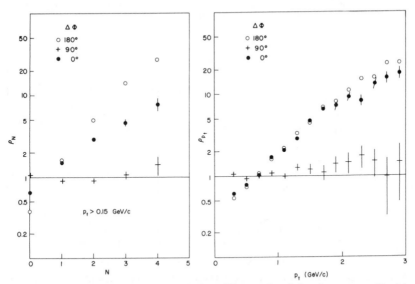

Figure 11 The structure of LPTH events in three different azimuthal wedges separated by $\Delta\phi$ from the large transverse momentum π^0 trigger, as studied by the CS Collaboration at the ISR. The data are normalized to "normal" collisions. The multiplicity (N) and transverse momentum (p_T) distributions out of the trigger plane are not affected. The trigger jet multiplicity is depressed relative to that of the away-jet by the trigger-bias effect.

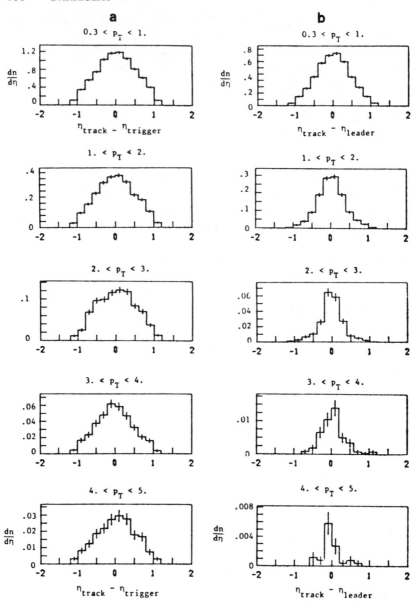

Figure 12 Data obtained by the CCOR Collaboration in the study of charged particles associated with a π^0 having more than 7 GeV/c transverse momentum and produced in the opposite azimuthal hemisphere. (*a*) Rapidity distributions are referred to the π^0 rapidity. The away-jet structure is washed out by the wide rapidity span of the away jet. (*b*) Rapidity distributions are referred to the rapidity of the away particle of highest momentum (leader). The away-jet structure is now clearly evidenced, the collimation increasing with the transverse momentum p_T of the charged particle.

angular correlation is observed with the trigger, which belongs to the other jet, but a clear correlation with the away-side particle having the largest transverse momentum (the leader) is in evidence. The jet collimation is observed to increase with the leader transverse momentum indicating that jet fragmentation proceeds with limited transverse momentum with respect to the jet axis.

The general features of the event structure have been successfully reproduced by a number of phenomenological analyses based on the parton model (see Section 1.5). The introduction of scaling violation in the structure functions (Contogouris & Gaskell 1977, Contogouris, Gaskel & Nicolaidis 1978b) and of QCD effects (Feynman, Field & Fox 1978) improve the agreement with the data.

Having briefly illustrated the main features of LPTH events I now turn to a more quantitative analysis of their properties.

2.3   Jet Fragmentation

Let us first consider internal jet properties depending only upon their mode of fragmentation, not of formation. The away-side jet, partner of the trigger jet, is often best suited for such studies. It is free of trigger bias, while in the trigger jet resonance production is enhanced and may produce large effects even if it corresponds to a small branching fraction.

Global information on the fragmentation mechanism is provided by Fermilab measurements (CLFCI, FLPW) of the ratio of jet to single-particle cross section. This is a large quantity, typically two orders of magnitude, increasing with x_T. It is particularly sensitive to the detailed behavior of the fragmentation function near $z = 1$. The data (Figure 13) are in good agreement with the results of a QCD calculation.

Measurements of the jet multiplicity, including both neutral and charged fragments, have been performed at the CERN ISR (CS). They are compared in Figure 14 with data from e^+e^- annihilation and νp interactions. A similar dependence upon energy, approximately logarithmic, is observed in the three cases.

It is experimentally difficult to identify all particles in an LPTH final state: information on particle ratios within a jet is rather crude. The positive-to-negative-charge ratio, R, has been measured in several experiments and observed to increase with z; charge correlations among the jet fragments have also been studied. In the absence of correlation, the charge combinations $(+ +), (\pm \mp), (- -)$ would be populated in the proportion $R^2 : 2R : 1$. The ratio $\rho = (\pm \mp)/[2\sqrt{(+ +)(- -)}]$ is therefore a measure of the charge correlation, $\rho = 1$ corresponding to no correlation. For the pair of leading charged fragments in each jet the values $\rho = 1.52 \pm 0.05$ and $\rho = 1.60 \pm 0.03$ have been reported for the trigger jet and away-side jet respectively. This observation that opposite charge pairs are favored in the fragmentation chain agrees with any model in which hadronization proceeds via production of successive q\bar{q} pairs (Field & Feynman 1978, Andersson, Gustafson & Peterson 1978b) and suggests similarities between the quark and hadron fragmentation mechanisms (Hoyer et al 1979, Anderson, Gustafson & Peterson 1977, 1978a, Hayot & Jadach 1978).

The BFS Collaboration has performed extensive studies of trigger jets in which the trigger particle is identified. The mean associated momentum for charged secondaries is small because of the trigger-bias effect. Its dependence upon p_T is shown in Figure 15. Such jet configurations are well suited

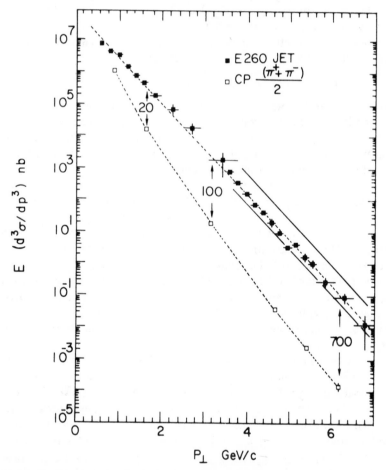

Figure 13 The inclusive jet-production cross section, as measured by the CLFCI Collaboration, is compared to inclusive single π production. The solid lines correspond to two different QCD predictions [quoted by Jacob (1979) in his recent review].

to study the role of resonances, which are relatively enhanced. Data on ρ, K*, and Δ production have been reported.

Having reviewed some global properties of the fragmentation mechanism, let us consider in more detail the behavior of the fragmentation functions. It is customary to relate the fragment momenta to the total jet momentum $\mathbf{P_J}$:

$$\mathbf{p}_i = z_i\ \mathbf{P_J} + \mathbf{q}_{Ti}, \qquad \mathbf{q}_{Ti} \cdot \mathbf{P_J} = 0.$$

The evaluation of the total jet momentum, and therefore of z and $\mathbf{q_T}$, involves an estimate of the contribution of soft fragments, which may be confused with the spectator debris: the higher P_{TJ}, the smaller the corresponding correction and the more confidence in the calculation. The

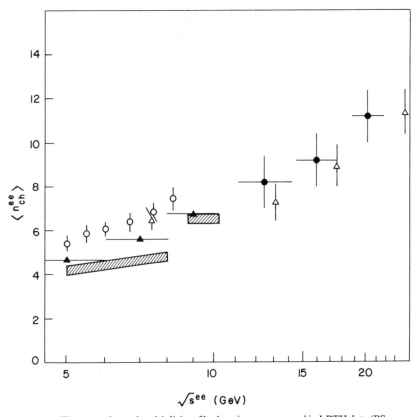

Figure 14   The mean charged multiplicity of hadron jets as measured in LPTH data (BS: *open circles*, CS: *full circles*), in neutrino-proton interactions (*full triangles*, Van der Welde 1979), in e^+e^- annihilations at Spear and Doris (*crosshatched zones*) and at PETRA (*open triangles*, Wolf 1979). All data are adjusted to the equivalent e^+e^- energy.

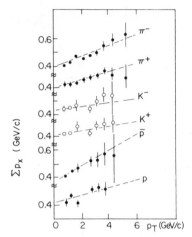

Figure 15 Total transverse momentum of charged secondaries in trigger jets with an identified single-particle trigger of transverse momentum p_T (data from the BFS Collaboration).

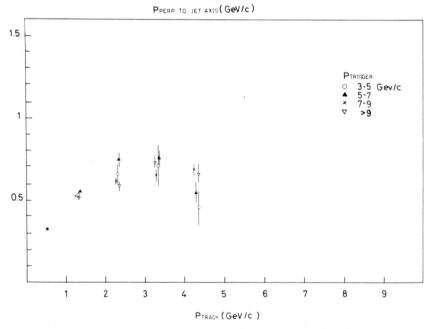

Figure 16 The mean transverse momentum distribution of jet fragments with respect to the jet axis as measured by the CCOR Collaboration for different values of the trigger transverse momentum.

CCOR and CS Collaborations report

$$\langle q_T \rangle = 0.55 \pm 0.05 \text{ GeV}/c$$

for P_{TJ} in the range between 5 and 10 GeV/c (Figures 16 and 17). The CCOR analysis also includes a clear demonstration of the axial symmetry of the jet around its axis.

From QCD arguments we should expect two kinds of jets, quark jets and gluon jets, with two different modes of fragmentation. On very general grounds (relative to quarks, the higher color charge carried by gluons, and the importance of the 3-gluon vertex), theoretical studies of gluon jet fragmentation indicate that they should generate high multiplicity final states with soft fragments and no leading flavor (Brodsky, de Grand & Schwitters 1978, Konishi, Ukawa & Veneziano 1979a,b, Einhorn & Weeks 1978, Kane & Yao 1978, Shizuya & Tye 1978, 1979). We should also expect $\langle q_T \rangle$ (or, better, $\langle q_T^2 \rangle$) to increase with P_J and develop a high q_T tail. A quark jet should even occasionally be accompanied by a radiated gluon jet. None of these is apparent in the data, which are not accurate enough and do

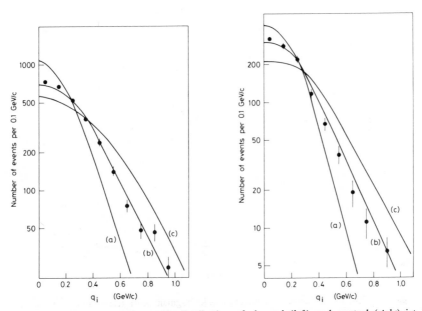

Figure 17 Transverse momentum distribution of charged (*left*) and neutral (*right*) jet fragments with respect to the jet axis (CS Collaboration). Rather than q_T, it is its component on the plane of the beams, $\tilde{q}_T$, that is measured. Solid lines correspond to Gaussian distributions corrected for detector acceptance and having $\langle \tilde{q}_T^2 \rangle^{1/2}$ = (*a*) 0.3, (*b*) 0.45, and (*c*) 0.60 GeV/c. Using the CCOR result on the isotropy of the jet fragmentation around its axis, these data correspond to $\langle q_T \rangle \simeq 0.55$ GeV/c.

not span a large enough P_J range to reveal such details. It is nonetheless interesting to compare the LPTH value of $\langle q_T \rangle$—even if it only contains crude information—with the equivalent quantity measured in e^+e^- annihilation and in νp interactions (Figure 18). Hadron jets from these three different sources are observed to exhibit very similar transverse fragmentation once soft fragments are ignored.

The z dependence of fragmentation functions has been the subject of many measurements. However, the z (of away-side fragments) is often referred to the trigger transverse momentum rather than to P_J, in which case it is usually called x_e. This introduces additional complications involving the transverse momentum correlation between the two jets (see Section 2.4). To a good approximation, z distributions can be described by an exponential form with the same slope as for e^+e^- jets (Figure 19).

An essential feature of the fragmentation function is its approximate

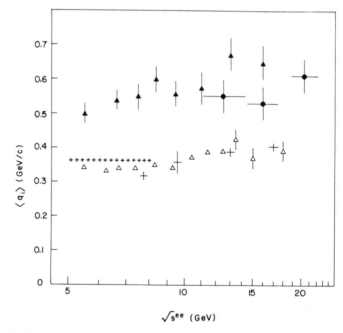

Figure 18   The mean transverse momentum of jet fragments with respect to the jet axis. All data are adjusted to the equivalent e^+e^- energy $(s^{ee})^{1/2}$. LPTH data $(p_T > 800 \text{ MeV}/c)$ are shown as black circles (CS). The e^+e^- data (*crosses*) and neutrino data (*open triangles*) are for all fragments. Selecting the harder fragments (Scott 1978) brings the neutrino data (*full triangles*) in agreement with LPTH data. Similarly, for e^+e^- data measured at PETRA (Söding 1979), the selection $z > 0.2$ causes $\langle q_T \rangle$ to increase to about 0.45 GeV/c at $s^{1/2} \simeq 17$ GeV.

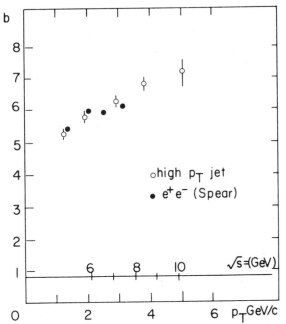

Figure 19 The slopes *b* obtained from exponential fits to the jet fragmentation function in the interval $0.2 < z < 0.8$ in e^+e^- annihilation (*full circles*) and LPTH data of the BS Collaboration (*open circles*).

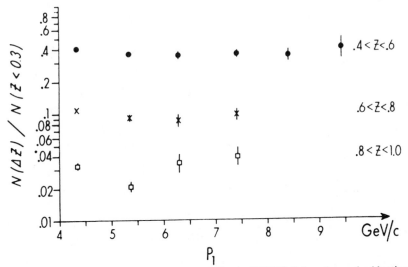

Figure 20 Normalized *z* distributions measured by the ABCS Collaboration and evidencing the scaling behavior of the fragmentation function: the fractional contribution of given *z* intervals is independent of the trigger momentum.

scaling behavior, namely its independence from P_J. Evidence that this is indeed the case at large enough P_J is overwhelming (Figure 20). A summary of LPTH data compared with e^+e^- annihilations and vp interactions is presented in Figure 21.

2.4 Jet-Jet Correlations

Correlation measurements between the two jets should provide information on the hard scattering subprocess and on the wave functions of the interacting partons, independently from the fragmentation mechanism.

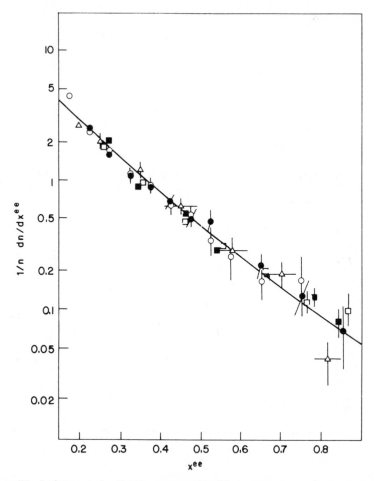

Figure 21   Jet fragmentation functions measured in different processes: v-p interactions (*open triangles*, Van der Welde 1979); e^+e^- annihilations (*solid line*, Hanson et al 1975); and pp collisions (*full circles* CS, $p_T < 6$ GeV/c, *open circles* CS, $p_T > 6$ GeV/c, *full squares* CCOR, $p_T > 5$ GeV/c, *open squares* CCOR, $p_T > 7$ GeV/c).

Quark-quark scattering should produce two jets having uncorrelated charges. This is indeed what is observed in the highest available range of transverse momentum: the charge correlation coefficient ρ defined in Section 2.3 has been measured between the leading charged fragments of jets produced in p-p collisions and found equal to 1.00 ± 0.03 (CS).

At lower transverse momenta some charge compensation between the trigger and the away-jet is however still present (BFS). This again indicates that constituent interchange may play a significant role at moderate p_T (Chase 1978a). It is worth noting that a measurement of ρ in e^+e^- annihilations at PETRA and PEP should reveal the quark-antiquark nature of the final state; to my knowledge this has not yet been observed.

Direct information on the pion structure functions is provided by a Fermilab experiment (FLPW) that compares the jet production cross sections for p-p and π-p collisions. The ratio between these two quantities is displayed in Figure 22 together with the similar ratio measured for inclusive

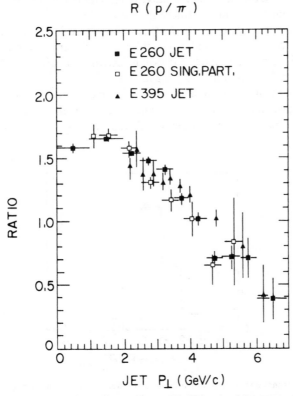

Figure 22   Fermilab data on jet production at 90°. The ratio of the cross sections $p + p \rightarrow$ jet $+ \ldots$ to $\pi + p \rightarrow$ jet $+ \ldots$ is displayed vs the jet transverse momentum, evidencing the increased pion efficiency at larger values of p_T.

pion production. It shows that pions are more efficient than protons at producing large transverse momentum jets: pion constituents carry a larger momentum fraction than proton constituents do. This interpretation is confirmed by the angular dependence of the cross-section ratio, which decreases in the forward direction.

Much effort has been devoted to collecting information on the transverse motion of the colliding partons. The observation that internal transverse momenta k_T of the order of 1 GeV/c were necessary was an outstanding conclusion of early phenomenological analyses.

In the QCD language one is led to distinguish between the "primordial" k_T (associated with the hadron size) and the k_T component acquired from gluon radiative corrections when p_T increases. Their distributions have been the subject of many theoretical studies, usually in connection with large p_T production of massive lepton pairs in hadron collisions. The situation is somewhat controversial and is commented upon in Section 3:3. For the time being I shall retain the naive parton model definition of k_T as a convenient means to compare different data.

Direct measurements of the transverse momentum imbalance between the two jets have been reported at Fermilab (FLPW) and at the ISR (CS); they indicate an approximately Gaussian distribution with ~ 1.2 GeV/c half width at half maximum (Figure 23).

In experiments using a single particle as a large transverse momentum trigger, the distribution of the momentum component of the away-jet fragments normal to the trigger plane, called p_{out}, is customarily measured.

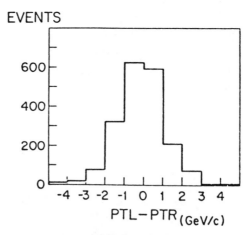

Figure 23 The transverse momentum imbalance between two 90° jets produced in p-p collisions (FLPW). Similar data have been obtained by the CS Collaboration at the ISR. The observed width corresponds to an effective k_T with a mean value of about 1 GeV/c.

The distribution depends not only upon the transverse motion of the incident partons but also on the transverse fragmentation of the away-jet. In fact $\langle p_{out} \rangle$ is expected not only to increase trivially with z, the momentum fraction of the away-jet carried by the fragment, but also with the total jet momentum because of the increased gluon radiation. This is indeed what is observed (Figure 24) again yielding $\langle k_T \rangle$ values of the order of 1 GeV/c. Similar values are deduced from the analysis of dilepton production in p-p collisions. I return to this result in Section 3.3.

Additional information on jet-jet correlations is provided by measurements of the inclusive cross section for production of symmetric pion pairs in p-p collisions (CFS, CCOR). The restriction to pions having approximately opposite transverse momenta simplifies the interpretation of the data (Baier, Cleymans & Petersson 1978): the role played by k_T is reduced. Fermilab data on $\pi^+ \pi^-$ pairs are compared to the result of a QCD calculation in Figure 25a. The same calculation, extended to ISR energies, is compared in Figure 25b to similar data for π^0 pairs. The agreement is quite encouraging.

2.5 The Spectators

The debris of the colliding hadrons, after amputation of the two partons taking part in the hard scattering, populate the large rapidity regions. Their observation in the vicinity of the incident beam (or beams) is not always easy. Several interesting results have nontheless been collected and deserve brief mention.

The density of spectator fragments decreases with increasing transverse

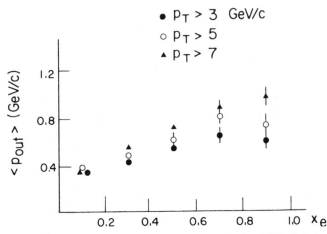

Figure 24   The dependence of $\langle p_{out} \rangle$ upon x_e as measured by the CCOR Collaboration for different values of the trigger transverse momentum p_T.

momentum of the trigger, more rapidly than could be expected from pure energy-momentum conservation. They are observed to carry a transverse momentum opposite to that of the trigger, typically 100 MeV/c per particle (BFS). This provides additional information on the transverse motion of incident partons and yields an independent measurement of the transverse momentum imbalance between the two jets. Fontannaz, Pire & Schiff (1978), using diquark fragmentation functions obtained from deep inelastic leptoproduction data (in the target fragmentation region), show that the BFS results are consistent with the following simple picture: one of the proton valence quarks is kicked out at large p_T and the left-over diquark

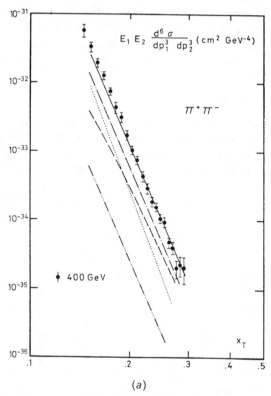

(a)

Figure 25   Invariant cross sections for the production of large mass pion pairs: (a; above) $\pi^+\pi^-$ from the CFS Collaboration (b; *right*) $\pi^0\pi^0$ from the CCOR Collaboration. The solid lines are the results of a QCD calculation by Baier, Engels & Petersson (1979). The contributions of various terms are indicated in (a): $q+q \rightarrow q+q$(----), $q+g \rightarrow q+g$(-·-·-), $g+g$ $\rightarrow g+g$(······), and $g+g \rightarrow q+\bar{q}$(---·---·).

travels along and fragments according to the same mechanism as in leptoproduction.

Data taken with a forward large transverse momentum trigger provide a means to identify the incident hadron from which the trigger jet originates (CCHK). They show strong correlations between the trigger and its parent proton but none with the other incident proton. Moreover these correlations, when studied for different types of trigger particles, are well understood from simple dimensional-counting arguments. The distribution of charged particles in reduced longitudinal momentum x_L is well described by a power law $(1-x_L)^n$ with n as predicted by dimensional counting rules (Drijard et al 1979).

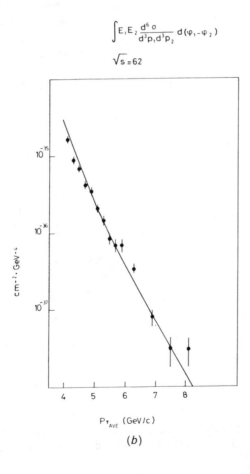

$$\int E_1 E_2 \frac{d^6 \sigma}{d^3 p_1 d^3 p_2} d(\varphi_1 - \varphi_2)$$

$$\sqrt{s} = 62$$

(b)

3   LPTH AND OTHER DEEP HADRONIC PROBES

3.1   *Future Prospects*

In the course of the preceding section we had many opportunities to compare LPTH data either with the results of perturbative QCD calculations or with measurements performed in e^+e^- and νp experiments. Emphasis has been put on the successes of the QCD approach, at the risk of appearing somewhat biased in its favor. The motivation in doing so was not to oversimplify a rather complex experimental reality. But the difficult LPTH experiments, with complicated multibody final states, require guidance to suggest relevant measurements and efficient modes of data presentation. Such a guidance is presently best provided by the QCD approach, which is the most successful at describing existing data and possesses a strong predictive power. The hope of reaching in the QCD framework a coherent and unified description of strong interaction dynamics is an important stimulus to encourage further experimental efforts.

There are many future LPTH prospects. The use of higher performance

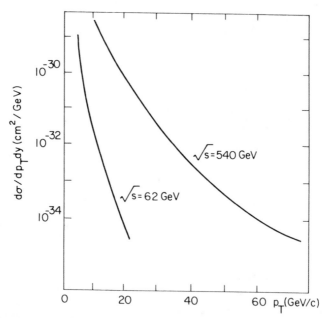

Figure 26   Anticipated jet yields in p$\bar{\text{p}}$ collisions at ISR, $s^{1/2} = 62$ GeV, and at the collider, $s^{1/2} = 540$ GeV as calculated in the QCD leading log approximation by Horgan (quoted by Jacob 1979).

detectors, such as the axial field magnet at the ISR, should help refine the present picture. Whenever possible experiments testing a specific aspect of the interaction should be preferred over those involving its whole complexity. Section 2 gave us many examples of that kind: ratios of inclusive cross sections measured either with different beams or for different types of produced particles, jet fragmentation studies without reference to the mode of formation, measurements of the transverse momentum imbalance between the two jets independently from their mode of fragmentation, etc. A particularly rich source of information, as yet almost unexploited, is the study of final states containing a large transverse momentum direct photon, the analysis of which should be greatly simplified.

Improvements of the existing machines, namely the availability of pp̄ collisions at the ISR and the increase of the Fermilab energy will help in successfully achieving these tasks.

In addition, a new unexplored energy domain will be within reach when the SPS pp̄ collider in Europe and Isabelle and the Fermilab collider in the United States begin to operate. Anticipated jet yields are very large at such machines (Figure 26) and spectacular effects are expected. The availability of measurable cross sections up to transverse momenta of the order of 50–100 GeV/c will allow one to probe hadronic structure much deeper than was previously possible. Even in the small x_T region, accessible with low luminosities, a copious production of gluon jets is expected.

3.2 Deep Hadronic Probes in the Naive Parton Model

Experience demonstrates that for LPTH experiments to progress they must be constantly confronted with the results of other measurements using different probes: e^+e^- annihilation, deep inelastic leptoproduction, Drell-Yan dilepton production (Drell & Yan 1970, 1971). In the preceding section such comparisons were made whenever possible and striking similarities were noted. It is instructive at this point to briefly assess the merits of LPTH in this context. A crude investigation in the framework of a naive quark parton model will help in drawing the general lines. The more realistic QCD picture is treated in the next subsection.

Four processes are considered in Table 1: (a) e^+e^- annihilation into hadrons; (b) Drell-Yan production of lepton pairs; (c) leptoproduction experiments in the deep inelastic region (with incident neutrinos, electrons, or muons); (d) LPTH. In terms of naive quark diagrams, a unified description of these processes involves three independent quantities: the structure functions describing the quark content of hadrons, the fragmentation functions describing the hadron content of quarks, and the scattering amplitudes of colliding quarks. The latter can only be inferred from (d) and

Table 1 Deep hadron probes

Process	Naive parton model	Some relevant QCD terms (excluding scaling violations)
a) e^+e^- annihilation		$\gamma \rightarrow q\bar{q}$ (leading) $\gamma \rightarrow$ onium $\rightarrow 3g$ $\gamma \rightarrow q\bar{q}\, g$
b) Drell-Yan pairs		$q\bar{q} \rightarrow \gamma$ (leading) $q\bar{q} \rightarrow \gamma g$ $qg \rightarrow q\gamma$
c) Deep inelastic leptoproduction (γ, Z, W)		$\gamma q \rightarrow q$ (leading) $\gamma g \rightarrow q\bar{q}$
d) LPTH		$qq \rightarrow qq$ $qg \rightarrow qg$ $gg \rightarrow gg$ $gg \rightarrow q\bar{q}$ constituent interchange $q+g \rightarrow q+\gamma$ (direct photons)

are irrelevant to the other processes. However, the former are best measured in (a) and (b), to which they contribute uniquely. Disentangling quark-quark scattering amplitudes from (d) is therefore likely to require prior knowledge of the structure and fragmentation functions obtained from simpler processes. It is also apparent from the naive quark picture that the richer information provided by (d) is obtained at the price of a higher complexity of the final state. While only two colored jets have to annihilate their unphysical quantum numbers in the first three processes, four colored jets are present in (d), thereby requiring a more complex set of final state interactions.

In this too-naive picture LPTH appears as the process involving the greatest complication, but it provides some information unobtainable from other processes. Similar conclusions apply when considering its role in the QCD framework.

3.3   Deep Hadronic Probes in the QCD Framework

A first, major, question to answer is to what extent does QCD preserve the validity of the naive parton model picture: does the general LPTH event structure persist, does it still make sense to talk about coplanar large transverse momentum jets, about fragmentation and structure functions, about spectator fragments, etc? This has been the subject of intense theoretical studies during the past three years, which have converged to a set of very important results. A first step was to recognize that the Q^2 dependence of the moments of the structure functions measured in leptoproduction experiments is amenable to a simple parton model interpretation (Altarelli & Parisi 1977, Kim & Schilcher 1978, Abad & Humpert 1978). The effects of gluon corrections, such as gluon bremsstrahlung from a quark parton, factor out from the hard scattering between the target and the virtual probe (mass squared $= Q^2$). This results simply in a redefinition of the structure functions, which acquire a Q^2-dependence inducing scaling violation, but it preserves the naive parton-model picture. Politzer (1977a) suggested that such a factorization mechanism could occur for processes more complicated than leptoproduction (for which the hadron structure is probed directly by a local current). His conjecture was confirmed in particular for quark-quark scattering (Sachrajda 1978a, Furmanski 1978) and for the Drell-Yan process (Politzer 1977b, Sachrajda 1978b). The general proof was soon established (Ellis et al 1979, Amati, Petronzio & Veneziano 1978a,b, Libby & Sterman 1978a,b) that, for any hard process which admits a parton model interpretation, the effects of QCD interactions to all orders in perturbation theory are simply absorbed in a universal renormalization of the structure and fragmentation functions. In the meantime, different theoretical approaches (Sterman & Weinberg

1977, Wosiek & Zalewski 1979, Furmanski 1979, Llewellyn Smith 1978b, Dokshitzer, D'yakonov & Troyan ˙1978a,b) had pointed to similar conclusions.

The factorization of the hard scattering subprocess preserves the validity of the parton model picture. In particular, for LPTH, we need only use the running coupling constant $\alpha_s(Q^2) \propto (\log Q^2/\Lambda^2)^{-1}$, $\Lambda \sim 500$ MeV, and the Q^2-dependent structure and fragmentation functions measured in leptoproduction and e^+e^- annihilation experiments. This is precisely the approach followed in most of the QCD calculations mentioned in the preceding sections and with which LPTH data were confronted.

We may now return to our evaluation of the role of LPTH among other deep hadronic probes and feel justified in continuing to use the parton model language.

A first remark concerns the hard scattering subprocess: in addition to quark-quark scattering, quark-gluon and gluon-gluon scattering contribute to leading order (Cutler & Sivers 1978, Hagiwara 1979). For each of these terms we need to know the Q^2-dependent structure and fragmentation functions. Factorization does not imply that the structure function effectively measured in a given hard scattering process can be blindly plugged in a calculation involving another hard scattering process. Gluons, sea, and valence quarks of various flavors may play a different role in the two cases and, since each of these is expected to have different structure functions, with different Q^2 dependences, they will combine differently in the two processes. This trivial remark, which of course applies as well to fragmentation functions, indicates potential difficulties in comparing the reactions listed in Table 1.

In particular, gluons are not directly probed in leptoproduction experiments and reliable measurements of their densities are difficult to obtain (Altarelli & Martinelli 1978, Anderson, Matis & Myrianthopoulos 1978, Glück & Reya 1979, Wu Ki Tung 1978, Mendez & Weiler 1979). LPTH data, if the effect of gluon terms could be easily disentangled, would provide a more convenient measure of gluon densities. A particularly interesting channel in this context is direct photon production, which is expected to be dominated by the gluon + quark → photon + quark term. Meanwhile most QCD calculations of LPTH data have used educated guesses of the gluon densities, relating them to the well-known structure functions of the valence quarks (see for example Glück & Reya 1977).

Quark structure functions are available for valence and sea quarks of different flavors as a result of the recent accumulation of accurate leptoproduction data (Gordon et al 1978, Bosetti et al 1978, de Groot et al 1979a,b, Bodek et al 1979, Fernandez et al 1979, Benvenuti et al 1979). Their Q^2 dependence is found consistent with QCD predictions; in fact the observations that the relation between different moments is in quantitative agreement with QCD (Bosetti et al 1978, de Groot et al 1979a,b) has been considered by many as a spectacular success of the theory although the relevance of this result has recently been contested (Harari 1979, Abbott & Barnett 1979).

Structure functions of particles other than nucleons are not accessible to leptoproduction experiments: their measurement is the privileged domain of LPTH and Drell-Yan processes. A first step in this direction has been taken by the FLPW Collaboration (Dris et al 1979). More, and more accurate, data should become available in the not too distant future.

Fragmentation functions should be most directly obtained from e^+e^- annihilation data (Hanson et al 1975, 1978, Berger et al 1979a,b). However, the flavor content of e^+e^- jets is completely different from that of LPTH quark jets: the latter are expected to be mostly of u and d parentage owing to the dominance of these flavors in the initial state. In e^+e^- annihilations, where copious production of heavy flavors is expected as soon as the corresponding thresholds are open, the experimental problem of identifying the parent flavor is very difficult. As a result, little is known on the flavor dependence of the fragmentation functions. Leptoproduction experiments are potentially a rich source of information in this context. The jet-structure of their final state (quark + diquark) has been observed and detailed measurements of their structure are under way. A possible indication of QCD effects is provided by the observation that $\langle q_T^2 \rangle$ seems to increase with jet momentum in neutrino data (Bosetti et al 1978, Mazzanti, Odorico & Roberto 1979).

Much less is known about the fragmentation of gluons. LPTH is the only hard process to which they contribute to leading order, but available data do not provide any evidence for gluon jets different from quark jets. However, e^+e^- annihilations may be a rich source of gluon jets, either as decay products of bound states of heavy quark-antiquark pairs, or as companions of quark-antiquark jet pairs (Ellis, Gaillard & Ross 1976, de Grand, Ng & Tye 1977). Recent PETRA experiments (Barber et al 1979, Brandelik et al 1979) have provided evidence for their production; if such an interpretation is confirmed, gluon fragmentation functions might soon be measured at PETRA and PEP.

There have been several very recent attempts to perform QCD calculations of quark and gluon fragmentation functions (Altarelli et al 1979, Konishi, Ukawa & Veneziano 1979a,b, Shizuya & Tye 1979, Bassetto, Ciafaloni & Marchesini 1979). They confirm the properties of the fragmentation mechanism that were intuitively assumed in earlier parton model analyses, in particular the connection between structure and fragmentation functions. However, when it comes to detailed confrontations between theory and experiment, the situation is increasingly complicated by the presence of different flavors and the role of resonance formation (Dias de Deus & Sakai 1979).

After these comments we may be surprised by the similarity observed between LPTH and other jets (Figures 14 and 18). It may be accidental or the data may be too crude to reveal the expected differences. More refined measurements are necessary to assess the role of gluon jets in LPTH final states. The presently explored p_T range may be too low to easily distinguish gluon from quark jets (Shizuya & Tye 1979, Einhorn & Weeks 1978).

Another essential ingredient in the QCD description of LPTH processes is the transverse momentum distribution (k_T) of the incident partons. If it does not contribute much to single-particle inclusive cross sections, it obviously plays an essential role in the description of two-particle correlations in the back-to-back configuration and of transverse momentum imbalance between the two jets (Owens 1979b, Owens & Kimel 1978, Alonso et al 1979, Fontannaz & Schiff 1978, Chase 1978b). Some information on this quantity is in principle available from the lack of collinearity of the quark and diquark jets in leptoproduction final states, but it is in Drell-Yan production that it has been most extensively studied. Experimentally (Lederman 1978) the transverse momentum distribution of muon pairs produced at Fermilab and at the ISR is observed to be very wide, corresponding to an effective value of $\langle k_T \rangle$ of at least 1 GeV/c in the ISR energy range. Even if part of the effect at Fermilab arises from nuclear complications, this result has caused some surprise and its interpretation has led to much confusion and controversy. The confusing issue is the presence of two short-distance scales in the problem (Kajantie & Lindfors 1978, Caswell, Horgan & Brodsky 1978): the dimuon mass and its transverse momentum, or, in the case of LPTH, the mean transverse momentum of the jets and their p_T imbalance. How does factorization work in such a case?

The Drell-Yan term, $q + \bar{q} \to \gamma^*$, accounts well for the total dimuon production cross section once scaling violations are introduced in the structure functions according to the factorization prescription (Altarelli, Ellis & Martinelli 1979, Kubar-Andre & Paige 1979). Higher order terms are dominated by the $q + \bar{q} \to \gamma^* + g$ correction, which contributes significantly.

For large transverse momentum pairs, however, terms such as $q + g \to q + \gamma^*$ are expected to play an important role (Fritzsch & Minkowski 1978, Halzen & Scott 1979, Altarelli, Parisi & Petronzio 1978) even if their contribution to the total cross section is minor (Kripfganz & Contogouris 1979, Contogouris & Kripfganz 1979). In this context the experimental observation of a hadron jet balancing the dilepton transverse momentum would be very informative and should provide a way to single out gluon jets (Brucker et al 1978).

Georgi (1979) has recently given a clear analysis of the problem in which he shows that k_T effects can be included in factorized QCD inclusive cross sections but that k_T is a very complicated variable from the point of view of factorization. The moments of the distribution functions do not factor out at fixed k_T. Recent QCD calculations (Glück & Reya 1978, Soper 1979, Dokshitzer, D'yakonov & Troyan 1978a,b, Parisi & Petronzio 1979) achieve good agreements with the data. Contrary to the suggestions of earlier analyses they do not require a large primordial k_T; a value of the order of 200 MeV/c is sufficient, the broad distribution of the dilepton transverse momentum being well accounted for by the dynamical recoils.

The extension of the Drell-Yan results to LPTH is not straightforward since different hard scattering terms are involved both for low and for large p_T imbalances between the two jets. The similarity noted between the effective $\langle k_T \rangle$ values deduced from LPTH data and dilepton production experiments may well disappear when better quality data will allow for a more accurate comparison.

A last comment concerns the role played by constituent interchange terms, which in QCD language correspond to higher-twist corrections. Section 2 included several indications (e.g. baryon production, quantum number correlations between the two jets) that they may play a role in LPTH, in particular at moderate values of p_T where they are enhanced by the trigger-bias effect. Such terms are not expected to contribute significantly to the other processes listed in Table 1.

This elementary evaluation of the role of LPTH among other deep hadronic probes suggests the following concluding remarks:

1. QCD has reached a state where it can efficiently guide the choice of future LPTH measurements and it offers clear schemes in which to present and analyze their results.

2. Most available LPTH data are in agreement with the expectations of QCD calculations, an encouraging stimulus for future work.

3. LPTH is the most dynamically complex deep hadronic probe. On one hand this implies that its analysis requires the previous understanding of other simpler processes such as e^+e^- annihilation, leptoproduction, or Drell-Yan dileptons. On the other hand LPTH offers a unique access to quantities not directly probed in other processes, in particular those related to gluons.

4. Wherever possible LPTH results should be formulated in a simple form allowing for specific tests of the theory. Factorization is very suggestive in this context as it simplifies the interpretation of experiments that differ only in a few of their aspects (cross-section ratios, etc).

5. Direct photon production is a rich potential source of information. Its relatively simple description in the QCD framework (inasmuch as it is not related to the more complicated π^0 production) should encourage detailed measurements of the structure of associated final states.

6. LPTH and the Drell-Yan process benefit from the availability of $\bar{p}$, K and π beams at fixed-target machines. This advantage over leptoproduction experiments gives them the unique potential to measure the structure functions of these particles.

7. In the not too distant future, new machines (CERN SPS $p\bar{p}$ collider, Isabelle, Fermilab collider) will be available to explore a new range of transverse momenta, up to $\sim 100 \text{ GeV}/c$. LPTH and the Drell-Yan process will have the unique opportunity to probe much smaller distances than previously possible. Even at low luminosity copious production of gluon jets is anticipated.

8. The progress of our understanding of strong interaction dynamics requires a constant confrontation between the theory and the results of experiments in which the deep hadronic structure can be probed. LPTH has a major role to play in this context and it should never be "the neglected stepchild of the quark-parton model" as Cutler & Sivers (1977) once called it. Higher order QCD effects are known to be important, for example in the Drell-Yan process where they induce a renormalization of the cross section by a factor of almost two. LPTH offer another laboratory in which they can be measured.

ACKNOWLEDGMENTS

I am deeply indebted to L. Di Lella, J. Ellis, and M. Jacob, who have contributed very useful comments to this review. In particular I owe much to the recent rapporteur talks of L. Di Lella (1979) and M. Jacob (1979) and to illuminating discussions with their authors.

I wish to thank Michèle Jouhet for her help to bring this report to its present form.

APPENDIX: EXPERIMENTAL ASPECTS

Table 2 briefly summarizes the main characteristics of the detectors used in LPTH studies. The following abbreviations have been used: SAS for single-arm magnetic spectrometer, DAS for double-arm magnetic spectrometer, SFMD for split-field magnet detector. Solid angle, polar angle, azimuth, and rapidity are denoted Ω, θ, ϕ, and y respectively. Collaborations are referred to in the first column by their acronym; their detailed denomination is given in Table 3.

Table 3 lists publications issued by the experimental collaborations. Contributions to conferences have usually been ignored; they are only retained if not covered by a subsequent journal publication.

Table 2   Detectors

Collaboration	Beam + Target	Trigger	Secondaries
ACHM	p + p	γ, $\Delta\Omega = 0.3$ sr, $\theta = 53°, 90°$ lead glass array	Charged hadrons, no momentum analysis, $\Delta\Omega \sim 4\pi$ streamer chambers
ABCS	p + p	γ, $(\Delta\phi = 45°, \Delta\theta = 90°) \times 4$ Four liquid argon calorimeters (Pb) in various azimuthal configurations	
BFS	p + p	$p^{\pm}, \pi^{\pm}, K^{\pm}, \Delta\Omega = 4$ msr, $\theta = 90°$ SAS with Cerenkov counters	Charged particles in the SFMD
BS	p + p	$p^{\pm}, \pi^{\pm}, K^{\pm}, \Delta\Omega = 12$ msr, $\theta = 36$ to 90° SAS with Cerenkov counters	
BCB	$\pi^{\pm}$, p + p	γ, $\theta = 15$ to 110° Fine grain lead-scintillator sandwich	
CLFCI	$\pi^{\pm}, K^{\pm}$, p + p, Al, Be	Hadron jets, $\theta = \pm 90°$, $\Delta\Omega = 2 \times 1.5$ sr Double-arm hadron calorimeter	Magnetic spectrometer covering the trigger acceptance (charged particles).
C	p + p	γ, $\Delta\Omega = 0.1$ sr, $\theta = 90°$ lead glass array	Charged particles in the SFMD
CCHK	p + p	h^{+}, $\Delta\phi = 20°$, $\Delta y = 0.8$, $\theta = 9$ to 21°, 45° geometry selection on straight tracks $\pi^{\pm}, K^{-}$, p, Cerenkov counters, $\theta = 20°$	Charged particles in the SFMD
CCOR	p + p	γ, $\Delta\Omega \sim 2 \times 1$ sr, $\theta = \pm 90°$ Double-arm lead glass arrays	Charged particles in drift chamber detector in superconducting solenoid, $\Delta\phi = 2\pi$, $\Delta y = 1.4$
CCR	p + p	γ, $\Delta\Omega \sim 2 \times 0.6$ sr, $\theta = \pm 90°$ Double-arm lead glass arrays	Charged particles, no magnetic analysis, in spark chambers covering the trigger acceptance
CCRS	p + p	γ, $\Delta\Omega \sim 2 \times 0.6$ sr, $\theta = \pm 90°$ Double-arm lead glass arrays	Charged particles in DAS covering the trigger acceptance
CRB	p + p	γ, $\Delta\Omega \sim 0.5$ sr, $\theta = 90°$, lead glass array	
CS	p + p	γ, $\Delta\Omega \sim 2 \times 0.8$ sr, $\theta = \pm 90°$, Double-arm lead glass arrays	Charged particles in DAS covering the trigger acceptance
CP	p + p, N	$\pi^{\pm}, p^{\pm}, K^{\pm}, \theta \sim 90°$, $\Delta\Omega = 20\mu$ sr lab, SAS with Cerenkov counters	
CF	p + N	γ, $\theta = 65, 93°$ lead glass array	
CFS	p + Be, W	Hadron pairs in DAS with Cerenkov counters for particle identification $\theta = \pm 90°$, $\Delta\Omega = 2 \times 0.063$ sr	
DLR	p + p	$\pi^{\pm}, K^{\pm}, p^{\pm}, \theta = 90°$, $\Delta\Omega \simeq 12$ msr in SAS with Cerenkov counters	Crude charged-particle detection without momentum analysis (spark chambers + scintilators) over $\Delta\phi = 2\pi$, $\Delta y = 3$.
FJH	p + Be	Double-arm lead glass spectrometer	
FLPW	$\pi^{\pm}$, p + p	Hadron jets, $\theta = \pm 90°$, $\Delta\Omega \simeq 1.5 + 2$ sr Double-arm hadron calorimeter	Charged particles in drift chambers covering the trigger acceptance (no momentum analysis)
FMP	p + N	$K^{\pm}$ in single arm of DAS equipped with Cerenkov counters	
NN	p + p, C (internal)	γ, $\theta = 40$ to 110°, $\Delta\Omega = 9.5$ μsr lab lead glass array	
PSB	p + p	γ, $\theta = 90°, 17.5°$, $\Delta\Omega = 0.3$ sr lead glass array	Crude charged-particle detection without momentum analysis over $\Delta\Omega \sim 4\pi$ (scintillators)
SS	p + p	$h^{\pm}$, $\theta = 90°$, $\Delta\Omega = 85$ msr, SAS	

Table 3   Experimenter Collaborations: Publications

Aachen–CERN–Heidelberg–München (ACHM)

Eggert et al 1975a. A study of high transverse momentum π^0's at ISR energies

Eggert et al 1975b. Angular correlations in proton-proton collisions producing a large transverse momentum π^0

Athens–Brookhaven–CERN–Syracuse (ABCS)

Cobb et al 1978. Azimuthal correlations of high transverse momentum π^0 pairs

Kourkoumelis et al 1979a. Study of resolved high-p_T neutral pions at the CERN ISR

Kourkoumelis et al 1979b. Inclusive π^0 production at very large p_T at the ISR

Kourkoumelis et al 1979c. Inclusive η production at high p_T at the ISR

Kourkoumelis et al 1979d. Azimuthal correlation of a high $p_T \pi^0$ and η with a second π^0 produced in p-p collisions at the ISR

Kourkoumelis et al 1979e. Correlations of high transverse momentum π^0 pairs produced at the CERN ISR

Kourkoumelis et al 1979f. Measurements of π^0 fragments from jets produced in p-p collisions at the CERN ISR

Diakonou et al 1979. Direct production of high p_T single photons in p-p collisions at the CERN ISR

British–French–Scandinavian (BFS)

Albrow et al 1978a. Studies of p-p collisions at the CERN ISR with an identified charged hadron of high transverse momentum at 90°. (I) On forward particles in high p_T reactions

Albrow et al 1978b. Ibid. (II) On the distribution of charged particles in the central region

Albrow et al 1979. Ibid. (III) Jet-like structures

British–Scandinavian (BS)

Alper et al 1973a. Production of high transverse momentum particles in p-p collisions in the central region at the CERN ISR

Alper et al 1973b. Particle composition at high transverse momenta in p-p collisions in the central region at the ISR

Alper et al 1975a. The production of charged particles with high transverse momentum in p-p collisions at the CERN ISR

Alper et al 1975b. Production spectra of $\pi^\pm$, $K^\pm$, $p^\pm$ at large angles in proton-proton collisions in the CERN ISR

Brookhaven–Caltech–Berkeley (BCB)

Donaldson et al 1976. Inclusive π^0 production at large transverse momentum from $\pi^\pm p$ and pp interactions at 100 and 200 GeV/c

Donaldson et al 1978a. Angular dependence of high transverse momentum π^0 production in $\pi^\pm p$ and pp interactions

Donaldson et al 1978b. Comparison of high transverse momentum π^0 production from π^-, K^-, p and $\bar{p}$ beams

Caltech–UCLA–Fermilab–Chicago Circle–Indiana (CLFCI)

Bromberg et al 1977. Observation of the production of jets of particles at high transverse momentum and comparison with inclusive single-particle reactions

Bromberg et al 1978. Production of jets and single particles at high p_T in 200 GeV hadron-beryllium collisions

Bromberg et al 1979a. Jets produced in π^-, π^+ and proton interactions at 200 GeV on hydrogen and aluminium targets

Table 3 (continued)

Bromberg et al 1979b. Production and correlations of charged particles with high p_T in 200 GeV/c $\pi^\pm$p, K$^-$p and pp collisions
Bromberg et al 1979c. Experimental tests of quantum chromodynamics in high p_T jet production in 200 GeV/c hadron-proton collisions

CERN (C)
Darriulat et al 1976a. Large transverse momentum photons from high energy proton-proton collisions
Darriulat et al 1976b. Structure of final states with a high transverse momentum π^0 in p-p collisions

CERN–College de France–Heidelberg–Karlsruhe (CCHK)
Cottrell et al 1975. Measurement of large transverse momentum positive particles produced at medium angles at $\sqrt{s} = 52.5$ GeV
Della Negra et al 1975a. Observation of leading particles in p-p interactions with large transverse momentum secondaries
Della Negra et al 1975b. Composition of particles emitted at large p_T and medium angles in pp collisions at $\sqrt{s} = 52.5$ GeV
Della Negra et al 1976. Study of events with a positive particle of large transverse momentum emitted near the forward direction in p-p collisions at $\sqrt{s} = 52.5$ GeV
Della Negra et al 1977. Observation of jet structure in high p_T events at the ISR and the importance of parton transverse momentum
Drijard et al 1979. Quantum number effects in events with a charged particle of large transverse momentum. (I) Leading particles in single and diquark jets
Drijard et al 1980. Ibid. (II) Charge correlations in jets

CERN–Columbia–Oxford–Rockefeller (CCOR)
Angelis et al 1978. A measurement of inclusive π^0 production at large p_T from p-p collisions at the CERN ISR
Angelis et al 1979. A study of final states containing high p_T π^0's at the CERN ISR

CERN–Columbia–Rockefeller (CCR)
Büsser et al 1973. Observation of π^0 mesons with large transverse momentum in high energy proton-proton collisions
Büsser et al 1974a. Correlations between large transverse momentum π^0 mesons and charged particles at the CERN ISR
Büsser et al 1974b. Correlation between two large transverse momentum π^0 mesons at the CERN ISR

CERN–Columbia–Rockefeller–Saclay (CCRS)
Büsser et al 1975. A study of high transverse momentum η and π^0 mesons at the CERN ISR
Büsser et al 1976. A study of inclusive spectra and two-particle correlations at large transverse momentum

CERN–Rome–Brookhaven (CRB)
Amaldi et al 1978. Search for single photon direct production in pp collisions at $\sqrt{s} = 53.2$ GeV
Amaldi et al 1979a. Comparison of direct photon production in pp collisions at $\sqrt{s} = 30.6$ GeV and 53.2 GeV
Amaldi et al 1979b. Single direct photon production in pp collisions at $\sqrt{s} = 53.2$ GeV in the p_T interval 2.3 to 5.7 GeV/c
Amaldi et al 1979c. Inclusive η production in pp collisions at ISR energies

Table 3 (continued)

CERN–Saclay (CS)

Clark et al 1978a. Inclusive π^0 production from high energy p-p collisions at very large transverse momentum

Clark et al 1978b. Large transverse momentum π^0 production in pp, dp, and dd collisions at the CERN ISR

Clark et al 1979. Large transverse momentum jets in high energy pp collisions

Chicago–Princeton (CP)

Cronin et al 1973. Production of hadrons with large transverse momentum at 200 and 300 GeV

Cronin et al 1975. Production of hadrons at large transverse momentum at 200, 300 and 400 GeV

Antreasyan et al 1977a. Production of π^+ and π^- at large transverse momentum in p-p and p-d collisions at 200, 300 and 400 GeV

Antreasyan et al 1977b. Production of kaons, protons and antiprotons with large transverse momentum in p-p and p-d collisions at 200, 300 and 400 GeV

Antreasyan et al 1979. Production of hadrons at large transverse momentum in 200, 300 and 400 GeV p-p and p-nucleus collisions

Frisch et al 1980. The relative production of $\pi^\pm$, $K^\pm$, p and $\bar{p}$ at large transverse momentum in 200 and 300 GeV π-p collisions

Columbia–Fermilab (CF)

Appel et al 1974. Hadron production at large transverse momentum.

Columbia–Fermilab–Stony Brook (CFS)

Kephart et al 1977. Measurement of the dihadron mass continuum in p-Be collisions and a search for narrow resonances

McCarthy et al 1978. Nucleon number dependence of the production cross sections for massive dihadron states

Fisk et al 1978. Correlations between two hadrons at large transverse momenta

Jöstlein et al 1979a. Scaling properties of high mass symmetric hadron- and pion-pair production in proton-beryllium collisions

Jöstlein et al 1979b. Inclusive production of large transverse momentum hadrons and hadron pairs

Daresbury–Liverpool–Rutherford (DLR)

Alper et al 1976. Multiplicities associated with the production of pions, kaons or protons of high transverse momentum at the ISR

Alper et al 1978. Multiplicity distributions associated with charged hadron production over a range of transverse momentum and production angle at ISR energies

Fermilab–Johns Hopkins (FJH)

Cox 1979. A search for direct photon production at Fermilab energies and comparison with direct photon measurements at ISR energies

Fermilab–Lehigh–Pennsylvania–Wisconsin (FLPW)

Corcoran et al 1978. Comparison of high p_T events produced by pions and protons

Corcoran et al 1979. A two-jet calorimeter experiment at Fermilab

Corcoran et al 1980. Evidence that high p_T jet pairs give direct information on parton-parton scattering

Dris et al 1979. π^+ structure information from a jet experiment

Fermilab–Michigan–Purdue (FMP)

Akerlof et al 1977. Measurements of ϕ production in proton-nucleus collisions at 400 GeV/c

Table 3 (continued)

NAL–Northern Illinois (NN)

Carey et al 1974a. Production of large transverse momentum γ rays in p–p collisions from 50 to 400 GeV

Carey et al 1974b. Inclusive π^0 production in p–p collisions at 50–400 GeV/c

Carey et al 1974c. Unified description of single particle production in p–p collisions

Pisa–Stony Brook (PSB)

Finocchiaro et al 1974. Measurement of charged particle multiplicities associated with large transverse momentum photons in proton-proton collisions

Kephart et al 1976. Charged particle multiplicities associated with large transverse momentum photons

Saclay–Strasbourg (SS)

Banner et al 1973. Large transverse momentum particle production at 90° in p–p collisions at the ISR

Literature Cited

Abad, J., Humpert, B. 1978. *Phys. Lett. B* 77:105

Abbott, L., Barnett, R. M. 1979. *SLAC preprint SLAC-PUB-2325*

Akerlof, C. W. et al. 1977. *Phys. Rev. Lett.* 39:861

Albrow, M. G. et al. 1978a. *Nucl. Phys. B* 135:461

Albrow, M. G. et al. 1978b. *Nucl. Phys. B* 145:305

Albrow, M. G. et al. 1979. *Nucl. Phys. B* 160:1

Alonso, J. L. et al. 1979. *Nucl. Phys. B* 157:498

Alper, B. et al. 1973a. *Phys. Lett. B* 44:521

Alper, B. et al. 1973b. *Phys. Lett. B* 44:527

Alper, B. et al. 1975a. *Nucl. Phys. B* 87:19

Alper, B. et al. 1975b. *Nucl. Phys. B* 100:237

Alper, B. et al. 1976. *Nucl. Phys. B* 114:1

Alper, B. et al. 1978. *Nucl. Phys. B* 141:189

Altarelli, G. et al. 1979. *Nucl. Phys. B* 160:301

Altarelli, G., Ellis, R. K., Martinelli, G. 1979. *Nucl. Phys. B* 157:461

Altarelli, G., Martinelli, G. 1978. *Phys. Lett. B* 76:89

Altarelli, G., Parisi, G. 1977. *Nucl. Phys. B* 126:298

Altarelli, G., Parisi, G., Petronzio, R. 1978. *Phys. Lett. B* 76:351

Amaldi, E. et al. 1978. *Phys. Lett. B* 77:240

Amaldi, E. et al. 1979a. *Phys. Lett. B* 84:360

Amaldi, E. et al. 1979b. *Nucl. Phys. B* 150:326

Amaldi, E. et al. 1979c. *Nucl. Phys. B* 158:1

Amati, D., Petronzio, R., Veneziano, G. 1978a. *Nucl. Phys. B* 140:54

Amati, D., Petronzio, R., Veneziano, G. 1978b. *Nucl. Phys. B* 146:29

Anderson, H. L., Matis, H. S., Myrianthopoulos, L. C. 1978. *Phys. Rev. Lett.* 40:1061

Andersson, B., Gustafson, G., Peterson, C. 1977. *Phys. Lett. B* 71:337

Andersson, B., Gustafson, G., Peterson, C. 1978a. *Phys. Lett. B* 72:503

Andersson, B., Gustafson, G., Peterson, C. 1978b. *Nucl. Phys. B* 135:273

Angelis, A. L. S. et al. 1978. *Phys. Lett. B* 79:505

Angelis, A. L. S. et al. 1979. *Phys. Scr.* 19:116

Antreasyan, D. et al. 1977a. *Phys. Rev. Lett.* 38:112

Antreasyan, D. et al. 1977b. *Phys. Rev. Lett.* 38:115

Antreasyan, D. et al. 1979. *Phys. Rev. D* 19:764

Appel, J. A. et al. 1974. *Phys. Rev. Lett.* 33:719, 722

Baier, R. et al. 1977. *Nucl. Phys. B* 118:139

Baier, R., Cleymans, J., Petersson, B. 1978. *Phys. Rev. D* 17:2310

Baier, R., Engels, J., Petersson, B. 1979. *Univ. Bielefeld. Rep. BI-TP/79/10*

Banner, M. et al. 1973. *Phys. Lett. B* 44:537

Barber, D. P. et al. 1979. *Phys. Rev. Lett.* 43:830

Bassetto, A., Ciafaloni, M., Marchesini, G. 1979. *Phys. Lett. B* 86:366

Benvenuti, A. et al. 1979. *Phys. Rev. Lett.* 42:149

Berger, C. et al. 1979a. *Phys. Lett. B* 78:176

Berger, C. et al. 1979b. *Phys. Lett. B* 81:411

Berman, S. M., Jacob, M. 1970. *Phys. Rev. Lett.* 25:1683

Berman, S. M., Bjorken, J. D., Kogut, J. 1971. *Phys. Rev. D* 4:3388

Bjorken, J. D. 1969. *Phys. Rev.* 179 : 1547
Bjorken, J. D. 1973. *Phys. Rev. D*.8 : 4098
Bjorken, J. D. 1974. *Acta. Phys. Pol. B* 5 : 145
Bjorken, J. D. 1975. *High* P_T *dynamics.* Delivered at SLAC Summer Inst. on Part. Phys.
Bjorken, J. D. 1971. *SLAC Rep. SLAC-PUB-974*
Bjorken, J. D., Paschos, E. A. 1969. *Phys. Rev.* 185 : 1975
Blankenbecler, R., Brodsky, S. J., Gunion, J. F. 1972a. *Phys. Lett. B* 39 : 649
Blankenbecler, R., Brodsky, S. J., Gunion, J. F. 1972b. *Phys. Rev. D* 6 : 2652
Blankenbecler, R., Brodsky, S. J., Gunion, J. F. 1973a. *Phys. Lett. B* 42 : 461
Blankenbecler, R., Brodsky, S. J., Gunion, J. F. 1973b. *Phys. Rev. D* 8 : 287
Blankenbecler, R., Brodsky, S. J., Gunion, J. F. 1978. *Phys. Rev. D* 18 : 900
Bodek, A. et al. 1979. *Phys. Rev. D* 20 : 1471
Bosetti, P. C. et al. 1978. *Nucl. Phys. B* 142 : 1
Brandelik, R. et al. 1979. *Phys. Lett. B* 86 : 243
Breidenbach, M. et al. 1969. *Phys. Rev. Lett.* 23 : 935
Brodsky, S. J., Farrar, G. 1973. *Phys. Rev. Lett.* 31 : 1153
Brodsky, S. J., Farrar, G. 1975. *Phys. Rev. D* 11 : 1309
Brodsky, S. J., de Grand, J., Schwitters, R. 1978. *Phys. Lett. B* 79 : 255
Bromberg, C. et al. 1977. *Phys. Rev. Lett.* 38 : 1447
Bromberg, C. et al. 1978. *Nucl. Phys. B* 134 : 189
Bromberg, C. et al. 1979a. *Phys. Rev. Lett.* 42 : 1202
Bromberg, C. et al. 1979b. *Phys. Rev. Lett.* 43 : 561
Bromberg, C. et al. 1979c. *Phys. Rev. Lett.* 43 : 565
Brucker, J. M. et al. 1978. *Phys. Lett. B* 78 : 630
Büsser, F. W. et al. 1973. *Phys. Lett. B* 46 : 471
Büsser, F. W. et al. 1974a. *Phys. Lett. B* 51 : 306
Büsser, F. W. et al. 1974b. *Phys. Lett. B* 51 : 311
Büsser, F. W. et al. 1975. *Phys. Lett. B* 55 : 232
Büsser, F. W. et al. 1976. *Nucl. Phys. B* 106 : 1
Carey, D. C. et al. 1974a. *Phys. Rev. Lett.* 32 : 24
Carey, D. C. et al. 1974b. *Phys. Rev. Lett.* 33 : 327
Carey, D. C. et al. 1974c. *Phys. Rev. Lett.* 33 : 330
Caswell, W. E., Horgan, R. R., Brodsky, S. J. 1978. *Phys. Rev. D* 18 : 2415
Chase, M. K. 1978a. *Phys. Lett. B* 79 : 114
Chase, M. K. 1978b. *Nucl. Phys. B* 145 : 189

Chen, C. K. 1978. *Phys. Rev. D.* 18 : 3303
Clark, A. G. et al. 1978a. *Phys. Lett. B* 74 : 267
Clark, A. G. et al. 1978b. *Nucl. Phys. B* 142 : 189
Clark, A. G. et al. 1979. *Nucl. Phys. B* 160 : 397
Cobb, J. H. et al. 1978. *Phys. Rev. Lett.* 40 : 1420
Combridge, B. L. 1978. *Phys. Rev. D.* 18 : 734
Combridge, B. L., Kripfganz, J., Ranft, G. 1977. *Phys. Lett. B* 70 : 234
Contogouris, A. P., Gaskell, R. 1977. *Nucl. Phys. B.* 126 : 157
Contogouris, A. P., Gaskell, R., Nicolaidis, A. 1978a. *Phys. Rev. D* 17 : 839
Contogouris, A. P., Gaskell, R., Nicolaidis, A. 1978b. *Phys. Rev. D* 17 : 2992
Contogouris, A. P., Gaskell, R., Papadopoulos, S. 1978. *Phys. Rev. D* 17 : 2314
Contogouris, A. P., Kripfganz, J. 1979. *Phys. Rev. D* 19 : 2207
Contogouris, A. P., Papadopoulos, S., Hongoh, M. 1979. *Phys. Rev. D* 19 : 2607
Corcoran, M. D. et al. 1978. *Phys. Rev. Lett.* 41 : 9
Corcoran, M. D. et al. 1979. *Phys. Scr.* 19 : 95
Corcoran, M. D. et al. 1980. *Phys. Rev. Lett.* 44 : 514
Cottrell, R. et al. 1975. *Phys. Lett. B* 55 : 341
Cox, B. 1979. *Fermilab Rep. Conf. 79/85 EXP*
Cronin, J. W. et al. 1973. *Phys. Rev. Lett.* 31 : 1426
Cronin, J. W. et al. 1975. *Phys. Rev. D* 11 : 3105
Cutler, R., Sivers, D. 1977. *Phys. Rev. D* 16 : 679
Cutler, R., Sivers, D. 1978. *Phys. Rev. D* 17 : 196
Dalitz, R. H. 1965. In *High Energy Physics, Les Houches*, p. 253. New York : Gordon & Breach
Darriulat, P. et al. 1976a. *Nucl. Phys. B* 107 : 365
Darriulat, P. et al. 1976b. *Nucl. Phys. B* 110 : 429
Deden, H. et al. 1975. *Nucl. Phys. B* 85 : 269
de Grand, J., Ng, Y. J., Tye, S. H. H. 1977. *Phys. Rev. D* 16 : 3251
de Groot, J. G. H. et al. 1979a. *Phys. Lett. B* 82 : 45
de Groot, J. G. H. et al. 1979b. *Z. Phys.* C143
Della Negra, M. et al. 1975a. *Phys. Lett. B* 59 : 401
Della Negra, M. et al. 1975b. *Phys. Lett. B* 59 : 481
Della Negra, M. et al. 1976. *Nucl. Phys. B* 104 : 365
Della Negra, M. et al. 1977. *Nucl. Phys. B* 127 : 1

Diakonou, M. et al. 1979. *Phys. Lett. B* 87:292

Dias de Deus, J., Sakai, N. 1979. *Phys. Lett. B* 86:321

Di Lella, L. 1979. Presented at *10th Int. Symp. on Multipart. Dynamics, Goa, India, 25–27 September 1979; CERN Intern. Rep. CERN-EP/79–145*, 19 November 1979

Dokshitzer, Yu. L., D'yakonov, D. I., Troyan, S. I. 1978a. *Phys. Lett. B* 78:290

Dokshitzer, Yu. L., D'yakonov, D. I., Troyan, S. I. 1978b. *Phys. Lett. B* 79:269

Donaldson, G. et al. 1976. *Phys. Rev. Lett.* 36:1110

Donaldson, G. et al. 1978a. *Phys. Lett. B* 73:375

Donaldson, G. et al. 1978b. *Phys. Rev. Lett.* 40:917

Drell, S. D., Yan, T. M. 1970. *Phys. Rev. Lett.* 25:316

Drell, S. D., Yan, T. M. 1971. *Ann Phys.* 66:555

Drijard, D. et al. 1979. *Nucl. Phys. B* 156:309

Drijard, D. et al. 1980. *Nucl. Phys.* In press

Dris, M. et al. 1979. *Phys. Rev. D* 19:1361

Duke, D. W. 1977. *Phys. Rev. D* 16:1375

Eggert, K. et al. 1975a. *Nucl. Phys. B* 98:49

Eggert, K. et al. 1975b. *Nucl. Phys. B* 98:73

Eichten, T. et al. 1973. *Phys. Lett. B* 46:274

Einhorn, M. B., Weeks, B. G. 1978. *Nucl. Phys. B* 146:445

Ellis, J., Gaillard, M. K., Ross, G. G. 1976. *Nucl. Phys. B* 111:253

Ellis, R. K. et al. 1979. *Nucl. Phys. B* 152:285

Ellis, S. D., Jacob, M., Landshoff, P. V. 1976. *Nucl. Phys. B* 108:93

Ellis, S. D., Kislinger, M. B. 1974. *Phys. Rev. D* 9:2027

Ellis, S. D., Stroynowski, R. 1977. *Rev. Mod. Phys.* 49:753

Escobar, C. O. 1975. *Nucl. Phys. B* 98:173

Escobar, C. O. 1977. *Phys. Rev. D* 15:355

Escobar, C. O. 1979. *Phys. Rev. D* 19:844

Farrar, G. R. 1977. *Phys. Lett. B* 67:357

Farrar, G. R., Frautshi, S. C. 1976. *Phys. Rev. Lett.* 36:1017

Fernandez, E. et al. 1979. *Phys. Rev. Lett.* 43:1975

Feynman, R. P. 1969. *Phys. Rev. Lett.* 23:1415; see also in *Photon-Hadron Interactions* New York: Benjamin

Feynman, R. P., Field, R. D., Fox, G. C. 1977. *Nucl. Phys. B* 128:1

Feynman, R. P., Field, R. D., Fox, G. C. 1978. *Phys. Rev. D* 18:3320

Field, R. D. 1978. *Phys. Rev. Lett.* 40:997

Field, R. D., Feynman, R. P. 1977. *Phys. Rev. D* 15:2590

Field, R. D., Feynman, R. P. 1978. *Nucl. Phys. B* 136:1

Finocchiaro, G. et al. 1974. *Phys. Lett. B* 50:396

Fishbach, E., Look, G. W. 1977a. *Phys. Rev. D* 15:2576

Fishbach, E., Look, G. W. 1977b. *Phys. Rev. D* 16:1369

Fishbach, E., Look, G. W. 1977c. *Phys. Rev. D* 16:1571

Fisk, R. J. et al. 1978. *Phys. Rev. Lett.* 40:984

Fontannaz, M., Pire, B., Schiff, D. 1978. *Phys. Lett. B* 77:315

Fontannaz, M., Schiff, D. 1978. *Nucl. Phys. B* 132:457

Frazer, W., Gunion, J. F. 1979. *Phys. Rev. D* 20:147

Frisch, H. J. et al. 1980. *Phys. Rev. Lett.* 44:511

Fritzsch, H., Gellman, M., Leutwyler, H. 1973. *Phys. Lett. B* 47:365

Fritzsch, H., Minkowski, P. 1977. *Phys. Lett. B* 69:316

Fritzsch, H., Minkowski, P. 1978. *Phys. Lett.* 73:80

Furmanski, W., 1978. *Phys. Lett. B* 77:312

Furmanski, W. 1979. *Nucl. Phys. B* 158:429

Georgi, H. 1979. *Phys. Rev. Lett.* 42:294

Glück, M., Reya, E. 1977. *Nucl. Phys. B* 130:76

Glück, M., Reya, E. 1978. *Nucl. Phys. B* 145:24

Glück, M., Reya, E. 1979. *Nucl. Phys. B* 156:456

Gordon, B. A. et al. 1978. *Phys. Rev. Lett.* 41:615

Gross, D. J., Wilczek, F. A. 1973. *Phys. Rev. Lett.* 30:1343

Gustafson, G., Månsson, O. 1979. *Phys. Lett. B* 81:75

Hagiwara, K. 1979. *Phys. Lett. B* 84:241

Halzen, F., Ringland, G. A., Roberts, R. G. 1978. *Phys. Rev. Lett.* 40:991

Halzen, F., Scott, D. M. 1978a. *Phys. Rev. Lett.* 40:1117

Halzen, F., Scott, D. M. 1978b. *Phys. Rev. D* 18:3378

Halzen, F., Scott, D. M. 1979. *Phys. Rev. D* 19:216

Hanson, G. et al. 1975. *Phys. Rev. Lett.* 35:1609

Hanson, G. et al. 1978. *Jets in e^+e^- annihilations, SLAC Intern. Rep. SLAC-PUB-2118*

Harari, H. 1979. *Nucl. Phys. B* 161:55

Hayot, F., Jadach, S. 1978. *Phys. Rev. D* 17:2307

Hoyer, P. et al. 1979. *Nucl. Phys. B* 151:389

Hwa, R. C., Spiessbach, A. J., Teper, M. J. 1976. *Phys. Rev. Lett.* 36:1418

Hwa, R. C., Spiessbach, A. J., Teper, M. J. 1978. *Phys. Rev. D* 18:1475

Jacob, M. 1979. Presented at *EPS Int. Conf. on High Energy Phys., Geneva 1979*, p. 473. Geneva: CERN

Jacob, M., Landshoff, P. V. 1976. *Nucl. Phys. B* 113:395

Jacob, M., Landshoff, P. V. 1978. *Phys. Reports* 48:285

Jones, D., Gunion, J. F. 1979. *Phys. Rev. D* 19:867

Jones, D., Rückl, R. 1979. *Phys. Rev. D* 20:232

Jöstlein, H. et al. 1979a. *Phys. Rev. Lett.* 42:146

Jöstlein, H. et al. 1979b. *Phys. Rev. D* 20:53

Kajantie, K., Lindfors, J. 1978. *Nucl. Phys. B* 146:465

Kane, G. L., Yao, Y. P. 1978. *Nucl. Phys. B* 137:313

Kephart, R. D. et al. 1976. *Phys. Rev. D* 14:2909

Kephart, R. D. et al. 1977. *Phys. Rev. Lett.* 39:1440

Kim, K. J., Schilcher, K. 1978. *Phys. Rev. D* 17:2800

Konishi, K., Ukawa, A., Veneziano, G. 1979a. *Phys. Lett. B* 80:259

Konishi, K., Ukawa, A., Veneziano, G. 1979b. *Nucl. Phys. B* 157:45

Kourkoumelis, C. et al. 1979a. *Phys. Lett. B* 83:257

Kourkoumelis, C. et al. 1979b. *Phys. Lett. B* 84:271

Kourkoumelis, C. et al. 1979c. *Phys. Lett. B* 84:277

Kourkoumelis, C. et al. 1979d. *Phys. Lett. B* 85:147

Kourkoumelis, C. et al. 1979e. *Nucl. Phys. B* 158:39

Kourkoumelis, C. et al. 1979f. *Phys. Lett. B* 86:391

Kripfganz, J., Contogouris, A. P. 1979. *Phys. Lett. B* 84:473

Kripfganz, J., Ranft, G. 1977. *Nucl. Phys. B* 124:353

Kripfganz, J., Schiller, A. 1978. *Phys. Lett. B* 79:317

Kubar-Andre, J., Paige, F. E. 1979. *Phys. Rev. D* 19:221

Landshoff, P. V., Polkinghorne, J. C. 1973a. *Phys. Rev. D* 8:927

Landshoff, P. V., Polkinghorne, J. C. 1973b. *Phys. Rev. D* 8:4159

Landshoff, P. V., Polkinghorne, J. C. 1973c. *Phys. Lett. B* 45:361

Lederman, L. M. 1978. *Proc. 19th Int. Conf. on High Energy Phys., Tokyo*, p. 720

Levin, E. M., Ryskin, M. G. 1976. *Sov. Phys. JETP* 42:783

Libby, S. B., Sterman, G. 1978a. *Phys. Lett. B* 78:618

Libby, S. B., Sterman, G. 1978b. *Phys. Rev. D* 18:3252

Llewellyn Smith, C. H. 1978a. *Phys. Lett. B* 79:83

Llewellyn Smith, C. H. 1978b. *Acta. Phys. Austr.*, Suppl. 19, p. 331

Matveev, V. A., Muradyan, R. M., Tavkhelidze, A. V. 1973. *Nuovo Cimento Lett.* 7:719

Mazzanti, P., Odorico, R., Roberto, V. 1979. *Phys. Lett. B* 81:219

McCarthy, R. L. et al. 1978. *Phys. Rev. Lett.* 40:213

Mendez, A., Weiler, T. 1979. *Phys. Lett. B* 83:221

Owens, J. F. 1979a. *Phys. Rev. D* 19:3279

Owens, J. F. 1979b. *Phys. Rev. D* 20:221

Owens, J. F., Kimel, J. D. 1978. *Phys. Rev. D* 18:3313

Owens, J. F., Reya, E., Glück, M. 1978. *Phys. Rev. D* 18:1501

Panofsky, W. K. H. 1968. *Proc. 14th Int. Conf. on High Energy Phys., Vienna.* CERN

Parisi, G., Petronzio, R. 1979. *Nucl. Phys. B* 154:427

Politzer, H. D. 1973. *Phys. Rev. Lett.* 30:1346

Politzer, H. D. 1977a. *Phys. Lett. B* 70:430

Politzer, H. D. 1977b. *Nucl. Phys. B* 129:301

Ranft, J., Ranft, G. 1976. *Nucl. Phys. B* 110:493

Rückl, R., Brodsky, S. J., Gunion, J. F. 1978. *Phys. Rev. D* 18:2469

Sachrajda, C. T. 1978a. *Phys. Lett. B* 76:100

Sachrajda, C. T. 1978b. *Phys. Lett. B* 73:185

Scott, W. G. 1978. In *Neutrino 1978, Int. Conf. on Neutrino Phys., Purdue Univ.*, ed. J. Fowler

Selove, W. 1972. *CERN-NP Intern. Rep. 72–25*

Selove, W. 1979. Presented at *14th Rencontre de Moriond, Les Arcs, France, March 1979; Univ. Penn. Rep. UPR-70E, May 1979*

Shizuya, K., Tye, S. H. H. 1978. *Phys. Rev. Lett.* 41:787

Shizuya, K., Tye, S. H. H. 1979. *Phys. Rev. D* 20:1101

Sivers, D., Blankenbecler, R., Brodsky, S. J. 1976. *Phys. Rep. C* 23:1

Söding, P. 1979. *Jet analysis, EPS Int. Conf. on High Energy Phys., Geneva*

Soper, D. E. 1979. *Phys. Rev. Lett.* 43:1847

Sterman, G., Weinberg, S. 1977. *Phys. Rev. Lett.* 39:1436

Tung, W. K. 1978. *Phys. Rev. D* 17:738

Van der Welde, J. C. 1979. *Phys. Scr.* 19:173

Wolf, G. 1979. *DESY Intern. Rep. 79141*

Wosiek, J., Zalewski, K. 1979. *Nucl. Phys. B* 161:294

Wu, T. T., Yang, C. N. 1965. *Phys. Rev. B* 137:708

Ann. Rev. Nucl. Part. Sci. 1980. 30:211–52

PROTON MICROPROBES ✸5617
AND PARTICLE-INDUCED
X-RAY ANALYTICAL
SYSTEMS

T. A. Cahill

Department of Physics, University of California, Davis, California 95616

CONTENTS

1 INTRODUCTION

It was in 1970 that the mating of energy dispersive x-ray detectors with beams of low energy (MeV) ions for the purpose of elemental analysis was first reported by Johansson et al (1970). Early enthusiasm for the technique, fueled by claims of high sensitivity for the method, pressing environmental concerns, and under-utilization of numerous low energy accelerators, led to several years of rapid development of the technique and wide application to physical and biological problems. The purposes of this review are to identify the applications that have proved to be viable, isolate the factors

211

0163-8998/80/1201-0211$01.00

that make the successful applications competitive with alternate methodologies, and project present trends so as to predict those areas that will benefit from the method in the future.

The experience of the groups that have used particle-induced x-ray emission (PIXE) for analytical purposes can be summarized as follows:

1. PIXE is a multielement, nondestructive technique, with all elements between sodium and uranium detectable, in principle, in a single irradiation, with relatively uniform sensitivity from element to element.
2. PIXE can be accurate to about $\pm 5\%$ and precise to about $\pm 2\%$ for thin targets, with the only serious corrections involving x-ray attenuation in the lighter elements (sodium through calcium).
3. PIXE operates with total target masses that range between about 10^{-3} g for standard systems to 10^{-12} g for proton microprobes, achieving fractional mass sensitivities of about one part in 10^6 in a few minutes.

These capabilities successfully meet the needs of atmospheric chemistry and several other areas.

It is important to understand what PIXE is not—it is not a method of ultra trace analysis like neutron activation analysis (NAA) or some forms of mass spectroscopy (MS). Its sensitivity for a given element depends strongly on the nature of both the sample and the substrate upon which it is usually placed. In some cases, e.g. when only medium and heavy elements are of interest and samples of abundant mass are available over a large enough area, PIXE may not have any particular advantage over modern x-ray fluorescence (XRF) systems.

Since about 1975, the basic components of the PIXE systems in regular use have changed rather little. Energy dispersive Si(Li) x-ray detectors have not significantly improved in energy resolution, crystal thickness, area, or the ability to handle high count rates. The consensus that had developed by 1973 that protons between 1 and 4 MeV were optimum for most purposes has not been shaken, although α particles of equivalent velocity are also widely used. Yet, the systems in use today are enormously more powerful as usable analytical instruments.

Innovations such as extraction of the ion beam into an air or helium atmosphere, on-demand pulsing of the ion beam by the x-ray detector, critical edge and pierced ("funny") filters, location of the detector at back angles, automation of sample handling and data acquisition, and, especially, fast on-line data reduction codes designed specifically for PIXE spectra have allowed the method's potential to be realized in routine practice. In addition, mature PIXE laboratories are extending their capabilities by developing or adopting other analytical methods: ion beam nuclear activation (ACT) and ion scattering (RBS, for Rutherford backscattering in this review) for very light elements; energy dispersive or wavelength dispersive x-ray fluorescence (XRF, λ-XRF) for medium mass elements in thick targets; and other chemical, optical, or nuclear techniques.

The ultimate expression of PIXE capabilities is now being realized in proton microprobes that yield elemental information at mass levels approaching one attogram (10^{-18} g), in quantitative nondestructive analyses, providing information difficult or impossible to obtain by alternative techniques. The appreciation of this capability has fueled explosive growth in proton microprobe systems with PIXE and nuclear analytical capabilities in the past three years.

1.1 Other Reviews

1.1.1 PIXE ANALYTICAL SYSTEMS   Since the articles by Johansson et al (1970), Cookson & Poole (1970), Needham et al (1970), Duggan & Spaulding (1970), an extensive review literature has developed regarding PIXE and its analytical applications. This includes articles by Valkovic (1973), Owers & Shalgosky (1974), Folkmann (1975), and Johansson & Johansson (1976). In addition, the work of Johansson (1977) contains numerous articles in all areas of PIXE, while method summaries, tables, and references are included in works by Cahill (1975) and Mitchell & Ziegler (1977).

This review utilized a literature search (to November 1979), a listing of PIXE grants from the International Atomic Energy Agency, and surveys of 152 x-ray programs from the Smithsonian Science Information Exchange. In addition, a survey was sent to the major PIXE groups, with emphasis on those not recently represented in the literature with descriptions of their present methods. The 55 responses aided greatly in establishing present practice and future plans for PIXE systems (Table 1). This was doubly important because PIXE publications extend into journals in the fields of the applications, journals not widely read by the physicists and nuclear chemists who are most active in PIXE development. This literature is so extensive that only selected references can be cited. Yet the literature of the fields of application is the best measure of the method's value as it becomes a significant factor in the field of analytical chemistry.

1.1.2 PROTON MICROPROBES   A limited review literature was available on proton microprobes prior to recent articles by Cookson (1979, Cookson et al 1979) and Martin & Nobiling (1979). These are most welcome as 16 of 21 proton microprobes have become operational since 1976. This review relies heavily upon the comprehensive review by Cookson (1979), which contains detail far beyond the limited scope of the present article. The technology is changing rapidly as the new facilities come on-line, and applications of many of the systems have not yet appeared in print.

Table 1 PIXE programs

Laboratory	Ion, Energy (MeV)	Primary application	Comments[a]	References
AUSTRALIA				
Canberra	p, 3.0	Mineral	ACT	Clark 1976
Lucas Heights	p, 3.0	Archeology	RBS	Duerden et al 1980
Melbourne	p, 2.5	Biological	RBS, XRF, ACT	Chaudri et al 1980[c]
		Materials science	RBS	Legge et al 1980[c]
AUSTRIA				
Linz	p, α 0.1–0.8	Materials science	High I	Benka et al 1978
BANGLADESH				
Dacca	p, α 2.5	Fluids	EB	Khan et al 1979
BELGIUM				
Geel	p, α 1–3	Aerosols	RBS, XRF, ACT	Barfoot et al 1980[c]
Gent	p, 2.4	Medical	NAA, AA	Maenhaut et al 1979
Liege	p, α 3; 18	Minerals		Roelandts et al 1978
Namur	p, α 0.5–3.2	Biological	EB, **RBS**, ACT, λ	Deconninck 1977
BRAZIL				
Rio de Janeiro	p, 3.0			Montenegro et al 1979
Sao Paulo	p	Aerosols		Orsini et al 1977
CANADA				
Chalk River	p, 0.8	Materials science	RBS	I. Mitchell[b]
Guelph	p, 2.0	Biological	RBS	Campbell et al 1977
Manitoba	p, 20–50	Aerosols	ACT	
Montreal	p	Medical		LeComte et al 1978
CHILE				
Santiago	p, 6	Aerosols		Morales 1979[b]
CHINA (PRC)				
Fudan	p, 3.0	Archeology	EB	Chen et al 1980
Shanghai Institute	p, 3.5	Water		Li et al 1979
CHINA (ROC)				
Lungan	p, 3.0	Materials science	XRF	Lin et al 1978
DENMARK				
Aarhus	p, 0.5–5	Materials science	RBS, ACT	Folkmann et al 1980[c]
Copenhagen	p, 2.0–3.0	Aerosols	RBS	Kemp & Palmgren-Jensen 1977

	Beam	Application	Techniques	Reference
FINLAND				
Helsinki	p, 2.0	Aerosols	RBS, XRF, ACT	Raunemaa et al 1980[c]
Turko	p, d, α 5–21	Biological	ACT	Dahlbacka et al 1980[c]
FRANCE				
Saclay	p, α 4–22; 10–30	Materials science		Poncet & Engelmann 1978
Strasbourg	p, 4; HI	Materials science		Pape et al 1978
				Heitz et al 1978
GERMANY				
Bochum	p, 1.5–4	Biological	EB	Wilde et al 1978
Bonn	α, 30	Materials science	RBS	Bauer et al 1978
Frankfurt		Aerosols		Garten et al 1980[c]
Marburg	p, 2–4	Aerosol	RBS, ACT	Hasselmann et al 1977
GREAT BRITAIN				
Birmingham	p, 1–3	Biological		Saied et al 1980[c]
Hammersmith	p, 2.8	Medical	EB	Huda 1979
Harwell	p, 3.0	Materials science		Cairns & Cookson 1980
Manchester	p, 2.0	Materials science	RBS	Jarjis 1979
Surrey	p, α 0.3–2	Materials science	RBS	P. Hemmeut[b]
GREECE				
Athens	p, 2.0	Medical	EB	Katsanos et al 1976
HUNGARY				
Budapest	p, d, α 1.4–4		RBS, XRF, ACT	Szökefalvi-Nagy et al 1980[c]
Debrecen	p, α, HI 0.8–5		RBS, XRF, ACT	Vegh et al 1978
INDIA				
Bhudaneswar		Forensic		Sen et al 1980[c]
Kanpur	p, 2.0	Fluids		Varier et al 1980[c]
IRAN				
Tehran	p, 0.5	Water	EB	Lodhi & Sioshansi 1977
ISRAEL				
Weizmann	H.I., 34	Materials science	RBS	Blahir et al 1980[c]
ITALY				
Catania	p, α 1.0–2.5	Aerosols	TkTg, RBS, AA, ES	Baeri et al 1978
Milan	p, 2.8	Aerosols	XRF	Milazzo et al 1980[c]

Table 1 —*continued*

Laboratory	Ion, Energy (MeV)	Primary application	Comments[a]	References
JAPAN				
Nagoya	p, 0.15, 2			Moriya et al 1978
Osaka	p, α 1.0, 2.0	Materials science	RBS	Takai et al 1977, 1978
Sendai	p, 3.5			Kaji et al 1977
Tokyo	α 30	Biological		Uemura et al 1978
Tsukuba	p, Tandem	Materials science		Shima 1979
Yokohama	p, 2.5	Fluids		Hashimoto et al 1980[c]
NETHERLANDS				
Amsterdam	p, 3.0	Biological	RBS, ACT	Vis 1975
Eindhoven	p, 3.0	Fluids	XRF	Kivits et al 1979
NEW ZEALAND				
Lower Hutt	p, 2.5	Biological	RBS, XRF, ACT	Whitehead 1979
POLAND				
Cracow	p, 2.9	Biological	ACT	Hrynkiewicz et al 1980
Warsaw	p, 2–3	Minerals	RBS	Babinski et al 1980[c]
PORTUGAL				
Sacavem	p, 0.2–2			Lopes et al 1979
RUMANIA				
Bucharest	p, 4	Biological		Badica et al 1978
SOUTH AFRICA				
Johannesburg	p, 3	Aerosols	High *I*	J. P. F. Sellschop 1979[b]
Pelindaba	p, 2.4	Aerosols		Mingay & Barnard 1978
Faure	p, 1.5–3.5	Biological		Peisach et al 1973
SWEDEN				
Lund	p, 2.5	Aerosols	RBS, XRF, ACT	Johansson & Johansson 1976
Uppsala	α, 1.8	Art	RBS	Tove et al 1980

Location	Energy	Application	Methods[a]	Reference
USA				
Bell Labs	α 2–4	Materials science	RBS, XRF, ES	Feldman et al 1973
Brigham Young	p, 2	Aerosols	Chemical	Mangelson et al 1977
Brooklyn	p, 3		EB	Bauman et al 1979
Colorado	p, 4.3	Biological	XRF, ACT	Rudolph et al 1972
Davis (Univ. Calif.)	α, 18, 4.5	Aerosols	EB, RBS, XRF	Cahill 1975
Duke	p, 3	Biological	TkTg, λ	Walter et al 1977
Florida State	p, 4	Aerosols	RBS	Johansson et al 1975
Florida Univ.	p, 3.8	Biological	TkTg	Van Rinsvelt et al 1977
Geneseo, NY	p, 0.4–2.0			Chen 1977
Ill. Inst. Technol.	p, α, <2.5	Materials science	EB, RBS, ACT	Rickards & Ziegler 1975
IBM, NY	p	Aerosol	XRF	Mandler & Semmler 1974
Kentucky	p, 2, 4	Minerals		Pepper 1979[b]
Mississippi	p, 3	Gasses		Wolfe 1979[b]
Naval Research Lab	p, 3	Material science	RBS, ACT, XRF	Knudsen 1978
Newport News	p, 3.8	Biological	High I	Jolly et al 1978
Nevada	p	Aerosols		Hoffer et al 1979
Purdue	p, 2.5–5	Water	XRF, ACT	Rickey et al 1977
Rice	p, 3	Biological		Valkovic et al 1975
Rutgers	p, 1.65	Mineral		Kugel & Herzog 1977
Tallahassee	p, 3	Aerosols	EB	H. Kaufmann 1980[b]
Texas A&M	α, HI, 0.5–1.0 MeV/amu	Biological	ACT	Cross et al 1977
YUGOSLAVIA				
Ljubljana	p, 0.9	Biological	ACT	Kump et al 1977

[a] EB—external beams in air, helium, etc.
TkTg—targets thick enough to stop the ion beam.
High I—beam currents in excess of 1 μA.
XRF—x-ray fluorescence.
λ—wavelength dispersive detection of x rays.
RBS—Rutherford backscattering or other scattering methodology.
ACT—activation analysis for light element detection.

NAA—neutron activation analysis.
AA—atomic absorption.
MS—mass spectroscopy.
ES—electron microprobe or Auger analysis.
[b] Private communication (survey information).
[c] 2nd Int. Conf. on PIXE, Lund, Sweden (1980).

2 PARTICLE-INDUCED X-RAY ANALYTICAL SYSTEMS

The application of ion-induced characteristic x rays for elemental analysis demands that one create and then cross the interface between basic atomic physics and the well-developed discipline of analytical chemistry. This review first treats the question of PIXE methodologies as seen by the physicist or nuclear chemist, and then discusses the difficult interface to more traditional areas of analytical chemistry.

Analytical application of ion beams via production of characteristic x rays is based on the inversion of the standard relationship used to relate detected events N_x to incident ion flux N_I:

$$(\rho t)_Z = N_x/N_I(4\pi A_z/N_0\sigma_i^{emis}\Omega\varepsilon_d),\qquad 1.$$

where $(\rho t)_Z$ is the areal density (g cm^{-2}) of the element with atomic number Z and mass A_z, ε_d is the detection efficiency at energy E_x by a detector of solid angle Ω, N_0 is Avogadro's number, and σ_i^{emis} is the characteristic x-ray production cross section for a given transition. The natural units (g cm^{-2}) are often irrelevant to analytical chemistry, which normally expresses elemental results in fractional mass units such as parts per million (ppm). For a fixed incident ion condition, one can write the simple relationship for each transition,

$$(\rho t)_Z = (N_x/N_I)F_{z,i},\qquad 2.$$

subject only to x-ray attenuation effects that depend on the samples.

The characteristic x-ray production cross section, σ_i^{emis}, is a product of the probability for vacancy formation σ_v and the fluorescence yield ω_i for each transition. Theoretical models for production such as the plane wave Born approximation (Merzbacher & Lewis 1958, with recent summaries in Rice et al 1977 and Benka et al 1978), or the binary encounter approximation (Garcia et al 1973) agree with experimental data to within 5–30% in the proton energy region between about 1 and 5 MeV. Cross sections are very sensitive to ion velocity in this energy region, and scale approximately as (Cross et al 1977)

$$\sigma_k = f(Z_1^2 V_1^4 Z_2^{-12})\qquad 3.$$

for ions of atomic number Z_1 and velocity V_1 on targets with atomic number Z_2, although the effective Z_2 dependence falls to about Z_2^{-6} for light elements.

Fluorescence yield ω_i was reviewed by Bambynek et al (1972); radiative decay rates are available from Rosner & Bhalla (1970) and Scofield (1974b) via the relativistic Hartree-Fock-Slater theory. Agreement between theoretical and experimental radiative decay rates is excellent for a wide variety of vacancy production modes (Criswell & Gray 1974, Close et al 1973, Abrath & Gray 1974, Wyrick & Cahill 1973), but theoretical production cross sections are not accurate enough for use in PIXE analyses. Summaries of experimental production cross sections for K and L transitions are given, respectively, by Rutledge & Watson (1973), Hardt & Watson (1976), and Gardner & Gray (1978). Parametrization of cross-section data was done by Johansson & Johansson (1976), and Figure 1 is drawn from the fits. The similarities of the K and L cross sections for transitions of similar x-ray energy are striking, and the L cross sections for heavy elements, Z_H, are approximately the same as the K_α cross sections for light elements with atomic numbers $Z_L \simeq 0.37 Z_H$.

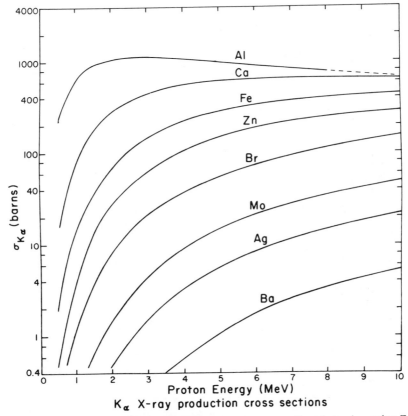

K_α X-ray production cross sections

Figure 1a K_α x-ray production cross sections vs proton energy E_p and atomic number Z_2 generated from a fit to data by Johansson & Johansson (1976). For heavier ions of equivalent velocity, cross sections scale approximately as Z_i^2.

One consequence of the nature of x-ray production by MeV ions is that probabilities of emission are large, usually greater than 10^2 barns, for all elements, either through the K or L shell x rays. Another is that probabilities vary slowly from element to element, unlike thermal neutron capture cross sections used in NAA.

Large cross sections are, of themselves, inadequate for good analytical sensitivities, as detection limits are always based upon a comparison of a characteristic x-ray peak and its background continuum x-rays and/or interfering x- or γ-ray lines. In a typical case, one has a sample superimposed on a supporting foil (Figure 2). The matrix contributes N_B, while the sample contributes both N_x for element Z and a continuum $N_{B,x}$. Thus, the nature of the continuum must be evaluated in order to establish the minimum detectable limit of mass (*MDL*), or sensitivity (*S*) in fractional mass units. Most of the early papers in PIXE including Johansson et al

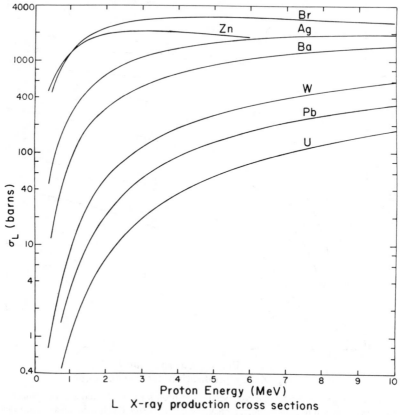

L X-ray production cross sections

Figure 1b   L x-ray production cross sections vs proton energy E_p and atomic number Z_2.

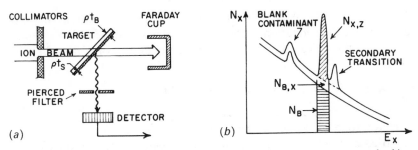

Figure 2   A schematic of a PIXE analysis of a target composed of a sample ρt_s on a backing ρt_B is shown in (*a*). A spectrum is shown schematically in (*b*), with the contribution of a sample ρt_s adding to that of a contaminated blank ρt_B.

(1970), Needham et al (1970), Campbell et al (1971), Watson et al (1971), Jolly & White (1971), Flocchini et al (1972), Cookson et al (1972), Deconninck (1972), Duggan et al (1972), Gordon & Kraner (1972), Lochmuller et al (1972), Verba et al (1972), Demortier et al (1972), Rudolph et al (1972), Pape et al (1972), Saltmarsh et al (1972), Umbarger et al (1973), Young et al (1973), Barnes et al (1973), and Johansson et al (1972) focused on the nature of the peak-to-background ratio, isolating the desirability of thin, low Z backings and low velocity ion beams.

The comprehensive work of Folkmann et al (1974a,b) advanced this work on a sound theoretical and experimental footing. They showed that most of the x rays below the energy of the maximum energy transfer from the ion of mass M to a free electron of mass m, $E_x \simeq 4mE/M$, were due to free electron bremsstrahlung, with a contribution due to bound electron bremsstrahlung that extended to somewhat higher energies than E_x (Figure 3). This part of the background scales as $(Z_1)^2$, so that ions of equal velocity are roughly equivalent in this region as shown in direct comparisons of equal velocity ions (Cahill et al 1974). Above the energy E_x, which is about 8 keV for 4-MeV protons or 16-MeV α particles, the background is due to projectile bremsstrahlung, Compton events in the detector from γ rays, and electrons striking the detector or a highly charged insulating target. Some of these effects are more severe for ions at higher energies, and require careful γ shielding and secondary electron suppression.

With information on both characteristic x rays and continuum background processes, S and MDL can be calculated for situations in which closely interfering lines are absent. The definition of the minimum number of counts that meet a 95% confidence level is approximately

$$N_x \geqq 3(N_B + N_{Bx})^{1/2}, \qquad\qquad\qquad 4.$$

although for small numbers of events, the student-t test modifies this to N_x

$\geqq 2 + 2[2(N_B + N_{Bx}) + 1]^{1/2}$ (Ziegler 1975). From these equations, one can derive the scaling equation for S in fractional mass units for thin samples, defined as those degrading incident ion energy or exit x-ray flux less than 10%:

$$S \propto \Delta E^{1/2}(Q\Omega\rho t_s)^{-1/2}(\sigma_B/\sigma_X^2)^{1/2}(1 + \rho t_B/\rho t_s)^{1/2}, \qquad 5.$$

where ΔE is the detector energy resolution, Q is total charge of ions, Ω is the detector solid angle, ρt_s is the areal density of sample material superimposed on a low Z, clean backing of areal density ρt_B, and the effective cross sections for the background counts and characteristic x-ray counts are σ_B and σ_X, respectively. This equation exhibits those parameters that can and have been used to optimize fractional mass sensitivity.

Combining the x-ray production cross sections with the probability of generation of the background continuum, one can estimate MDL as a function of proton energy E_p and atomic number Z. In order to do this,

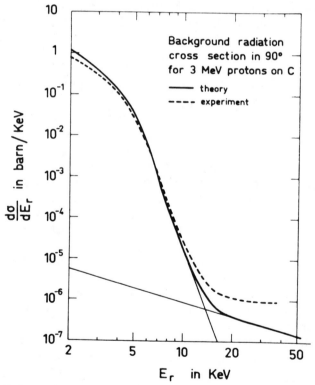

Figure 3   Background spectrum, experimental and theoretical, for 3-MeV protons on thin clean substrates (Folkmann et al 1974a).

assumptions must be made concerning the thickness of the supporting matrix; for Figure 4 these included detector energy resolution of 165 eV, solid angle of $0.003 \times 4\pi$, collected charge of 10 μC of protons, and target thickness of 0.1 mg cm^{-2}. Detection of low Z elements is through K x rays, and of high Z elements through L x rays.

2.1   Optimization of PIXE Systems

The inherent simplicity of PIXE analyses is firmly based upon the inherent simplicity of inner shell ionization processes, so that with an adequate x-ray detector one can perform satisfactory analyses using standard techniques. However, advances made in most aspects of the method have increased accuracy, precision, and sensitivity, and decreased cost, especially for applications where alternative nonaccelerator methodologies are available.

2.1.1 ION BEAMS   While most survey respondents reported using protons with energies averaging 2.8 MeV, some active programs use energies ranging from 0.2 to 50 MeV, α particles between 1 and 30 MeV, and heavy ions. The most notable advances are in control of the ion beam by the x-ray detector, on-demand beam pulsing, and extraction of the ion beam into air or helium.

On-demand beam pulsing was first proposed by Jaklevic & Goulding (1972) for electron beams in x-ray tubes. The detection of an x ray switches off the electron beam, minimizing the chance for a second x ray to arrive and pile-up on the first while the charge is being integrated. While space charge effects make realization of this idea difficult with electron beams, no such problems exist for ion beams, and a system was promptly built at the University of California at Davis by Thibeau et al (1973) using electrostatic

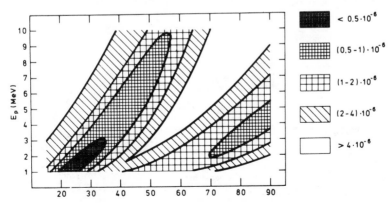

Figure 4 Sensitivity of PIXE analyses vs proton energy E_p and atomic number Z_2, based upon typical parameters given in the text (Johansson & Johansson 1976).

deflection plates operated at 5 kV, a pulse-off time of 0.5 μs, and a hold-off duration of 40 μs. The benefits of this simple device include reduced pulse pile-up events, increased count rate, and reduced dead time. Irradiation damage of the target is also reduced, since the damage is due to the average flux I_{ave}, which is now related to the instantaneous flux I_{inst} by $I_{ave} = I_{inst}$ $(1 - R T_R)$, where R is the count rate and T_R is the hold-off duration (Cahill 1975). Many of the larger programs now use on-demand beam pulsing (Koenig et al 1977, and survey results).

A second innovation in PIXE ion beam involves extraction of the beam into air or He (Figure 5). Beams have been extracted into air through a Ni foil (Seaman & Shane, 1975), through a Be foil (Katsanos et al 1976, Lodhi & Sioshansi 1977), through an Al foil (Deconninck 1977), through Al and Kapton foils (Huda 1979), and through a W foil (F. C. Yung, private communication, 1979). Beams have been extracted into helium through a Be foil (Modjtahed-Zadeh et al 1975) and through a W foil (Bauman et al 1979). A study of exit foils and analytical sensitivity is included in Raith et al (1980), who show comparable sensitivity for analyses in air and vacuum. The advantages of analysis in air include better sample cooling and lower damage rates in air and helium, which is important for biological samples. It also allows one to analyze liquids without vessels, eliminates charging of insulating samples, and handles large and awkward objects including archeological artifacts such as ancient swords. It reduces cost of automation through elimination of the vacuum chamber. The disadvantages are associated with the exit window backgrounds, the effective thickness of the air blank (about 2.6 mg cm^{-2} for a 2-cm path viewed by the detector), and the Ar interference (about 26 μg cm^{-2}).

In addition to incorporating these innovations, many groups have reduced beam areas to as low as 0.01 cm^2 in order to gain mass sensitivity.

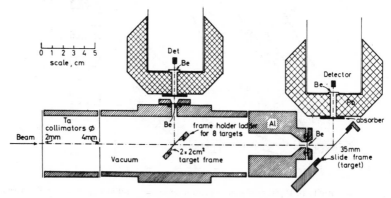

Figure 5   External beam system (Katsanos et al 1976).

In this regard, PIXE gains a major advantage over XRF systems, as the latter normally operate over about 10 cm². Thus, for equivalent fractional mass sensitivity S (Perry & Brady 1973, Goulding & Jaklevic 1977), PIXE requires only 10^{-3} as much mass as XRF. This aspect has been used with great effect for particle-sizing air samplers, as available mass is strictly limited (Section 3.1.2).

2.1.2 TARGETS   Many PIXE groups have adopted 5-cm × 5-cm plastic 35-mm slide frames as the standard automated sampling system; thin Mylar or Kapton foils of 0.6 mg cm^{-2} are widely used as backings. Thinner foils allow greater sensitivity, but at the expense of fragility and difficulties in automatic sample handling. Thus, many groups have relinquished some sensitivity for the reliability and convenience necessary for large numbers of analyses. Advances in sample preparation techniques are treated in Section 3.

Techniques for handling thick targets have been advanced by many of the users (Chemin et al 1974, Reuter et al 1975, Ahlberg 1975, Van de Kam et al 1977, Ahlberg 1977, Willis et al 1977a,b, Kropf 1977, Benka et al 1977, Van Rinsvelt et al 1977, Bauer et al 1978, Vegh et al 1978, Mommsen et al 1978, Huda 1979). Heating and charging effects must be handled carefully, normalization to beam current may have to be done by ion scattering or other indirect methods, and, in addition to severe x-ray attenuation corrections, one must consider refluorescence corrections normally negligible in thin targets. Thus, quantitative information on light elements is difficult to obtain, and one has surrendered some of the advantages of PIXE. The difficulty of obtaining elemental distributions vs depth has caused a decline of PIXE use in some industrial laboratories working in materials science. Thus, while 2/3 of all groups reported increasing PIXE activity, several groups specializing in thick targets reported declining use or termination of thick target PIXE programs in favor of other methods such as XRF. Much thick target PIXE work is being done with external beams, which reduces several of the problems while adding other advantages; more and more groups routinely use XRF for samples not well suited to PIXE.

2.1.3 FILTERS AND COLLIMATORS   Sensitivity has been increased through use of special types of x-ray filters such as the pierced or funny filter and the critical edge filter. Detector shielding and collimation are improved through layered collimators and integral electron deflection magnets. The pierced filter represents an attempt to gain sensitivity for heavy elements while cutting down high count rates for light elements (Cahill 1975, Kaufmann et al 1977). A filter capable of blocking soft x rays has a small hole in the center, typically about 10% of the detector solid angle. Thus, elements from Na through Cl have 10% of the solid angle for elements Ca and heavier; this sacrifices some sensitivity for abundant elements while gaining sensitivity for heavier elements, aids dead time correction circuits that handle pulses from soft x rays poorly, and increases detector resolution for light elements through collimation. The filter is now used by most aerosol groups, and is as close to "something for nothing" as one gets in PIXE systems.

Critical edge filters are also being used, especially in studies in materials science involving detection of a small mass in the presence of large nearby constituents and for thick samples with high proton fluxes (Gordon & Kraner 1972, Pabst & Schmid 1975, Ishii et al 1975, Badica et al 1978). A similar goal is approached through use of the proton beam to generate characteristic x rays, which in turn fluoresce a sample seen by the detector through a critical edge filter (Lin et al 1978).

2.1.4 DETECTORS AND ELECTRONICS   While only modest advances have been made in energy dispersive x-ray detectors, count rate capabilities are being raised to the point that 20,000 per second are not unusual for groups analyzing medium and heavy elements (see also Section 2.1.1).

In addition, more groups are moving the detector to backward angles to reduce backgrounds (Tawara et al 1976). The problem of sensitive detection of medium mass elements, $Z = 40$–70, caused by low cross sections for K lines and strong interferences for L lines with abundant light elements, continues to raise the possibility of wavelength dispersive x-ray detectors. These have resolutions that can be better by a factor of 50 than those of Si(Li) detectors for soft x rays, $E_x < 4$ keV, allowing resolution of L x rays of elements such as Cd and Sn from K x rays of K. The price one pays is lack of multielement detection and poor solid angles, resulting in roughly equivalent sensitivities for most light elements but real advantages for selected species. No active group is heavily involved with this methodology at this time, but the potential exists, especially in biological and geological samples.

2.1.5 DATA ACQUISITION AND REDUCTION   One of the areas of PIXE that is now evolving rapidly is that of automatic or semiautomatic data reduction. Three approaches are being widely used: (a) semiautomatic reduction with peaks and backgrounds established by means such as cursors and light pens; (b) automatic peak finding and integrating routines derived from γ-ray codes such as SAMPO or written entirely for x-ray work; and (c) automatic x-ray specific codes written to embody known x-ray relationships, blank background shapes, and detector imperfections such as escape peaks, radiative Auger transitions, and compton edges.

The second approach was used by Harrison & Eldred (1973), Nielson et al (1976), Willis & Walter (1977), and Nass et al (1978). As written and modified by Harrison & Eldred, the code RACE proceeds by (a) recording a blank foil and storing in memory (a physical blank, including contaminants, etc); (b) subtracting the blank channel by channel from a spectrum; (c) extracting residual counts due to N_{Bx} via a polynomial; (d) finding all statistically significant peaks with a correlation function; (e) integrating each peak twice, once with a linear sum and once via a Gaussian fit free in location, height, and width. Only then does the code examine the possibilities for each peak, recalibrate the energy, if necessary, to ± 5 eV, and extract areal density values. Thus, all primary and secondary lines are separately shown, and lines with no known x-ray association are still retained. At that point, the code performs logic on the spectrum to resolve interferences, calculate attenuation factors, eliminate sum peaks, etc.

The third approach is represented by the code HEX (Kaufmann & Akselsson 1975, Kaufmann et al 1977, LeComte et al 1978, Van Espen et al 1979) and its derivatives. In it, the theoretical spectrum shape is parametrized, as are all primary and secondary x-ray transitions, escape peaks, etc. The fit is then made to the spectrum using all available data in the entire spectrum, and one result is given for each element (Figure 6). HEX-type codes probably have slightly better precision, since they use all lines for every element in the fit, but they are dependent on a sophisticated library and cannot handle unusual spectra as RACE-type codes can. Both have

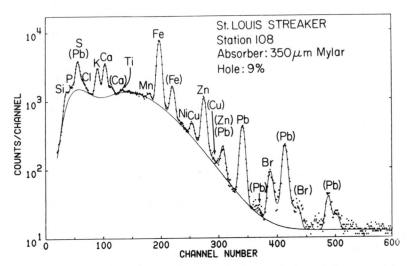

Figure 6   Computer fit by program HEX to a sample of atmospheric particulate matter taken with a Florida State "streaker." (Kaufmann et al 1977).

performed very well in interlaboratory comparisons. More laboratories are undertaking the considerable effort involved in full automation of data acquisition and on-line data reduction (Harrison & Eldred 1973, Cahill 1977, LeComte et al 1978) with advantages in quality assurance as well as convenience.

Extraction of the areal densities of the elements by the codes must be followed by corrections applied to evaluate x-ray attenuation and re-fluorescence effects due to finite target thickness. For targets thin with respect to energy loss of the ion beam, the two major corrections arise from a layering of material on the substrate and from the finite size of individual particles in the sample (Rhodes & Hunter 1972). Both become very important for soft x rays. Yet, it is in these elements between Na and Ca that PIXE analyses are most competitive with XRF, so that effort has been expended in making such corrections well. For example, in the code RACE, Harrison & Eldred (1973) calculated an approximate effective thickness for every sample by adding the normal fraction of light elements (such as oxides) associated with the observed heavier elements. This gives a value ρt_s that is used to calculate absorption for each element in the sample. The code also calculates separately particle size corrections based upon 16 possible particle size spectra and the normal composition in each size range for typical aerosols. This is an example of specialization based upon a specific type of program, but within the scope of its assumptions the approach has proven sound.

2.2   Calibration, Validation, and Verification

Three factors are necessary for the establishment and acceptance of the accuracy and precision of PIXE results.

1. Calibration, involving the relationship between the number of ions N_{I}, the number of characteristic x rays N_x, and the areal density ρt_z of element Z present in the sample. Systems for thin samples can be calibrated directly from Equation 1, as all parameters can be measured, but most groups use thin gravimetric elemental standards to find directly the calibration factor $F_{z,i}$ in Equation 2. By using polynomial fits, the nominal $\pm 5\%$ accuracy quoted by the manufacturer can be reduced to $\pm 3\%$ if care is taken to evaluate nonuniformity in the standard foils (Cahill 1975). Equations 1 and 2 are not directly relevant to thick samples, however, and such groups often calibrate using standard reference materials (SRMs), such as those from the US National Bureau of Standards or the Geological Survey, or added tracer elements such as yttrium.

2. Validation involves checking the calibration over a range of realistic samples and analytical conditions, and this step should be independent of calibration. Again, use of SRMs and spiked samples is common, as is

Table 2 Results of formal interlaboratory intercomparisons—ratios to standards

Method	Number of groups reporting data[a]	Solution[b] standards	Rock[c] standards	Aerosol[d] standards	Aerosol[e] samples
PIXE	7	1.03 ± 0.16	0.99 ± 0.29	0.99 ± 0.19	0.98 ± 0.08; 1.01 ± 0.16
XRF	8	0.97 ± 0.12	1.07 ± 0.20	1.03 ± 0.14	0.97 ± 0.08; 1.08 ± 0.15
λ-XRF	3	1.19 ± 0.34	1.12 ± 0.47	1.37 ± 0.50	—
AA, ES[h]	3	0.88 ± 0.17	0.40 ± 0.31	0.47 ± 0.29	1.04^f, 0.84^g
ACT	—	—	—	—	1.04^f, 0.84^g
NAA	1	0.97 ± 0.08	—	—	0.76 ± 0.15^g

[a] Camp et al 1975; each result represents the mean and standard deviation for all laboratories using the method for all elements quoted.
[b] Two samples, each including Al, S, K, V, Cr, MnFe, Zn, Cd, Au.
[c] Two samples, each including Al, Si, K, Ca, Ti, Mn, Fe.
[d] Two samples, each including Al, Si, S, K, Ca, Ti, Mn, Fe, Cu, Zn, Se, Br, Pb.
[e] Camp 1979; three samples or more, including up to 20 elements, of which S, Ca, Ti, Fe, Cu, Zn, Se, Br, and Pb are intercompared. Each result represents a single laboratory with the result being the mean and standard deviation for each element as compared to the referees.
[f] Laboratory reported S and Pb only.
[g] Laboratory reported S only.
[h] Atomic absorption; emission spectroscopy.

reanalysis of samples analyzed previously by the PIXE system or other compatible methods such as XRF. Informal exchange of samples between laboratories is very useful in this regard. Validation results are often included as part of publications on PIXE applications.

3. External verification involves participation in formal intermethod and interlaboratory comparisons of unknown samples with results published by a third party. The success of PIXE programs in such comparisons has greatly aided acceptance of the method (Table 2).

One other step remains before PIXE can be compared with other methods of analytical chemistry. Results in areal densities (g cm^{-2}) are desirable for aerosol samples, as one can establish a conversion factor related to the air volume pulled through the samples per unit area of substrate (cm^2 m^{-3}) and transfer the analytical results into the preferred atmospheric units (g m^{-3}). For other types of material, the preferred units are fractional mass S (in ppm) and to get areal densities into fractional mass units requires knowledge of total sample areal density ρt_s. The most direct way to obtain this value is to weigh a uniform sample of known area, but the low mass loadings adequate for PIXE make weight determinations to about 1 μg necessary in many cases. Modern electrobalances can do this, but care must be taken of variables such as electrostatic charge, humidity, and sample uniformity (Engelbrecht et al 1980).

3 APPLICATIONS OF PIXE SYSTEMS

PIXE is in essence a subdiscipline of analytical chemistry, and the results of PIXE analyses are appearing in the literature of the fields of the applications. Thus, as PIXE results diffuse through the scientific literature, complete bibliographies become more difficult to obtain and more irrelevant to this review. Results of the informal survey of most of the larger groups, combined with recent publications, yields an approximate evaluation of PIXE applications in many fields. These are described by discipline in order of annual numbers of analyses.

3.1 Atmospheric Physics and Chemistry

From the earliest publications on PIXE, analyses have been made of particulate matter collected from the atmosphere (Johansson et al 1970). Presently, five of the six largest PIXE groups list such analyses as their most numerous product, more than 30,000 in 1979. These groups have published over 150 articles and reports since 1975. The reasons for such heavy use are numerous, but they appear to include the following.

1. The essential variability of the atmosphere in time and space makes a single measurement of all atmospheric components to ultimate sensitivity

at one place and time of limited use. Atmospheric conditions change radically on several characteristic timescales including: short (seconds to minutes) for conditions such as wind gusts; diurnal (24 hr) for day-to-night variations; synoptic (a few days) for the passage of weather systems; and seasonal. Likewise, source emissions vary radically by location, especially for anthropogenic sources. For these reasons, statistically valid measurements of atmospheric composition demand many measurements (Flocchini et al 1976, Nelson et al 1976, Vie le Sage et al 1978, Courtney et al 1978, Pilotte et al 1978, Everett et al 1979, Moschandreas et al 1979); statistical methods are widely employed in interpreting the results (Cahill et al 1977a).

2. All elements in atmospheric particles are potentially important. All elements contribute to the mass of atmospheric particles, and any compound in the proper size range can scatter light (Barone et al 1978). Many elements are potentially toxic, and analyzing separately for each by chemical or optical spectroscopic methods is expensive (Kemp & Palmgren-Jensen). Also, particles (unlike gases) retain elemental signatures characteristic of their sources, even at long distances, which aids source identification and transport evaluation (Feeney et al 1975, Prahm et al 1976, Cahill et al 1976, Winchester 1977, Meinert & Winchester 1977, Lannefors et al 1977, Berg & Winchester 1978, Lawson & Winchester 1979a,b).

3. Small amounts of mass are available for analysis. A typical urban aerosol concentration is about 80 μg m^{-3}, or less than 0.1 ppm by mass. While effective filtration can remove particles from atmospheric gases, large air pumps, large filters, and extended sampling durations are the rule, as traditional methods used about 0.2 g of material for analysis of a limited number of species. This lack of available sample mass is compounded if particle size information is desired.

4. Information on particle size is important for understanding particulate matter. One of the major realizations of the past decade in atmospheric chemistry was that particulate matter occurs in characteristic size modes, generally peaked around 0.03, 0.3, and 30 μm diameter (Butcher & Charlson 1972). This trimodal distribution is important both in understanding aerosol sources and in evaluating effects such as lung capture (Desaedeleer et al 1977) and light scattering (Barone et al 1978). The modes are chemically distinct one from another (Feeney et al 1975, Flocchini et al 1976, Orsini et al 1977, Berg & Winchester 1978, Adams et al 1977). However, simultaneous measurement of size and chemistry requires devices such as impactors that can collect no more than a few monolayers of particles before they become clogged or start to make sizing errors. Three monolayers of 1-μm diameter particles amount to about 500 μg cm^{-2}, which, coincidentally, is about all the mass that can be tolerated by PIXE

systems when analyzing for light elements such as Si. PIXE's sensitivity to areal density, not collected mass, and the ability to match small area ion beams with small area particulate deposits on impactor stages has allowed the development of light, simple air sampling instrumentation designed for PIXE. This includes continuous time sequence filter samplers ("streakers") (Nelson et al 1976), rotating-drum monitoring impactors (Cahill et al 1976), six-stage single-orifice impactors (Van Grieken et al 1976), and tandem or stacked filter samplers (Cahill et al 1977b). In addition, manufacturers of filter media have responded to the demands of PIXE (and XRF) groups by developing thinner cleaner filters. The role of PIXE in size/composition measurements of atmospheric particulates is such that an estimated 90% of all such information in three or more size categories, worldwide, has been generated by PIXE programs.

3.2 Nonbiological Liquids

The second most active area of PIXE applications is the analysis of (nonbiological) liquids. A major effort of the US Environmental Protection Agency (EPA) in surveying drinking water is based mainly on PIXE (Rickey et al 1977), and several thousands of analyses are performed each year in the limited number of laboratories that possess such capabilities. Calibration of fluid systems is facilitated by easy application of tracer materials. The key problem is preparation of a suitable target from the fluid sample, either through liquid-specimen cells (Bertin 1978, Huda 1979, Khan et al 1979), external beams on liquid jets (Deconninck 1977), drying (Jolly & White 1971, Camp et al 1974, Baum et al 1977, Campbell et al 1977, Sioshansi et al 1977, Kivits et al 1979), or vapor filtration (Rickey et al 1977). The drying methods possess excellent sensitivities when deposits are placed on thin substrates, but they face the problem of nonuniform deposits during drying, problems handled, respectively, by aspirators, microdrops (two groups), a single drop within the beam, aspirators, and spinning filters (Kivits et al 1979). Vapor filtration is an elegant solution that involves pumping water through a semipermeable membrane. All these methods reach about 1 ppb for clean water samples. The specimen cell and external beam methods have simple preparation and high accuracy, but lose some sensitivity because large amounts of fluid act as an effective thick blank, ρt_B. They work very well for contaminated oils, slurries, etc. Multielemental analyses of fluids seem to possess characteristics that should make them prominent in PIXE analyses, despite the fact that they are already well handled in the liquid phase on an element-by-element basis by optical methods (AA, ES).

Another promising area is analysis of gases in cells (Katsanos & Hadjiantoniou 1978), which can reach high sensitivity when combined with cryogenic concentration (Cahill 1973, Wolfe 1979, private communication).

3.3   *Biology and Medicine*

Biological samples are analyzed in almost every PIXE laboratory, but few laboratories make very many analyses, probably because of the labor involved in sample preparation. The first publications on PIXE featured biological analyses such as blood, but widescale usage has been hampered by target preparation problems, sensitivity requirements for many elements below the nominal 1-ppm level attained by PIXE systems, and competition from well-established alternative methodologies. In addition, efforts to increase sensitivity often lead to thick targets that both forfeit quantitative information on the light elements handled so well by PIXE in thin samples and largely eliminate alternative ion beam methodologies for very light elements.

The techniques used in preparing a target from a biological sample are reviewed by Mangelson et al (1977, 1979) and Campbell (1977). Numerous groups have analyzed hair directly (Jolly et al 1974, Valkovic et al 1975, Valkovic 1977, Henley et al 1977, Whitehead 1979); some laboratories used proton microprobes (Table 5). Likewise, tooth enamel (Ahlberg et al 1975) and other such substances can be handled directly, although obtaining quantitative results is not easy because of nonuniformity. However, most work has gone into analyzing tissue samples, and target preparation is crucial. Thin tissue sections have been directly analyzed by Jundt et al (1974), Walter et al (1974), Kemp et al (1975), Campbell et al (1975), Vis et al (1977), Hasselmann et al (1977), and others. While results can be quantitative, sensitivities are marginal for many trace elements of interest since the original biological matrix is intact. Many kinds of biological samples can be effectively dried or lyophilized and ground into a powder. These can be dusted onto thin foils (Murray et al 1975) or pressed into pellets (Walter et al 1974, 1977, Mandler & Semmler 1974, Rupnick et al 1977), but again no concentration has been performed. Wet ashing has been used (Campbell et al 1975, Mangelson et al 1979), but reduction from the original sample mass (concentration) was no better than 0.58 and could be worse than unity, while contamination problems can be serious (Mangelson et al 1979). High temperature or heat ashing has been used to concentrate samples (Rudolph et al 1972, Van Rinsvelt et al 1977) but losses of volatile elements pose real problems despite excellent mass concentration. Work with low temperature or plasma ashing in oxygen (Bearse et al 1974, 1977, Sheline et al 1976, Zombola et al 1977, Mangelson et al 1979) has shown both good reduction of mass, to 4.6% of the original value (Mangelson et al 1979), and retention of potentially volatile metals. These advantages should make the method more popular in the future, as the factor of 20 concentration allows PIXE to reach sub-ppm trace levels in tissue while retaining its quantitative, multielemental characteristics. There may also be a role for inexpensive PIXE analyses to the 100-ppm level on ground plant material as a monitor of plant growth, response to salt stress (Murray et al 1975), and protein content (Kump et al 1977).

3.4   Geology and Soil Sciences

Geological samples can be easily prepared and analyzed by PIXE, but relatively few of the laboratories surveyed for this review do so. One reason is that localization of an analysis to a single microcrystalline type is so important that scanning electron microscope (SEM) and electron microprobe techniques are preferred over bulk analyses, even at the loss of any trace capability. Another reason is that the abundant aluminosilicate components of most rocks are of minor interest in many studies, and the heavier elements are better seen by wavelength dispersive XRF (λ-XRF) or, increasingly, XRF since abundant mass is available in bulk analyses. A third reason appears to be the poor reputation of all x-ray methods for absolute accuracy in thick samples. However, sensitivity is excellent for small crystals (Van Grieken et al 1975), the multielement nature aids studies (Roelandts et al 1978), and a major geological impact occurs in PIXE proton microprobes.

Promising areas involve soils (Shiokawa et al 1977) and sediment (Pilotte et al 1978). Recent work shows strong elemental chemical fractionation in soils vs particle size when suspended by wind. Laboratory analyses of soils subject to such erosional processes thus must involve resuspension and sizing for relevant measurements (Ashbaugh et al 1978), and such devices then resemble air samplers, giving PIXE numerous advantages (Section 3.1).

3.5   Materials Science, Archeology, Miscellaneous

PIXE analyses in materials science suffer from the problems of thick target analyses and the need for depth profiles in many industrial laboratories (Feldman et al 1973). Nevertheless, there are some elements for which no convenient nuclear reaction exists, and others in which a completely unknown contaminant is suspected as an impurity. Further, PIXE analyses conveniently use the same beams widely used for RBS analysis (Jarjis 1979). The major PIXE use for materials science may well be with proton microprobes, however.

There continue to be programs in archeology using PIXE beams (Gordon & Kraner 1972, Ahlberg et al 1975, Baijot-Stroobants & Bodart 1977, F. C. Yung, private communication, 1979, Boulle & Peisach 1979, Boulle et al 1979) and development of external beams will probably increase such uses since a much larger set of potential samples will be available for nondestructive analyses. Present miscellaneous uses of PIXE beams may be the major uses of tomorrow, from forensic analyses of gun residues (Barnes et al 1973), analysis of meteorites (Kugel & Herzog 1977), to marking mosquitoes with trace element coded fluorescent tags (McClelland et al 1972). Who knows?

4 PROTON MICROPROBES

While the first use of ion-induced x rays with highly focused or collimated proton beams dates from the early days of PIXE (Cookson & Poole 1970), rapid growth in facilities and utilization has occurred only within the past three years. These proton or nuclear microprobes use many analytical techniques besides PIXE, but the excellent sensitivity of PIXE lowers the amount of mass that can be detected in a nondestructive multielement analysis to below 10^{-18} g (Figure 7). The combination of trace sensitivity and low minimum detectable mass represents in many respects a capability unique to the method. In addition, use of other accelerator-based methods such as nuclear reactions and elastic ion scattering to extend elemental range to the lightest elements provides proton microprobes with capabilities absent from other electron beam devices, e.g. electron microprobes and scanning electron microscopes (SEMs) with x-ray capabilities. In this

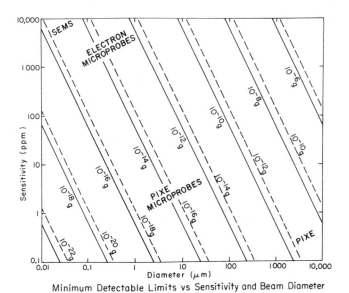

Minimum Detectable Limits vs Sensitivity and Beam Diameter

Figure 7 Minimum detectable mass vs beam diameter and analytical sensitivity. The dashed curve corresponds to an analysis made on a thin carbon backing foil $\rho t_B = 65\ \mu g\ cm^{-2}$ and a sample $\rho t_s = 35\ \mu g\ cm^{-2}$. The solid curves represent thick samples.

Table 3 Collimated nuclear microprobes[a]

Year	Laboratory	Ion, energy (MeV)	Minimum spot size (μm)[b]	Comments	References
1966	Harwell	p, 1.8	660	Later 25 μm	Pierce et al 1966
1966	Lucas Heights	p, 0.9	100	Later 60 μm	Mak et al 1966
1971	Los Alamos	p, 4.0	30	H^- ions	Armstrong & Wegner 1971
1973	Munich	α, 1.5	10		Bayerl & Eichinger 1978
1975	Heidelberg	p, 1.0	1	also ^{3}He	Nobiling et al 1975
1975	MIT	p, 2.5	25	EB[c]	Horowitz & Grodzins 1975
1976	Florida State	p, 4.7, 5.7	~30	Tandem	Cahill et al 1978
1978	Brookhaven	p, 3.5	25	EB[c]	Shroy et al 1978
1978	Grumman	d, 1.6	12	200 μm other dimension	Schulte & Kamykowski 1978

[a] Largely derived from Cookson (1979).
[b] FWHM of spot size.
[c] EB—external beam.

regard, however, groups using proton microprobes with PIXE are mimicking the trend of standard PIXE groups: they seek smaller area beams for lowered minimum detectable mass levels, and they use nuclear methods to extend elemental range. Proton microprobes using PIXE thus appear to represent a natural evolution of the method, and most of the earlier discussion is relevant to them. The summary that follows therefore emphasizes the special problems and solutions associated with the use of small diameter beams. Much of the discussion is based on the recent comprehensive review by Cookson (1979), which provides detail and discussion beyond the scope of this review.

4.1   Preparation of Microbeams

Beams of MeV ions with diameters below 100 μm can be prepared either by collimation or by ion focusing. Collimation was the first method used in pioneering radiation dose studies by Zirckle & Bloom (1953). Since then, several laboratories, including those at Harwell, Heidelberg, and Florida State, have developed collimated microbeams prior to developing focused beams, while others have continued to use collimated beams (Table 3). Collimation is simpler, but for beams below 25 μm it's limited by low ion fluxes. The samples are often placed very close to the aperture, in order to maximize flux, which results in very tight geometries.

More recently, groups at MIT (Horowitz & Grodzins 1975) and Brookhaven (Shroy et al 1978) developed collimated microbeams that are used outside of the vacuum and exit via foils or differential pumping. Numerous advantages exist in performing irradiations in air or other atmospheres, including additional cooling that may prove to be especially useful for biological samples.

Focused microbeams were first developed at Harwell in 1970 (Cookson et al 1972); they used a focusing system of four magnetic quadruples called a "Russian quadruplet" on a 3-MeV single-ended Van de Graaff accelerator (Figure 8). This focusing arrangement allows equal magnification in both planes from an object slit and placed 4.6 m from the specimen, which is 0.21 m beyond the last focusing lens. Careful analysis of focusing aberrations allowed a minimum beam spot of 3 μm diameter at full width half maximum. Typical current capability was 30 pA μm^{-2}. This system has been operational for a decade, and similar systems have been built at Manchester, Surrey, Namur, Melbourne, Studsvik, and Bochum (Table 4). The Harwell system can also operate with the second lens shut down, which creates a beam with rectangular beam spot narrower in one dimension but with current densities of five times the current density of the Russian quadruplet, or about 150 pA μm^{-2}. Other microprobes using a single stage of demagnification include the electrostatic triplet at Bell Labs, and the doublets of Heidelberg, Karlsruhe, and Florida State. Systems using two stages of demagnification have been built at ETH Zurich and UCLA; both possess an intermediate focus. Other systems are being constructed using still more configurations. In all these microprobes, the performance is characterized by higher beam currents than those obtainable with collimated microprobes, with the Harwell three-lens system reaching 150 pA μm^2, Karlsruhe doublets at 60 pA μm^2, and Russian quadruplet systems at around 20–30 pA μm^2 when operated with single-ended Van de Graaf accelerators. The limited results with tandem accelerators to date indicate that the lower current densities of these machines result in currents of $\lesssim 2$ pA μm^{-2}, although energies can be higher. For details of the focusing and performance of these machines, see Cookson (1979) and Martin & Nobiling (1979).

Special problems exist for proton microprobes in establishing the exact location of the point of irradiation. One of the great advantages of electron-based microprobes and scanning microscopes is that we can use electrons from the sample to provide an image that can then be correlated with the elemental analyses. Most proton microprobes use optical methods such as scintillators to locate the point of analysis, but a secondary electron imaging system was recently reported for the Harwell system (Younger & Cookson 1979). The beam can be rastered electrostatically over the sample, in a manner similar to SEMs, and good images have been obtained without deposition of coatings often used in electron microscopy. This advance appears to be a key one for large-scale use of proton microprobes, as it matches one of the most attractive features of electron systems while allowing PIXE trace sensitivities and light-element detection available only with ion beams.

The energy and current capabilities of proton microprobes make

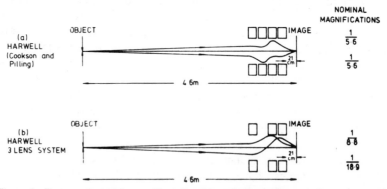

Figure 8 Proton microprobes at Harwell, with excitations corresponding to a Russian quadruplet (*a*) and Harwell three-lens system (*b*) (Cookson 1979).

Table 4 Focused proton microprobes[a]

Year	Laboratory	Ion energy (MeV)	Minimum spot size[b] (μm)	Comments	References
1970	Harwell	p, 3	3	Mag. Russian quadruplet. 2 μm later using only 3 lenses	Cookson et al 1972
1974	ETH Zurich	p, 6, T[c]	11 × 26	2 mag. doublets, intermediate focus	Bonani et al 1978
1974	UCLA	p, 2	40	2 mag. doublets, intermediate focus, later 0.01 nA into 8 μm	Singh et al 1976
1976	Heidelberg	p, 2, p, 6, T	2	Magnetic doublet	Nobiling et al 1977
1977	Bell Labs		12	Electrostatic triplet	Augustyniak et al 1978
1977	Lower Hut	0, 3	18	Russian quadruplet	Coote et al 1978
1977	Studsvik	p, 5.5	25	Russian quadruplet	Brune et al 1977
1977	Bochum	p, 4	~50	Unsymmetric quadruplet	Wilde et al 1978
1977	Karlsruhe	p, 3	2.5	Magnetic doublet	Heck 1978
1978	Melbourne	p, 5	10	Russian quadruplet	Legge 1978
1978	Manchester	p, 5	<10	Russian quadruplet	Calvert 1979[d]
≧1977	Kingston	p, 4		Mag. doublet, 10 μm aim	MacArthur et al 1977
1978	Surrey	p, 2	30 so far	Russian quadruplet	Hemment 1978[d]
1978	Namur	p, 3.2	20	Russian quadruplet	Deconninck 1979[d]
1980	Oxford	p, 6, T	4	Two doublets available	Watt et al 1979[d]
≧1979	Darmstadt	HI		Mag. triplet, heavy ions	Fisher 1977
1979	Florida State	p, 4, T	25	Magnetic doublet	Fletcher et al 1979[e]
1980	Eindhoven	p, 3.5 cyc	10	Quadruplet	M. Prims 1980[e]
≧1980	Amsterdam	cyc		Doublet	den Ouden et al 1980[e]

[a] Largely derived from Cookson (1979).
[b] FWHM of spot size.
[c] Tandem Van de Graaff.
[d] Quoted in Cookson (1979)
[e] Private communication.

questions of target damage more important than for larger area PIXE systems. This is especially important when nuclear reactions are used for light-element analysis, as the lower cross sections put a premium on high current fluxes. Energy densities are on the order of 0.5–2 kW mm^{-2}; this limits use to samples that can stand such energy densities, samples often drawn from problems in metallurgy or geology. Even so, samples can be easily vaporized or highly modified by such beams. This problem has always existed with electron beams, however, which operate at about 1 kW mm^{-2} (Bosch et al 1978). The work of Martin and co-workers suggests that

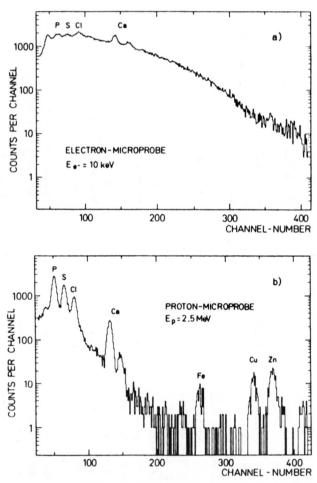

Figure 9 Comparison of PIXE and electron microprobe analyses of a biological sample (Bosch et al 1980).

high energy heavy ions may actually be less damaging than light ions per unit information in soft x-rays (Martin 1978a,b).

4.2   Analytical Methods

Most work with proton microprobes has involved elemental analyses of physical and biological samples. Ion beams are used for this purpose rather than easily obtainable fine electron beams for two main reasons: (a) nuclear methods can detect very light elements unobtainable by SEMs and electron microprobes, and (b) PIXE analyses allow trace sensitivities to better than 1 ppm, in contrast to typical values of 1000 ppm for electron beams. The reaction methods used with nuclear microprobes are given by Cookson (1979) and include detection of light elements via $^1H(^{11}B, \alpha)2\alpha$, $^2H(^3He, p)$, $^7Li(p, \alpha)\alpha$, $^9Be(\alpha, n)$, and $^9Be(d, p)$, and a variety of $(p, p'\gamma)$, (p, d), (d, p), $(d, p\gamma)$, and (d, α) reactions. In addition, the well-developed field of Rutherford backscattering analysis yields both elemental and depth information, especially in layered structures. PIXE, however, is particularly suitable for use in proton microprobes, because of its large cross sections, nondestructive trace analyses, quantitative nature, and inclusive detection of any element of $Z = 11$ or higher. Si(Li) detectors can be positioned very close to the beam allowing very high solid angles, high absolute detector efficiencies, and relatively lower damage rates to samples than for nuclear methods. A comparison of PIXE results to those obtainable with electron beams is shown in Figure 9 (Bosch et al 1980). Clearly, sensitivities are greatly improved by the use of PIXE so that a number of elements obscured in the electron beam analysis are easily seen by PIXE.

It should also be mentioned that the proton microprobes have not yet used many of the techniques adopted in mature standard PIXE systems, so that further improvement can be expected. Most groups have not used on-demand beam pulsing, although it would reduce sample damage in biological samples. Use of this technique would be fairly easy for collimated microprobes. Likewise, although some work is done with critical edge filters, pierced filters that would aid biological and pollution analyses have not been incorporated in such systems. There is no reason, in principle, that sensitivities should be any lower in PIXE microprobes than in standard PIXE systems.

4.3   Applications of Proton Microprobes Using PIXE

Applications of proton microprobes using PIXE are increasing rapidly as many laboratories have entered the field since 1976. Table 5, largely from Cookson (1979), displays results published to date, but descriptions of many more applications are in press or under preparation at this time.

One factor that stands out in the targets chosen for study is that, with the

Table 5 Microbeam analysis using PIXE[a]

Specimens	Main elements	Resolution (μm)	Comments	Laboratory[b] and reference
Biological				
Human hair	As, Pb, etc	7	Linear scan of diameter	H, Cookson & Pilling 1975
Hamster cell	Si; P, S, Cl, K	2	Linear scan in air	H, Pierce 1978
Rose petal	Z ≦ 20	—	Scanning maps	H, Legge 1976
Human lung	Al	25	Line scans and maps	Mi, Grodzins et al 1976
Rat's eye	S, Cl, K, Ca	100	Line scans and maps	Mi, Horowitz et al 1976
Human hair	Cl, Ca, Fe, Zn, As, Hg	100	Line scans along length	Mi, Horowitz et al 1976
Wheat	Many, Si-Pb	10	Computer maps	Me, Mazzolini & Legge 1978
Ant's head	Ca, Fe	10	Computer maps	Me, Legge 1978
Chick embryo	K, Ca, Cr, Cu, Zn	100	Step-by-step profile	B, Wilde et al 1978
Rat's tooth	P, S, Cl, K, Ca, Zn	20	Step; beam in air	Br, Shroy et al 1978
Pollen tube	P, S, Ca (Ca at 16×10^{-15} g)	2	Linear scan	He, Bosch et al 1980
Metallurgical[c]				
Steels	Cr, Fe, Ni	—	Linear scan	H, Pierce et al 1974
Nuclear fuel	U	—	Linear scan	H, McMillan et al 1976[a]
Steels	Fe, Cr	—	Maps	H, Pierce & Huddleston 1977
Steels	Ti, Cr	13	Linear and area scans	H, McMillan et al 1976[b]
Alloys	Ti-Mo	4	Linear and area scans	H, McMillan et al 1979

Geological

Copper pyrites	Si, S, K, Fe, Cu	5	Spectra from areas	H, Pierce 1973
Sphalerite	Ca, Fe, Zn	25	Maps	Mi, Horowitz & Grodzins 1975
Zircons	Hf, Zr, Tl, Pb	15	Linear scans	H, Clark et al 1979
Monazite, U/Th halo	P, Sc, Y, rare earths, Th, U, Pb; no superheavy elements	30	3 samples	F, Cahill et al 1978
Monazite, Giant halo	P, Sc, Y, rare earths, Th, U, Pb plus Br, Sb, superheavy elements?	30	4 samples	F, Gentry et al 1976
Monazite, Giant halo	Y, rare earths, Th, U, Pb; no Br, Sb, or superheavy elements	10	One of above specimens	H, Cookson et al 1978
Monazite, U/Th halo	Y, rare earths, Th, U; no superheavy elements		—	Z, Bonani et al 1978
Perthite	K, Fe, Ga, Sr, Rb, Pb	11	Linear scan, few ppm	Z, Bonani et al 1978
Lunar ilmenite	Ti, Fe, Zr, Y, Nb	3	Linear	H, Bosch et al 1980

Miscellaneous

Furnace slag	Si, Fe, Cu, Zn, Pb	5	Maps	H, Cookson et al 1972
Electronic components	Si, Ge	5	Linear scans	H, Peisach et al 1973
Corrosion layers	Fe, Cr	5	Depth on tapered sections	H, Singleton & Hartley 1979
Soot	V, Fe, Ni	10	Linear scans	H, McKenzie 1976[a]
MgO doped with Al	P, Ti, Cr, Fe	140	Map	Mi, Grodzins et al 1976
Fe implanted in C	Fe	30	Map	Z, Bonani et al 1978

[a] Largely derived from Cookson (1979).
[b] Key to laboratories: B = Bochum; Br = Brookhaven; F = Florida State; H = Harwell; He = Heidelberg; Me = Melbourne; Mi = MIT; Z = Zurich.
[c] Many PIXE results were compared with activation data for light elements.

exception of some biological samples, most must be considered thick targets. This raises all the problems mentioned in Section 2.1.2, and consequently most groups rely on semiquantitative analyses of spatial variation of mass, spatial variations of elemental ratios, or wholly qualitative pattern displays and maps. As in electron beam devices, this may be perfectly adequate to come to a sound conclusion, but it does make reproducibility difficult unless an entire sample has been analyzed in area and depth. Depth discrimination can be difficult, since at low ion energies, PIXE x-ray production cross sections fall very sharply with decreasing energy, resulting in a sensitivity to near surface regions in most samples. Thus, a small amount of material at the surface yields identical results to larger amounts at depth, and beyond a certain depth (tens of micrometers) no sensitivity exists at all. Nuclear reaction studies, on the other hand, utilize energy loss and resonant reaction rates to obtain depth information.

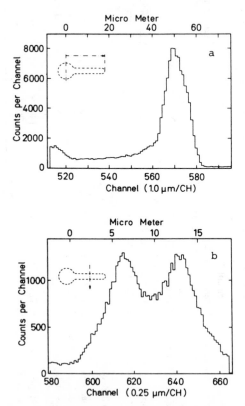

Figure 10 Linear scans of pollen tubes by the Heidelberg proton microprobe (Bosch et al 1980). The calcium content in one of the specimens was 0.016 pg, and the resolution was 1–3 μm. A schematic of the pollen tubes and the axes of analysis are shown in each figure.

Depth information can be gained by turning the sample over and scanning a diameter, as in a human hair (Cookson & Pilling 1975).

Spatial maps must contend with the trade-off between sensitivity and spatial resolution for a given dimension. To reach ppm sensitivity requires long durations in each resolution cell, and as the length and width of the map increase, the number of cells increases as the length times the width. This quadratic behavior often results in use of linear scans when trace sensitivities are required, with an implied assumption about the unmapped areas. For this reason, many of the maps published using proton microprobes do not reach even close to ppm sensitivities, while linear scans have achieved such levels. A good example of such a scan is shown in Figure 10 (Bosch et al 1980) for a pollen tube used in studies of plant growth. The calcium levels shown correspond to about 0.016 pg, while sensitivities correspond to a few parts per million in mass. These authors also point out the care that must be taken in sample preparation. A further example of a linear scan at trace sensitivity is shown in Figure 11 (Clark et al 1979).

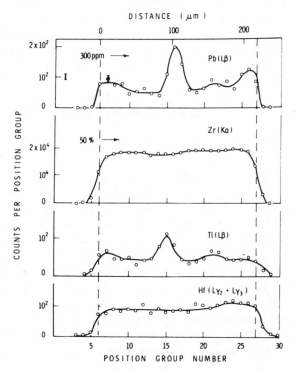

Figure 11   Linear scan of a zircon on the Harwell proton microprobe (Clark et al 1979). The existence of a "hot spot" on the order of 10 μm across is exhibited, shown by the Pb and Tl x rays and associated U fission tracks. Detection of the radiogenic lead and uranium (and hence the "hot spot") could not be accomplished with electron beams.

Zircons, typically 50×150 μm were scanned along various axes in order to determine the concentrations of Pb, Hf, Tl, and Zr, as well as their correlation to uranium seen in fission track analyses. The study was motivated by problems with dating the zircons through U-Pb isotopic data. The study shows high uranium "hot spots," with associated radiogenic lead that has not migrated appreciably, confirming a suggestion made by Steiger & Wasserburg (1966) that could account for the erroneous dates. Electron microprobes (and optical microscopes) could not detect the "hot spot." A further example of the essential role of spatial resolution in analyses of microcrystals is given in Cookson et al (1978), in which sequential analyses made without moving the proton beam more than a few micrometers still resulted in analyses differing by 20%, while abundant amounts of Br and Sb seen in previous analyses of the same 60-μm crystal (Cahill et al 1978) were totally absent. These results also illustrate the role of proton energy, as the center of the crystal was not investigated at the lower proton energies at Harwell.

Quantitative results with microprobes are possible when spatial variation is either absent or adequately known. PIXE analyses were able to determine elements as light as Si in the monazite (La, Ce, Th) PO_4 crystal, and oxygen could be inferred from the phosphorus content. The results are consistent with other analyses, and even allow dating the inclusions through U/Th/Pb ratios to 7×10^8 yr (Cahill et al 1978). The aforementioned work also graphically illustrates that it is far easier to obtain a microbeam than it is to control its location on a microsample.

5   CONCLUDING REMARKS

The steady expansion of PIXE capabilities and its successes in numerous disciplines have largely overcome the reluctance of many scientists to depend upon an analytical method based upon a university-operated accelerator. PIXE capabilities have been integrated into many research programs, and instruments have been designed and built under the assumption that PIXE systems can perform the required analyses at a reasonable cost. PIXE capabilities have played a major role in the success of proton microprobes, and they, in turn, have become uniquely capable in a very important sub-area of analytical chemistry. Yet in their essence, these advances must be based upon both the simple elegance of inner shell atomic physics and the willingness of numerous physicists and nuclear chemists to exploit these processes in areas far removed from their original training.

ACKNOWLEDGMENTS

I would like to express my appreciation for the abundant response to the survey by many PIXE groups, and express my regret that so little could be included in this short article. I would also like to thank my wife and many members of the Air Quality Group at Davis for suggestions and editorial assistance, and Charlene Curry, Dwight Wholgemuth, and Charles Goodart for preparation of the manuscript.

Literature Cited

Abrath, F., Gray, T. J. 1974. *Phys. Rev. A* 10:1157

Adams, F., Dams, R., Guzman, L., Winchester, J. W. 1977. *Atmos. Environ.* 11:629–34

Ahlberg, M. 1975. *Trace analysis of thick samples by proton induced x-ray emission.* Thesis. Univ. Lund, Sweden. 84 pp.

Ahlberg, M. S., Akselsson, R., Folkman, B. 1975. *Archeometry* 18:39

Ahlberg, M. S. 1977. *Nucl. Instrum. Methods* 142:61–65

Armstrong, D. D., Wegner, H. E. 1971. *Rev. Sci. Instrum.* 42:40

Ashbaugh, L. L., Cahill, T. A., Mix, R. 1978. *Proc. 10th Aerosol Technol. Symp.* Alburquerque, N. Mex.

Augustyniak, W. M., Betteridge, D., Brown, W. L. 1978. *Nucl. Instrum. Methods* 149:665–68

Badica, T., Ciortea, C., Petrovici, A., Popescu, I. 1978. *Nucl. Instrum. Methods* 156:537–40

Baeri, P., Campisano, S. U., Immé, G., Rimini, E., Della Mea, G. 1978. *Nucl. Instrum. Methods* 149:435–40

Baijot-Stroobants, J., Bodart, F. 1977. *Nucl. Instrum. Methods* 142:293–300

Bambynek, W., Craseman, B., Fink, R. W., Freund, H. U., Mark, H., Swift, C. D., Price, R. E., Rao, P. V. 1972. *Rev. Mod. Phys.* 44:716

Barnes, B. K., Beghian, L. E., Kegel, G. H. R., Mathur, S. C., Quinn, P. 1973. *Radioanal. Chem.* 15:13

Barone, J. B., Cahill, T. A., Eldred, R. A., Flocchini, R. G., Shadoan, D. J., Dietz, T. M. 1978. *Atmos. Environ.* 12:2213–21

Bauer, K. G., Fazly, Q., Mayer-Kuckuk, T., Mommsen, H., Schurkes, P. 1978. *Nucl. Instrum. Methods* 148:407–13

Baum, R. M., Willis, R. D., Walter, R. L., Gutknecht, W. F., Stiles, A. R. 1977. *X-ray Fluorescence of Environmental Samples,* ed. T. Dzubay. Ann Arbor Press, Mich., P. 165

Bauman, S. E., Williams, E. T., Finston, H. L., Bond, A. H., Lesser, P. M. S. 1979. *Nucl. Instrum. Methods* 165:57–62

Bayerl, P., Eichinger, P. 1978. *Nucl. Instrum. Methods* 149:663–64

Bearse, R. C., Close, D. A., Malanify, J. J., Umbarger, C. J. 1974. *J. Anal. Chem.* 46:499–503

Bearse, R. C., Burns, C. E., Close, D. A., Malanify, J. J. 1977. *Nucl. Instrum. Methods* 142:143–50

Benka, O., Geretschlager, M., Paul, H. 1977. *Nucl. Instrum. Methods* 142:83–84

Benka, O., Kropf, A. 1978. *At. Data Nucl. Data Tables* 22:219–33

Benka, O., Geretschlager, M., Kropf, A. 1978. *Nucl. Instrum. Methods* 149:441–44

Berg, W. W. Jr., Winchester, J. W. 1978. *Chem. Oceanogr.* 7:173–231

Bertin, E. P. 1978. *Introduction to X-ray Spectrometric Analysis.* p. 413

Bonani, G., Suter, M., Jung, H., Stroller, C., Wolfi, W. 1978. *Nucl. Instrum. Methods* 157:55–63

Bosch, F., El Goresy, A., Kratschmer, W., Martin, B., Povh. B., Nobiling, R., Travel, K., Schwalm. D. 1976. *Phys. Rev. Lett.* 37:1515

Bosch, F., El Goresy, A., Martin, B., Povh, B., Nobiling, R., Schwalm, D., Traxel, K. 1978. *Science* 199:765

Bosch, F., El Goresy, A., Herth, W., Martin, B., Nobiling, R., Povh., B., Reiss, H. D., Traxel, K. 1980. *Nucl. Sci. Appl.* 1:1–39

Boulle, G. J., Peisach, M. 1979. *J. Radioanal. Chem.* 50:205

Boulle, G. J., Peisach, M., Jacobson, L. 1979. *S. Afr. J. Sci.* 75:215

Brune, D., Lindh, U., Lorenzen, J. 1977. *Nucl. Instrum. Methods* 142:51–54

Butcher, S. S., Charlson, R. J. 1972. *An Introduction to Air Chemistry.* New York: Academic

Cahill, T. A. 1973. *Rep. Univ. Calif. Davis-Crocker Nucl. Lab. No. 162,* pp. 1–65

Cahill, T. A., Flocchini, R. G., Feeney, P. J.,

Shadoan, D. J. 1974. *Nucl. Instrum. Methods* 120 : 193–95

Cahill, T. A. 1975. In *New Uses for Low Energy Accelerators*, ed. J. Ziegler, pp. 1–72 New York : Plenum

Cahill, T. A., Flocchini, R. G., Eldred, R. A., Feeney, P. J., Lange, S., Shadoan, D., Wolfe, G. 1976. *Accuracy in Trace Analysis : Sampling, Sample Handling, and Analysis*, 422 : 1119–35. US Natl. Bur. Stand.

Cahill, T. A. 1977. *Nucl. Instrum. Methods* 142 : 1–3

Cahill, T. A., Eldred, R. A., Flocchini, R. G., Barone, J. B. 1977a. *Nucl. Instrum. Methods* 142 : 259–61

Cahill, T. A., Ashbaugh, L. L., Barone, J. B., Eldred, R. A., Feeney, P. J., Flocchini, R. G., Goodart, C., Shadoan, D., Wolfe, G. W. 1977b. *J. Air. Pollut. Control Assoc.* 27 : 675–78

Cahill, T. A. 1978. In Proc. 3rd Int. Conf. on Ion Beam Analysis, Washington DC. *Nucl. Instrum. Methods* 149 : 431–33

Cahill, T. A., Fletcher, N. R., Medsker, L. R., Nelson, J. W., Kaufmann, H. 1978. *Phys. Rev. C* 17 : 1183–95

Cairns, J. A., Cookson, J. A. 1980. *Nucl. Instrum. Methods* 168 : 511–16

Camp, D. C., Cooper, J. A., Rhodes, J. R. 1974. *X-Ray Spectrom.* 3 : 47

Camp, D. C., VanLehn, A. L., Rhodes, J. R., Pradzynski, A. H. 1975. *X-Ray Spectrom.* 4 : 123–37

Camp, D. C. 1979. *Environ. Sci. Technol.* In press

Campbell, A. B., Sartwell, B. D. 1978. *Nucl. Instrum. Methods* 149 : 496 (Abstr.)

Campbell, J. L., O'Brien, P., McNelles, L. A. 1971. *Nucl. Instrum. Methods* 92 : 269

Campbell, J. L., Orr, B. H., Herman, A. W., McNelles, L. A., Thomson, J. A., Cook, W. B. 1975. *Anal. Chem.* 47 : 1452

Campbell, J. L. 1977. *Nucl. Instrum. Methods* 142 : 263–73

Campbell, J. L., Orr, B. H., Noble, A. C. 1977. *Nucl. Instrum. Methods* 142 : 289–91

Chemin, J. F., Mitchell, I. V., Saris, F. W. 1974. *J. Appl. Phys.* 45 : 532, 537

Chen, J., Li, H., Ren, C., Tang, G., Wang, X., Yang, F., Yao, H. 1980. *Nucl. Instrum. Methods* 168 : 437–40

Chen, J. R. 1977. *Nucl. Instrum. Methods* 142 : 9–19

Chen, J. R., Franciso, R. B., Miller, T. E. 1977. *Science* 196 : 906

Chu, T. C., Navarette, V. R., Kaji, N., Izawa, G., Shiokawa, T., Ishii, K., Morita, S., Tawara, H. 1978. *J. Radioanal. Chem.* 36 : 195–207

Clark, G. 1976. *Proc. 4th Conf. Sci. Ind. Appl. Small Accel., Denton, Tex.,* ed. T. Duggan, I. Morgan, J. Martin, p. 147.

New York : IEEE

Clark, G. J., Gulson, B. L., Cookson, J. A. 1979. *Geochim. Cosmochim. Acta* 43 : 905

Close, D. A., Bearse, R. C., Malanify, J. J., Umbarger, C. J. 1973. *Phys. Rev. A* 8 : 1873–79

Cookson, J. A., Poole, M. 1970. *New Scientist,* p. 404

Cookson, J. A., Ferguson, A. T. G., Pilling, F. D. 1972. *J. Radioanal. Chem.* 12 : 39

Cookson, J. A., Pilling, F. D. 1975. *Phys. Med. Biol.* 20 : 1015

Cookson, J. A., Pilling, F. D. 1976. *Phys. Med. Biol.* 21 : 965

Cookson, J. A., Fletcher, N. R., Kemper, K. W., Medsker, L. R., Cahill, T. A. 1978. *Proc. Int. Symp. on Super Heavy Elements,* ed. M. Lodhi, pp. 164–69. New York : Pergamon

Cookson, J. A. 1979. *Nucl. Instrum. Methods* 165 : 477–508

Cookson, J. A., McMillan, J. W., Pierce, T. B. 1979. *J. Radioanal. Chem.* 48 : 337–57

Coote, G. E., Sparks, R. J., Pohl, K. P., West, J. G., Purcell, C. R. 1978. Inst. Nucl. Sci., DSIR Lab., Lower Hutl, New Zealand, *Semiann. Rep. Jan. 1978, INS-R-242,* p. 10 ; as quoted in Cookson 1979

Courtney, W. J., Rheingrover, S., Pilotte, J., Kaufmann, H. C., Cahill, T. A., Nelson, J. W. 1978. *J. Air Pollut. Control. Assoc.* 28 : 224–28

Criswell, T. L., Gray, T. J. 1974. *Phys. Rev. A* 10 : 1145–50

Cross, J. B., Zeisler, R., Schweikert, E. A. 1977. *Nucl. Instrum. Methods* 142 : 111–19

Deconninck, G. 1972. *J. Radioanal. Chem.* 12 : 157

Deconninck, G. 1977. *Nucl. Instrum. Methods* 142 : 275–84

Deconninck, G., Bodart, F. 1978. *Nucl. Instrum. Methods* 149 : 609–14

Demortier, G., Lefebre, J., Gillet, C. 1972. *J. Radioanal. Chem.* 12 : 277

Desaedeleer, G. G., Winchester, J. W., Akselsson, K. R. 1977. *Nucl. Instrum. Methods* 142 : 97–09

Duerden, P., Bird, R., Scott, M. D., Clayton, E., Russell, L. H. 1980. *Nucl. Instrum. Methods* 168 : 447–52

Duggan, J. L., Spaulding, J. 1970. *Proc. 2nd Conf. on Use of Small Accel. in Teach. Res., AEC Conf. 700322*

Duggan, J. L., Beck, W. L., Albrecht, L., Munz, L., Spaulding, J. D. 1972. *Adv. X-Ray Anal.* 15 : 407

Engelbrecht, D., Cahill, T. A., Feeney, P. J. 1980. *J. Air Pollut. Control Assoc.* 30 : 391–92

Everett, R. G., Hicks, B. B., Berg, W. W., Winchester, J. W. 1979. *Atmos. Environ.* 13 : 931–34

Feeney, P. J., Cahill, T. A., Flocchini, R. G.,

Eldred, R. A., Shadoan, D. J., Dunn. T. 1975. *J. Air. Pollut. Control Assoc.* 25:1145–47

Feldman, L. C., Poate, J. M., Ermanis, F., Schwartz, B. 1973. *Thin Solid Films* 19:81

Fisher, B. E. 1977. *Prog. Rep. GSI-J-I-78 GS I, Darmstadt*, as quoted in Cookson 1979

Flocchini, R. G., Feeney, P. J., Sommerville, R. J., Cahill, T. A. 1972. *Nucl. Instrum. Methods* 100:397–402

Flocchini, R. G., Cahill, T. A., Shadoan, D. J., Lange, S., Eldred, R. A., Feeney, P. J., Wolfe. G. W., Simmeroth, D. C., Suder, J. K. 1976. *Environ. Sci. Technol.* 10:76–82

Folkmann, F., Gaarde, C., Huus, T., Kemp, K. 1974a. *Nucl. Instrum. Methods* 116:487–99

Folkmann, F., Borggren, J., Kjeldgaard, A. 1974b. *Nucl. Instrum. Methods* 119:117–23

Folmmann, F. 1975. *J. Phys. E* 8:429

Garcia, J. D., Fortner, R. J., Kavanaugh, T. M. 1973. *Rev. Mod. Phys.* 45:111

Gardner, R. K., Gray, J. J. 1978. *At. Data Nucl. Data Tables* 21:515–36

Gentry, R. V., Cahill, T. A., Fletcher, N. R., Kaufmann, H. C., Medsker, L. R., Nelson, J. W., Flocchini, R. G. 1976. *Phys. Rev. Lett.* 37:11

Gordon, B. M., Kraner, H. W. 1972. *J. Radioanal. Chem.* 12:181

Goulding, F. S., Jaklevic, J. M. 1977. *Nucl. Instrum. Methods.* 142:323–32

Grodzins, L., Horowitz, P., Ryan, J. 1976. See Clark 1976, p. 75

Hansen, L. D., Ryder, J. F., Mangelson, N. F., Hill, M. W., Fancette, K. J., Eatough, D. J. 1980. *Anal. Chem.* 52:821–24

Hardt, T. L., Watson, R. L. 1976. *At. Data Nucl. Data Tables* 17:107–25

Harrison, J. F., Eldred, R. A. 1973. *Adv. X-Ray Anal.* 17:560

Hasselmann, I., Koenig, W., Richter, F. W., Steiner, U., Watjen, U., Bode, J. C., Ohta, W. 1977. *Nucl. Instrum. Methods* 142:163–70

Heck, D. 1978. *Karlsruhe Rep. KFK* 2680:115; as quoted in Cookson 1979

Heitz, C., Kwadow, M., Tenorio, D. 1978. *Nucl. Instrum. Methods* 149:483–88

Henley, E. C., Kassouny, M. E., Nelson, J. W. 1977. *Science* 197:277–78

Hoffer, T., Kliwer, J., Moyer, J. 1979. *Atmos. Environ.* 13:619–27

Horowitz, P., Grodzins, L. 1975. *Science* 189:795–96

Horowitz, P., Aronson, M., Grodzins, L., Ladd, W., Ryan, J., Merriam, G., Lechene, C. 1976. *Science* 194:1162

Huda, W. 1979. *Nucl. Instrum. Methods* 158:587–94

Hyrnkiewicz, A. Z., Szymczyk, S., Kajfosz,

J., Olech, M. 1980. *Nucl. Instrum. Methods* 126:75

Ishii, K., Morita, S., Tawara, H., Chu, T. C., Kaji, H., Shiokawa T. 1980. *Nucl. Instrum. Methods* 168:517–21

Jaklevic, J. M., Goulding, F. S. 1972. *IEEE Trans. Nucl. Sci.* NS–19:384

Jarjis, R. A. 1979. *Nucl. Instrum. Methods* 160:457–60

Johansson, S. A. E., Johansson, T. B. 1976. *Nucl. Instrum. Methods* 137:473–516

Johansson, S. A. E. 1977. *Proc. Int. Conf. Part. Induced X-Ray Emission. Its Anal. Appl. Nucl. Instrum. Methods* 142:1–338

Johansson, T. B., Akselsson, R., Johansson, S. A. E. 1970. *Nucl. Instrum. Methods* 84:141–43

Johansson, T. B., Akselsson, R., Johansson, S. A. E. 1972. *Adv. X-Ray Anal.* 15:373

Johansson, T. B., Van Grieken, R. E., Nelson, J. W., Winchester, J. W. 1975. *Anal. Chem.* 47:855–60

Jolly, R. K., White, M. B. 1971. *Nucl. Instrum. Methods* 97:103

Jolly, R. K., Pehrson, G. R., Gupta, S. K., Buckle, D. C., Aceto, H. 1974. *Proc. 3rd Conf. Appl. Small Accel. in Teach. Res. Denton, Tex.* p. 203

Jolly, R. K., Kane, J. R., Buckle, D. C., Randers-Pehrson, G., Teoh, W., Aceto, H. 1978. *Nucl. Instrum. Methods* 151:183–88

Jundt, F. C., Purser, K. H., Kubo, H., Schenk, E. A. 1974. *J. Histochem. Cytochem.* 22:1; Quoted in Mangelson et al. 1979

Kaji, H., Shiokawa, T., Ishii, K., Morita, S., Kamiya, M., Sera, K. 1977. *Nucl. Instrum. Methods* 142:21–26

Katsanos, A., Xenoulis, A., Hadjiantoniou, A., Fink, R. W. 1976. *Nucl. Instrum. Methods* 137:119–24

Katsanos, A., Hadjiantoniou, A. 1978. *Nucl. Instrum. Methods* 149:469–73

Kauffman, R. L., Feldman, L. C., Chang, R. P. H. 1978. *Nucl. Instrum. Methods* 149:619–22

Kaufmann, H. C., Akselsson, R. 1975. *Adv. X-Ray Anal.* 18:353

Kaufmann, H. C., Akselsson, R., Courtney, W. J. 1977. *Nucl. Instrum. Methods* 142:251–57

Kemp, K., Jensen, F. P., Moller, J. T. 1975. *Phys. Med. Biol.* 20:834; reported in Mangelson et al 1979

Kemp, K., Palmgren-Jensen, F. P. 1977. *Nucl. Instrum. Methods* 142:101–3

Khan, A. H., Khaliguzzam, M., Husain, M., Abdullah, M. 1979. *Nucl. Instrum. Methods* 165:253–59

Kivits, H. P. M., De Rooij, F. A. J., Wijnhoven, G. P. J. 1979. *Nucl. Instrum. Methods* 164:225–29

Knudsen, A. R. 1978. *Nucl. Instrum. Methods* 149 : 445–49

Koenig, W., Richter, F. W., Steiner, U., Stock, R., Thielmann, R., Watjen, U. 1977. *Nucl. Instrum. Methods* 142 : 225

Kropf, A. 1977. *Nucl. Instrum. Methods* 142 : 79–81

Kugel, H. W., Herzog, G. F. 1977. *Nucl. Instrum. Methods* 142 : 301–5

Kump, P., Rupnik, P., Budnar, M., Kreft, I. 1977. *Nucl. Instrum. Methods* 142 : 205–8

Lannefors, H. O., Johansson, T. B., Granat, L., Rudell, B. 1977. *Nucl. Instrum. Methods* 142 : 105–10

Lawson, D. R., Winchester, J. W. 1979a. *Atmos. Environ.* 13 : 925–30

Lawson, D. R., Winchester, J. W. 1979b. *Science* 205 : 1267–69

LeComte, R., Paradis, P., Monaro, S., Barrette, M., Lamoureux, G., Menard, H. A. 1978. *Nucl. Instrum. Methods* 150 : 289–99

Legge, G. J. F. 1976. *Proc. Aust. Conf. Nucl. Technol., Lucas Heights*, p. 93

Legge, G. J. F. 1978. *Proc. 2nd Aust. Conf. Nucl. Technol. Anal.*, p. 18

Li, M. et al. 1979. *Kexue Tongbao* 24 : 19

Lin, T., Luo, C., Chou, J. 1978. *Nucl. Instrum. Methods* 151 : 439–44

Lochmuller, C. H., Galbraith, I., Walter, R. 1972. *Anal. Lett.* 5 : 943

Lodhi, A. S., Sioshansi, P. 1977. *Nucl. Instrum. Methods* 142 : 45–47

Lopes, J. S., Jesus, A. P., Ramos, S. C. 1979. *Nucl. Instrum. Methods* 164 : 369–74

MacArthur, J. D., Ewan, G. T., Green, B. J. 1977. *Res. Nucl. Phys., Queen's Univ., Kingston*, p. 23 ; quoted in Cookson 1979

MacDonald, G. L. 1978. *Anal. Chem.* 50 : 135

Maenhaut, W., DeReu, L., Van Rinsvelt, H. A., Cafmeyer, J., Van Epsen, P. 1979. 4th Int. Conf. on Ion Beam Analysis, Aarhus. *Nucl. Instrum. Methods.* 168 : 551–56

Mak, B. W., Bird, J. R., Sabine, T. M. 1966. *Nature* 211 : 738

Mandler, J. W., Semmler, R. A. 1974. See Jolly et al 1974, p. 173

Mangelson, N. F., Hill, M. W., Nielson, K. K., Eatough, D. J., Christensen, J. J., Izatt, R. M., Richards, D. O. 1979. *Anal. Chem.* 51 : 1187–94

Mangelson, N. F., Hill, M. W., Nielson, K. K., Ryder, J. F. 1977. *Nucl. Instrum. Methods* 142 : 133–42

Martin, B., Nobiling, R. 1979. In *Applied Charged Particle Optics*, ed. A. Septier, New York : Academic. In press

Martin, F. W. 1973. *Science* 179 : 173–75

Martin, F. W. 1978a. *Ultramicroscopy* 3 : 75–79

Martin, F. W. 1978b. *Nucl. Instrum. Methods* 149 : 475–81

Mazzolini, A. P. J., Legge, G. J. F. 1978. See Legge 1978 ; p. 93

McClelland, G. A. G., McKenna, R. J., Cahill, T. A. 1972. Mark-release-recapture studies of biting insects. In *Proc. Biting Fly Control Environ. Qual., Alberta, Canada*, pp. 49–52

McGinley, J. R., Stock, G. J., Schweikert, E. A., Gross, J. B., Zeisler, R., Zikovsky, L. 1978. *J. Radioanal. Chem.* 43 : 411–20

McMillan, J. W., Pierce, T. B. 1976b. *Conf. Ion Beam Surface Layer Anal.*, ed. J. Mayer et al, New York : Plenum, 901 pp.

McMillan, J. W., Hirst, P. M., Pummery, F. C. W., Huddleston, J., Pierce, T. B. 1978. *Nucl. Instrum. Methods* 149 : 83

McMillan, J. W., Pummery, F. C. W., Huddleston, J., Pierce, T. B. 1979. To be published (from Cookson 1979)

Meinert, D. L., Winchester, J. W. 1977. *J. Geophys. Res.* 82 : 1778–82

Merzbacker, E., Lewis, H. W. 1958. In *Encyclopedia of Physics*, ed. S. Flugge, 34 : 166. Berlin : Springer

Mingay, D. W., Barnard, E. 1978. *Nucl. Instrum. Methods* 157 : 537–44

Mitchell, I. V., Ziegler, J. F. 1977. In *Ion Beam Handbook for Material Analysis*, ed. J. W. Mayer, E. Rimini, pp. 311–484. New York : Academic

Modjtahed-Zadeh, R., Rastegar, B., Gallmann, A., Guillaume, G., Jundt, F., Sioshansi, P. 1975. *Nucl. Instrum. Methods* 131 : 563–65 (in French)

Mommsen, H., Bauer, K. G., Fazly, Q. 1978. *Nucl. Instrum. Methods* 157 : 305–9

Montenegro, E. C., Baptista, G. B., Barros Leite, C. V., de Pinho, A. G., Pashoa, A. S. 1979. *Nucl. Instrum. Methods* 164 : 231–34

Moriya, Y., Ato, Y., Miyagawa, S. 1978. *Nucl. Instrum. Methods* 150 : 523–28

Moschandreas, D. J., Winchester, J. W., Nelson, J. W., Burton, R. M. 1979. *Atmos. Environ.* 13 : 1413–18

Murray, S. A., Cahill, T. A., Paulson, K. N., Spurr, A. R. 1975. *Commun. Soil Sci. Plant Anal.* 6 : 33–34

Nass, M. J., Lurio, A., Ziegler, J. F. 1978. *Nucl. Instrum. Methods.* 154 : 567–71

Needham, P. B., Sartwell, J. R., Sartwell, B. D. 1970. *Adv. X-Ray Anal.* 14 : 184

Nelson, J. W., Desaedeleer, G. G., Akselsson, K. R., Winchester, J. W. 1976. *Adv. X-Ray Anal.* 19 : 403–14

Nielson, K. K., Hill, M. W., Mangelson, N. F. 1976. *Adv. X-Ray Anal.* 19 : 511–20

Nobiling, R., Civelekoglu, Y., Povh, B., Schwalm, D., Traxel, K. 1975. *Nucl. Instrum. Methods* 130 : 325–34

Nobiling, R., Traxel, K., Bosch, F., Civelekoglu, Y., Martin, B., Povh, B., Schwalm, D. 1977. *Nucl. Instrum. Methods* 142 : 49–50

Orsini, C. Q., Boueres, L. C. 1977. *Nucl. Instrum. Methods* 142 : 27–32

Orsini, C. Q., Kaufmann, H. C., Akselsson, K. R., Winchester, J. W., Nelson, J. W. 1977. *Nucl. Instrum. Methods* 142 : 91–96

Owers, M. J., Shalgosky, H. I. 1974. *J. Phys. E* 7 : 593

Pabst, W., Schmid, K. 1975. *X-Ray Spectrum* 4 : 85

Pape, A., Sens, J. C., Fintz, P., Gallmann, A., Gove, H. E., Guillaume, G., Stupin, D. M. 1972. *Nucl. Instrum. Methods* 105 : 161–62

Pape, A., Chevallier, A., Chevallier, J. 1978. See Cookson et al 1978, pp. 170 : 71

Peisach, M., Newton, D. A., Peck, P. F., Pierce, T. B. 1973. *J. Radioanal. Chem.* 16 : 445

Perry, S. K., Brady, F. P. 1973. *Nucl. Instrum. Methods* 108 : 389

Pierce, T. B., Peck, P. F., Cuff, D. R. A. 1966. *Nature* 211 : 66

Pierce, T. B. 1973. *J. Radioanal. Chem.* 17 : 55

Pierce, T. B., McMillan, J. W., Peck, P. F., Jones, I. G. 1974. *Nucl. Instrum. Methods* 118 : 115

Pierce, T. B., Huddleston, J. 1977. *Nucl. Instrum. Methods* 144 : 231

Pierce, T. B. 1978. *Proc. Adv. Group. Meet., Costa Rica*, p. 33. Vienna : IAEA; as quoted in Cookson 1979

Pilotte, J. O., Winchester, J. W., Glassen, R. C. 1978. *Water, Air, Soil Pollut.* 9 : 363–68

Poncet, M., Engelmann, C. 1978. *Nucl. Instrum. Methods* 149 : 461–67

Prahm, L. P., Torp, U., Stern, R. M. 1976. *Tellus* 28 : 355–72

Raith, B., Roth, M., Gollner, K., Gonsior, B., Ostermann, H., Uhlorn, C. D. 1977. *Nucl. Instrum. Methods* 142 : 39

Raith, B., Wilde, H. R., Roth, M., Stratmarn, A., Gonsior, B. 1980. *Nucl. Instrum. Methods* 168 : 251–58

Reuter, W., Lurio, A., Cardone, F., Ziegler, J. F. 1975. *J. Appl. Phys.* 46 : 3194

Reuter, W., Lurio, A. 1978. *Nucl. Instrum. Methods* 149 : 495 (Abstr.); Private communication

Rhodes, J. R., Hunter, C. B. 1972. *J. X-Ray Spectrum.* 1 : 113–17

Rice, R., Basbas, G., McDaniel, F. D. 1977. *At. Data Nucl. Data Tables* 20 : 503–11

Rickards, J., Ziegler, J. F. 1975. *Appl. Phys. Lett.* 27 : 707

Rickey, F. A., Mueller, K., Simms, P. C., Michael, B. D. 1977. In *X-Ray Fluorescence Analysis of Environmental Samples*, ed. T. Dzubay, pp. 135–143. Ann Arbor Sci., Mich.

Robertson, R. 1977. *Nucl. Instrum. Methods* 142 : 121–26

Roelandts, I., Weber, G. Y., Quaglia, L.

1978. *Nucl. Instrum. Methods* 157 : 141–46

Rosner, H. R., Bhalla, C. P. 1970. *Z. Phys.* 231 : 347

Rudolph, H., Kliwer, J. K., Kraushaar, J. J., Ristinen, R. A., Smythe, W. R. 1971. *Proc. 18th Ann. ISA Anal. Instrum. Symp.*, San Francisco : ISA. 151 pp.

Rudolph, H., Kliwer, J. K., Kraushaar, J. J., Ristinen, R. A., Smythe, W. R. 1972. *Anal. Instrum.* 10 : 151–56

Rupnik, P., Kump, P., Budnar, M., Kreft, I. 1977. *Nucl. Instrum. Methods* 142 : 209–11

Rutledge, C. H., Watson, R. L., 1973. *At. Data Nucl. Data Tables* 12 : 195–216

Salem, S. I., Panossian, S. L., Krause, R. A. 1974. *At. Data Nucl. Data Tables* 14 : 91–109

Saltmarsh, M. J., Van de Woude, A., Ludemann, C. A. 1972. *Appl. Phys. Lett.* 21 : 64

Scheer, J., Voet, L., Watjen, U., Koenig, W., Richter, F. W., Steiner, U. 1977. *Nucl. Instrum. Methods* 142 : 333–38

Schulte, R. L., Kamykowski, E. A. 1978. *J. Radioanal. Chem.* 43 : 233

Scofield, J. H. 1974a. *Phys. Rev. A* 9 : 1041–49

Scofield, J. H. 1974b. *At. Data Nucl. Data Tables* 14 : 121–37

Seaman, G. C., Shane, K. C. 1975. *Nucl. Instrum. Methods* 126 : 473–74

Sheline, J., Akselsson, R., Winchester, J. W. 1976. *J. Geophys. Res.* 81 : 1047–50

Shima, K. 1979. *Nucl. Instrum. Methods* 165 : 21–26

Shiokawa, T., Chu, T. C., Navarette, V. R., Kaji, H., Izawa, G., Ishii, K., Morita, S., Tawara, H. 1977. *Nucl. Instrum. Methods* 142 : 199–204

Shroy, R. E., Kraner, H. W., Jones, K. W. 1978. *Nucl. Instrum. Methods* 157 : 163–68

Singh, M., Melvin, J., Cho, Z. H. 1976. *IEEE Trans. Nucl. Sci.* 23 : 657

Singleton, J. F., Hartley, N. E. W. 1979. *J. Radioanal. Chem.* 48 : 317

Sioshansi, P., Lodhi, A. S., Payrovan, H. 1977. *Nucl. Instrum. Methods* 142 : 285–87

Steiger, R., Wasserburg, G. J. 1966. *J. Geophys. Res.* 71 : 6065–90

Takai, M., Gamo, K., Masuda, K., Namba, S., Mizobuchi, A. 1977. *Jpn. J. Appl. Phys.* 16 : 1647

Takai, M., Gamo, K., Yagita, H., Masuda, K., Namba, S. 1978. *Nucl. Instrum. Methods* 149 : 457–59

Tawara, H., Ishii, K., Morita, S. 1976. *Nucl. Instrum. Methods* 132 : 503

Thibeau, H., Stadel, J., Cline, W., Cahill, T. A. 1973. *Nucl. Instrum. Methods* 111 : 615–17

Tove, P. A., Sigurd, D., Patersson, S. 1980. *Nucl. Instrum. Methods* 168 : 441–46

Uemura, Y. J., Kuno, Y., Koyama, H., Yamazaki, T., Kienle, P. 1978. *Nucl. Instrum. Methods* 153 : 573–79

Umbarger, C. J., Bearse, R. C., Close, D. A., Malanify, J. J. 1973. *Adv. X-Ray Anal.* 16 : 102

Valkovic, V. 1973. *Contemp. Phys.* 14 : 415

Valkovic, V. 1977. *Nucl. Instrum. Methods* 142 : 151

Valkovic, V., Liebert, R. B., Zabel, T., Larson, H. T., Mijanic, D., Wheeler, R. W., Phillips, G. C. 1974. *Nucl. Instr. Methods* 114 : 573

Valkovic, V., Rendic, D., Phillips, G. C. 1975. *Environ. Sci. Technol.* 9 : 1150

Van de Heide, J. A., Kivits, H. P. M., Beuzekom, D. C. 1979. *X-Ray Spectrosc.* 8 : 63

Van der Kam, P. M. A., Vis, R. D., Verheul, H. 1977. *Nucl. Instrum. Methods* 142 : 55–60

Van Espen, P., Nullens, H., Maenhaut, W. 1979. *Microbeam Analysis 1979*, ed. D. Newbury, pp. 265–667. San Francisco Press

Van Grieken, R. E., Johansson, T. B., Winchester, J. W., Odom, L. A. 1975. *Z. Anal. Chem.* 275 : 343–48

Van Grieken, R. E., Johansson, T. B., Akselsson, K. R., Winchester, J. W., Nelson, J. W., Chapman, K. R. 1976. *Atmos. Environ.* 10 : 571–76

Van Rinsvelt, H. A., Lear, R. D., Adams, W. R. 1977. *Nucl. Instrum. Methods* 142 : 171–80

Vegh, J., Berenyi, D., Koltay, E., Kiss, I., Seif el-Nasr, S., Sorkadi, L. 1978. *Nucl. Instrum. Methods* 153 : 553–55

Verba, J. W., Sunier, J. W., Wright, B. T., Slaus, I., Holman, A. B., Kulleck, J. G. 1972. *J. Radioanal. Chem.* 12 : 181

Vie le Sage, R., Berg, W. W., Cohn, S. L., Winchester, J. W., Nelson, J. W. 1978. *Pollut. Atmos.* 77 : 6–16

Vis, R. D. 1975. *J. Radioanal. Chem.* 27 : 447

Vis, R. D., Van der Kam, P. M. A., Verheul, H. 1977. *Nucl. Instrum. Methods* 142 : 159—62

Vis, R. D., Wiedersphan, K. J., Verheul, H. 1978. *J. Radioanal. Chem.* 45 : 407–16

Walter, R. L., Willis, R. D., Gutknecht, W. F., Joyce, J. M. 1974. *Anal. Chem.* 46 : 843

Walter, R. L., Willis, R. D., Gutknecht, W. F., Shaw, R. W. Jr. 1977. *Nucl. Instrum. Methods* 142 : 181–97

Watson, R. L., Sjurseth, J. R., Howard, R. W. 1971. *Nucl. Instrum. Methods* 93 : 69–76

Watson, R. L., Leeper, A. K., Sonobe, B. I. 1977. *Nucl. Instrum. Methods* 142 : 311–16

Whitehead, N. E. 1979. *Nucl. Instrum. Methods* 164 : 381–88

Wilde, H. R., Roth, M., Unlorn, C. D., Gonsior, B. 1978. *Nucl. Instrum. Methods* 149 : 675–78

Wilk, S. F. J., McKee, J. S. C., Randell, C. P. 1977. *Nucl. Instrum. Methods* 142 : 33–38

Willis, R. D., Walter, R. L., Shaw, R. W. Jr., Gutknecht, W. F. 1977a. *Nucl. Instrum. Methods* 142 : 67–77

Willis, R. D., Walter, R. L. 1977. *Nucl. Instrum. Methods* 142 : 231–42

Willis, R. D., Walter, R. L., Doyle, B. L., Shafroth, S. M. 1977b. *Nucl. Instrum. Methods* 142 : 317–21

Winchester, J. W. 1977. *Nucl. Instrum. Methods* 142 : 85–90

Wyrick, R. K., Cahill, T. A. 1973. *Phys. Rev.* A 8 : 2288–91

Young, F. C., Roush, M. L., Berman, P. G. 1973. *Int. J. Appl. Radioisotopes* 24 : 153

Younger, P. A., Cookson, J. A. 1979. *Nucl. Instrum. Methods* 158 : 193–98

Ziegler, J. 1975. Private communication and in Cahill 1975

Zirckle, R. E., Bloom, W. 1953. *Science* 117 : 487

Zombola, R. R., Witos, P. A., Bearse, R. C. 1977. *Anal. Chem.* 49 : 2203–5

Ann. Rev. Nucl. Part. Sci. 1980. 30:253–98

RELATIVISTIC CHARGED PARTICLE IDENTIFICATION BY ENERGY LOSS

✕5618

W. W. M. Allison
Nuclear Physics Laboratory, University of Oxford, Oxford, United Kingdom OX1 3RH

J. H. Cobb*
Department of Physics, University of Lancaster, Lancaster, United Kingdom

CONTENTS

*Current address: EP Division, CERN, 1211 Geneva 23, Switzerland.

253

0163-8998/80/1201-0253$01.00

1   INTRODUCTION

1.1   *Relativistic Particle Identification*

Although the physical mechanisms underlying the phenomenon of energy loss by a fast charged particle passing through matter have been understood for many years (Bohr 1913, Bethe 1930), only recently has it become technically feasible to use energy-loss measurements for the identification of single particles in the relativistic region. Energy-loss measurements can provide single particle identification in the velocity range $5 \lesssim \beta\gamma \lesssim 300$[1] and therefore the technique is applicable both at fixed-target accelerators and at storage rings where interactions produce a high multiplicity of secondary particles with momenta of a few tens of GeV/c. The more conventional methods of particle identification by time-of-flight measurement and Cerenkov detectors fail or become difficult to use in this region, especially when implemented over a large solid angle. Inability to identify the secondary particles prohibits a complete analysis of an interaction; Lorentz invariant quantities cannot be constructed, the production and exchange of quantum numbers cannot be studied, and the analysis can only be performed in terms of laboratory variables.

The secondary particles to be identified are the leptons (electron and muon) and the long-lived stable hadrons (π meson, K meson, and proton), although the improbable but interesting observation of free fractionally charged quarks should not be overlooked. Specific methods exist for the identification of the leptons: muons, which are not greatly different in mass from pions, can be identified by their ability to penetrate great thicknesses of matter without interacting strongly; electrons, since their mass is so small, can be identified by their ability to initiate electromagnetic showers in a high-Z material. Hadrons must be identified by mass separation: a precision measurement of momentum in a magnetic spectrometer combined with an estimate of velocity yields the mass via the relation $\beta\gamma = P/\mu c$. The separation between particle species that can be achieved depends on the precision with which $\beta\gamma$ is estimated.

The particle velocity, β, may be deduced directly by measuring the time of flight over a fixed path of length L, the time difference between two species of masses μ_1 and μ_2 being $t_1 - t_2 \sim (L/2c) \times (\mu_1^2 - \mu_2^2)c^2/P^2$. If we assume a time resolution of a fraction of a nanosecond, then a path length of roughly P^2 meters is required to separate π and K mesons at a momentum of P GeV/c. This technique clearly becomes impractical when typical momenta

---

[1] Here β is the velocity, v/c, of the particle; γ is the Lorentz factor, $(1-\beta^2)^{-1/2}$. The product $\beta\gamma$ is equal to $P/\mu c$, where μ is the mass of the particle and P is its momentum. The symbol m in this paper will be used for the mass of the electron.

exceed a few GeV/c. Alternatively, a Cerenkov counter may be used. Cerenkov radiation is emitted when a particle of velocity βc moves through a medium of refractive index $[\varepsilon(\omega)]^{1/2}$, where $\varepsilon(\omega)$ is the dielectric constant. This is free radiation of frequency ω emitted at an angle $\theta = \text{arc cos } \beta^{-1}[\varepsilon(\omega)]^{-1/2}$ when β exceeds the threshold velocity $\beta_{\bar{\gamma}} = [\varepsilon(\omega)]^{-1/2}$ (see appendix). The mean number of photons emitted per unit path length is $dN/dx = \alpha/c[1 - 1/\beta^2\varepsilon(\omega)]d\omega$ and rises rapidly from threshold to 90% of the asymptotic value when $\beta\gamma \sim 3\beta_{\bar{\gamma}}\gamma_{\bar{\gamma}}$. A Cerenkov counter therefore provides a threshold rather than a velocity measurement. Usually practical considerations limit the spectrum to the optical region and, in order to obtain a refractive index sufficiently close to unity over a large frequency range, the radiator medium must be a gas. The photon flux is low and the path-length in the radiator must be rather long, up to several meters; it increases quadratically with the desired threshold. Several independent Cerenkov counters are required in order to provide sufficient thresholds to cover the entire spectrum of interesting momenta. Furthermore, to handle high multiplicity events, each has to be subdivided into many cells with the consequent difficulties of separating the light emitted at finite angles by different particles. Cerenkov counters have been fully reviewed by Litt & Meunier (1973).

The transition radiation emitted in the x-ray region when a charged particle crosses a discontinuity in dielectric constant may be detected and used for velocity discrimination (Garibyan 1960, Harris et al 1973, Allison 1977). This method is applicable only in the extreme relativistic region. With careful optimization a threshold of $\beta\gamma$ as low as 300–500 may be reached (Cobb et al 1977, Camps et al 1975, Commichau et al 1979). Relatively compact detectors can therefore be built to provide electron identification for momenta above 500 MeV/c and π meson identification above 50 GeV/c but there seems to be little practical possibility of achieving lower thresholds.

There is therefore a region where it is difficult to use the more conventional methods of particle identification. Recent attention has turned to the use of ionization measurements.

By ionization measurements we mean an indirect estimate of the energy loss, Δ, of a charged particle as it passes through a thin absorber, specifically by measuring the number of electrons liberated in a gas-filled proportional counter. The energy loss of a charged particle rises with $\beta\gamma$ and a measurement of it may be used to estimate the velocity. This proportional counter can be the same device that provides the spatial coordinates for momentum measurement and pattern recognition, so that particle identification will be available over the entire solid angle covered by the detector.

The use of energy-loss measurements for the identification of relativistic particles was first discussed in detail by Alikhanov et al (1956). Measurements of ionization are commonly made in nuclear emulsions and in bubble chambers for the identification of nonrelativistic particles. With recent technical developments of proportional counters, the method can be extended to the relativistic region; several large ionization detectors are currently operating or under construction (see Table 2). Nevertheless the required mass resolution is not easily achieved because of the weak dependence of the ionization on velocity and the intrinsic statistical fluctuations of the ionization process. Therefore, to optimize the performance of such detectors, the underlying physics has to be understood. This review discusses the physics of ionization by a relativistic charged particle and how this is related to the mass resolution of a possible detector.

1.2   The Energy-Loss Method

The energy-loss or ionization method complements Cerenkov counters and x-ray transition radiation detectors by using the broad spectral range of virtual photons associated with the incident charged particle, from the soft ultraviolet to the x-ray region, with the atoms of the absorber medium acting as detectors. The large bandwidth and indirect means of observation of these photons lead to the large ("Landau") fluctuations in the signals from a proportional counter. Figure 1 shows the spectrum of signals observed when protons of $\beta\gamma = 3$ pass through a 2.3-cm thick proportional counter filled with argon at atmospheric pressure (Walenta et al 1979). The data for electrons with $\beta\gamma = 4000$ show a similar spectrum with the peak shifted upwards by $\sim 50\%$. The spectrum describes indirectly the probability $F(\Delta)$ of an energy loss Δ when the particle passes through the counter. Each Δ is the sum of energy transfers E_i in 50–100 distinct collisions between the particle and the gas atoms. The observation is indirect because what is measured is the number of ionization electrons in the gas rather than the energy loss itself. The collisions include the excitation of bound and unbound atomic states and "hard" scatters between the charged particle and quasi-free atomic electrons. The spectrum of energy transfers extends from a few eV up to the kinematic limit, $E_{max} \sim 2m\beta^2c^2\gamma^2$, for a collision between a heavy charged particle and a free electron of mass m. The cross section for these hard collisions is described by the familiar Rutherford formula, $d\sigma/dE \sim 1/(\beta^2 E^2)$, and consequently changes insignificantly with $\beta\gamma$ in the relativistic region ($\beta\gamma \gtrsim 4$). From the point of view of particle identification these hard collisions are a nuisance for they contribute to the "tail" extending to high energies (Figure 1). Individual collisions hard enough to eject a relativistic electron are rare and well beyond the dynamic range of typical ionization measurements. For this reason we may restrict the discussion to nonrelativistic electrons. A fortiori the energy transfer is negligible compared with the incident particle energy. The peak region of the distribution is largely the result of soft collisions in which the atom as a whole absorbs a virtual photon, producing ionization. The energies involved in such collisions are characterized by the atomic structure of the material and the cross-section peaks in the region of the photoabsorption edges.

As discussed in the appendix, the electromagnetic field of a moving charged particle can be described as expanding in the transverse dimension as the velocity approaches the phase velocity of light in the medium, and the energy-loss cross section grows logarithmically with the $\beta\gamma$ of the particle. The peak of the energy-loss spectrum, or any other estimator of the soft collisions, therefore rises with log $\beta\gamma$. This dependence of the ionization on velocity enables us to estimate the particle mass in the relativistic region. The rise would continue indefinitely in a medium of zero density. In a medium of finite density the dielectric properties modify the electromagnetic field limiting its expansion (Fermi 1940, Crispin & Fowler 1970). At

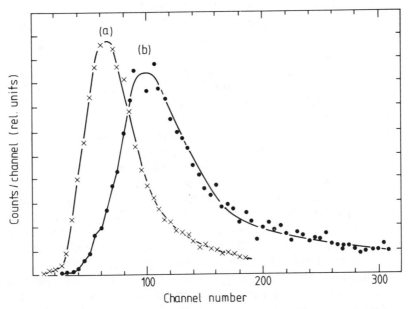

Figure 1   The observed ionization pulse height distributions in 2.3 cm of argon/10% CH_4 at 1 atm for (*a*) protons 3 GeV/*c* and (*b*) electrons 2 GeV/*c* (Walenta et al 1979). The peaks of the distributions correspond to an energy loss of about 4 keV. The lines are drawn simply to guide the eye.

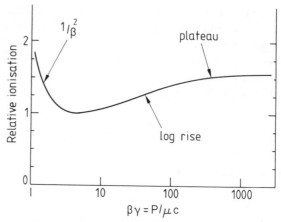

Figure 2   The typical dependence of ionization on $\beta\gamma$.

large values of $\beta\gamma$ the peak energy loss becomes independent of velocity[2]. The dependence of the energy loss on $\beta\gamma$ is shown schematically in Figure 2. The nonrelativistic region where the energy loss falls as $1/\beta^2$ is well known. The region of the "relativistic rise" starts at $\beta\gamma \sim 4$ and the energy loss rises logarithmically with $\beta\gamma$ until the Fermi plateau is reached. The magnitude of the relativistic rise depends upon the atomic structure and density of the material. It is a few percent in solids and liquids and 50–70% in high-Z noble gases at atmospheric pressure, reaching the plateau for $\beta\gamma \sim 200$–500. A gas is therefore the only practical choice for the sampling medium of an ionization detector.

Many independent measurements or samples of ionization are necessary to resolve K and π mesons of the same momentum. Typically the ionization must be measured to 5–6% (FWHM) by collecting a spectrum such as Figure 1a for *each track*. Because of the distribution tail, resolution is lost if, instead, a smaller number of thicker samples is used. Usually ionization detectors make 100–300 measurements of the energy loss of a single particle in a total thickness of 100–1000 cm atm. The single sample thickness is chosen to represent a statistically significant number of soft collisions but to keep the probability of a hard collision small. The information gained from the multiple sampling of the energy-loss distribution must then be analyzed carefully to obtain a normally distributed estimator of ionization with a resolution of a few percent. A common procedure is to take the mean of the

---

[2] This applies to any form of restricted energy loss; the true mean energy loss continues to rise even in the Fermi plateau region, simply through an extension of the upper kinematic limit. It is not, however, an experimentally accessible parameter.

lowest 40–60% of the measurements. Figure 3 (Lehraus et al 1978) shows an example of the mass resolution that may be achieved in this way.

In this review we show how the mass resolution of a detector may be calculated from the detector parameters and the atomic structure of the medium through the well-defined intermediate steps of the differential energy-loss cross section $d\sigma/dE$ and the energy-loss distribution in a finite material thickness $F(\Delta)$. The discussion therefore falls naturally into three sections.

In Section 2 we derive the relationship between $d\sigma/dE$ and the atomic inelastic structure function $f(k,\omega)$. The latter, more generally known as the generalized oscillator strength density, may be modeled using the experimental photoabsorption cross section of the medium, $\sigma_\gamma(E)$. Knowing $d\sigma/dE$, including its velocity dependence, we may calculate the mass resolution for particles of known momentum incident on an ideal detector in which data on every collision that occurs are available for analysis. Such detectors are not realizable; practical detectors have a coarse grain structure and measure $F(\Delta)$ rather than $d\sigma/dE$.

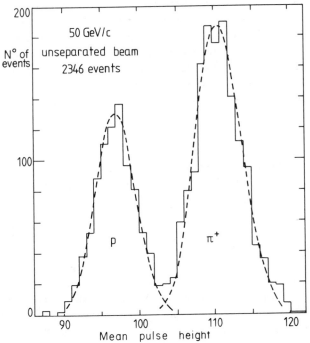

Figure 3   The separation of pions and protons of 50 GeV/c obtained by Lehraus et al (1978) using the EPI detector. The mean of the lowest 40% of 128 × 6-cm samples is plotted. The resolution shown by the dashed curves is 6% FWHM. The gas is argon/5% CH_4 at STP.

In Section 3 we discuss how $d\sigma/dE$ may be folded to give $F(\Delta)$. We comment on the empirical relationship between $F(\Delta)$ and the observed ionization distribution and finally compare experimental data with theoretical predictions.

In Section 4 we show how data from practical devices may be analyzed and how the ionization and the mass resolution depend on the detector parameters and the choice of gas. We develop some simple formulae that, with some qualifications, describe the resolution of a range of devices. We mention some of the practical difficulties that are encountered in the design of such devices and summarize the properties of existing detectors.

2   THE ATOMIC CROSS SECTION

2.1   Kinematics

As it passes through a medium, a fast charged particle loses energy in a number of independent collisions each with its own energy and momentum transfer $(E,\mathbf{p})$ from the particle to the medium[3]. The number of such collisions follows a Poisson distribution with mean given by $N_a\sigma_a x$, where N_a is the number of atoms per unit volume, σ_a is the atomic cross section, and x the thickness of the medium. The distribution of the energy loss Δ in x due to all these collisions (the energy loss distribution) is therefore completely defined once the cross section is known as a function of E and $\mathbf{p}$.

The first constraint on a collision comes from the fact that the incident fast particle (velocity βc and mass μ) is not excited by the collision. This reduces to

$$E(1 - E/2\gamma\mu c^2) = \boldsymbol{\beta} \cdot \mathbf{p}\, c - p^2/2\gamma\mu. \qquad\qquad 1.$$

For reasons explained in the introduction we are only concerned with soft collisions ($E \ll \gamma\mu c^2$, $p \ll \beta\gamma\mu c$) for which Equation 1 simplifies to

$$E = \beta pc \cos \psi. \qquad\qquad 2.$$

We therefore consider the cross section as a function of E and the modulus $|\mathbf{p}|$. Figure 4 shows the E vs $|\mathbf{p}|$ plane with lines of constant ψ, which is the angle between $\mathbf{p}$ and $\boldsymbol{\beta}$. Since we are concerned with energy-loss processes, E as defined above is positive and ψ lies in the first quadrant. The limit $\psi = 0$ gives for an energy E a minimum momentum transfer equal to $E/\beta c$. The boundary of the physical scattering region is therefore a function of β, and Figures 4 and 5 show this for $\beta = 0.9$. If E, p, and ψ are measured for a single

---

[3] In a single crystal the collisions may not be independent. This phenomenon, known as channeling, was reviewed by Gemmell (1974) and is not discussed here.

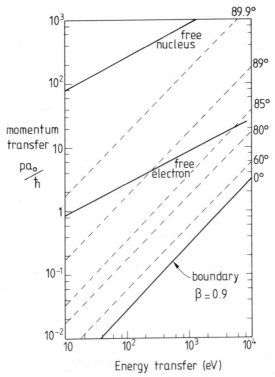

Figure 4   The kinematic region for soft energy and momentum transfer in natural units (a_0 = Bohr radius) for $\beta = 0.9$. Dashed lines indicate the recoil angle ψ. Elastic collisions with a free stationary electron or nucleus ($A = 4$) follow the solid lines.

energy-loss "collision," Equation 2 shows that β may be determined[4]. Normally data on energy loss contain no information on p or ψ; we must integrate the cross section over these variables.

Further information on the cross section in the (E, p) plane requires some knowledge of the target medium. Consider collisions with a single isolated electron or nucleus. For a free stationary electron the constraint is $E = p^2/2m$. This line and an equivalent line for typical nuclear collisions ($A = 4$) are shown on Figure 4. Nuclear collisions contribute negligibly to the energy-loss process even for large values of momentum transfer (small impact parameter). Such collisions are important in giving rise to multiple scattering. We may ignore radiative nuclear collisions (Bremsstrahlung) as exceptional processes in absorbers whose thickness is 10^{-3} or less of a

---

[4] This is the principle of the "ring imaging" Cerenkov counter (Seguinot & Ypsilantis 1977). However, being sensitive only to β, the method loses resolution at high values of $\beta\gamma$.

radiation length. We conclude that the dominant inelastic process in the collision of fast particles with atoms will be quasi-elastic collisions with constituent electrons.

Consider now a medium containing *bound* electrons with binding energy E_1 and internal momentum $\mathbf{q}$. Typically $\mathbf{q}$ is of order $(2mE_1)^{1/2}$. Kinematics of the electron requires $E = E_1 + (\mathbf{p} + \mathbf{q})^2/2m$. This traces out a broad band on the (E, p) plane according to the magnitude and direction of $\mathbf{q}$. The band consists of two parts: the resonance region where $p^2/2m$ is small compared with E_1, and the quasi-free region where the constituent electron behaves almost as if it were free and stationary (Rutherford scattering). This is illustrated in Figure 5 for the example of a single atomic shell of binding energy 30 eV. Our problem is to evaluate the cross section at each point on this plane; we expect that most of the cross section lies within such bands.

Since the interaction involves the exchange of a virtual photon, the range and, therefore, the cross section increase as the photon becomes more nearly real. Figure 5 shows the relation between E and p for real photons in the absence of dispersion ($p = E/c$). This line lies just outside the physical

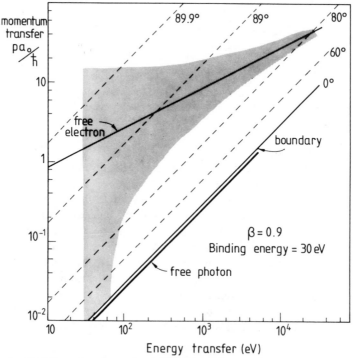

Figure 5 The kinematic region as described in Figure 4. The shading describes qualitatively the region for collisions with a bound electron, binding energy 30 eV. The energy-momentum relation for the absorption of free photons in the low density limit ($\varepsilon = 1$) is also shown.

region for energy-loss collisions. As β tends to one, this boundary of the physical region asymptotically approaches the real photon condition giving the cross section increase. Note that all the increase comes at very low p in the resonance region. At finite densities the photon condition is given by $p = E\varepsilon^{1/2}/c$ and the asymptotic behavior is modified. If ε, the dielectric constant, is greater than one, the physical region may cross the photon line and Cerenkov radiation results; if ε is less than one, the increase of the cross section with β saturates since the physical region never reaches the photon line (see appendix).

We write the complex dielectric constant of a medium, $\varepsilon = \varepsilon_1 + i\varepsilon_2$, where ε_1 and ε_2 describe its polarization and absorptive properties respectively. Indeed the atomic photoabsorption cross section, σ_γ, is directly related to ε_2 and the electron density N by

$$\sigma_\gamma = \frac{Z\omega}{Nc} \frac{\varepsilon_2}{(\varepsilon_1)^{1/2}} \simeq \frac{Z\omega}{Nc} \varepsilon_2, \qquad\qquad 3.$$

where σ_γ and ε are functions of $\omega(= E/\hbar)$ and Z is the atomic number. The second form of Equation 3 applies for low density media ($\varepsilon_1 \sim 1$): ε_1 is given in terms of ε_2 by the Kramers-Kronig relation (see, for example, Jackson 1975). Just as $\varepsilon(\omega)$ comprises all the necessary knowledge of the medium in the restricted kinematic region for free photons, we may introduce $\varepsilon(k, \omega)$, the generalized dielectric constant ($k = p/\hbar$), to describe the response of the medium over the whole scattering region for inelastic collisions. The energy-loss cross section may be derived in terms of $\varepsilon(k, \omega)$ and in turn $\varepsilon(k, \omega)$ may be related to atomic matrix elements.

For free photons the difference between a perfectly homogeneous medium and an inhomogeneous medium of the same mean density gives rise to the incoherent scattering away from the forward direction known as Rayleigh scattering. For the virtual photons associated with an incident charged particle, scattering from inhomogeneities in the dielectric gives rise to transition radiation (Garibyan 1960). In fact, such scattering only occurs from macroscopic inhomogeneities even at x-ray wavelengths. We may therefore ignore the effect of inhomogeneities in the medium and work with a continuum model.

2.2   The Mean Energy Loss in a Medium[5]

For collisions in which E is very small compared with the incident particle energy, we may calculate the cross section treating the electromagnetic field semiclassically. We derive the electric field $\mathbf{E}$ at the position of the particle, $\mathbf{r} = \beta ct$, and describe the mean energy loss per unit time as the effect of this

[5] The reader who is not concerned with the details of the cross-section derivation may wish to skip Sections 2.2 and 2.3.

electric field doing work on the particle. In Section 4 we reinterpret this energy loss as a probability of loss of discrete energy $\hbar\omega$ in each frequency range; this is the quantization method originally introduced by Planck to describe black body radiation and generally known as semiclassical radiation theory. In this approach the density effect is included from the beginning rather than treated as a correction (Sternheimer 1952). This is appropriate because the effect is significant for particle identification. Derivations that overlap in part with the next two sections were given by Landau & Lifshitz (1960) and in the major review of energy loss by Fano (1963).

In a nonmagnetic dielectric medium Maxwell's equations may be written (using Gaussian units)

$$\text{div } \mathbf{H} = 0; \qquad \text{curl } \mathbf{E} = -\frac{1}{c}\frac{\partial \mathbf{H}}{\partial t} \tag{4.}$$

$$\text{div } (\varepsilon \mathbf{E}) = 4\pi\rho; \qquad \text{curl } \mathbf{H} = \frac{1}{c}\frac{\partial(\varepsilon \mathbf{E})}{\partial t} + \frac{4\pi}{c}\mathbf{j}. \tag{5.}$$

The incident charge velocity βc gives the driving terms;

$$\rho = e\delta^3 \, (\mathbf{r} - \beta ct); \qquad \mathbf{j} = \beta c \rho. \tag{6.}$$

We analyze the problem in terms of potentials in the Coulomb gauge;

$$\mathbf{H} = \text{curl } \mathbf{A}; \qquad \text{div } \mathbf{A} = 0; \qquad \mathbf{E} = -\frac{1}{c}\frac{\partial \mathbf{A}}{\partial t} - \text{grad } \phi. \tag{7.}$$

Equations 4 are satisfied identically. Equations 5 become

$$\mathbf{V} \cdot (\varepsilon \mathbf{V}\phi) = -4\pi \, e \, \delta^3 \, (\mathbf{r} - \beta ct); \tag{8.}$$

$$-\nabla^2 \mathbf{A} = -\frac{1}{c^2}\frac{\partial}{\partial t}\left(\varepsilon \, \frac{\partial \mathbf{A}}{\partial t}\right) - \frac{1}{c}\frac{\partial}{\partial t}(\varepsilon \, \mathbf{V}\phi) + 4\pi e \, \beta\delta^3(\mathbf{r} - \beta ct).$$

These equations are solved by expressing all fields in terms of their Fourier transforms;

$$F(\mathbf{r}, t) = \frac{1}{(2\pi)^2}\int d^3k \, d\omega \, F(\mathbf{k}, \omega) \exp i(\mathbf{k} \cdot \mathbf{r} - \omega t). \tag{9.}$$

One obtains

$$\phi(\mathbf{k}, \omega) = 2e\delta(\omega - \mathbf{k} \cdot \beta c)/k^2\varepsilon;$$

$$\mathbf{A}(\mathbf{k}, \omega) = 2e \, \frac{(\omega \mathbf{k}/k^2 c - \beta)}{(-k^2 + \varepsilon\omega^2/c^2)} \, \delta(\omega - \mathbf{k} \cdot \beta c). \tag{10.}$$

The electric field in space and time may then be expressed as follows

$$\mathbf{E}(\mathbf{r}, t) = (1/2\pi)^2 \int\int [i\omega/c\,\mathbf{A}(\mathbf{k}, \omega) - i\mathbf{k}\phi(\mathbf{k}, \omega)] \exp i(\mathbf{k}\cdot\mathbf{r} - \omega t)\,d^3k\,d\omega$$

11.

The mean energy loss per unit length, known conventionally as $\langle dE/dx\rangle$, is due to the longitudinal component of $\mathbf{E}$ at the point $\mathbf{r} = \boldsymbol{\beta} ct$:

$$\langle dE/dx\rangle = \frac{e\mathbf{E}(\boldsymbol{\beta} ct, t)\cdot\boldsymbol{\beta}}{\beta}$$

12.

$$\langle dE/dx\rangle = \frac{e^2 i}{\beta 2\pi^2}\int\int \left[\frac{\omega}{c}\left(\frac{\omega\mathbf{k}\cdot\boldsymbol{\beta}}{k^2} - \beta^2\right)\Big/\left(-k^2 + \frac{\varepsilon\omega^2}{c^2}\right) - \frac{\mathbf{k}\cdot\boldsymbol{\beta}}{k^2\varepsilon}\right]$$

$$\times\ \delta(\omega - \mathbf{k}\cdot\boldsymbol{\beta} c)\exp i(\mathbf{k}\cdot\boldsymbol{\beta} c - \omega)t\,d^3k\,d\omega$$

13.

$$= \frac{e^2 i}{\beta\pi}\int dk\int_{-\infty}^{\infty} d\omega\,\frac{k}{\beta}$$

$$\times\left[\omega\left(\frac{\omega^2}{k^2 c^2} - \beta^2\right)\Big/(-k^2 c^2 + \varepsilon\omega^2) - \frac{\omega}{k^2\varepsilon c^2}\right].$$

14.

In the last step we have re-expressed the d^3k integration as $2\pi k^2\,dk\,d(\cos\psi)$ and integrated over $\cos\psi$. In doing so we have assumed that ε is isotropic. Note that the time dependence has dropped out but that the integration is still over both positive and negative ω. Since $\varepsilon(-\omega) = \varepsilon^*(\omega)$ we may combine positive and negative frequencies:

$$\langle dE/dx\rangle = -\frac{2e^2}{\beta^2\pi}\int_0^{\infty} d\omega\int_{\omega/v}^{\infty} dk$$

$$\times\left[\omega k(\beta^2 - \omega^2/k^2 c^2)\,\mathrm{Im}\left(\frac{1}{-k^2 c^2 + \varepsilon\omega^2}\right) - \frac{\omega}{kc^2}\,\mathrm{Im}\left(\frac{1}{\varepsilon}\right)\right],$$

15.

where the integration is now over the physical scattering region of Figure 5 with $E = \hbar\omega$ and $p = \hbar k$. The upper limits of integration written formally as infinity will not concern us in our study of thin absorbers.

2.3 The Generalized Oscillator Strength Density

The only unknown quantity in Equation 15 is the complex dielectric constant, which is a function of k as well as of ω, as discussed at the end of Section 2.1. It is convenient to express the imaginary part, ε_2, in terms of the generalized oscillator strength density $f(k, \omega)$ thus

$$\varepsilon_2(k, \omega) = \frac{2\pi^2 Ne^2}{m\omega}f(k, \omega).$$

16.

Fano & Cooper (1968) reviewed the importance of the concept of oscillator strength in atomic physics and its relation to the simple classical model. Its generalization as a function of k as well as of ω was discussed by Inokuti (1971). The function $f(k, \omega)$ dω is a measure of the fraction of electrons coupling to the field between ω and $\omega + d\omega$ when the wave number is k : $f(k, \omega)$ is related to the atomic matrix elements between the ground state and the complete set of orthogonal atomic states n with excitation $(E_n - E_0)$ as follows

$$f(k, \omega) = \frac{1}{Z} \sum_n \frac{2m(E_n - E_0)}{\hbar^2 k^2} \, \delta\left(\omega - \frac{E_n - E_0}{\hbar} \right) \left| \langle n | \sum_{j=1}^{z} \exp\left(i\mathbf{k} \cdot \mathbf{r}_j \right) |0\rangle \right|^2.$$

17.

The sum over n includes both discrete and continuum states.

For atomic hydrogen $f(k, \omega)$ may be calculated rigorously; such calculations were reviewed by Inokuti (1971). They show a resonance region and a quasi-free region as expected from our qualitative discussion based on constituent scattering (Figure 5). For other atoms reasonably reliable calculations by Hartree-Fock and other methods have been made (McGuire 1971, Manson 1972, Berg & Green 1973, Inokuti et al 1978). Experimental investigations of the energy and momentum transfer of low energy electrons in thin targets have been used to investigate $f(k, \omega)$ for many atoms and molecules (see, for example, Bonham et al 1978).

In the context of relativistic charged particle identification, it is not clear that the finer details of the atomic structure are significant. The energy loss measured in an experiment with, for example, a proportional chamber may be discussed adequately with a simplified model of the generalized oscillator strength density. Such a model may be constructed from the photoabsorption cross section using the dipole approximation in the resonance region and point-like scattering in the quasi-free region.

Where ka is small, the dipole approximation is valid (a is the size of the ground-state orbital). This means that the exponential in Equation 17 may be expanded, retaining only the term linear in k. Then the only states that contribute to the sum over n are those that correspond to optically "allowed" transitions, and $f(k, \omega)$ is independent of k. In this region, therefore, $\varepsilon_2(k, \omega)$ is independent of k and equal to its value for free photons. Outside the dipole region the generalized oscillator strength is still constrained by the Bethe sum rule (Bethe 1930, Bethe & Jackiw 1968):

$$\int_0^\infty f(k, \omega) \, d\omega = 1.$$

18.

This sum rule[6] says that the area under every slice of the $f(k, \omega)$ surface at constant k is unity[7]—in fact, if $f(k, \omega)\, d\omega$ represents the fraction of electrons coupling to the field in the range ω to $\omega + d\omega$, f should integrate to unity. This result is quite general and goes beyond the dipole approximation. At high k many of the contributing atomic states have nonvanishing higher multipole matrix elements and are different from those contributing in the dipole-dominated low-k region.

We can now make a model of $f(k, \omega)$. We assume that we may extend the dipole approximation to describe the entire resonance region, defined as the area below and to the right of the free electron line in Figure 5. Thus in this region, $f(k, \omega)$ is independent of k and given in terms of the atomic photoabsorption cross section of free photons, $\sigma_\gamma(\omega)$ by Equations 3 and 16:

$$f(k, \omega) = \frac{mc}{2\pi^2 e^2 Z} \left[\sigma_\gamma(\omega)\right].$$ 19.

The quantity $\sigma_\gamma(\omega)$ is a convenient experimentally available measure of oscillator strength density. At fixed k we therefore have a total oscillator strength in the resonance region given by

$$\int_{res} f(k, \omega)\, d\omega = \frac{mc}{2\pi^2 e^2 Z} \int_{\hbar k^2/2m}^{\infty} \left[\sigma_\gamma(\omega)\right] d\omega.$$ 20.

The remaining oscillator strength necessary to satisfy the Bethe sum rule must be attributed to the quasi-free region. We approximate this by a δ function on the free electron line, ignoring the Fermi motion of bound electrons:

$$f(k, \omega) = \frac{mc}{2\pi^2 e^2 Z} \delta\left(\omega - \frac{\hbar k^2}{2m}\right) \int_0^\omega \sigma_\gamma(\omega')\, d\omega'.$$ 21.

The normalization is determined uniquely by the Bethe sum rule. The integral in Equation 21 describes the number of electrons that are effectively free for an energy transfer, $\hbar\omega$. We note that the photoabsorption cross section satisfies the Thomas-Reiche-Kuhn sum rule,

$$\int_0^\infty \sigma_\gamma(\omega)\, d\omega = 2\pi^2 e^2 Z/mc.$$ 22.

[6] In a relativistic theory the sum rule applies at constant $q^2 = k^2 - \omega^2/c^2$. On or near the free electron line where we make use of the Bethe sum rule $\omega \ll kc$ and the nonrelativistic form (Equation 18) is a good approximation.

[7] This normalization follows from the choice of N as the electron density adhered to throughout this paper. This is convenient for handling gas mixtures.

This model may be written formally as the following general expression for $\varepsilon_2(k, \omega)$:

$$\varepsilon_2(k, \omega) = \frac{Nc}{\omega Z}\left[\sigma_\gamma(\omega)H\left(\omega - \frac{\hbar k^2}{2m}\right) + \int_0^\omega \sigma_\gamma(\omega')\,d\omega'\,\delta\left(\omega - \frac{\hbar k^2}{2m}\right)\right], \quad 23.$$

where $H(x)$ is the step function, $H = 1$ for $x > 0$ and $H = 0$ otherwise. The approximation of the oscillator strength in the quasi-free region by a δ function is adequate since we are really only concerned with the projection of the cross section on the ω or energy transfer axis. Detailed calculations for argon, for example, (see Section 2.5) show that the contribution of the quasi-free region to the energy-loss process is very smooth, broad, and not large, even assuming the δ-function form (see Figure 6b).

Before we can integrate over k in Equation 15 we need $\varepsilon_1(k, \omega)$. The first term of Equation 15 contains a denominator $-k^2c^2 + \varepsilon\omega^2$, which represents the photon propagator and is therefore large only near the photon line—that is, at the lower part of the resonance region where the dipole approximation is good. So for this term we ignore quasi-free electron scattering and write

$$\varepsilon_2(k, \omega) = \frac{Nc}{\omega Z}\sigma_\gamma(\omega) = \varepsilon_2(\omega) \qquad\qquad 24.$$

for all k. The real and imaginary parts of ε are related by the Kramers-Kronig relation,

$$\begin{aligned}
\varepsilon_1(\omega) - 1 &= \frac{2}{\pi}P\int_0^\infty \frac{x\varepsilon_2(x)}{x^2 - \omega^2}\,dx \\
&= \frac{2}{\pi}\frac{Nc}{Z}P\int_0^\infty \frac{\sigma_\gamma(x)\,dx}{x^2 - \omega^2},
\end{aligned} \qquad 25.$$

where P indicates that the principal value of the integral is to be taken. Therefore ε_1 is independent of k for the first term of Equation 15 while for the second term it may be assumed to be unity for low density media.

We are now in a position to integrate over k in Equation 15. After some straightforward manipulation we get

$$\begin{aligned}
\langle dE/dx \rangle = -\int_0^\infty d\omega\, \frac{e^2}{\beta^2 c^2 \pi}\Bigg[&\frac{Nc}{Z}\sigma_\gamma(\omega)\ln\left[(1-\beta^2\varepsilon_1)^2 + \beta^4\varepsilon_2^2\right]^{-1/2} \\
&+ \omega\left(\beta^2 - \frac{\varepsilon_1}{|\varepsilon|^2}\right)\Theta + \frac{Nc}{Z}\sigma_\gamma(\omega)\ln\left(\frac{2m\beta^2 c^2}{\hbar\omega}\right) \\
&+ \frac{1}{\omega}\int_0^\omega \frac{\sigma_\gamma(\omega')}{Z}\,d\omega'\Bigg],
\end{aligned} \qquad 26.$$

where $\Theta = \arg(1 - \varepsilon_1\beta^2 + i\varepsilon_2\beta^2)$ and in terms 1, 3, and 4 we have dropped factors of $1/|\varepsilon|^2$, which are unimportant in low density media.

2.4   The Cross Section

We now reinterpret Equation 26 in terms of a number of discrete collisions with energy transfer $E = \hbar\omega$. With N as the number of electrons per unit volume, $d\sigma/dE$ the differential cross section per electron per unit energy loss, the average energy loss is given by

$$\left\langle \frac{dE}{dx} \right\rangle = - \int_0^\infty NE \frac{d\sigma}{dE} \hbar \, d\omega \qquad\qquad 27.$$

$$\frac{d\sigma}{dE} = \frac{\alpha}{\beta^2\pi} \frac{\sigma_y(E)}{EZ} \ln[(1 - \beta^2\varepsilon_1)^2 + \beta^4\varepsilon_2^2]^{-1/2} \; + \; \frac{\alpha}{\beta^2\pi} \frac{1}{N\hbar c} \left(\beta^2 - \frac{\varepsilon_1}{|\varepsilon|^2}\right) \Theta$$

$$+ \frac{\alpha}{\beta^2\pi} \frac{\sigma_y(E)}{EZ} \ln\left(\frac{2mc^2\beta^2}{E}\right) \; + \; \frac{\alpha}{\beta^2\pi} \frac{1}{E^2} \int_0^E \frac{\sigma_y(E')}{Z} \, dE', \qquad 28.$$

where α is the fine structure constant and ε_1, ε_2, and Θ are derived from σ_y using Equations 25, 24, and following 26 respectively.

In making this substitution we have implicitly assumed single photon exchange (per collision). A similar assumption enters if the first-order Born approximation is used explicitly and is valid for incident velocities much greater than the electron orbital velocities.

The first two terms of Equation 28 are referred to as the transverse cross section. They come from the magnetic vector potential term (in the Coulomb gauge) for which the electric field is transverse to the direction of 3-momentum transfer, $\hbar k$. Comparing the first term (with $\varepsilon_2 = 0$) with the simple model described in the appendix, we recognize the factor $\ln(\gamma'^2)$ responsible for the relativistic increase of the cross section as well as its saturation. In the limit that ε_2 vanishes, the second term describes the emission of Cerenkov radiation; Θ jumps from near 0 below Cerenkov threshold to almost π above:

$$N\left(\frac{d\sigma}{d\omega}\right)_{\check{z}} = \frac{\alpha}{c}\left(1 - \frac{\varepsilon_1}{\beta^2|\varepsilon|^2}\right) \simeq \frac{\alpha}{c} \sin^2 \theta_{\check{z}}. \qquad 29.$$

This is the familiar result for the Cerenkov flux in terms of the Cerenkov angle $\theta_{\check{z}}$. Above ionization threshold ε_2 does not vanish and it is impossible to distinguish this component of the cross section as responsible for Cerenkov radiation; indeed this term may be negative. Chechin et al (1972) used a cross section identical with Equation 28 except in the form of this term. Their derivation depends on certain assumptions about the transparency of the medium that were first made by Budini & Taffara (1956).

Their formula has recently been used in calculations of energy loss by Lapique & Puiz (1980).

The third and fourth terms of Equation 28 are known as the longitudinal cross section. They come from the electrostatic term in the Coulomb gauge, which has the electric field parallel to the momentum transfer. In the nonrelativistic theory they are the only terms. Their sole dependence on velocity is through the $1/\beta^2$ factor common to all terms and so they become effectively constant in the relativistic region. The third term comes from the resonance region while the fourth represents Rutherford scattering from those electrons that are quasi-free for an energy transfer E. For low density media the longitudinal cross section does not depend on either ε_1 or density. The transverse cross section does depend on density and in the limit of zero density becomes

$$\frac{d\sigma_T}{dE} = \frac{\alpha}{\beta^2 \pi} \left[\frac{\sigma_\gamma(E)}{EZ} (\ln \gamma^2 - \beta^2) \right]. \qquad 30.$$

This cross section increases linearly with $\ln \gamma$ without saturation due to the density effect.

It is instructive to consider the limit $\beta = 1$ of the cross section in Equation 28. Let us approximate the dielectric constant, $\varepsilon(\omega)$, in terms of the plasma frequency, $\omega_p^2 = 4\pi N e^2/m$, thus

$$\varepsilon_1 - 1 = \omega_p^2 \, P \int_0^\infty \frac{f(\omega') \, d\omega'}{\omega'^2 - \omega^2} \simeq \frac{\omega_p'^2}{\omega^2}, \qquad 31.$$

where ω_p' is the effective plasma frequency due to those electrons that are effectively free; $\omega_p'^2(\omega) = 4\pi N e^2/m \times \int_0^\omega f(\omega') \, d\omega'$. The second equality in Equation 31 is valid for values of ω near which $f(\omega)$ is smoothly varying[8]. With this approximation and insofar as ε_2 is small compared with $(\varepsilon_1 - 1)$, the transverse cross section becomes

$$\frac{d\sigma_T}{dE} = \frac{\alpha}{\beta^2 \pi} \frac{\sigma_\gamma(E)}{EZ} \left[\ln \left(\frac{\omega^2}{\omega_p'^2} \right) - 1 \right]. \qquad 32.$$

Comparing this with Equation 30 we expect the cross section to saturate at a velocity of order $\gamma = \omega/\omega_p'$. This result shows that the relativistic rise extends over the widest range for gases (low ω_p) with high ionization potentials (large ω) and is the reason for the choice of a gas, and a noble gas in particular, as the working media in relativistic dE/dx detectors. The

---

[8] This relation depends on the approximation that

$$P \int_0^\infty f(\omega') \left[\left(\frac{\omega'}{\omega} \right)^2 - 1 \right]^{-1} d\omega' \text{ is equal to } \int_0^\omega f(\omega') \, d\omega'.$$

approximation in Equation 31 is made here for the purpose of illustration. In practice, the full expression of Equation 28 may be used.

To use Equation 28 we need only know the photoabsorption cross section per electron (σ_γ/Z) for the medium of interest. With the advent of powerful sources of synchrotron radiation have come reliable experimental photoabsorption data. References have been listed by Way (1978). For the noble gases see West & Marr (1976) and West & Morton (1978). For simple molecular gases including hydrocarbons and CO_2 see Lee et al (1973, 1977). Theoretical calculations for the elements are in fair agreement and may also be used (McGuire 1968). For mixtures of gases the cross section per electron may be constructed by independent addition. Where data for molecular gases are not available, the cross section may be approximated by independent addition of the elements, although this will clearly be a poor guess for the outer electrons whose energies are affected by chemical binding. Although calculated energy-loss distributions are not too sensitive to the assumed photoabsorption spectrum, it is important to ensure that for each element or molecule the spectrum satisfies the sum rule (Equation 22) including the contribution of discrete absorption lines. In the absence of firm data the sum rule may be satisfied artificially by adding the necessary absorption below threshold to account for discrete excitation. With σ_γ/Z known, ε_2 and ε_1 follow from Equations 3 and 25 respectively. We call calculations using the full Equation 28 together with photoabsorption cross sections the photo absorption ionization (PAI) model.

2.5   A Calculation for Argon and the "Ideal" Detector

Using the PAI model we explore the dependence of the energy-loss cross section for argon on energy transfer E, velocity β, and gas pressure P. Complete and reliable data on σ_γ are available (West & Marr 1976). Figure 6a shows $E\sigma_\gamma$ plotted against E on a log scale. Equal areas under this curve represent equal contributions to the optical oscillator strength density. The K, L, and M shells are clearly visible although there are contributions at intermediate energies owing to correlation effects (Chang & Fano 1976). Figure 6b shows $E^2 \, d\sigma/dE$ evaluated according to Equation 28 for $\beta = 1$ in argon at normal density. Equal areas under the curve represent equal contributions to the energy loss. The contributions from the Rutherford term, the transverse term (shaded), and the longitudinal term in the resonance region respectively are shown above one another. 1% of the total energy loss due to collisions of less than 50 keV is in the form of Cerenkov radiation below ionization threshold. An important difference between the shapes of Figures 6a and b is due to the slowly rising Rutherford or quasi-free term that forms a background under the resonant structures.

In the relativistic region the transverse cross section increases with $\ln \gamma$

until limited by the density effect. The combined cross section (Equation 28) normalized to its value at $\beta\gamma = 4$ is shown as a function of $\beta\gamma$ for three different values of energy transfer E in Figure 7. For particle identification we are interested in the slope of the relativistic increase and at what value of $\beta\gamma$ it saturates. We see that both of these quantities vary with E. The value of

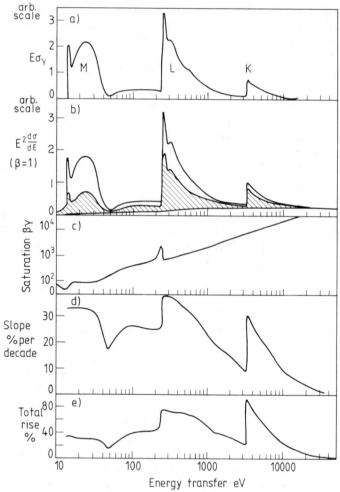

Figure 6 Results of a calculation based on the PAI model for argon at normal density as a function of energy transfer, E. (*a*) The input photoabsorption spectrum. (*b*) The calculated energy-loss cross section for $\beta = 1$. Upper unshaded area is the resonance region longitudinal term; shaded area is the transverse term; lower unshaded area is the quasi-free or Rutherford term. (*c*) The value of $\beta\gamma$ at which the slope of the log rise of the cross section for energy transfer E saturates (slope falls to 5% of maximum). (*d*) The slope of the log rise in the absence of density effect. (*e*) The total rise relative to minimum at normal density.

$\beta\gamma$ at which saturation is effectively complete is shown in Figure 6c. The behavior is very close indeed to a line $\beta\gamma = 2\omega/\omega'_p$ except at optical frequencies and just below the L-shell edge. This dependence was expected on the basis of Equation 32 and clearly gives a convenient way of estimating the saturation point. The density effect becomes significant at about 40% of this saturation value as shown in Figure 7. For the outer shell of electrons of argon this corresponds to $\beta\gamma \sim 40$ at normal density. As the density of argon is varied, the saturation point follows a (density)$^{-1/2}$ dependence.

The slope of the relativistic rise, which depends largely on the relative contribution of transverse and longitudinal terms, is shown as a function of E in Figure 6d. It is independent of density. The slope is given in percent per decade so that, for instance, the percentage difference in the cross section due to incident K and π mesons of the same momentum is half the figure shown. The total rise relative to the ionization minimum is shown in Figure 6e for argon at normal density.

In comparing the possible contribution made by different shells to particle identification we must consider statistical effects. There are about 30 collisions with M-shell electrons per centimeter of relativistic track in argon at normal density, between one and two collisions with L-shell electrons per centimeter, and one collision with a K-shell electron every 30 cm. This shows that, even in an "ideal" detector in which one could imagine recording every individual collision in the gas and its energy, K-shell ionization can make no effective contribution to individual particle identification. On the other hand, L-shell ionization is effective in a few meters, while in the absence of the density effect M-shell ionization would be effective in 15–20 cm.

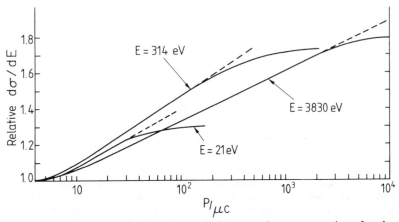

Figure 7 The calculated $\beta\gamma$ dependence of the energy-loss cross sections for three representative energies for argon at normal density normalized to $\beta\gamma = 4$. The dashed lines show the dependence in the low density limit.

In Figure 8 the solid curve shows the mass resolution obtainable from a knowledge of the spectrum of individual energy transfers of a particle with known momentum traversing one meter of pure argon at normal density. The resolution is calculated by a maximum likelihood analysis of this spectrum, using as likelihood the function $Nx(d\sigma/dE)$, which depends on ln (p/μ). For this reason the resolution is expressed as the logarithm of the resolvable mass ratio. For example the K to π meson mass ratio represents a difference in ln μ of 1.25. It is possible in principle to separate K and π mesons by the Rayleigh criterion (one FWHM) in the range $\beta\gamma = 4$–150 with such an "ideal" detector of length one meter. Whether this level of separation is sufficient depends on the circumstances of the experiment, in particular on the relative populations of the different masses and on the degree of confidence required. (In this illustrative discussion we have ignored the fact that K and π mesons of the same momentum have sufficiently different $\beta\gamma$ values that the resolution is not the same in the two cases). Alternatively we can consider the perfect "primary ionization" detector, which resolves the number of collisions without measuring their energies. Since such collisions are predominantly M-shell processes the

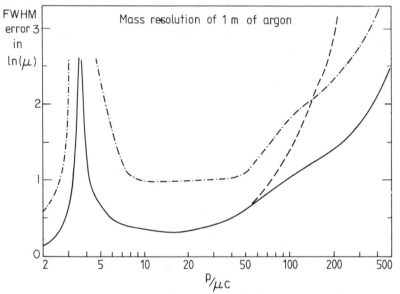

Figure 8   The potential and actual mass resolution inherent in the energy loss in 1 m of pure argon at normal density (excluding Cerenkov radiation below ionization threshold). Solid line: the energy transfer in every collision is known. Dashed line: only the number of such collisions is known (primary ionization). Dotted and dashed line: the integrated energy loss in each of 62 × 1.6-cm samples is known (ISIS1).

resolution is rapidly curtailed by the density effect above $\beta\gamma$ equal to 50. This is shown by the dashed line. At lower velocities all electron shells share in the log rise and $1/\beta^2$ behavior so that measurement of primary ionization is as effective as the ideal detector.

Neither of the above detectors is yet realizable. For comparison we show as a dash-dot curve the resolution of a practical device, ISIS1, described briefly in Section 4, which measures the energy loss integrated in each of 62 × 1.6-cm samples. Its resolution is typically worse than ideal by a factor of two to three although the difference is much less pronounced at high $\beta\gamma$. This is because the sample length is of the order of the mean free path between L-shell collisions, which are the source of the rise in this region. At nonrelativistic velocities the mass resolution is excellent, simply on account of the steepness of the energy-loss curve (see Figure 2). Near $\beta\gamma = 3.5$ there is a "cross-over" region where particle identification is not possible (see, however, Footnote 13 in Section 4.1). We postpone further discussion on mass resolution to Section 4.

3   THE ENERGY-LOSS DISTRIBUTION

In Section 3.1 we consider how the cross section of Section 2 may be used to derive the energy-loss distribution in a finite thickness of material, and in Section 3.2 how to relate this to actual measurements. In Section 3.3 we compare the results of various calculations of energy loss with available data.

3.1   Calculation of the Energy-Loss Distribution

The probability distribution $F(x,\Delta)$ of energy loss Δ by particles of given velocity traversing a thickness of material x is uniquely determined by the cross section $d\sigma/dE$ and the number of electrons in the layer Nx. Since $d\sigma/dE$ has a significant value over a wide range of E, the distribution $F(x,\Delta)$ is both broad and skew. This problem was first described by Williams (1929) and recently was discussed fully by Bichsel & Saxon (1975). There are two approaches, the convolution method and the Laplace transform method.

The convolution method considers a very small thickness δx such that the chance of two collisions within δx is negligible. Then

$$F(\delta x,\Delta) = (1-\tau)\,\delta(\Delta) + N\delta x\frac{d\sigma}{dE}(\Delta) + O(\tau^2), \qquad 33.$$

where τ, the chance of one collision, is $N\delta x \int_0^\infty d\sigma/dE\ dE$. From this the distribution for any given thickness may be built up by repeated application of the formula

$$F(2x,\Delta) = \int_0^\Delta F(x,\Delta-E)\,F(x,E)\mathrm{d}E, \qquad\qquad 34.$$

which follows from the statistical independence of the energy loss in different layers. Bichsel & Saxon (1975) used this method.

Alternatively the folding may be done by Monte Carlo method. The mean number of collisions, τ, in the sample is calculated from the integrated cross section. Then for each trial the actual number is chosen from a Poisson distribution with mean τ. Finally the energy loss is just the energy sum of such collisions chosen according to the normalized $\mathrm{d}\sigma/\mathrm{d}E$ spectrum. This or related methods have been used by Ispiryan et al (1974), Cobb et al (1976), and Ermilova et al (1977).

The Laplace transform method was introduced by Landau (1944). He considered the change in $F(x,\Delta)$ as a result of passing through a thin elemental layer δx:

$$F(x+\delta x,\Delta)-F(x,\Delta) = -N\delta x \int_0^\infty \frac{\mathrm{d}\sigma}{\mathrm{d}E}(E)\,F(x,\Delta)\mathrm{d}E$$

$$+ N\delta x \int_0^\Delta F(x,\Delta-E)\frac{\mathrm{d}\sigma}{\mathrm{d}E}(E)\,\mathrm{d}E. \qquad 35.$$

The first term describes the probability that the energy loss was already equal to Δ before entering δx where a further collision increased the energy beyond Δ. The second term corresponds to the case in which the energy loss at x was $\Delta-E$ but a collision with energy E occurred in the element δx. This equation may be put in the form of a transport equation

$$\frac{\partial F}{\partial x}(x,\Delta) = \int_0^\infty N\frac{\mathrm{d}\sigma}{\mathrm{d}E}(E)\,[F(x,\Delta-E)-F(x,\Delta)]\mathrm{d}E. \qquad 36.$$

This may be solved by taking Laplace transforms of both sides with respect to Δ and solving for $\bar{F}(x,s)$, the transform of $F(x,\Delta)$:

$$\bar{F}(x,s) = \exp\left[-x \int_0^\infty N\frac{\mathrm{d}\sigma}{\mathrm{d}E}(E)(1-e^{-sE})\mathrm{d}E\right] \qquad 37.$$

where use has been made of the boundary condition:

$$F(x=0,\Delta) = \delta(\Delta); \qquad \bar{F}(x=0,s) = 1. \qquad 38.$$

Inverting the Laplace transform (σ is very small but positive)

$$F(x,\Delta) = \frac{1}{2\pi i}\int_{-i\infty+\sigma}^{i\infty+\sigma} \mathrm{d}s \, \exp\left[s\Delta-x\int_0^\infty N\frac{\mathrm{d}\sigma}{\mathrm{d}E}(E)(1-e^{-sE})\mathrm{d}E\right]. \qquad 39.$$

This result is exact. On the other hand realistic models of the cross section are not simple and the integration must be carried out numerically. Talman (1979) recently used this method.

Of course, given the same cross section, $d\sigma/dE$, the convolution and Laplace transform methods must give the same energy-loss distributions. The important difference between the work of different authors has been the approximations made for $d\sigma/dE$ and the range of validity of the resulting energy-loss distributions, as discussed in Section 3.3. The convolution method is the easiest to handle for thin absorbers where the number of collisions is not large and acceptable approximations in the form of the cross section are few. For the new results reported in this paper we have used this method in conjunction with the PAI model.

3.2   The Ionization Distribution

Up to this point we have been discussing the distribution of energy loss in a thin absorber by a relativistic particle. Unfortunately this is not directly measurable[9]. The most faithful manifestation of the energy loss is the deposited ionization, but others, such as bubble density (Chechin et al 1972, Fisher et al 1975) and emulsion grain density (reviewed by Crispin & Fowler 1970), have been discussed. Except for the case of ionization these show poor velocity resolution and are not suitable for the identification of single particles. Therefore in the following we confine our attention to gas-filled proportional counters.

We assume that the number of electrons n_i liberated in the absorber is related to the energy deposited Δ by the relation

$$\Delta = n_i W \qquad\qquad 40.$$

and that, on average, the energy per ion pair W is a constant independent of Δ. This assumption is essentially statistical and depends on the random nature of the secondary mechanisms responsible for thermalization of the deposited energy. The situation was discussed recently by Inokuti (1975). Experimental values of W vary from 26 eV for argon to 36 eV for nitrogen; for a review see ICRU (1979). For the energy deposited by relativistic charged particles, we are concerned with the linearity of Equation 40, not only for the combined energy loss Δ, but also at the level of individual energy transfers, E. Insofar as this is not so, the velocity dependence of ionization may not follow that of energy loss. Although this linearity must break down for E close to the first ionization potential, for mixtures involving molecular gases with low ionization potentials, departures from

---

[9] By contrast the energy loss of low energy electrons is measured directly in the study of the generalized oscillator strength density (Bonham et al 1978).

linearity should be smaller than for pure noble gases. Photon data for an argon/methane mixture are consistent with constant W to within 3% at least above $E = 250$ eV. In propane W is constant to within 8% down to $E = 80$ eV (Srdoc 1973).

To calibrate absolutely an ionization distribution, the response to a known energy deposition must be measured. This usually takes the form of monochromatic x rays. However technical differences between x-ray and charged particle data, such as the effects of triggering or local saturation of the gas amplification, can make absolute calibration unreliable. Most data are uncalibrated or shown relative to other charged particle data at minimum ionization, $\beta\gamma \sim 4$.

While Equation 40 may apply statistically, for a fixed value of Δ there are fluctuations of n_i (Fano 1947). There are further statistical fluctuations in the gas amplification process of the n_i electrons at the collecting anode. This effect has been summarized by Charpak (1970) and the broad conclusion is that the rms of the combined fluctuations from both sources σ_p in the observed amplified signal P is

$$\sigma_p/P \sim n_i^{-1/2} \sim 0.17\, \Delta^{-1/2} \qquad\qquad 41.$$

where Δ is in keV. This relation is followed to within $\pm 10\%$ for photons in propane (Srdoc 1973). It represents the intrinsic energy resolution of a proportional counter. The fluctuations represented by Equation 41 are small compared to the fluctuations in the energy deposited itself. Consequently the intrinsic energy resolution is not an important parameter of an energy-loss detector.

Some of the energy loss may be carried away from the region of the track in the form of radiation. At intermediate energies the absorption length of the gas is too short but at x-ray energies fluorescent photons from K-shell vacancies may travel several centimeters. These are too rare to have an effect on the ionization loss distribution but may be useful for calibration purposes (Allison et al 1979). As far as the Cerenkov energy is concerned, we remark that all proportional chamber gas mixtures are chosen to have short ultraviolet photoabsorption lengths that prevent Geiger discharges in the gas amplification region. So that, while 1% of the energy loss may propagate away from the track in pure argon, in a practical proportional chamber gas the figure is smaller. The proportional chamber quenching gas with its low ionization potential also plays a major role converting metastable states into ionization. Without this mechanism the linear dependence of Equation 40 would be less plausible for collisions with outer electrons.

Energy is also carried away from the region of the track in the form of δ-ray electrons. Such electrons come from collisions of several tens of keV or

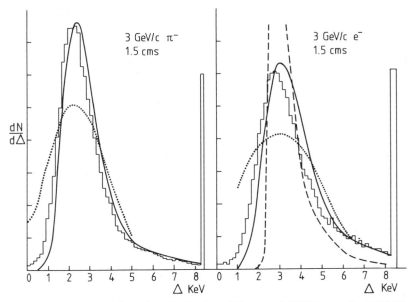

Figure 9 Experimental energy-loss distributions of Harris et al (1973) for π and e at 3 GeV/c in 1.5 cm of argon/7% CH_4 at normal density. The dashed and dotted curves are calculations using the model of Landau (1944) with corrections of Maccabee & Papworth (1969) and Blunck & Leisegang (1950) respectively. The solid curves are the predictions of the PAI model.

more and are in the tail of the energy-loss distribution. They are infrequent and their escape is not important.

3.3 *Comparison with Data*

The experimental data with which calculations may be compared consist of energy-loss distributions[10] $F(\Delta)$ and their dependence on velocity, pressure, thickness, and composition. Rather few clean energy-loss distributions have appeared in the literature (Dimcovski et al 1971, Harris et al 1973, Kopot et al 1976, Bunch 1976, Walenta et al 1979, Allison et al 1979). However, there are many measurements of widths and peaks of distributions, for instance Jeanne et al (1973), Onuchin & Telnov (1974), and other work cited by them. In Figure 9 we show the distributions obtained by Harris et al (1973) for the energy loss in 1.5 cm of argon/7% CH_4 at normal density. Data on the relativistic rise in argon and xenon are shown in Figures 10 and 11. Reliable data for propane are also available. The

---

[10] For convenience we use this term for the experimental as well as the theoretical distributions, remembering that the former is really the ionization distribution as discussed in Section 3.2.

pressure dependence of the rise has been studied extensively by Walenta et al (1979), Fancher et al (1979), and Hasebe et al (1978).

The dashed lines on Figures 9–11 are the predictions of the most general model, due to Sternheimer (1952) based on the work of Landau (1944). The basic assumption of Landau's model is that the Rutherford term in the cross section is the only significant source of statistical fluctuation responsible for the distribution width and that the contribution of the cross section in the resonance region may be evaluated in the low density limit by using the oscillator strength sum rule to give a "restricted mean energy loss." This depends only on the mean ionization potential defined by $\ln I = \int \ln(\hbar\omega) f(\omega) d\omega$, and not on details of the optical oscillator strength, $f(\omega)$. The density effect is described by a correction term (Sternheimer 1952, Sternheimer & Peierls 1971 and other references cited there). The model provides a poor description of the distributions (Figure 9) and the relativistic rise (Figures 10, 11). In particular the calculated rise is some 10–

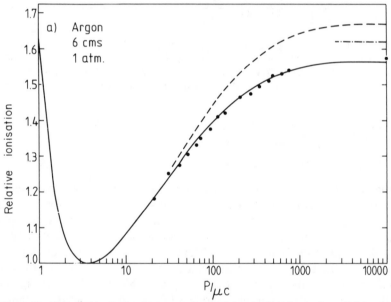

Figure 10   Values of ionization in argon at normal density (relative to $\beta\gamma \sim 4$). Measured points from Lehraus et al (1978) refer to the mean of the lowest 40% of 128 × 6-cm samples of argon/5% CH_4 and, with the exception of the highest point, were taken with incident hadrons. The experimental errors are slightly larger than the points shown. The dash line and dash-dot line are the predictions of the Sternheimer (1952) and Ermilova et al (1977) models respectively. The solid line is the most probable energy loss in 4.5 cm of pure argon calculated with the PAI model.

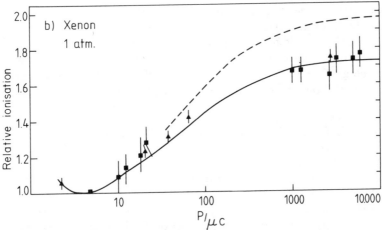

Figure 11 Relative ionization in xenon at normal density taken from Fischer et al (1975) (*triangles*), Walenta et al (1979) (*squares*), "most probable" values for 1.27 cm of xenon/18% methylal, and 2.3 cm of xenon/7.5% CH_4/5% C_3H_8 respectively. The data above $\beta\gamma = 1000$ were taken with incident electrons; the remainder with hadrons. Dashed line is the prediction of the Sternheimer (1952) model and solid line is that of Cobb et al (1976) for pure xenon.

15% too large for the noble gases[11]. Recently Chechin & Ermilova (1976) discussed the range of validity of the Landau model. Landau's basic assumption is expected to fail for thicknesses x such that $\xi/I < 100$ where $\xi = 2\pi Ne^4x/m\beta^2c^2$. For relativistic particles in argon, xenon, and propane this corresponds to a minimum thickness of about 170, 150, and 30 cm atm respectively. On the other hand, for efficient particle identification, ionization must be measured in a number of samples, each thin enough to distinguish the spread due to the large fluctuations of the Rutherford contribution from the resolution due to fluctuations in the resonance region. This requires $\xi/I \sim 1$, incompatible with the basic assumption of the Landau model. An early attempt to improve the description of the distributions included a first-order atomic shell effect (Blunck & Leisegang 1950), but the model is not applicable for $\xi/I < 10$ (Chechin & Ermilova 1976).

More recent models have developed along the lines discussed in Section 2. Simple models of this type, by including the shell structure of the atoms describe the energy-loss distribution with fair success; the detailed

---

[11] A conjecture by Garibyan & Ispiryan (1972) that this is a transition effect due to the chamber windows has not been confirmed by the experiments of Smith & Mathieson (1975) or Allison et al (1976).

treatment of the dielectric constant is necessary to achieve a good description of the relativistic rise. For example, Ispiryan et al (1974) and Talman (1979) used as a crude model of the photoabsorption spectrum a number of δ functions representing the fraction f_i of electrons in each shell at an energy $E_i = 1.3 \times$ the energy of the shell edge. The cross section in Equation 28 then becomes in the low density limit

$$\frac{d\sigma}{dE} = \frac{2\pi e^4}{mc^2\beta^2}\frac{f_i}{E_i}\delta(E-E_i)\left(\ln\frac{2mc^2\beta^2\gamma^2}{E_i}-\beta^2\right)+\frac{2\pi e^4}{mc^2\beta^2}\frac{f_i}{E^2}H(E-E_i), \qquad 42.$$

where H is a unit step function.

The continuous curves superimposed on Figure 9 are the result of a calculation using the PAI model for pure argon. Use of Equation 42 produces similar results at least for 3-GeV/c pions. Taking into account the quoted energy normalization uncertainty of 20% in the data, the agreement is fair although there is an excess of signals at small energy loss also noted in other data (Allison et al 1979). The experimental resolution function (Equation 41) has not been included in the calculation but has a small effect.

Ermilova et al (1977) calculated the energy loss on the Fermi plateau using the empirical photoabsorption spectrum, although this was done earlier in connection with the track density in bubble chambers by Chechin et al (1972). Substituting the approximate Formula 32 for the first two terms of Equation 28 yields

$$\frac{d\sigma}{dE} = \frac{\alpha}{\beta^2\pi}\frac{\sigma_\gamma(E)}{EZ}\left[\ln\left(\frac{\omega 2mc^2\beta^2}{\omega_p'^2\hbar}\right)-1\right]+\frac{\alpha}{\beta^2\pi}\frac{1}{E^2}\int_0^E\frac{\sigma_\gamma(E')}{Z}dE' \qquad 43.$$

for the cross section. In their treatment Ermilova et al (1977) took the logarithmic term to be the only contribution to the restricted energy loss. With these approximations they overestimated the relativistic rise of argon as shown on Figures 10 and 11. Cobb (Cobb 1975, Cobb et al 1976, Bunch 1976) was the first to use the full cross section (Equation 28), although he and his colleagues approximated the photoabsorption cross section by a series of Lorentzians or δ functions similar to the model of Ispiryan et al (1974). The energies chosen were equal to the values of the absorption edges. This choice was criticized by Ermilova et al (1977). However, the calculations gave very good agreement with available data on the relativistic rise for both argon and xenon (Figure 11). (Increasing the ionization energies by 30% as suggested by Ermilova et al increased the calculated relativistic rise by about 3% for argon.) Figure 10 shows the results of new calculations with the PAI model for pure argon. They are in agreement with the data of Lehraus et al (1978).

We conclude that (a) for thin absorbers the Landau-Sternheimer model is inapplicable; and (b) the PAI model provides a good description of the data although further examples, including the effects of quenching gases, remain to be calculated. Possible residual differences between calculation and experimental data, for instance in the width of the spectra, may be due to a breakdown of the assumed correspondence between energy loss and measured ionization.

4 MULTIPLE SAMPLING DETECTORS

4.1 Statistical Analysis of Signals

Let us consider how the ionization resolution and thence velocity and mass resolution may be obtained in practice. This problem was first considered by Alikhanov et al (1956). As discussed in the introduction a detector capable of distinguishing between different mass assignments to a track of known momentum must make a large number n of measurements Δ_i. The most probable energy loss is a very poor estimator for a spectrum with limited statistics. In principle the most efficient use of the data to distinguish between two mass hypotheses is to compute the likelihood ratio

$$L_{12} = \prod_{i=1}^{n} F_1(\Delta_i)/F_2(\Delta_i) \qquad\qquad 44.$$

where F_1 and F_2 are the normalized pulse height distributions expected for two masses μ_1 and μ_2 and L_{12} is the likelihood ratio of μ_1 to μ_2. The distributions F vary little in shape with velocity[12]. For example Figure 12 shows the calculated dependence of the relative width on the most probable energy loss for argon according to the PAI model. The small variation with $\beta\gamma$ is due to the dependence on the statistics of both the total number and the relative numbers of hard and soft collisions.

For practical detectors we may assume that the shape of F is velocity invariant[13] and find the most likely values and error of the ionization scale

---

[12] To the extent that this is true the dependence of different ionization estimators on velocity is the same for a given sample thickness. This is assumed in Figures 10 and 11 where ratios of "most probable" values and truncated mean values are compared indiscriminately.

[13] The loss of information associated with this simplification is not serious. It has been suggested (Nygren 1976, Talman 1979) that kaons and pions might be separated by the different widths of their distributions in the region of minimum ionization. Careful study shows that this is a false hope (Allison et al 1978a). A likelihood ratio analysis using the PAI model shows that more than 5000 × 1.6-cm atm samplings in argon would be required to separate K and π by two standard deviations. Talman (private communication) has agreed that this identification is impractical. His original paper erroneously assumes that the log likelihood function is parabolic in the neighborhood of the double solution.

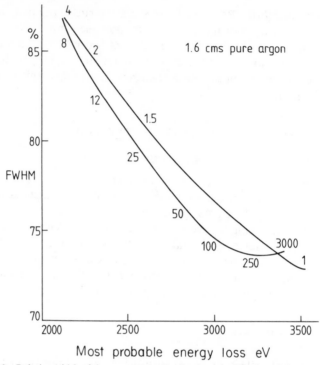

Figure 12   Relative width of the energy-loss distribution as a function of the most probable energy loss for 1.6 cm of pure argon at normal density calculated with the PAI model. The figures on the graph are the corresponding values of $\beta\gamma$.

parameter λ which maximizes

$$L(\lambda) = \prod_{i=1}^{n} F(\Delta_i/\lambda) \qquad\qquad 45.$$

for each track. Our simulation studies show that $\ln \lambda$ is normally distributed and that the relative values of λ obtained for different tracks are insensitive to the precise form of the function used. By monitoring the average value of λ fitted to, for instance, beam tracks of known velocity, a normalized ionization estimate and error for each track can be determined[14]. The probability of a certain mass assignment to a track may then be calculated by comparing the expected ionization (based on the known momentum, the

---

[14] In a practical detector the sample length depends on track inclination. If it is assumed that the shape F is unaffected, the likelihood method may be used as described and the fitted value of λ corrected by a simple $\cos \theta$ factor. This approximation is probably satisfactory for inclinations up to about 45°.

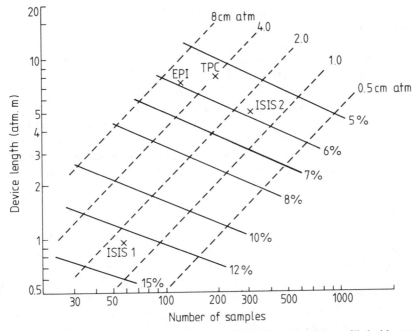

Figure 13   The ionization resolution (% FWHM) of a multisampling detector filled with pure argon calculated with the PAI model for $\beta\gamma = 100$. The dashed lines are loci of constant sample thickness. The devices EPI, ISIS1, ISIS2, and TPC are described in Table 2.

relativistic rise curve, and the hypothesized mass) with the observed ionization value and error using normal error statistics.

Although the likelihood method is efficient, some authors have considered it too time consuming in practice and have studied other methods. If the energy-loss distribution were Gaussian, it would be sufficient simply to take the mean of all measurements. Unfortunately this is not possible because of the long tail. However, this difficulty may be overcome by taking a fixed fraction r of the signals with the smallest amplitudes and evaluating their mean. This procedure has been studied by Alikhanov et al (1956) and many workers in the field since. They find the resolution insensitive to the value of r chosen in the range 0.4–0.6, and little worse than that obtained by the likelihood method. However, we believe that the likelihood method should not be ignored; there is a negligible loss of information if the data are histogrammed in coarse bins[15] and, if the median of the distribution is used

---

[15] This implies that the "least count" in the measurement of the ionization may also be coarse; 4–5 bits of precision are more than sufficient. However, the stability must be much better than this and 8 bits or more are usually used.

to derive a good starting value, a maximum likelihood fit can be carried out rapidly.

Figure 13 is a plot of the calculated ionization resolution obtainable from a likelihood analysis of data from devices using pure argon at $\beta\gamma = 100$. Lines of constant resolution (FWHM) are shown as a function of the device length and number of samples. Crosses show the length and number of samples for four devices currently operational or under construction. The gas in these devices is not pure argon but this has a small effect. The external particle identifier (EPI) has achieved 6% (FWHM) ionization resolution (Lehraus et al 1978) and ISIS1 between 12 and 14% (Allison et al 1979). Both of these figures are in good agreement with the predictions shown on Figure 13.

4.2   The Choice of Detector Parameters

The parameters that determine the mass resolution of an ionization detector are the number of samples n, the sample size x, the gas composition, and the gas pressure P. These influence the ionization resolution, the slope of the relativistic rise, and the $\beta\gamma$ value at which the density effect becomes important.

A study of a wide variety of gases and conditions has not yet been carried out using the PAI model. Instead, in Table 1 we show the results of calculations with an earlier model by Cobb et al (1976) for some gases at normal density. These show that, while the ionization resolution for the rare gases is comparable, the resolution for low-Z molecular gases is better.

Table 1   Calculations of resolution in 5 m of gas at normal density divided into 1.5-cm samples (from Cobb et al 1976)

Gas	Relativistic rise	Ionization resolution (% FWHM)	K/π[a] limit (GeV/c)
Helium	1.41	5.2	45
Neon	1.54	5.4	50
Argon	1.58	5.3	55
Krypton	1.60	5.2	55
Xenon	1.70	5.5	95
Methane	1.36	3.9	30
Ammonia	1.39	4.2	45
Nitrogen	1.48	4.6	45
Carbon dioxide	1.45	3.9	50
Argon/20% CO_2	1.55	4.9	55

[a] The K/π limit corresponds to the momentum at which K and π mesons are separated by $1.8 \times$ FWHM resolution.

However, this apparent advantage is more than offset by the smaller relativistic rise. Among the rare gases the rise is largest for high Z. The behavior of both the resolution and the relativistic rise is a direct consequence of the size of energy transfer in a typical collision, as discussed in Section 2.5. The experimental data of Walenta et al (1979), who compare propane, argon, and xenon, confirm these trends. Therefore to get the best mass resolution over the widest range of $\beta\gamma$, heavy noble gases are preferred. Xenon has not yet been used; energy-loss detectors are usually filled with argon and a small component of a molecular gas to ensure their operation as drift or proportional chambers.

We now consider more generally whether there are any simple optimum choices of detector parameters and gas composition. To do this we derive some approximate formulae based on the PAI model and supported by experimental data. They are intended as a guide only; they are not a substitute for a calculation using the photoabsorption data for the gas mixture concerned.

For pure argon the ionization resolution $R(\%$ FWHM), shown in Figure 13, is adequately described by the formula

$$R = 96n^{-0.46}(xP)^{-0.32}. \qquad \qquad 46.$$

One expects the power of n to be -0.5; Walenta et al (1979) found a power -0.43 empirically using a truncated mean analysis rather than a maximum likelihood method. Figure 14 shows experimental data for the relative widths of single sample distributions ($n = 1$) as a function of the dimensionless scaled value of x, namely $\xi/I = 2\pi Ne^4x/m\beta^2c^2I$. Numerically ξ

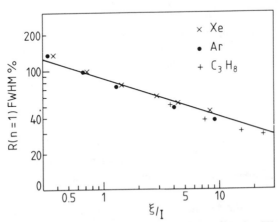

Figure 14   Experimental data on the width of the energy-loss distribution (Walenta et al 1979) as a function of the scaled sample thickness ξ/I. The solid line is the extrapolation of Equation 46 for pure argon to a single sample ($n = 1$).

$= 6.83vxP/\beta^2$ where v is the mean number of electrons per molecule and xP is in cm atm. I is the mean ionization potential defined in Section 3.3. Equation 46 gives a fair description of the data, not only for argon, but for xenon and propane as well in the case $n = 1$. Extending the formula to $n \neq 1$ for all gases by working in terms of ξ/I in place of xP, we find

$$R = 81n^{-0.46}(\xi/I)^{-0.32}. \qquad 47.$$

Judging from Figure 14, this formula is good to 10–20% in the range $0.5 < \xi/I < 10$.

The slope of the log rise below the onset of the density effect depends on the gas mixture and on the sample thickness. This may be seen by examining Equation 26 in the low density limit ($\varepsilon_1 = 1$),

$$\langle dE/dx \rangle = -\int_0^\infty d\omega \, \frac{e^2}{\beta^2 c^2 \pi} \left[\frac{Nc\sigma_\gamma(\omega)}{Z} \left(\ln \frac{2m\beta^2\gamma^2 c^2}{\hbar\omega} - \beta^2 \right) \right.$$
$$\left. + \frac{1}{\omega} \int_0^\infty \frac{\sigma_\gamma(\omega')}{Z} \, d\omega' \right]. \qquad 48.$$

The dependence of the slope on gas composition is through the $\hbar\omega$ denominator in the logarithm, which results in a form

$$S = \frac{200 \ln 10}{\ln(2m\beta^2\gamma^2 c^2) - 1 - \ln I + \chi} = \frac{460}{15.6 - \ln I + \chi} \qquad 49.$$

where S is the slope in percent per decade (of $\beta\gamma$) relative to minimum ionization ($\beta\gamma = 4$); χ describes the contribution of the last term in Equation 48, the Rutherford or quasi-free term, which exhibits no log rise. Its magnitude is about 0.5 for ξ/I near 1 and increases slowly with sample thickness as further contributions from quasi-free scattering get folded into the main peak of the energy-loss distribution. This reduces the slope of the relativistic rise for thicker samples. The effect is very clear in the data of Walenta et al (1979) for propane, argon, and xenon. These data may be described empirically by the expression

$$S = (\xi/I)^{-0.2} \frac{430}{16 - \ln I}, \qquad 50.$$

which is close to the expected form (Equation 49). The first factor describes the dependence on sample width over the range covered in Figure 14 and shows that the slope increases by 15% if the sample thickness is halved. The second factor gives the dependence on gas mixture; for example, the slope is 10% greater in xenon than argon. For molecular gases ξ/I is larger and, in general, $\ln I$ is smaller than for rare gases so that the slopes are lower on both counts.

The density effect starts to modify the slope as soon as the velocity dependence of collisions with the outer shells of electrons saturate. As shown in Section 2.5 this happens over the $\beta\gamma$ range ε/E_p to $2\varepsilon/E_p$, where ε is related to the outer shell binding energy and E_p is the plasma energy ($E_p = 0.19\,(\nu P)^{1/2}$ eV). At higher $\beta\gamma$ the slope falls progressively. No simple formula is available and the detailed behavior must be calculated for the conditions concerned. Nevertheless the choice of low pressure (small E_p) noble gases (large ε) is favored.

For velocities below the onset of the density effect we may calculate the mass resolution for particles of known momentum using Equations 47 and 50:

$$\delta(\ln\mu) = \delta(\ln\beta\gamma) = \frac{\lambda}{\lambda_0}\,R\,\ln 10/S \qquad\qquad 51.$$

$$= 0.43\,\frac{\lambda}{\lambda_0}\,n^{-0.34}(n\xi/I)^{-0.12}(16-\ln I),$$

where λ/λ_0 is the ionization relative to minimum. This formula shows that in this region the mass resolution is very insensitive to the gas mixture (I), rather insensitive to the total mass of gas ($n\xi$), and only depends significantly on n. We conclude that the only real way to improve the mass resolution of an energy-loss detector below the onset of the density effect is to sample the ionization as finely as possible.

For very thin samples where ξ/I is much less than one, the shape of the energy-loss distribution will show structures due to the energy spectrum of single collisions. It follows that the shape of the distribution will not scale as assumed in Section 4.1 and the analysis discussed above will not be applicable. The velocity and mass resolution of such fine grain ionization data is ultimately limited by the statistics of the energy-loss collisions rather than by imperfect sampling. Such an ideal detector was discussed at the end of Section 2.5. There are practical problems in the measurement of fine grain energy-loss data in drift chambers. These have been studied experimentally by Rehak & Walenta (1980) and in simulation by Lapique & Puiz (1980). The main problems are the loss of resolution due to electron diffusion in the gas and the large fluctuations of the gas amplification for single electrons. Primary ionization has also been studied in other ways; by counting streamers in low density gases in a streamer chamber (Davidenko et al 1969, Eckardt et al 1977), by measuring the efficiency of spark chambers (Söchting 1979), and by observing the time-jitter of proportional chamber signals (Rehak & Walenta 1980). The streamer chamber method has been developed for individual particle identification, and Eckardt et al (1977) achieved an ionization resolution of 19% (FWHM) on a 90-cm track length in a helium-neon mixture. The method is currently limited by the difficulty of resolving clusters from different collisions.

4.3   Practical Considerations

So far we have discussed only the principles of energy-loss detectors without considering how they may be put into practice. In this section we mention briefly some of the practical considerations, discussed in greater detail in the references cited.

In general an energy-loss detector will be large, 15 m^3 or more, its length being determined by the path length necessary to achieve a given mass resolution and its aperture by the geometrical acceptance. The practical problem is to instrument this large volume. In experiments at storage rings, at least, the detector will be inside the magnetic field volume and perform the dual function of particle identification and momentum analysis. The many samplings of the particle "track" will provide a high degree of information redundancy for pattern recognition. Although there is no fundamental conflict between obtaining coordinate and energy-loss information, an energy-loss detector is seldom optimized for its performance as a particle identifier alone. The practical solution is a detector employing drift chamber techniques but using electron drift paths that are long compared with conventional drift chambers, the length being related to the expected particle flux through the detector.

In such a detector each track is sampled by drifting the ionization electrons to a uniformly spaced array of anode wires under the influence of a high uniformity electric field. Proportional gas amplification then takes place at each wire, and the ionization sample is identified with the charge, which is the integral of the current signal from the wire. The shape of this signal varies with the angle of the track and the drift distance through the effect of electron diffusion. As a result, to avoid bias, it is necessary to measure a shape-independent integral (Pleming 1977). The shorter the sample size, the smaller the component of pulse width due to track inclination and the better the multitrack resolution of the detector. The ability to handle high multiplicities and high rates means that considerable attention must be paid to the elimination of baseline shifts in the electronics (Brooks et al 1978).

For each track the pulse heights of many signals must be collected with systematic errors of less than 1%. The extent of cross talk or correlation between these signals, which are treated as independent, is of outstanding importance. The average cross talk between *any* two signals needs to be less than about 3×10^{-4}; this affects the design of the electronics, power supplies, and grounding. The correlation between signals from neighboring samples in the detector can be larger ($\sim 10^{-1}$) without affecting the resolution. Delta-ray electrons passing from one sample to another are too few to cause problems in detectors with good spatial resolution; such low energy electrons are quickly separated from the parent track by multiple scattering. In the presence of a magnetic field they are trapped in the original sample. Capacitative coupling between wires may be reduced by the use of intermediate field wires between collecting anodes (Allison et al 1976). Cross coupling due to electron diffusion is important for long drift-path detectors but can be minimized by a careful choice of gas mixture (Allison et al 1974). Although the effect of *linear* cross talk on the average pulse height is zero, it does change the shape of the asymmetric ionization distribution slightly. Practical reasons against a choice of very small sample width include the electromechanical stability of large arrays of close wires, the mechanical tolerances needed for uniform gas amplification, the small signal amplitudes, and the cost of the electronics.

In order that individual track signals may be sensed and integrated without bias there must be an adequate signal-to-noise ratio. This depends on the bandwidth of the electronics, the electronic noise, the number of ionization electrons, and the gas amplification factor (Pleming 1977). The gas amplification cannot be increased beyond about 10^4 without the risk of angle-dependent saturation effects (Frehse et al 1978, Allison et al 1978c). More significantly, for long drift-path geometries the production of positive ions in the gas amplification produces a volume space charge that distorts the drift field (Allison et al 1974, 1979, Friedrich et al 1979). To avoid this effect, the flux, gain, drift distance, or duty cycle of the device must be limited. Considerable attention must be paid to gas purity, especially with the longer drift paths, if the ionization electrons are not to be lost by attachment to electronegative molecules. Typically the concentration of oxygen, as the dominant electronegative impurity, must be kept in the range 0.1–1.0 parts per million (Allison et al 1974, 1979).

In Section 4.2 it was shown that, at velocities for which the density effect is unimportant, the mass resolution is more or less independent of the choice of gas (at constant mass of gas). In practice consideration of systematic effects modifies this conclusion. For gases with higher I the variation of ionization with velocity is larger and the ionization resolution worse than for those with lower I (Equations 47 and 50). Gases with higher I are therefore to be preferred since the mass resolution is less sensitive to systematic and calibration errors. Sources of these are associated with the density sensitivity of the gas amplification factor and mechanical tolerances of the chamber. Some care is required to reduce these errors to the level of 1%.

Data from a single event with 10 or more tracks and several hundred signals per track amount to typically 10^5 bits of information. These data

Table 2 Relativistic energy-loss particle identifiers (May 1980)

Name	Gas	Samples (cm)	Acceptance	Physics objective[g]	Status
EPI[a]	Ar/5% CH_4 1 atm	128 × 6	2m × 1m	diffraction dissociation with BEBC (CERN)	Data 1978
ISIS1[b]	Ar/20% CO_2 1 atm	80 × 1.6	4m × 2m	Strong interaction and charm physics	Data early 1980
ISIS2[b]	Ar/20% CO_2 1 atm	320 × 1.6	4m × 2m	with EHS (CERN)	Construction
CRISIS[c]	Ar/20% CO_2 1 atm	192 × 1.6	1m × 1m	Hadron physics with FHS (FNAL)	Construction
JADE[d]	Ar/C_2H_6 4 atm	48 × 1	$\sim 4\pi$	e^+e^- at PETRA	Data 1979
TPC[e]	Ar/20% CH_4 10 atm	192 × 0.4	$\sim 4\pi$	e^+e^- at PEP	Construction
UA1[f]	Ar/C_2H_6 1 atm	200 × 0.8	$\sim 4\pi$	$\bar{p}p$ collider at CERN	Construction

[a] Jeanne et al 1973, Lehraus et al 1978, and Figure 13.
[b] Allison et al 1974, 1978b, 1978c, 1979, and Figure 13.
[c] Wadsworth et al 1979.
[d] Barber et al 1976, Farr et al 1978, Wagner et al 1980.
[e] Nygren 1976, Fancher et al 1979, and Figure 13.
[f] Astbury et al 1978.
[g] BEBC = Big European Bubble Chamber; EHS = European Hybrid Spectrometer; FHS = Fermilab Hybrid Spectrometer.

collected by drift chambers in a few microseconds represent a peak rate to the electronics of about 10^{10} Hz. The design of the electronics must aim at performing a parallel-to-serial conversion so that the data reach a computer at a manageable rate. Indeed the use of long drift paths in the chambers is the first stage of such a parallel-to-serial conversion. Analogue storage techniques on discrete capacitors (Brooks et al 1978, Farr & Heintze 1978) or on charge-coupled devices (Fancher et al 1979) have been used; other possibilities such as "flash" analogue-to-digital converters were reviewed by Dhawan & Ludlum (1978).

4.4   Existing Detectors

Following the initial idea of a relativistic particle identifier based on energy loss (Alikhanov et al 1956), and the exploratory work of Ramana Murthy & Demeester (1967) and Dimcovski et al (1971), Jeanne et al (1973) proposed a full-scale device known as the external particle identifier (EPI) to operate behind the BEBC bubble chamber at CERN. The device consists of 4096 independent rectangular proportional counters defined by planes of cathode wires. Its acceptance is modest and its multitrack capability is limited by the small number of transverse cells. Its mass resolution, however, is excellent as shown in Figure 3. Table 2 gives a list of devices now operating or under construction. Most of these have been reviewed by Heintze (1978). Apart from EPI they all make extensive use of long electron

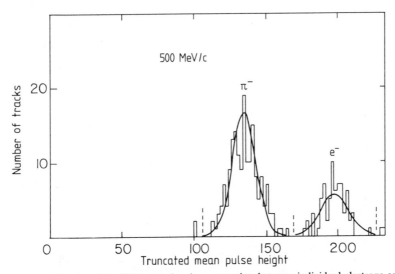

Figure 15   Results of the ISIS1 test showing separation between individual electrons and pions at 500 MeV/c. The truncated mean is the mean of the lowest 36 of the 60 samples. The curves represent ionization resolutions of 12 and 14% (FWHM) respectively.

drift paths. JADE, the TPC, and the UA1 detectors are designed for colliding-beam machines and have cylindrical geometries within the magnetic field volume where it is difficult to achieve the necessary track length and the design must be optimized for momentum as well as mass resolution. The relativistic particle identification capabilities of JADE and UA1 are marginal. By working with argon at 10 atm, the TPC is expected to have good mass resolution up to 15 GeV/c.

The situation is easier for fixed-target experiments where space allows the use of normal pressure gases and dedicated particle identifiers. The ISIS2 chamber is expected to achieve good mass resolution in a fiducial volume of 40 m³ with only 320 channels of electronics by using drift paths of up to 2 m. A similar but much smaller device, CRISIS, is under construction for use at FNAL (Wadsworth et al 1979). Tests with ISIS1 have shown the expected mass resolution (Figure 15; Allison et al 1979). Figure 16 shows a plot of the position information for drifted tracks recorded by ISIS1. The high density of spatial information, the clear identification of background signals (δ rays

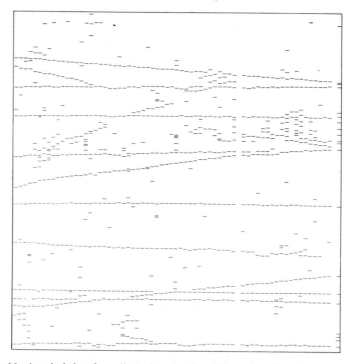

Figure 16   A typical plot of coordinate data showing a fair number of tracks passing through ISIS1 during a test. The ordinate is the drift time and the abscissa is the sample or wire number. Pulse height information is also recorded for each "hit" shown.

etc), and the good multitrack resolution are apparent. As a result, clear sections of each track, free of contamination by other signals, can be distinguished and unbiased ionization estimates obtained for a high multiplicity of tracks.

5 SUMMARY

Spectacular progress has been made since Alikhanov et al (1956) first described in detail the formidable practical problem of identifying relativistic charged particles by energy loss. Recent tests with large detectors indicate that these problems have been overcome and that we may now look forward to the successful application of the method to provide particle identification in high energy physics experiments.

With the advances in practical application have come a better theoretical understanding and more thorough calculations. These are now able to give a good description of data for energy loss in thin samples ($\xi/I \sim 1$, equivalent to about 1 cm atm of argon). More clean data would be welcome for a variety of gases at different densities and thicknesses and for a range of $\beta\gamma$ values. Nevertheless it is now possible to predict the mass resolution of a detector with confidence. Calculations point to argon or, better, xenon as the best gas, with the choice of density governed by a compromise between the constraint of detector length and mass resolution at high $\beta\gamma$. The only way to improve the resolution within this constraint is to sample more finely.

Further efforts continue to develop much finer grain ionization detectors although these present significant practical difficulties. We note that the mass resolution of the "ideal" detector, which resolves the energy loss in individual atomic collisions, is a factor 2.5 or so better than current detectors at lower values of $\beta\gamma$ (less than 50 for argon). At higher values of $\beta\gamma$ where an improved resolution would be most welcome, the factor is significantly smaller.

ACKNOWLEDGMENTS

We would like to thank Drs. L. Lyons, M. Inokuti, and W. G. Scott for their valuable comments and careful reading of drafts of the manuscript. Finally we thank Miss Daphne Pollard for her patience and skill in typing the manuscript.

Appendix

A picture of the field of a relativistic charged particle in a medium

The main features of the logarithmic rise of the energy-loss cross section, the density effect, and their relationship with Cerenkov radiation depend only on the virtual wave structure of the electromagnetic field of the incident particle and may be understood with the aid of a simple model.

Consider a scalar field ϕ coupled to a particle moving with constant velocity v along the x axis through a dispersive medium in which the phase velocity of the field is $u(\omega)$, a function of frequency. In the rest frame of the particle the field is static. In the frame of the medium it is a wave packet that is neither dispersed nor attenuated and that moves along x with constant amplitude and phase with respect to the particle. Thus the phase velocity along x of each frequency component must equal v since it is a component of a static field as seen by an observer moving with the particle, $v = \omega/k_x$ (k is the wave number of the Fourier component of frequency ω). By definition of the phase velocity, $u = \omega/|k|$. Therefore the space-time dependence of each frequency component is described in the 2-dimensional case[16] by the phase factor $\exp i(k_x x + k_y y - \omega t) = \exp i \, \omega/v \, [x - vt + (v^2/u^2 - 1)^{1/2} y]$. The dimensionless velocity $v/u(\omega)$ may be denoted by $\beta'(\omega)$. If β' is less than unity, the transverse dependence of the field will be an evanescent wave of range $y_0(\omega) = v/\omega (1 - \beta'^2)^{-1/2} = \beta'\gamma'\lambda$ where $\gamma'(\omega) = (1 - \beta'^2)^{-1/2}$ and λ is the free wavelength over 2π. For larger velocities β' gets closer to unity and the range increases linearly with $\beta'\gamma'$. If v becomes greater than u, the field is a traveling wave and there is a flux away from the source at an angle arc cos $(1/\beta')$. These features of wave motion are common to, for example, acoustic and surface waves.

For the electromagnetic field of a charged particle moving in vacuum, $u = c$ for all frequencies $\beta' = v/c$ and $\gamma' = \gamma$. The expansion of the electromagnetic field with $\beta\gamma$ responsible for the relativistic rise is therefore seen as a general consequence of wave motion. In a medium, on the other hand, $u = c/(\varepsilon)^{1/2}$ where $\varepsilon(\omega)$ is the dielectric constant. There are two cases:

(a) $\varepsilon(\omega) > 1$. Below the velocity $\beta = 1/(\varepsilon)^{1/2}$ the transverse waves are evanescent. Above this velocity free radiation is emitted insofar as the medium is transparent. This is Cerenkov radiation and occurs mainly in the region of ω below ionization threshold.

(b) $\varepsilon(\omega) < 1$. In this case, the range is now $y_0(\omega) = v/\omega(1 - \beta^2\varepsilon)^{-1/2}$, which has a limiting value of $\lambda(1 - \varepsilon)^{-1/2}$. Ninety percent of this range is achieved by a $\beta\gamma$ value of $2(1 - \varepsilon)^{-1/2}$ where the field saturates, becoming insensitive to

---

[16] In the 3-dimensional problem the exponent in y is replaced by a Bessel function in the cylindrical polar coordinate r. However, the qualitative features are unaltered.

any further increase of velocity. Using a crude model of the dielectric constant, $\varepsilon(\omega) = 1 - \omega_p^2/\omega^2$, we see that the energy-loss cross section is expected to saturate at $\beta\gamma = 2\omega/\omega_p$. This expectation is confirmed by proper calculation and comparison with experiment in this paper.

Alternatively the quanta of the electromagnetic field in the medium may be considered as having an "effective mass," similar to the concept of effective mass ascribed to electrons in the band theory of solids. Normal modes of the field satisfy the equation $c^2k^2 - \omega^2\varepsilon(\omega) = 0$ whence $m(\omega) = \hbar\omega/c^2 [1 - \varepsilon(\omega)]^{1/2}$. The exchange of such photons represents a short range force limited by their Compton wavelength, which is equal to $\lambdabar(1 - \varepsilon)^{-1/2}$, the same result as obtained above.

Literature Cited

Alikhanov, A. I. et al. 1956. *Proc. CERN Symp. on High Energy Accelerators Pion Phys.* 2 : 87–98

Allison, W. W. M. et al. 1974. *Nucl. Instrum. Methods* 119 : 499–507

Allison, W. W. M. et al. 1976. *Nucl. Instrum. Methods* 133 : 325–34

Allison, W. W. M. 1977. Presented at *Int. Symp. on Transition Radiat. High Energy Part. Yerevan, USSR.* Oxford, OUNPL33/77

Allison, W. W. M., Bunch, J. N., Cobb, J. H. 1978a. *Nucl. Instrum. Methods* 153 : 65–67

Allison, W. W. M. et al. 1978b. *Nucl. Instrum. Methods* 156 : 169–70

Allison, W. W. M. et al. 1978c. The design of ISIS and the results of tests with ISIS1. *CERN/EP/EHS/PH78-10*

Allison, W. W. M. et al. 1979. *Nucl. Instrum. Methods* 163 : 331–41

Astbury, A. et al. 1978. A 4π solid angle detector for the SPS used as a proton-antiproton collider at a centre of mass energy of 540 GeV. *Proposal CERN/SPSC/P92*

Barber, D. P. et al. 1976. A proposal for a compact magnetic detector at PETRA. Hamburg: DESY

Berg, R. A., Green, A. E. S. 1973. *Adv. Quantum Chem.* 7 : 277–88

Bethe, H. A. 1930. *Ann. Phys.* 5 : 325–400

Bethe, H. A., Jackiw, R. W. 1968. *Intermediate Quantum Mechanics*, pp. 302–5. New York: Benjamin. 2nd ed.

Bichsel, H., Saxon, R. P. 1975. *Phys. Rev. A* 11 : 1286–96

Blunck, O., Leisegang, S. 1950. *Z. Phys.* 128 : 500–5

Bohr, N. 1913. *Philos. Mag.* 25 : 10–31

Bonham, R. A. et al. 1978. *Adv. Quantum Chem.* 11 : 1–32

Brooks, C. B. et al. 1978. *Nucl. Instrum. Methods* 156 : 297–99

Budini, P., Taffara, L. 1956. *Nuovo Cimento* 4 : 23–45

Bunch, J. N. 1976. D.Phil. thesis. Univ. Oxford. Available from Rutherford Laboratory, *HEP/T/70*

Camps, C. et al. 1975. *Nucl. Instrum. Methods* 131 : 411–16

Chang, T. N., Fano, U. 1976. *Phys. Rev. A* 13 : 263–81

Charpak, G. 1970. *Ann. Rev. Nucl. Sci.* 20 : 195–254

Chechin, V. A. et al. 1972. *Nucl. Instrum. Methods* 98 : 577–87

Chechin, V. A., Ermilova, V. C. 1976. *Nucl. Instrum. Methods* 136 : 551–58

Cobb, J. H. 1975. D.Phil. thesis. Univ. Oxford. Available from Rutherford Laboratory, *HEP/T/55*

Cobb, J. H., Allison, W. W. M., Bunch, J. N. 1976. *Nucl. Instrum. Methods* 133 : 315–23

Cobb, J. H. et al. 1977. *Nucl. Instrum. Methods* 140 : 413–27

Commichau, V. et al. 1979. *CERN/EP/EHS/PH79-3*

Crispin, A., Fowler, G. N. 1970. *Rev. Mod. Phys.* 42 : 290–316

Davidenko, V. A. et al. 1969. *Nucl. Instrum. Methods* 67 : 325–30

Dhawan, S., Ludlam, T. 1978. *IEEE Trans. Nucl. Sci.* NS-25 : 944–51

Dimcovski, Z. et al. 1971. *Nucl. Instrum. Methods* 94 : 151–55

Eckardt, V. et al. 1977. *Nucl. Instrum. Methods* 143 : 235–39

Ermilova, V. C., Kotenko, L. P., Merzon,

G. I. 1977. *Nucl. Instrum. Methods* 145:555–63

Fancher, D. et al. 1979. *Nucl. Instrum. Methods* 161:383–90

Fano, U. 1947. *Phys. Rev.* 72:26–29

Fano, U. 1963. *Ann. Rev. Nucl. Sci.* 13:1–66

Fano, U., Cooper, J. W. 1968. *Rev. Mod. Phys.* 40:441–507

Farr, W. et al. 1978. *Nucl. Instrum. Methods* 156:283–86

Farr, W., Heintze, J. 1978. *Nucl. Instrum. Methods* 156:301–9

Fermi, E. 1940. *Phys. Rev.* 57:485–93

Fischer, J. et al. 1975. *Nucl. Instrum. Methods* 127:525–37

Fisher, C. M., Guy, J. G., Venus, W. A. 1975. *Nucl. Instrum. Methods* 133:29–34

Frehse, H. et al. 1978. *Nucl. Instrum. Methods* 156:87–96

Friedrich, D. et al. 1979. *Nucl. Instrum. Methods* 158:81–88

Garibyan, G. M. 1960. *Sov. Phys. JETP* 10:372–76

Garibyan, G. M., Ispiryan, K. A. 1972. *JETP Lett.* 16:413–15

Gemmell, D. S. 1974. *Rev. Mod. Phys.* 46:129–227

Harris, F. et al. 1973. *Nucl. Instrum. Methods* 107:413–22

Hasebe, N. et al. 1978. *Nucl. Instrum. Methods* 155:491–501

Heintze, J. 1978. *Nucl. Instrum. Methods* 156:227–44

ICRU. 1979. *Int. Comm. on Radiat. Units Measure. Ave. energy required to produce an ion pair. Rep. 31*

Inokuti, M. 1971. *Rev. Mod. Phys.* 43:297–347

Inokuti, M. 1975. *Radiat. Res.* 64:6–22

Inokuti, M., Itikawa, Y., Turner, J. E. 1978. *Rev. Mod. Phys.* 50:23–35

Ispiryan, K. A., Margarian, A. T., Zverev, A. M. 1974. *Nucl. Instrum. Methods* 117:125–29

Jackson, J. D. 1975. *Classical Electrodynamics*, pp. 310–12 New York: Wiley. 848 pp. 2nd ed.

Jeanne, D. et al. 1973. *Nucl. Instrum. Methods* 111:287–300

Kopot, E. A. et al. 1976. *Sov. Phys. JETP* 70:387–96; transl. 43:200–4

Landau, L. 1944. *J. Phys. USSR* 8:201–5

Landau, L., Lifshitz, E. M. 1960. *Electrodynamics of Continuous Media*, pp. 256–62, 344–59. New York: Pergamon

Lapique, F., Puiz, F. 1980. *CERN/EF 79–4* (submitted to *Nucl. Instrum. Methods*)

Lee, L. C. et al. 1973. *J. Quant. Spectrosc. Radiat. Transfer* 13:1023–31

Lee, L. C. et al. 1977. *J. Chem. Phys.* 67:1237–46

Lehraus, I. et al. 1978. *Nucl. Instrum. Methods* 153:347–55

Litt, J., Meunier, R. 1973. *Ann. Rev. Nucl. Sci.* 23:1–43

Maccabee, H. D., Papworth, D. G. 1969. *Phys. Lett. A* 30:241–42

Manson, S. T. 1972. *Phys. Rev. A* 5:668–77

McGuire, E. J. 1968. *Phys. Rev.* 175:20–30

McGuire, E. J. 1971. *Phys. Rev. A* 3:267–79

Nygren, D., spokesman. 1976. Proposal for a PEP facility based on the time projection chamber. *PEP-4 proposal.*

Onuchin, A. P., Telnov, V. I. 1974. *Nucl. Instrum. Methods* 120:365–68

Pleming, R. W. 1977. D.Phil. thesis. Univ. Oxford. Available from Rutherford Laboratory, *HEP/T/69*

Ramana Murthy, P. V., Demeester, G. D. 1967. *Nucl. Instrum. Methods* 56:93–105

Rehak, P., Walenta, A. H. 1980. *IEEE Trans Nucl. Sci.* NS-27:54–58

Seguinot, J., Ypsilantis, T. 1977. *Nucl. Instrum. Methods* 142:377–91

Smith, G. C., Mathieson, E. 1975. *Nucl. Instrum. Methods* 131:13–15

Söchting, K. 1979. *Phys. Rev. A* 20:1359–65

Srdoc, D. 1973. *Nucl. Instrum. Methods* 108:327–32

Sternheimer, R. M. 1952. *Phys. Rev.* 88:851–59

Sternheimer, R. M., Peierls, R. F. 1971. *Phys. Rev. B* 3:3681–92

Talman, R. 1979. *Nucl. Instrum. Methods* 159:189–211

Wadsworth, B. F. et al. 1979. *IEEE Trans. Nucl. Sci.* NS-26:120–28

Wagner, A. et al. 1980. *Nucl. Instrum. Methods.* In press

Walenta, A. H. et al. 1979. *Nucl. Instrum. Methods* 161:45–58

Way, K. 1978. *At. Nucl. Data Tables* 22:125–30

West, J. B., Marr, G. V. 1976. *Proc. R. Soc. Ser. A* 349:397–421; *At. Nucl. Data Tables* 18:497–508

West, J. B., Morton, J. 1978. *At. Nucl. Data Tables* 22:103–7

Williams, E. J. 1929. *Proc. R. Soc. Ser. A* 125:420–45

Ann. Rev. Nucl. Part. Sci. 1980. 30 : 299–335

THE TAU LEPTON[1]

✕5619

Martin L. Perl

Stanford Linear Accelerator Center, Stanford University, Stanford, California 94305

CONTENTS

1   INTRODUCTION

In the last few years there has been an important addition to the known elementary particles—the tau (τ) lepton. It is an important addition first because, to the best of our knowledge, the tau is a fundamental particle.

---

[1] Work supported by the Department of Energy, contract DE-AC03-76SF00515.

0163-8998/80/1201-0299$01.00

That is, unlike most of the so-called elementary particles such as the proton or pion, the tau is not made up of simpler particles or constituents (see Section 1.1). Second, the tau is important because all its measured properties agree with its designation as a lepton. Hence it joins the very small lepton family of particles; a family that prior to the tau's discovery had only four members: the electron (e), its associated neutrino (v_e), the muon (μ), and its associated neutrino (v_μ). Third, the tau is important because, along with the discovery of the fifth quark, it appears to confirm some general theoretical ideas about the connection between leptons and quarks. A brief discussion on quarks and the lepton-quark connection is presented in Sections 1.1 and 2.2.

This article reviews the experimental work done on the tau, why we believe it is a lepton, the measured properties of the tau, and the experimental work still to be done on the tau. I present just enough theory to provide a framework for discussing the experimental results. Correspondingly, I present a full set of experimental references (up to November 1979), but only a few general theoretical references.

The history of the discovery of the tau was reviewed by Feldman (1978a); I only outline it here. It is an old idea to look for leptons with masses greater than that of the electron or muon—the so-called heavy leptons. The first searches for a heavy charged lepton using electron-positron collisions were carried out by Bernardini et al (1973) and by Orioto et al (1974) at the ADONE e^+e^- storage ring. They looked for the electromagnetic production process

$$e^+ + e^- \rightarrow l^+ + l^-, \qquad 1.$$

where l represents the new lepton. The ADONE storage ring did not have enough energy to produce the tau.

The first evidence for the tau was obtained (Perl 1975, Perl et al 1975) in 1974 at the SPEAR e^+e^- storage ring by using the reaction in Equation 1. Subsequent experiments at SPEAR, which is at the Stanford Linear Accelerator Center (SLAC), and at the DORIS e^+e^- storage ring at the Deutsches Elektronen-Synchrotron (DESY) confirmed this discovery and measured the properties of the tau. Recent reviews have been given by Kirkby (1979), Flügge (1979), Feldman (1978a), and Tsai (1980).

1.1 The Definition of a Lepton

The definition of a lepton is based upon our experience with the electron, muon, and their associated neutrinos. We use these criteria to classify a particle as a lepton (Perl 1978):

1. The lepton does not interact through the strong interaction. Thus the lepton is differentiated from the hadron particle family, such as the pion, proton, and ψ/J.

2. The lepton interacts through the weak interactions and, if charged, through the electromagnetic interaction.

3. The lepton has no internal structure or constituents. I shall call a particle without internal structure or constituents a point particle. This is, of course, always a provisional definition since going to higher energy may reveal the internal structure or the constituents of a particle. However, the requirement on a lepton is to be understood in contrast to the properties of hadrons. That is, hadronic properties such as electromagnetic form factors are explained in terms of the hadron's internal constituents—the quarks. The form factor concept provides a quantitative test of whether or not a particle has internal structure (Section 3.2).

The leptons that we now know share two additional properties that may not be central to the definition of a lepton (Perl 1978):

4. The known leptons have spin 1/2. We can, however, conceive of particles that have properties 1–3 listed above and yet have other spins, zero for example (Farrar & Fayet 1980).

5. All the known leptons obey a lepton conservation law. This is defined formally in Section 2.1. I will give an intuitive definition here. A lepton, such as the e⁻, possesses an intrinsic property called lepton number, which cannot disappear. This property can either be transferred to an associated neutrino (transferred from e⁻ to v_e) or it can be cancelled by combining the lepton with its antilepton (e⁻ combined with e⁺). As with the property of spin-1/2 we do not know if this is intrinsic to all leptons or only an accidental property of the known leptons.

1.2 The Tau Lepton

The tau lepton has the crucial lepton defining properties 1–3 listed in the previous section. It also has property 4, namely spin 1/2; and very probably it has property 5, lepton conservation. It is easiest to get a general picture of the properties of the τ by comparing it with the e and μ (Table 1).

The astonishing property of the tau is its large mass of about 1782 MeV/c^2, 3600 times the electron mass and 17 times the muon mass. Until the tau was discovered many physicists held the vague idea that the simplicity and lack of structure of the leptons were associated with their relatively small mass. The masses of the electron, muon, and their associated neutrinos are all smaller than the mass of the lightest hadron, the neutral pion, which has a mass of 135 MeV/c^2. Indeed, the word lepton

Table 1 The known leptons

Properties	Electron	Muon	Tau
Charged lepton symbol	e^-, e^+	μ^-, μ^+	τ^-, τ^+
Charged lepton mass (MeV/c^2)	0.51	105.7	1782^{+3}_{-4}[a]
Charged lepton lifetime (s)	stable	2.20×10^{-6}	$< 2.3 \times 10^{-12}$[a]
Charged lepton spin	1/2	1/2	1/2
Associated neutrino	$\nu_e, \bar{\nu}_e$	$\nu_\mu, \bar{\nu}_\mu$	$\nu_\tau, \bar{\nu}_\tau$[b]
Associated neutrino mass	< 60 eV/c^2[c]	< 0.57 MeV/c^2[c]	< 250 MeV/c^2[a,c]

[a] The detailed discussion of these measurements appears in Section 4.

[b] All measured properties of the τ are consistent with it having a unique neutrino, ν_τ; however, more experimental work needs to be done on these properties, as discussed in Section 4.4.

[c] Could be 0.

comes from the Greek lepto meaning fine, small, thin, or light. However, this is certainly not descriptive of the tau whose mass is greater than that of many hadrons, almost twice the proton mass for example. Nevertheless the term lepton has been kept for the tau; often the oxymoron heavy lepton is used.

The relatively large mass of the tau allows it to decay to a variety of final states. Some of the decay modes that have been measured (Section 5) are

$$\tau^- \rightarrow \nu_\tau + e^- + \bar{\nu}_e$$
$$\tau^- \rightarrow \nu_\tau + \mu^- + \bar{\nu}_\mu$$
$$\tau^- \rightarrow \nu_\tau + \pi^-$$
$$\tau^- \rightarrow \nu_\tau + \rho^-$$
$$\tau^- \rightarrow \nu_\tau + \pi^- + \pi^+ + \pi^-$$
$$\tau^- \rightarrow \nu_\tau + \pi^- + \pi^+ + \pi^- + \pi^0$$

2.

An analogous set of decay modes occurs for the τ^+. Note that in all measured τ decays a neutrino is produced, which indicates that the τ obeys a lepton conservation rule.

2 THEORETICAL FRAMEWORK

By definition, a lepton interacts through the weak and electromagnetic forces but not through the strong interactions, and it has no internal structure or constituents. To proceed further, we need a more restrictive theoretical framework. I shall impose on the lepton conventional weak interaction theory (Bailin 1977, Zipf 1978) and some sort of lepton conservation rule, since these are restrictions that the τ obeys. However, the reader should keep in mind that there may exist leptons that do not obey these restrictions.

2.1 *Weak Interactions and Lepton Conservation*

Consider a charged and neutral lepton pair (L^-, L^0) with the same lepton number $n_L = +1$. Their antiparticles $(L^+, \bar{L}^0)$ have lepton number $n_L = -1$. Lepton conservation means that in all reactions the sum of the n_L's of all the particles remains unchanged.

Assuming (*a*) conventional weak interaction theory, (*b*) that the L^- is heavier than the L^0, and (*c*) that there is sufficient mass difference between the L^- and the L^0, the following sorts of decays will occur (Figure 1):

$$L^- \to L^0 + e^- + \bar{\nu}_e \qquad\qquad 3.$$

$$L^- \to L^0 + \mu^- + \bar{\nu}_\mu \qquad\qquad 4.$$

$$L^- \to L^0 + (\text{hadrons})^-. \qquad\qquad 5.$$

In Figure 1*c* the quark-antiquark pair $d\bar{u}$ replaces the lepton-neutrino pair, and the quarks convert to hadrons. If the L^0 is heavier than the L^- the reverse decays

$$L^0 \to L^- + e^+ + \bar{\nu}_e \qquad\qquad 6.$$

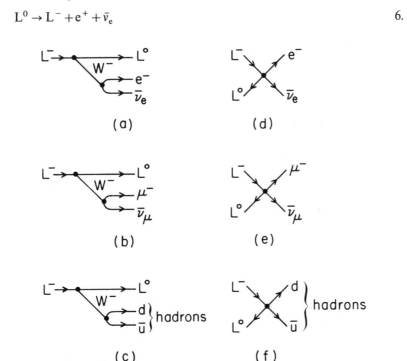

Figure 1 Diagrams for the decay of the τ into quarks or leptons: (*a–c*) intermediate boson view, (*d–f*) four-fermion coupling view.

$$L^0 \rightarrow L^- + (\text{hadrons})^+ \qquad\qquad 7.$$

will occur. We shall assume that lepton-W and quark-W vertices have conventional Weinberg-Salam theory couplings (Bailin 1977, Zipf 1978).

The W propagator diagrams of Figures 1a–c have become the conventional way to represent weak decays. However, when a lepton, such as the τ, has a mass much smaller than the proposed W mass (1.8 GeV/c^2 compared to roughly 100 GeV/c^2) the W propagator has no observable effect. Therefore for some of the discussions of the decays of the τ I shall use Figures 1d, e, which diagram the old four-fermion coupling of Fermi weak interaction theory (Bjorken & Drell 1964).

2.2 Simple Models for New Charged Leptons

2.2.1 SEQUENTIAL LEPTON MODEL   In this model (Perl & Rapidis 1974) a sequence of charged leptons of increasing mass is assumed, each lepton type having a separately conserved lepton number and a unique associated neutrino of smaller, but not necessarily zero, mass. That is, there is a particle (and antiparticle) sequence:

charged lepton	associated neutrino	
$e^-(e^+)$	$v_e(\bar{v}_e)$	
$\mu^-(\mu^+)$	$v_\mu(\bar{v}_\mu)$	
$1^-(1^+)$	$v_1(\bar{v}_1)$	8.
$\vdots$	$\vdots$	

Decays of the $1^\pm$ through the electromagnetic interaction such as $1^\pm \rightarrow e^\pm + \gamma$ or $1^\pm \rightarrow e^\pm + \gamma$ are forbidden. The $1^\pm$ can only decay through the weak interaction as described in Section 2.1, namely

$$
\begin{aligned}
1^- &\rightarrow v_1 + e^- + \bar{v}_e \\
1^- &\rightarrow v_1 + \mu^- + \bar{v}_\mu \\
&\vdots \\
1^- &\rightarrow v_1 + (\text{hadrons})^-.
\end{aligned}
\qquad 9.
$$

The vertical dots in Equation 9 indicate decays to all associated lepton-neutrino pairs of sufficently small masses. In Equation 9 and in the remainder of this paper we list only the decay modes of the negatively charged lepton; the decay mode of the positively charged lepton is obtained by changing every particle to its antiparticle. The neutrinos in this model are stable because their associated charged lepton has a larger mass and their lepton number is conserved. The search for the τ was based on this model (Perl 1975) and to the best of our knowledge the τ conforms to this model.

2.2.2 ORTHOLEPTON MODEL   In this model (Llewellyn Smith 1977) the new charged lepton, l^-, has the same lepton number as a smaller mass, same-sign, charged lepton, such as the e^-. Let us use the e^- example. We expect the dominant decay to occur through the electromagnetic interaction

$$l^- \to e^- + \gamma. \tag{10.}$$

However, current conservation forbids the l-γ-e vertex from having the usual form $\bar{\psi}_l \gamma^\mu \psi_e$ (Low 1965, Perl 1978), and this decay mode might be suppressed. Therefore decays through the weak interactions such as

$$
\begin{aligned}
l^- &\to e^- + e^+ + e^- \\
l^- &\to \nu_e + e^- + \bar{\nu}_e \\
l^- &\to e^- + (\text{hadrons})^0 \\
l^- &\to \nu_e + (\text{hadrons})^-
\end{aligned}
\tag{11.}
$$

could in principle be detected.

2.2.3 PARALEPTON MODEL   In this model (Llewellyn Smith 1977, Rosen 1978) the l^- has the same lepton number as a smaller mass, opposite-sign, charged lepton, such as the e^+. Electromagnetic decays such as $l^- \to e^- + \gamma$ are now forbidden. The decay modes through the weak interaction are

$$l^- \to \bar{\nu}_e + e^- + \bar{\nu}_e \tag{12.}$$

$$l^- \to \bar{\nu}_e + \mu^- + \bar{\nu}_\mu \tag{13.}$$

$$l^- \to \bar{\nu}_e + (\text{hadrons})^- \tag{14.}$$

In this illustration the l^- has the same lepton number as the e^+ and hence as the $\bar{\nu}_e$.

2.3 $e - \mu - \tau$ Universality

A special case falling within the sequential heavy lepton model is the model in which the e, μ, and τ differ only by having (a) different masses and (b) different and separately conserved lepton numbers. In this model the e, μ, and τ have the same spin (1/2), the same electromagnetic interactions, and the same weak interactions. They are all point particles and they are all associated with different massless, spin-1/2 neutrinos. Thus the comparative properties of the charged lepton depend only on the masses being different. We call this $e - \mu - \tau$ universality.

2.4 Leptons and Quarks

The Weinberg-Salam theory of the unification of weak and electromagnetic interactions (Bailin 1977, Zipf 1978) provides a quantitative model for new leptons that is related to the sequential lepton model. In its current form,

Weinberg-Salam theory classifies the leptons and quarks into left-handed doublets, containing at least

$$
\text{leptons} = \begin{pmatrix} v_e \\ e^- \end{pmatrix}_L, \begin{pmatrix} v_\mu \\ \mu^- \end{pmatrix}_L, \begin{pmatrix} v_\tau \\ \tau^- \end{pmatrix}_L,
$$

$$
\text{quarks} = \begin{pmatrix} u \\ d' \end{pmatrix}_L, \begin{pmatrix} c \\ s' \end{pmatrix}_L, \begin{pmatrix} t \\ b \end{pmatrix}_L,
$$

15.

and right-handed singlets, containing at least

$$
\text{leptons} = e_R, \mu_R, \tau_R
$$
$$
\text{quarks} = u_R, d'_R, c_R, s'_R, t_R, b_R.
$$

16.

This classification assumes that the t quark exists and that the v_τ is unique. The weak-electromagnetic interaction only connects particles to themselves or to the other member of the doublet. For leptons this is equivalent to lepton conservation. For the first two quark doublets there is only approximate conservation because the d' and s' quarks are mixed by the Cabibbo angle θ_c. That is

$$
d' = d \cos \theta_c + s \sin \theta_c
$$
$$
s' = -d \sin \theta_c + s \cos \theta_c
$$

17.

where d and c are pure quark states.

The τ plays an important role in this model, because with the τ there are three sets (usually called generations) of leptons and three sets of quarks (assuming the t quark exists). This theory does not require equal numbers of generations of leptons and quarks. But if it happens that the numbers of generations are equal, that is certainly very significant with respect to the connection between leptons and quarks.

Our immediate need for this theory is, however, more mundane. The theory predicts that the weak interactions between the members of a doublet are the same for all doublets. Hence from the $e - v_e$ or $\mu - v_\mu$ weak interactions we can predict the $\tau - v_\tau$ weak interactions if this theory is correct. Specifically, it predicts that (a) the $\tau - v_\tau$ coupling will be $V - A$ and (b) the coupling constant will be the universal Fermi weak interaction constant $G_F \approx 1.02 \times 10^{-5}/M^2_{proton}$. We discuss these predictions in Sections 4.2 and 4.4.

3   THE IDENTIFICATION OF THE TAU AS A LEPTON

The identification of the tau as a lepton is intertwined with all the properties of the tau. Therefore in a general sense the subject of this entire review is the demonstration that the tau is a lepton. However, it is useful to summarize this demonstration in this section.

3.1 Decay Process Signatures

In this discussion the tau is treated as a sequential lepton. The tau may be an electron-associated ortholepton with the decay $\tau^- \to e^- + \gamma$ strongly suppressed compared to the weak interaction decay modes (see Section 4.4), but this possibility does not alter the present discussion.

A crucial signature for identification of a particle as a sequential lepton is that it decays only via the weak interaction and that the various decay branching ratios are explained by the weak interactions. We can roughly calculate the weak interaction predictions for the τ decay by using Figure 1 and replacing the L, L^0 pair by the τ, ν_τ pair. Because the quark decay mode (Figure 1c) occurs in the different colors, there are five diagrams of equal weight. Therefore, we expect that the leptonic decays $\nu_\tau e^- \bar{\nu}_e$ or $\nu_\tau \mu^- \bar{\nu}_\mu$ will each occur 20% of the time and the semileptonic decays via the quark mode will occur 60% of the time.

A more precise calculation of the branching ratios uses (a) conventional Weinberg-Salam theory; (b) the masses of the τ and the final-state particles; (c) some theoretical concepts like the conserved vector current (CVC); and (d) some specific experimental parameters, for example, the pion lifetime, required to calculate the decay rate for $\tau^- \to \pi^- \nu_\tau$. Many of these

Table 2 Predictions for τ^- branching ratios

Mode	Branching ratio (%)	Additional input to calculation
$e^- \bar{\nu}_e \nu_\tau$	16.4–18.0	none
$\mu^- \nu_\mu \nu_\tau$	16.0–17.5	none
$\pi^- \nu_\tau$	9.8–10.6	π^- lifetime
$K^- \nu_\tau$	≈ 0.5	θ_{Cabibbo}
$\rho^- \nu_\tau$	20–23	CVC plus $e^+ e^-$ annihilation cross sections
$K^{*-} \nu_\tau$	0.8–1.5	Das-Mathur-Okubo sum rules plus θ_{Cabibbo}
$A_1^- \nu_\tau$	8–10	Weinberg sum rules
(2 or more π's, K's)$^- \nu_\tau$	23–25	quark model, CVC, $\sigma(e^+ e^- \to \text{hadrons})$, etc
$A_1^- \nu_\tau$ + (2 or more π's, K's)$^- \nu_\tau$ subtotal[a]	31–35	

[a] Does not include the ρ, K*, or A_1 modes.

calculations were first made by Thacker & Sakurai (1971) and by Tsai (1971). Table 2 gives the branching ratios, based on these references and on the work of Gilman & Miller (1978), Kawamoto & Sanda (1978), Pham, Roiesnel & Truong (1978), and Tsai (1980). We assume a massless v_τ, spin 1/2 for the τ and v_τ, V$-$A coupling, and the Weinberg-Salam weak interaction theory; and we use the additional inputs listed in the third column of Table 2. Two of the branching ratios are uncertain. The decay rate of the mode with three or more π's or K's, the multihadron decay mode, is difficult to calculate precisely (Section 5.3). The calculation of the A_1 decay mode depends upon knowing for certain that the A_1 exists, and on knowing the properties of the A_1 (Section 5.2). Since the total of the branching fractions must be one, any change in these decay rates will change all the branching fractions. In addition some of the calculations are uncertain because they depend on experimental data such as the total cross section for $e^+ + e^- \rightarrow$ hadrons. Therefore a range of theoretical predictions is given for some of the branching fractions in Table 2.

Note that the crude prediction using Figure 1 is quite good; the individual leptonic branching ratios are calculated to be 16–18% rather than 20%, so that the total semileptonic branching ratio prediction increases from 60% to 64–68%.

All of the decay modes listed in Table 2, except $\tau^- \rightarrow v_\tau + K^-$, have been seen and their measured branching ratios agree with the calculations (Section 5). Of equal importance, τ decays modes that would occur through the strong or electromagnetic interactions have not been found (Section 5.3). The tau's decay processes are thus consistent with it being a lepton and inconsistent with it being a hadron.

G. Feldman has remarked that in the W exchange model of τ decays (Figure 1a–c) all the decay modes of the τ are decay modes of the W if the v_τ is excluded. Hence the consistency of the measured with the predicted branching ratios may be thought of as repeated proof that the τ acts as a lepton in the $\tau - W - v_\tau$ vertex. The measurements themselves can be viewed as a study of the decay modes of a virtual W.

3.2 e^+e^- Production Process Signatures

3.2.1 THEORY   There are four general observations we can make about tau production in e^+e^- annihilation.

1. Taus should be produced in pairs via the one-photon exchange process (Figure 2a)

$$e^+ + e^- \rightarrow \gamma_{virtual} \rightarrow \tau^+ + \tau^- \qquad \qquad 18.$$

once the total energy, E_{cm}, is greater than twice the τ mass (m_τ).

2. For spin 0 or 1/2, the production cross section for point particles is

known precisely from quantum electrodynamics:

spin 0: $\qquad \sigma_{\tau\tau} = \dfrac{\pi\alpha^2\beta^3}{3s}$ $\qquad\qquad\qquad$ 19.

spin 1/2: $\qquad \sigma_{\tau\tau} = \dfrac{4\pi\alpha^2}{3s}\dfrac{\beta(3-\beta^2)}{2}$ $\qquad\qquad$ 20.

where $s = E_{cm}^2$; $\beta = v/c$, v being the cms velocity of the τ and c being the velocity of light; and α is the fine structure constant. The $\sigma_{\tau\tau}$ for higher spins was discussed by Tsai (1978), Kane & Raby (1980), and Alles (1979). I restrict further discussion in this section to spin 1/2, which is appropriate to the τ. It has become customary in e^+e^- annihilation physics to remove the $1/s$ dependence of cross sections (Equations 19 and 20) by defining

$$R = \sigma/\sigma_{e^+e^-\to\mu^+\mu^-},$$ $\qquad\qquad$ 21.

where

$$\sigma_{e^+e^-\to\mu^+\mu^-} = \frac{4\pi\alpha^2}{3s}$$ $\qquad\qquad$ 22.

Then for spin 1/2 we expect

$$R_\tau = \frac{\beta(3-\beta^2)}{2},$$ $\qquad\qquad$ 23.

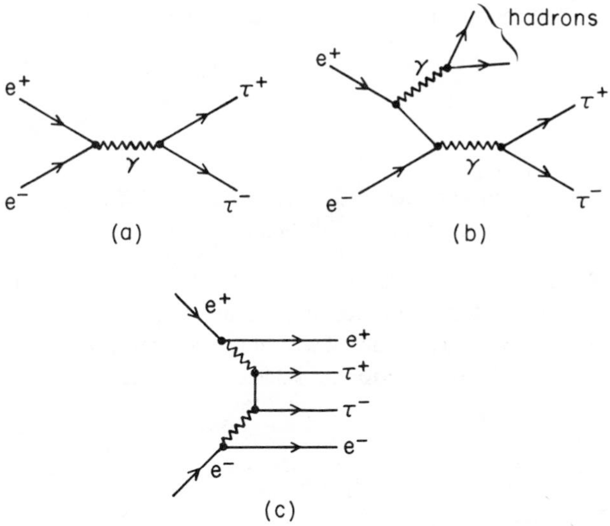

Figure 2 Diagrams for (*a*) $e^+e^- \to \tau^+\tau^-$ via one-photon exchange, (*b*) higher order production of a τ pair with hadrons, and (*c*) $e^+e^- \to e^+e^-\tau^+\tau^-$ via a two-virtual-photon process.

which has the simple property that $R_\tau \to 1$ as $\beta \to 1$; that is, as E_{cm} rises above the τ threshold. If the τ has internal structure, then Equation 20 is modified by a form factor $F(s)$

$$\sigma_{\tau\tau} = \frac{4\pi\alpha^2}{3s} \frac{\beta(3-\beta^2)}{2} |F(s)|^2. \qquad 24.$$

We expect that the internal structure will cause

$$|F(s)| \ll 1, \quad \text{when } E_{cm} \gg 2m_\tau. \qquad 25.$$

This is what happens in pair production of hadrons such as $e^+e^- \to \pi^+\pi^-$ or $e^+e^- \to p\bar{p}$. A point particle has $F(s) = 1$ for all s.

3. The production process

$$e^+ + e^- \to \tau^+ + \tau^- + \text{hadrons} \qquad 26.$$

should be very small compared to $e^+ + e^- \to \tau^+ + \tau^-$. This is because for a lepton the reaction in Equation 26 can only occur in a higher order process such as the one in Figure 2b, where two extra powers of α will appear in the cross section. On the other hand, for hadrons the reaction in Equation 26 is the common one. For example: in the region of several GeV the cross section for $e^+ + e^- \to K^+ + K^- + \text{hadrons}$ is much larger than the cross section for $e^+ + e^- \to K^+ + K^-$.

4. At sufficiently high energy, tau pairs should be produced in higher order electromagnetic processes (Figure 2c) such as

$$e^+ + e^- \to \tau^+ + \tau^- + e^+ + e^- \qquad 27.$$

and

$$e^+ + e^- \to \tau^+ + \tau^- + \gamma + \gamma \qquad 28.$$

These production processes are discussed in Section 6 (future studies of the τ) because there are not yet any published data on them.

3.2.2 EXPERIMENTAL RESULTS BELOW 8 GeV   Since the τ decays before detection, all production cross section measurements depend upon detection of some set of τ decay modes. Two sets have been used.

1. $e^\pm \mu^\mp$ events: the production and decay sequence

$$e^+ + e^- \to \tau^+ + \tau^-$$

$$\tau^+ \to e^+ + \nu_e + \bar{\nu}_\tau \qquad 29.$$

$$\tau^- \to \mu^- + \bar{\nu}_\mu + \nu_\tau$$

leads to $e^\pm \mu^\mp$ pairs being the only detected particles in the event. These two-prong, total charge zero, $e\mu$ events constitute a very distinctive signature, and thereby led to the discovery of the tau (Perl 1975, Perl et al 1975). The SPEAR data (Perl 1977a) on the energy dependence of the production of such events are shown in Figures 3 and 4. Figure 3 shows $R_{e\mu,\text{observed}}$, defined as

$$R_{e\mu,\text{observed}} = 2R_\tau B(\tau \to e\text{v's}) B(\tau \to \mu\text{v's}) A_{e\mu}, \qquad 30.$$

where the B's are branching fractions and $A_{e\mu}$ is the acceptance and efficiency of the apparatus. In the apparatus used for these data (the SLAC-LBL Mark I magnetic detector) $A_{e\mu}$ was almost independent of E_{cm}; hence the $R_{e\mu,\text{observed}}$ values are proportional to R_τ. Note that the sharp threshold at about 3.7 GeV and the leveling out of the R value as E_{cm} increases above 5 GeV are in agreement with Equation 23. This is shown explicitly in Figure 4 (Perl 1977a) where R_τ was calculated from Equation 30, and is in agreement with Equation 23—the theoretical curve.

2. $e^\pm$ hadron$^\mp$ and $\mu^\pm$ hadron$^\mp$ events: the production and decay

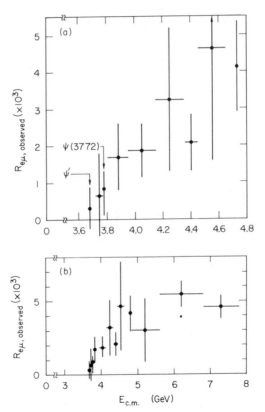

Figure 3 $R_{e\mu,\text{observed}}$ for (a) $3.6 \leqq E_{cm} \leqq 4.8$ GeV and (b) $3.6 \leqq E_{cm} \leqq 7.8$ GeV (Perl 1977a).

sequence

$$e^+ + e^- \rightarrow \tau^+ + \tau^-$$

$$\tau^+ \rightarrow e^+ + \nu_e + \bar{\nu}_\tau \qquad\qquad\qquad 31.$$

$$\tau^- \rightarrow \text{hadron}^- + \nu_\tau, \quad (\text{hadron} = \pi \text{ or } K)$$

leads to two-prong events consisting of one hadron and one opposite-sign electron. The restriction to events with one hadron is necessary to reduce the background (Brandelik et al 1978, Kirkby 1979) from charmed-particle production and decay processes such as

$$e^+ + e^- \rightarrow D^+ + D^-$$

$$D^+ \rightarrow e^+ + \nu_e + K^0$$

$$D^- \rightarrow \text{hadrons}$$

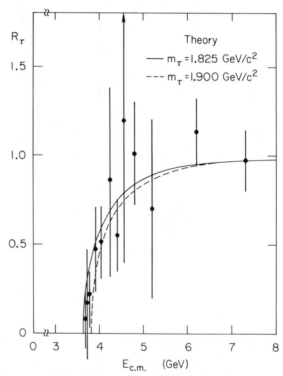

Figure 4 R_τ compared to theoretical curves for point-like, spin-1/2 particles for two τ masses (Perl 1977a). These older data indicated a τ mass between 1825 and 1900 MeV/c^2; the presently accepted value is 1782 MeV/c^2.

since charm-related events are predominantly multihadronic. Figure 5 shows early results from the DASP detector at DORIS (Brandelik et al 1978); more recent results from the DELCO detector at SPEAR are given in Figure 6 (Kirkby 1979). The data in both figures are consistent with Equation 23.

If the τ decays to μ + neutrinos instead of e + neutrinos in Equation 31, then $\mu^{\pm}$ hadron$^{\mp}$ events are produced. These μ-single-hadron events can also be separated from charmed-particle-related μ-multihadron events, and

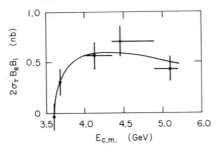

Figure 5   Cross section for e-hadron events from DASP (Brandelik et al 1978) compared to theoretical curve for point-like, spin-1/2 particles.

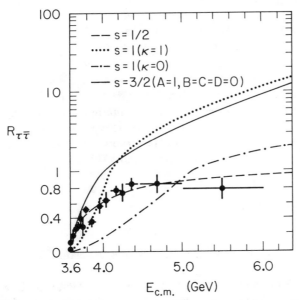

Figure 6 R_τ from DELCO (Kirkby 1979) compared to theoretical curves for point-like particles with spin 0, 1/2, 1, or 3/2. The constants A, B, C, and D are vertex parameters (Tsai 1978).

were very important in the early work on the τ (Cavalli-Sforza et al 1976, Feldman et al 1976, Burmester et al 1977a,b).

The $e\mu$, e-hadron, and μ-hadron data discussed above, and all other published data on e-hadron and μ-hadron events (e.g. Barbaro-Galtieri et al 1977, Bartel et al 1978), are consistent with Equation 23 and hence with Equation 20. Therefore all the production data are consistent with the τ being a point particle with spin 1/2, required properties for a lepton (Section 1.1). We can discuss this more quantitatively by examining how much the form factor $F_\tau(s)$ (Equation 24) is allowed to deviate from one. There are two possibilities: (a) We expect that this deviation will be largest at high E_{cm}. Looking at the higher energy data in Figure 4 we see that R_τ is consistent with $F(s) = 1$ within about $\pm 20\%$. Therefore these data allow a maximum deviation of $F(s)$ from one of about $\pm 10\%$ at these energies. (b) Alternatively, we can use a model for $F(s)$; the usual choice is (Hofstadter 1975, Barber et al 1979)

$$F(s) = 1 \mp \frac{s}{s - \Lambda_\pm^2}. \qquad\qquad 32.$$

Note that the larger Λ is, the less $F(s)$ deviates from one. This model has recently been applied to very high energy data from the PETRA e^+e^- colliding-beams facility at DESY.

3.2.3 EXPERIMENTAL RESULTS ABOVE 8 GeV   As E_{cm} increases in the PETRA and PEP energy range, ~ 10–40 GeV, τ pair events become increasingly distinctive for two reasons. (a) The increased energy of each of the τ's causes their respective decay products to move in opposite and roughly colinear directions. (b) Since the τ decays predominantly to one or three charged particles, the total charged multiplicity of τ events is small compared to the average charged multiplicity of hadronic events (≈ 12) (Wolf 1979). Figure 7 from the PLUTO group at PETRA is an example. A particularly striking signature (Barber et al 1979) is

$$e^+ + e^- \to \tau^+ + \tau^-$$

$$\tau^+ \to \mu^+ + \text{neutrinos} \qquad\qquad 33.$$

$$\tau^- \to \text{hadrons}^- \text{ or } \nu^- \text{ or } e^- + \text{neutrinos}.$$

The restriction to single hadron decays used in Section 3.2.2 is no longer necessary.

Figure 8 from the Mark-J Collaboration at PETRA (Barber et al 1979) shows the production cross section for τ pairs using the signature in Equation 33. The application of Equation 32 to these data yields, with 95% confidence (Barber et al 1979),

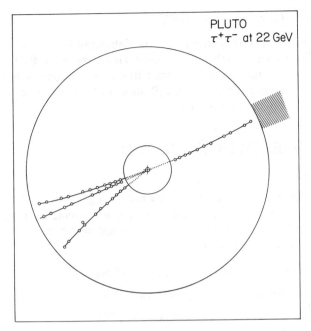

Figure 7   An example of a τ pair event at $E_{cm} = 22$ GeV obtained by the PLUTO group at PETRA. The single track is an electron, and the three tracks going in the other direction are pions.

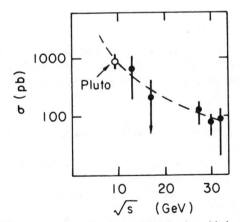

Figure 8   Comparison of the cross section for τ pair production with the theoretical curve for a point-like spin-1/2 particle. The data are from the Mark J detector at PETRA (Barber et al 1979), except for the point marked PLUTO obtained by the PLUTO group at DORIS.

$$\Lambda_- > 53 \text{ GeV}, \quad \Lambda_+ > 47 \text{ GeV}. \qquad 34.$$

These lower limits on Λ_τ mean that the data show no deviation from $F(s)$ = 1, as we can see directly from Figure 8. Hence these new PETRA data are consistent with the τ being a point particle. Very recent results from the TASSO Collaboration at PETRA (Brandelik et al 1980) are also consistent with the τ being a point particle.

4   PROPERTIES OF THE TAU

4.1   *Mass*

The τ mass (m_τ) is best determined by measuring the threshold for $e^+ + e^-$ $\rightarrow \tau^+ + \tau^-$. Table 3 summarizes three recent measurements, all of which are consistent. The DELCO measurement is based on the largest statistics and I use that value: 1782^{+3}_{-4} MeV/c^2.

4.2   *Spins, $\tau - v_\tau$ Coupling, and v_τ Mass*

The general form for the $\tau - v_\tau$ weak interaction four-current is

$$J^\lambda_{\tau v_\tau} = g_\tau(\psi^\dagger_{v_\tau} \mathcal{O} \psi_\tau)^\lambda, \quad \lambda = 0, 1, 2, 3 \qquad 35.$$

where g is a coupling constant, ψ_{v_τ} and ψ_τ are spin functions, and $\mathcal{O}$ is an operator. The decay of the τ is dependent on the current. For example, in the decay $\tau^- \rightarrow v_\tau + e^- + \bar{v}_e$ the four-particle matrix element (Figure 1d) is

$$T = J^\lambda_{\tau v_\tau} j_{\lambda, e\bar{v}_e}, \qquad 36.$$

where

$$j_{\lambda, e\bar{v}_e} = g\bar{u}_{v_e}\gamma_\lambda(1 - \gamma_5)u_e. \qquad 37.$$

Here the u's are the usual Dirac spinors and the γ's are the usual Dirac γ matrices (Bjorken & Drell 1964). Equation 37 is of course the conventional $V - A$ weak interaction current (Bjorken & Drell 1964). Also

$$\sqrt{2}g^2 = G_F = 1.02 \times 10^{-5}/M^2_{\text{proton}}. \qquad 38.$$

Hence the determination of $J^\lambda_{\tau v_\tau}$ involves the simultaneous determination of

Table 3   Measurements of the τ mass using the production threshold

Experiment	Mass (MeV/c^2)	Figure	Reference
DELCO	1782^{+3}_{-4}	9	Bacino et al 1978, Kirkby 1979
DASP	1807 ± 20	5	Brandelik et al 1978
DESY-HEIDELBERG	1787^{+10}_{-18}	10	Bartel et al 1978

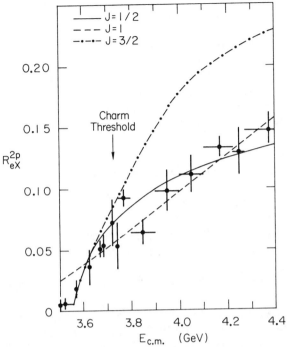

Figure 9   The threshold behavior of R_{ex} for e-hadron and $e\mu$ events from DELCO (Kirkby 1979).

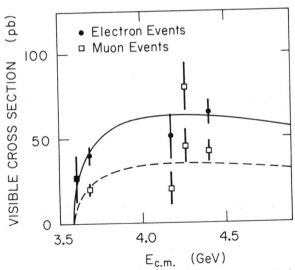

Figure 10   The threshold behavior of the cross section for e-hadron and μ-hadron events measured by the DESY-Heidelberg group (Bartel et al 1978). The theoretical curve is for a point-like, spin-1/2 particle.

g_τ, the τ spin, the ν_τ spin, the form of the operator $\mathcal{O}$, and even the determination of the ν_τ mass.

Fortunately, as discussed in Section 3.2.1, the τ pair production cross section depends upon the τ spin; hence we can independently determine that quantity. Equation 19 for spin 0 predicts a maximum R_τ value of 0.25; however, the measured maximum value is 1, hence spin 0 is excluded. Spin 1 and higher integral spins predict a β^3 factor (Tsai 1978). The behavior of $\sigma_{\tau\tau}$ near threshold (Figure 9) excludes a β^3 behavior and hence spin 1 or higher integral spins.

If the τ is not a point particle, spin 3/2 or higher half integral spins cannot at present be excluded. One can always select the arbitrary, energy-dependent parameters that occur in the τ-γ-τ vertex for spin 3/2, 5/2, ..., so that $\sigma_{\tau\tau}$ mimics the spin-1/2 case (Kane & Raby 1980). Such a model must also explain the $\tau^- \to \nu_\tau + \pi^-$ branching ratio, which requires spin 1/2 for a point particle (Alles 1979, Kirkby 1979). Further tests of the higher half integral spin proposal can be made; however, my strong instincts are to accept the spin-1/2 assignment.

The ν_τ spin is limited to half integral values by the existence of decay modes such as $\tau^- \to \nu_\tau + \pi^-$ and $\tau^- \to \nu_\tau + e^- + \bar{\nu}_e$. Furthermore, the measured value of the decay rate for $\tau^- \to \nu_\tau + \pi^-$ excludes spin 3/2 for a point-like τ or a point-like ν_τ (Alles 1979, Kirkby 1979). Higher half integral spins for the ν_τ have not been analyzed. Once again I use Ockham's razor and assign to the ν_τ a spin of 1/2 as the simplest hypothesis consistent with all observations on the ν_τ and τ.

Equation 35 now reduces to

$$J_{\tau\nu_\tau}^\lambda = g_\tau \bar{u}_{\nu_\tau} \mathcal{O}^\lambda u_\tau \qquad 39.$$

where the u's are Dirac spinors and $\mathcal{O}$ is some combination of scalar, pseudoscalar, vector, axial vector, and tensor current operators (Bjorken & Drell 1964, Rosen 1978). Insertion of Equation 39 into the matrix element (Equation 36) for

$$\tau^- \to \nu_\tau + e^- + \bar{\nu}_e$$

or

$$\tau^- \to \nu_\tau + \mu^- + \bar{\nu}_\mu \qquad 40.$$

shows that the e^- or μ^- momentum spectrum will depend on the form of $\mathcal{O}$. A similar situation occurred in the determination of the matrix element in μ decay

$$\mu^- \to \nu_\mu + e^- + \bar{\nu}_e \qquad 41.$$

(Marshak, Riazuddin & Ryan, 1969); except in that case the electron polarization was also measured, whereas in the τ case (Equation 40) the e^-

or μ^- polarization has not been measured. As discussed by Marshak, Riazuddin & Ryan (1969) the measurement of the e^- or μ^- momentum spectrum does not completely determine $\mathcal{O}$ in Equation 39. This dilemma has been resolved by all experimenters who have worked on the τ by (a) assuming that only vector (V) and axial vector (A) currents occur in Equation 39, as in all other weak interactions; and (b) allowing the relative strengths of the V and A currents to be fixed by measurement.

With these assumptions, Equation 39 reduces to

$$J^\lambda_{\tau\nu_\tau} = g_\tau \bar{u}_{\nu_\tau} \gamma^\lambda (v - a\gamma_5) u_\tau; \; v, a \text{ real}, \quad v^2 + a^2 = 1. \qquad 42.$$

Table 4 gives values of v and a for special choices of the current. In the τ rest system the normalized momentum distribution of the e or μ in Equation 40 is

$$dP/dy = y^2[6(v-a)^2(1-y) + (v+a)^2(3-2y)] \qquad 43.$$

where y = e or μ momentum/maximum e or μ momentum, and the e or μ mass and all neutrino masses are set to zero (Bjorken & Drell 1964). Of course, the e or μ momentum is measured in the laboratory system where the τ is in motion, and Equation 43 must be transformed properly.

Early studies of the v and a parameters were carried out using $e\mu$ events (Perl et al 1976, Barbaro-Galtieri et al 1977), e-hadron events (Yamada 1977, Brandelik et al 1978, Barbaro-Galtieri et al 1977), and μ-hadron events (Feldman et al 1976, Burmester et al 1977a,b). All these studies are consistent with $V-A$; Figure 11 from the PLUTO group is an example. Where the statistics are sufficient $V+A$ is excluded (Figure 12).

The most definitive study was carried out by the DELCO group (Kirkby 1979) (Figure 13), who determined the ρ Michel parameter (Michel 1950). With this parameter Equation 43 becomes

$$dP/dy = 4y^2[3(1-y) + (2\rho/3)(4y-3)] \qquad 44.$$

and

$$\rho = 3(v+a)^2/8. \qquad 45.$$

Table 4   Parameters for the $\tau - \nu_\tau$ current

Type	v	a	ρ
V–A	$\dfrac{1}{\sqrt{2}}$	$\dfrac{1}{\sqrt{2}}$	3/4
pure V	1	0	3/8
pure A	0	1	3/8
V+A	$\dfrac{1}{\sqrt{2}}$	$\dfrac{-1}{\sqrt{2}}$	0

They find $\rho = 0.72 \pm 0.15$, assuming the ν_τ mass is zero, which is in excellent agreement with V−A and in disagreement with V, A, or V+A (Table 4). Kirkby (1979) presents other evidence for V−A using $\langle p_e \rangle / E_{cm}$.

The effect of a nonzero ν_τ mass on the e or μ momentum spectrum is shown in Figure 12; the larger the ν_τ mass, the fewer the events with very energetic e's or μ's. Thus the ν_τ mass determination interacts with the V−A test. It has been shown that V+A is excluded for any value of the ν_τ mass (Bacino et al 1978, Kirkby 1979). However, the limited statistics of all the momentum spectrum measurements allow some deviation from V−A combined with some deviation of the ν_τ mass from zero. No study of the combined deviations has been done. Indeed, it has become conventional to assume that the $\tau - \nu_\tau$ current is precisely V−A and then to set an upper limit on the ν_τ mass. Table 5 gives three such determinations in historical order; all are consistent with a zero mass, although the limits are still quite

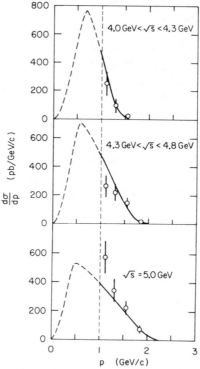

Figure 11   The momentum spectrum for muons in μ-hadron events obtained by the PLUTO group (Burmester et al 1977a). Muons with momenta less than the vertical dashed line could not be identified in the detector. The theoretical curve is for the decay $\tau^- \to \nu_\tau + \mu^- + \bar{\nu}_\mu$ with V−A coupling, all spins 1/2, and a massless ν_τ.

large because the spectra involve the ratio of the squares of the neutrino and tau masses.

To summarize this section: all measurements are consistent with

$$\tau \text{ spin} = 1/2$$

$$\nu_\tau \text{ spin} = 1/2$$

$$\tau - \nu_\tau \text{ current} = V - A$$ 46.

$$\nu_\tau \text{ mass} = 0$$

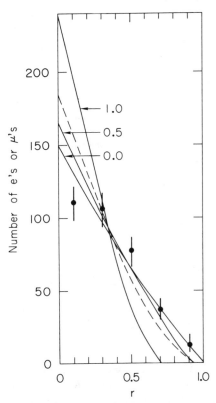

Figure 12   The momentum spectrum of e's or μ's in eμ events from the SLAC-LBL magnetic detector group (Perl et al 1976): $r = (p-0.65)/(p_{max}-0.65)$, where p is the momentum of the e or μ in GeV/c and p_{max} is its maximum value. The 0.65 constant is the lowest momentum at which e's and μ's could be identified in this detector. The solid theoretical curves are for the decays $\tau^- \rightarrow \nu_\tau + e^- + \bar{\nu}_e$ or $\tau^- \rightarrow \nu_\tau + \mu^- + \bar{\nu}_\mu$ with V $-$ A coupling, all spins 1/2, and a mass for the ν_τ indicated by the attached number in GeV/c^2. The dashed theoretical curve is for V $+$ A coupling and a massless ν_τ.

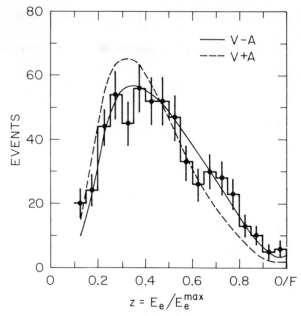

Figure 13   The normalized momentum spectrum for electrons from e-hadron and $e\mu$ events from DELCO (Kirkby 1979). The theoretical curves are for the decays $\tau^- \to \nu_\tau + e^- + \bar{\nu}_e$ with all spins 1/2, a massless neutrino, and the indicated coupling. The fits to $V - A$ (*solid*) and $V + A$ (*dashed*) give χ^2 per degrees of freedom of 15.9/17 and 53.7/17, respectively.

Table 5   Upper limits on ν_τ mass

Group	Upper limit on ν_τ mass (MeV/c^2)	Confidence (%)	Reference
SLAC-LBL	600	95	Perl et al 1977b
PLUTO	540	90	Knies 1977
DELCO	250	95	Kirkby 1979

Those are the standard set of parameters of the τ (Tsai 1980). They are, of course, compatible with e-μ-τ universality. As I have noted in this section, there has been a liberal use of Ockham's razor; measurements should continue on questions such as better limits on the ν_τ mass.

4.3   *Lifetime*

We have reduced the $\tau - \nu_\tau$ current to

$$J^\lambda_{\tau\nu_\tau} = g_\tau \bar{u}_{\nu_\tau} \gamma^\lambda (1 - \gamma_5) u_\tau \qquad\qquad 47.$$

but the value of g_τ is still undetermined. It cannot be determined by measurement of branching ratios or decay mode distributions because it appears as a constant multiplier in all decay mode matrix elements. One way to determine it is to measure the τ lifetime T_τ; other ways are discussed in Section 6. A specific example is helpful here. The decay rate for $\tau^- \to \nu_\tau + e^- + \bar{\nu}_e$ (Tsai 1971, Thacker & Sakurai 1971) is

$$\Gamma(\nu_\tau e^- \bar{\nu}_e) = \frac{2g_e^2 g_\tau^2 m_\tau^5}{192\pi^3}.$$ 48.

Note the analogy to the $\mu \to \nu_\mu + e^- + \bar{\nu}_e$ decay rate

$$\Gamma(\nu_\mu e^- \bar{\nu}_e) = \frac{G_F^2 m_\mu^5}{192\pi^3},$$ 49.

where $G_F^2 = 2g^4$. Then

$$T_\tau = \frac{B(\nu_\tau e^- \bar{\nu}_e)}{\Gamma(\nu_\tau e^- \bar{\nu}_e)} = B(\nu_\tau e^- \bar{\nu}_e)\left(\frac{g^2}{g_\tau^2}\right)\left(\frac{m_\mu^5}{m_\tau^5}\right)T_\mu,$$ 50.

where $B(\nu_\tau e^- \bar{\nu}_e)$ is the branching ratio for $\tau^- \to \nu_\tau + e^- + \bar{\nu}_e$ and T_μ is the μ lifetime. Using $B(\nu_\tau e^- \bar{\nu}_e) = 0.17$ (Section 5.1), Equation 50 predicts

$$T_\tau = 2.7 \times 10^{-13} \text{ s, for } g_\tau = g.$$ 51.

The measurement of T_τ requires a study of the average decay length of the τ's produced in $e^+ + e^- \to \tau^+ + \tau^-$; so far experiments have only been able to put an upper limit on this decay length. The smallest upper limit (Kirkby 1979, Bacino et al 1979a) is

$$T_\tau < 2.3 \times 10^{-12} \text{ s, (95\% confidence limit).}$$ 52.

Hence from Equation 50

$$g_\tau^2/g^2 > 0.12 \text{ (95\% confidence limit),}$$ 53.

or defining $G_\tau = \sqrt{2} g_\tau^2$,

$$G_\tau/G_F > 0.12 \text{ (95\% confidence limit).}$$ 54.

This is consistent with $G_\tau = G_F$, and hence with e-μ-τ universality and with Weinberg-Salam theory; but $G_\tau = G_F$ has not yet been proven.

4.4 Lepton Type and Associated Neutrino

Is the τ a sequential lepton with a unique lepton number, or does it share lepton number with the μ or e? Consider the μ first. If the τ were a μ-related ortholepton (Section 2.2.2) the reaction

$$\nu_\mu + \text{nucleus} \to \tau^- + \text{anything}$$ 55.

should occur. Analogously if the τ were a μ-related paralepton (Section 2.2.3)

$$\nu_\mu + \text{nucleus} \rightarrow \tau^+ + \text{anything} \qquad 56.$$

would occur. Cnops et al (1978) looked for the reactions in Equations 55 and 56 by using a neon-hydrogen-filled bubble chamber exposed to a ν_μ beam. They found no events; their upper limit on $G_{\tau-\nu_\mu}$ was $G_{\tau-\nu_\mu}/G_F < 0.025$ with 90% confidence. However, in these μ-related models $G_{\tau-\nu_\mu}$ is identical with what I call G_τ in Equation 54, and must have the value $G_{\tau-\nu_\mu}/G_F > 0.12$. This excludes the μ-related ortholepton and paralepton models.

The equivalent tests for e-related models have not been done because ν_e beams have not been built. Heile et al (1978) excluded the e-related paralepton model using an argument of Ali & Yang (1976). If the τ were an e-related paralepton its leptonic decay modes would be

$$\tau^- \rightarrow \bar{\nu}_e + e^- + \bar{\nu}_e$$
$$\tau^- \rightarrow \bar{\nu}_e + \mu^- + \bar{\nu}_\mu. \qquad 57.$$

The two identical $\bar{\nu}_e$'s in Equation 57 constructively interfere leading to the decay width ratio (Ali & Yang 1976)

$$\Gamma(\tau^- \rightarrow e^- + \text{neutrinos})/\Gamma(\tau^- \rightarrow \mu^- + \text{neutrinos}) \approx 2. \qquad 58.$$

The measurements of Heile et al (1978) show this ratio to be close to one; hence this model is wrong.

There is no way, using present data, to exclude the e-related ortholepton model for the τ. That is, it is possible that the τ^- and e^- have the same lepton number. The decay mode

$$\tau^- \rightarrow e^- + \gamma \qquad 59.$$

is then allowed. It has not been seen; the upper limit is 2.6% (Section 5.3). However, this does not exclude this model because the τ-γ-e vertex can be suppressed (Section 2.2.2).

More complicated models (Altarelli et al 1977, Horn & Ross 1977) such as the τ decaying into a mixture of ν_e and ν_μ have been considered and excluded. They are reviewed by Gilman (1978) and Feldman (1978b). However, one can always devise a complicated model that cannot be excluded. For example, suppose the τ shares its neutrino with a heavier τ'. That model cannot be excluded using present data, and if true the τ would not be a sequential lepton.

As we have done before, we select the simplest model consistent with all the data; that is the sequential model. It is, of course, consistent with e-μ-τ universality and with Weinberg-Salam theory.

5   DECAY MODES OF THE TAU

5.1   *Purely Leptonic Decay Modes*

Conventional weak interaction theory (Bjorken & Drell 1964) predicts that the decay width for

$$\tau^- \to v_\tau + 1^- + \bar{v}_1, \quad 1 = e \text{ or } \mu$$

is

$$\Gamma(v_\tau 1^- \bar{v}_1) = \frac{G_F^2 m_\tau^5}{192\pi^3}, \qquad\qquad 60.$$

where the mass of the 1 is neglected. This is a very basic and simple calculation and the only parameter is G_F; hence, the measurement of the purely leptonic branching fractions is crucial.

Feldman (1978b) reviewed all the data on B_e and B_μ, the branching fractions for

$$\tau^- \to v_\tau + e^- + \bar{v}_e$$

and $\qquad\qquad\qquad\qquad\qquad\qquad\qquad\qquad\qquad\qquad\qquad$ 61.

$$\tau^- \to v_\tau + \mu^- + \bar{v}_\mu$$

respectively. All measurements are consistent and their average values (Table 6) agree with the theoretical predictions from Table 2. Table 6 gives two sets of values for B_e and B_μ. One set assumes they are unrelated; the other set assumes they are connected via

$$\Gamma(v_\tau \mu^- \bar{v}_\mu)/\Gamma(v_\tau e^- \bar{v}_e) = 1 - 8y + 8y^3 - y^4 - 12y^2 \ln y = 0.972, \qquad 62.$$

where $y = (m_\mu/m_\tau)^2$ (Tsai 1971).

5.2   *Single Hadron or Hadronic Resonance Decay Modes*

In this section I consider the decay modes

$$\tau^- \to v_\tau + \pi^-$$

$$\tau^- \to v_\tau + K^-$$

$$\tau^- \to v_\tau + \rho^-$$

$$\tau^- \to v_\tau + K^*(890)^-$$

$$\tau^- \to v_\tau + A_1^-.$$

The first mode, $\tau^- \to v_\tau + \pi^-$, has the decay width (Tsai 1971)

$$\Gamma(v_\tau \pi^-) = \frac{G_F^2 f_\pi^2 \cos^2 \theta_c m_\tau^3}{16\pi} \left(1 - \frac{m_\pi^2}{m_\tau^2}\right)^2. \qquad 63.$$

Table 6 Purely leptonic decay mode branching fractions (in %)
(Feldman 1978b)

	B_e and B_μ free	$B_\mu = 0.972B_e$	Theory
B_e	16.5 ± 1.5	17.5 ± 1.2	16.4–18.0
B_μ	18.6 ± 1.9	17.1 ± 1.2	16.0–17.5

Here in comparison to Equation 60 two more parameters appear: θ_c, the Cabibbo angle, and f_π, the coupling constant that appears in the π decay width (Tsai 1971)

$$\Gamma(\pi^- \to \bar{v}_\mu + \mu^-) = \frac{G_F^2 f_\pi^2 \cos^2 \theta_c}{8\pi} m_\pi m_\mu^2 \left(1 - \frac{m_\mu^2}{m_\pi^2}\right)^2. \qquad 64.$$

However, $f_\pi^2 \cos^2 \theta_c$ can be evaluated experimentally from Equation 64, so the calculation of $\Gamma(v_\tau \pi^-)$ is firm.

Table 7 gives four published branching ratios for this decay mode. They are consistent with each other and with the theoretical prediction of 9.8–10.6% (Table 2). Dorfan (1979) recently gave a preliminary value of 10.7 $\pm 2.1\%$ based on a new analysis of SLAC-LBL Mark II data.

The second mode, $\tau^- \to v_\tau + K^-$, is Cabibbo suppressed, and weak interaction theory predicts (Tsai 1971)

$$\Gamma(v_\tau K^-)/\Gamma(v_\tau \pi^-) = \tan^2 \theta_c \frac{(1 - m_K^2/m_\tau^2)^2}{(1 - m_\pi^2/m_\tau^2)^2}. \qquad 65.$$

The smallness of this branching fraction, $\tan^2 \theta_c \approx 0.05$, and the difficulty of separating K's from the much larger π background has so far prevented the measurement of this mode.

Table 7 Branching fractions for $\tau^- \to v_\tau + \pi$

Experiment	Mode	Branching fraction[a] (%)	Reference
SLAC-LBL	xπ	$9.3 \pm 1.0 \pm 3.8$	Feldman 1978b
PLUTO	xπ	$9.0 \pm 2.9 \pm 2.5$	Alexander et al 1978a
DELCO	eπ	$8.0 \pm 3.2 \pm 1.3$	Bacino et al 1979b
SLAC-LBL	$\begin{cases} \text{x}\pi \\ \text{e}\pi \end{cases}$	$\left. \begin{matrix} 8.0 \pm 1.1 \pm 1.5 \\ 8.2 \pm 2.0 \pm 1.5 \end{matrix} \right\}$	Hitlin 1978
Average		8.3 ± 1.4	

[a] The first error is statistical, the second systematic.

The two measurements of $\tau^- \to v_\tau + \rho^-$ give

$$B(\tau^- \to v_\tau + \rho^-) = 20.5 \pm 4.1\% \quad \text{(Abrams et al 1979, Dorfan 1979)} \qquad 66.$$

$$B(\tau^- \to v_\tau + \rho^-) = 24 \pm 9\% \quad \text{(Brandelik et al 1979)}.$$

These values are consistent with theoretical predictions of 20–23% (Table 2). Figure 14 shows that the ρ momentum spectrum is consistent with the flat spectrum expected for this two-body decay.

The decay mode $\tau^- \to v_\tau + K^*(890)^-$, like $\tau^- \to v_\tau + K^-$, is Cabibbo suppressed and will be suppressed relative to $\tau^- \to v_\tau + \rho^-$ by a factor of $\tan^2 \theta_c \approx 0.05$. Hence a branching fraction of about 1% is expected. The first measurement of this mode was reported by Dorfan (1979) using SLAC-LBL Mark II data:

$$B[\tau^- \to v_\tau + K^*(890)^-] = 1.3 \pm 0.4 \pm 0.3\%, \qquad 67.$$

in good agreement with theory. In Equation 67 the first error is statistical and the second is systematic.

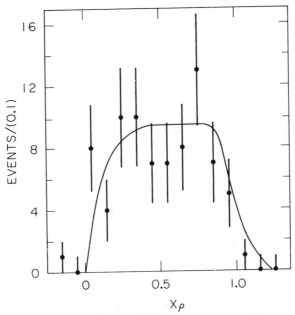

Figure 14   The distribution $X_\rho = (E_\rho - E_{min})/(E_{max} - E_{min})$ where E_ρ is the ρ energy and E_{max} and E_{min} are its maximum and minimum values from the SLAC-LBL magnetic detector (Abrams et al 1979). The theoretical curve is for $\tau^- \to v_\tau + \rho^-$ corrected for detector acceptance and measurement errors. Without these corrections the curve would be flat for $X_\rho = 0$ to 1 and zero above $X_\rho = 1$.

The search for the decay mode $\tau^- \to \nu_\tau + A_1^-$ is important (Sakurai 1975) because it may be the only way to establish the existence of the A_1 resonance, which is assumed to have mass ≈ 1100 MeV/c^2, $I^G = 1^-$, $J^P = 1^+$ (Bricman et al 1978). It is because of this importance that I consider the present data suggestive of the A_1's existence but not yet conclusive. The experimental analysis is difficult for two reasons: (a) The mode is found via

$$\tau^- \to \nu_\tau + A_1^-, \qquad A_1^- \to \pi^- + \pi^+ + \pi^- \qquad\qquad 68.$$

and this has to be separated from higher multiplicity $\pi^\pm$ and π^0 decay modes in which some π's are undetected; (b) After the $\pi^- \pi^+ \pi^-$ mode is separated out, one has to show that it contains a resonance with the expected properties (Basdevant & Berger 1978).

I only summarize the experimental situation here. The PLUTO group published two papers giving evidence for the A_1 (Alexander et al 1978b, Wagner et al 1980). Figure 15 shows their recent analysis (Wagner et al 1980) using

$$\tau^- \to \nu_\tau + \rho^0 + \pi^- \to \nu_\tau + \pi^- + \pi^+ + \pi^-. \qquad\qquad 69.$$

The SLAC-LBL group has also studied this decay mode (Jaros et al 1978); Figure 16 shows their data on Equation 68.

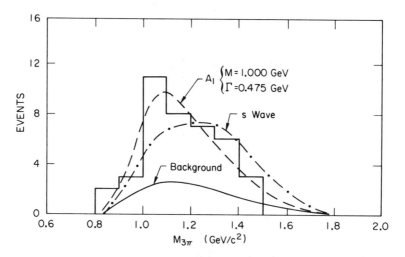

Figure 15 Invariant $\rho^0\pi$ mass distribution from the PLUTO group (Wagner et al 1980), compared to an s-wave without and with an imposed A_1 resonance of mass $= 1.0$ GeV/c^2 and width $= 0.475$ GeV/c (*dashed curve*), added to the expected background (*solid curve*).

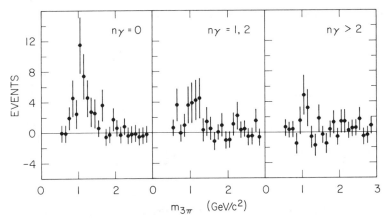

Figure 16 Invariant $(3\pi)^\pm$ mass distribution from four-prong events $\mu^\mp(3\pi)^\pm$ obtained by the SLAC-LBL magnetic detector (Jaros et al 1978). The number of photons accompanying the event is n_γ; hence (a) contains the possible $A^\pm \to (3\pi)^\pm$ signal, and (b) and (c) are measures of the possible feeddown background from events with π^0's.

5.3 *Multihadron Decay Modes*

The multihadron decay modes are

$$\tau^- \to \nu_\tau + (n_\pi \pi + n_K K)^-$$

$$n_\pi + n_K \geqq 2$$

70.

where the hadrons do not come from a single resonance such as the ρ or A_1. There is no general method for calculating the decay width $\Gamma(\nu_\tau + n_\pi \pi + n_K K)$ because that would require us to know how a W with a small virtual mass converts into hadrons, and of course we do not yet know this. Instead, several special methods have been used. Decay modes with even numbers of pions were considered by Gilman & Miller (1978). They used CVC to connect the vector hadronic weak interaction current to the vector hadronic electromagnetic current. We have experimental information on the latter since we know the multihadron cross sections for

$$e^+ + e^- \to \gamma_{\text{virtual}} \to \text{hadrons}.$$

71.

That is, Equation 71 is used to predict some of the rates for

$$W_{\text{virtual}} \to \text{hadrons}$$

72.

Pham, Roiesnel & Truong (1978) used current algebra and PCAC to calculate rates for the axial vector hadronic current; that is, for the decay modes with odd numbers of pions.

Table 8 Some measurements of multihadron decay modes, including the A_1 resonance if it exists

Mode for τ^-	Branching fraction (%)	Reference
$\nu_\tau + \pi^- + \pi^+ + \pi^- + \geq 0\pi^0$	18 ± 6.5	Jaros et al 1978
$\nu_\tau + \pi^- + \pi^+ + \pi^-$	7 ± 5	Jaros et al 1978
$\nu_\tau + \pi^- + \pi^+ + \pi^- + \pi^0$	11 ± 7	Jaros et al 1978
$\nu_\tau + \rho^0 + \pi^-$	5.4 ± 1.7	Wagner et al 1980
$\nu_\tau + \geq 3$ charged particles $\pm \geq 0\pi^0$	28 ± 6	Kirkby 1979

Unfortunately we do not yet have the required measurements to compare in detail with these calculations. This is because the experimental separation of the various multihadron modes is obscured (a) by the great difficulty of detecting π^0's in multihadron events; and (b) by the possibility of some charged π's escaping detection. Therefore, it is premature to attempt a precise comparison of theory and measurement of the multihadron decay modes. Table 8 lists existing measurements. All of them are compatible with calculations, given the large errors in the measurements and the uncertainties in the calculations. The branching fraction of $28 \pm 6\%$ for modes with at least two charged hadrons should be compared with the A_1 + multihadron subtotal in Table 2 of 31–35%. The latter are expected to be larger because they include some modes with one charged π and two or more π^0's.

5.4 Sequential Model Forbidden Decay Modes

If the τ lepton number were not perfectly conserved or if the τ, e, and μ lepton numbers were slightly mixed, the τ might decay electromagnetically into modes such as

$$
\begin{aligned}
\tau^- &\to e^- + \gamma \\
\tau^- &\to \mu^- + \gamma \\
\tau^- &\to e^- + e^+ + e^- \\
\tau^- &\to e^- + \mu^+ + \mu^- \\
\tau^- &\to \mu^- + \mu^+ + \mu^-
\end{aligned}
$$

73.

None of these decay modes have been found. Branching fraction upper limits are given in Table 9, along with upper limits on some other modes forbidden by the sequential model. The limits in Table 9 were obtained several years ago; increased statistics obtained since will allow searches that are five to ten times more sensitive.

Table 9   Upper limits on branching fractions for decay modes forbidden by the sequential model

Mode	Upper limit on branching fraction (%)	Confidence limit (%)	Experimental group or detector	Reference
$\tau^- \to e^- + \gamma$ $\tau^- \to \mu^- + \gamma$	12	90	PLUTO Group	a
$\tau^- \to e^- + \gamma$	2.6	90	LBL-SLAC lead glass wall	b
$\tau^- \to \mu^- + \gamma$	1.3	90	LBL-SLAC lead glass wall	b
$\tau^- \to$ (3 charged leptons)$^-$	1.0	95	PLUTO Group	a
$\tau^- \to$ (3 charged leptons)$^-$	0.6	90	SLAC-LBL magnetic detector	b
$\tau^- \to$ (3 charged particles)$^-$	1.0	95	PLUTO Group	a
$\tau^- \to \rho^- + \pi^0$	2.4	90	SLAC-LBL magnetic detector	b

[a] Flügge 1977.
[b] Perl 1977a.

6   FUTURE STUDIES OF THE TAU

6.1   *Future Studies of the Properties of the Tau*

As indicated throughout this review, there is more experimental work to be done on the properties of the τ for two reasons:

(*a*) Just as we continue to test the leptonic nature of the e and μ and the validity of the laws of lepton conservation, so we should continue to test these properties of the τ. The relatively large mass of the τ might be associated with deviation from $e - \mu - \tau$ universality. (*b*) Measurement of the properties of the τ can teach us about other areas of elementary particle physics, such as the existence of the A_1.

The main τ studies to be done are the following:

1. high energy, high statistics measurement of $\sigma(e^+e^- \to \tau^+\tau^-)$;
2. better determination of the limits on the ν_τ mass;
3. measurement of the τ lifetime and determination of G_τ;
4. discovery and measurement of $\tau^- \to \nu_\tau + K^-$;
5. elucidation of $\tau^- \to \nu_\tau + A_1$ and study of A_1 properties if it exists;
6. detailed study of multihadron decay modes of τ; and
7. better limits on sequential model forbidden decay modes.

6.2 *The Tau as a Decay Product*

Several known or proposed very heavy hadrons should decay to one or more τ's (Table 10). The measurement of these decay modes provides a separate measurement of G_τ, tests our standard model for the τ, such as the spin being 1/2, and provides information about the heavy hadron.

6.3 *Tau Production in Photon-Hadron Collisions*

We should be able to produce pairs of heavy charged leptons, such as τ pairs, by the very high energy Bethe-Heitler process (Kim & Tsai 1973, Tsai 1974, Smith et al 1977):

$$\gamma + \text{nucleus} \rightarrow \tau^+ + \tau^- + \text{anything.} \qquad 74.$$

We cannot learn much about the τ from this process, if our standard model of the τ is correct. However, this is a test of the model that should be done. An alternative method would use very high energy μ's via

$$\mu + \text{nucleus} \rightarrow \mu + \text{anything} + \gamma_{\text{virtual}}$$

$$\gamma_{\text{virtual}} \rightarrow \tau^+ + \tau^-. \qquad 75.$$

The detection of the τ pairs in the reactions of Equations 74 and 75 is difficult because the τ decay products are immersed in an enormous background of hadrons, e's, and μ's. The most promising signature appears to be $e^\pm \mu^\mp$ pairs.

6.4 *Tau Production in Hadron-Hadron Collisions*

Taus can be produced in hadron-hadron collisions in two types of processes: 1. Virtual photons from quark-antiquark annihilations can produce $\tau^+\tau^-$ pairs (Chu & Gunion 1975, Bhattacharya et al 1976). 2. Heavy hadrons such as those in Table 10 can be produced, and subsequently decay to $\tau\nu_\tau$ or $\tau^+\tau^-$ pairs.

Table 10   Some predicted decay modes of heavy hadrons to one or two τ's

Hadron	Quark composition	Predicted decay mode	Predicted branching fraction (%)	Reference
F	$c\bar{s}$	$F^+ \rightarrow \tau^+ + \nu_\tau$	≈ 3	Albright & Schrock 1979
B[a]	$b\bar{u},\, b\bar{d}$	$B^+ \rightarrow \tau^+ + \nu_\tau$	0.5–1.0	Ellis et al 1977
Υ	$b\bar{b}$	$\Upsilon \rightarrow \tau^+ + \tau^-$	≈ 3.5	Eichten & Gottfried 1977, Ellis et al 1977
T[b]	$t\bar{t}$	$T \rightarrow \tau^+ + \tau^-$	≈ 8	Ellis et al 1977

[a] The evidence for the existence of the B is scanty at present.
[b] This is a proposed particle; there is no evidence for its existence at present.

Unfortunately the detection of the τ's is a very difficult task because of the overwhelming background from e, μ, and hadron production. However, the second process can be used to produce ν_τ's in sufficient quantity to permit their detection.

6.5 Future Studies of the Tau Neutrino

The production and decay sequence

$$\text{proton} + \text{nucleon} \rightarrow F^- + \text{hadrons} \qquad 76.$$

$$F^- \rightarrow \tau^- + \bar{\nu}_\tau \qquad 77.$$

$$\tau^- \rightarrow \nu_\tau + \text{charged particles} \qquad 78.$$

can produce a ν_τ beam (Barger & Phillips 1978, Albright & Schrock 1979, Sciulli 1978). However, the neutrinos from π and K decay would overwhelm the ν_τ signal unless the majority of the π's and K's interact before they decay. Therefore the entire proton beam must be dumped in a thick target. There is still some problem with ν_e's and ν_μ's from D meson and other charmed-particle semileptonic decays, but the detection of the ν_τ appears feasible.

The detection of the ν_τ and the study of its interactions can provide the following information. (a) We can verify that the ν_τ is a unique lepton and is not a ν_e, thereby testing the electron-related ortholepton model (Section 4.4). (b) More generally we can determine if the ν_τ behaves conventionally with normal weak interactions. (c) We can measure the product of the cross section for Reaction 76 and the branching ratio for Reaction 77.

7 SUMMARY

All published studies of the τ are consistent with it being a point-like, spin-1/2, sequential charged lepton. These studies are consistent with $e - \mu - \tau$ universality, and they are consistent with conventional Weinberg-Salam theory. There is more experimental work to be done on measuring the properties of the tau and on using the tau to study other particles.

ACKNOWLEDGMENTS

In the writing of this review I was greatly aided by the reviews done by Gary Feldman, Gustave Flügge, Jasper Kirkby, and Y. S. (Paul) Tsai.

I wish to take this opportunity to thank my collaborators in the SLAC-LBL magnetic detector experiments at SPEAR; it was in these experiments that the first evidence for the existence of the tau was discovered. Two of these collaborators were particularly important to me. One is Gary Feldman who was a constant companion in my work on the tau. The other is Burton Richter; without his knowledge, experimental skill, and leadership SPEAR would not have been built.

Literature Cited

Abrams, G. S. et al. 1979. *Phys. Rev. Lett.* 43:1555

Albright, C. H., Schrock, R. E. 1979. *Phys. Lett. B* 84:123

Alexander, G. et al. 1978a. *Phys. Lett. B* 78:162

Alexander, G. et al. 1978b. *Phys. Lett. B* 73:99

Ali, A., Yang, T. C. 1976. *Phys. Rev. D* 14:3052

Alles, W. 1979. *Lett. Nuovo Cimento* 25:404

Altarelli, G. et al. 1977. *Phys. Lett. B* 67:463

Bacino, W. et al. 1978. *Phys. Rev. Lett.* 41:13

Bacino, W. et al. 1979a. *Phys. Rev. Lett.* 42:749

Bacino, W. et al. 1979b. *Phys. Rev. Lett.* 42:6

Bailin, D. 1977. *Weak Interactions.* London, England: Sussex Univ. Press. 406 pp.

Barbaro-Galtieri, A. et al. 1977. *Phys. Rev. Lett.* 39:1058

Barber, D. P. et al. 1979. *Phys. Rev. Lett.* 43:1915

Barger, V., Phillips, R. J. 1978. *Phys. Lett. B* 74:393

Bartel, W. et al. 1978. *Phys. Lett. B* 77:331

Basdevant, J. L., Berger, E. L. 1978. *Phys. Rev. Lett.* 40:994

Bernardini, M. et al. 1973. *Nuovo Cimento* 17:383

Bhattacharya, R., Smith, J., Soni, A. 1976. *Phys. Rev. D* 13:2150

Bjorken, J. D., Drell, S. D. 1964. *Relativistic Quantum Mechanics,* New York: McGraw Hill. 414 pp.

Brandelik, R. et al. 1978. *Phys. Lett. B* 73:109

Brandelik, R. et al. 1979. *Z. Phys. C* 1:233

Brandelik, R. et al. 1980. *Phys. Lett. B* 92:199

Bricman, C. et al. 1978. *Phys. Lett. B* 75:1. This is the Particle Data Group's review of particle properties.

Burmester, J. et al. 1977a. *Phys. Lett. B* 68:297

Burmester, J. et al. 1977b. *Phys. Lett. B* 68:301

Cavalli-Sforza, M. et al. 1976. *Phys. Rev. Lett.* 36:558

Chu, G., Gunion, J. F. 1975. *Phys. Rev. D* 11:73

Cnops, A. M. et al. 1978. *Phys. Rev. Lett.* 40:144

Dorfan, J. 1979. In *Particles and Fields—1979,* ed. B. Margolis, D. G. Stairs, p. 159. AIP Conf. Proc. 59, Particles and Fields Subseries No. 19. New York: Am. Inst. Phys. (Publ. 1980)

Eichten, E., Gottfried, K. 1977. *Phys. Lett. B* 66:286

Ellis, J., Gaillard, M. K., Nanopoulos, D. V.,

Rudaz, S. 1977. *Nucl. Phys. B* 131:285

Farrar, G. R., Fayet, P. 1980. *Phys. Lett. B* 89:191

Feldman, G. J. et. al. 1976. *Phys. Lett. B* 63:466

Feldman, G. J. 1978a. *Proc. Int. Meet. Frontier of Physics, Singapore.* In press; also issued as Stanford Linear Accel. Cent. *SLAC-PUB-2230*

Feldman, G. J. 1978b. In *Neutrinos—78,* ed. E. C. Fowler, p. 647. West Lafayette, Ind: Purdue Univ.

Flügge, G. 1977. In *Exp. Meson Spectrosc. 1977,* ed. E. Von Goeler, R. Weinstein, p. 132. Boston: Northeastern Univ.

Flügge, G. 1979. *Z. Phys. C* 1:121

Gilman, F. J. 1978. Stanford Linear Accel. Cent. *SLAC-PUB-2226*

Gilman, F. J., Miller, D. H. 1978. *Phys. Rev. D* 17:1846.

Heile, F. J. et al. 1978. *Nuc. Phys. B* 138:189

Hitlin, D. 1978. Quoted in *Proc. 19th Int. Conf. High Energy Phys.,* ed. S. Homma, M. Kawaguchi, H. Miyazawa, p. 784. Tokyo: Phys. Soc. Jpn.

Hofstadter, R. 1975. In *Proc. 1975 Int. Symp. Lepton Photon Interact. High Energies,* ed. W. T. Kirk, p. 869. Stanford, Calif: SLAC

Horn, D., Ross, G. G. 1977. *Phys. Lett. B* 67:460

Jaros, J. A. et al. 1978. *Phys. Rev. Lett.* 40:1120

Kane, G., Raby, S. 1980. *Phys. Lett. B* 89:203

Kawamoto, N., Sanda, A. I. 1978. *Phys. Lett. B* 76:446

Kim, K. J., Tsai, Y. S. 1973. *Phys. Rev. D* 8:3109

Kirkby, J. 1979. In *Proc. 1979. Int. Symp. Lepton Photon Interact. High Energies,* ed. T. B. W. Kirk, H. D. I. Arbarbanel, p. 107. Batavia, Ill: FNAL

Knies, G. 1977. In *Proc. 1977 Int. Symp. Lepton Photon Interact. High Energies,* ed. F. Gutbrod, p. 93. Hamburg: DESY

Llewellyn Smith, C. H. 1977. *Proc. R. Soc. London Ser. A* 355:585

Low, F. E. 1965. *Phys. Rev. Lett.* 14:238

Marshak, R. E., Riazuddin, Ryan, C. P. 1969. *Theory of Weak Interactions in Particle Physics,* New York: Wiley-Interscience. 761 pp.

Michel, L. 1950. *Proc. Phys. Soc. London Sect. A* 63:514

Orioto, S. et al. 1974. *Phys. Lett. B* 48:165

Perl, M. L., Rapidis, P. 1974. Stanford Linear Accel. Cent. *SLAC-PUB-1499*

Perl, M. L. 1975. In *Proc. Summer Inst. Part. Phys., SLAC-191,* ed. M. C. Zipf, p. 333. Stanford: SLAC

Perl, M. L. et al. 1975. *Phys. Rev. Lett.* 35:1489

Perl, M. L. et al. 1976. *Phys. Lett. B* 63:466
Perl, M. L. 1977a. See Knies 1977, p. 145
Perl, M. L. et al. 1977b. *Phys. Lett. B* 70:487
Perl, M. L. 1978. *New Phenomena in Lepton-Hadron Physics,* ed. D. E. C. Fries, J. Wess, p. 115. New York: Plenum. This reference discusses broader definitions of a lepton.
Pham, T. N., Roiesnel, C., Truong, T. N. 1978. *Phys. Lett. B* 78:623
Rosen, S. P. 1978. *Phys. Rev. Lett.* 40:1057
Sakurai, J. J. 1975. See Hofstadter 1975, p. 353
Sciulli, F. 1978. See Feldman 1978b, p. 863
Smith, J., Soni, A., Vermaseren, J. A. M. 1977. *Phys. Rev. D* 15:648
Thacker, H. B., Sakurai, J. J. 1971. *Phys. Lett. B* 36:103
Tsai, Y. S. 1971. *Phys. Rev. D* 4:2821
Tsai, Y. S. 1974. *Rev. Mod. Phys.* 46:815
Tsai, Y. S. 1978. Stanford Linear Accel. Cent. *SLAC-PUB-2105*
Tsai, Y. S. 1980. Stanford Linear Accel. Cent. *SLAC-PUB-2403*
Wagner, W. et al. 1980. *Z. Phys. C* 3:193
Wolf, G. 1979. See Kirkby 1979, p. 34
Yamada, S. 1977. See Knies 1977, p. 69
Zipf, M. C., ed. 1978. *Proc. Summer Inst. Part. Phys. SLAC-215, Stanford Linear Accel. Cent., Stanford Univ. Calif.*

Ann. Rev. Nucl. Part. Sci. 1980. 30:337–81

CHARMED MESONS PRODUCED IN e+e- ANNIHILATION

✖5620

Gerson Goldhaber

Department of Physics and Lawrence Berkeley Laboratory,
University of California, Berkeley, California 94720

James E. Wiss

Department of Physics, University of Illinois, Urbana, Illinois 61801

CONTENTS

1 INTRODUCTION

The study of high energy electron-positron annihilation has revealed a rich spectrum of new phenomena and led to the discovery of new, hadronically stable particles—the charmed mesons. These particles are particularly interesting because they are the lightest particles containing the fourth quark; thus they serve to test in a novel way many theoretical ideas about quarks and their interactions.

337

0163-8998/80/1201-0337$01.00

Before the existence of charm was generally accepted, elementary particles were assumed to be constructed from three constituents or quarks: up (u), down (d), and strange (s). Their quantum numbers are summarized in Table 1. This three-quark model for elementary particles, first proposed independently by Gell-Mann and Zweig in 1964 (Gell-Mann 1964, Zweig 1964) proved enormously successful in explaining the spectroscopy of the known hadrons, as well as many of the features of their interactions. Until the early 1970s there was little experimental need for a fourth quark. But then in 1970 Glashow, Illiopoulos & Maiani (GIM) demonstrated that the inclusion of a new quark, the charmed quark (c), would solve a problem with the Weinberg-Salam model of the weak interaction—the nonobservation of the strangeness-changing neutral current (Aronson et al 1970, Carithers et al 1973, Clark et al 1971, Klems et al 1970). The existence of a fourth quark, of course, implied the existence of numerous new particles. Using the quark model convention of constructing mesons from quark-antiquark pairs, one could anticipate the existence of $c\bar{u}$, $c\bar{d}$, $c\bar{s}$, and $c\bar{c}$ mesons. The lowest-lying states of these first three mesons, dubbed the D^0 ($c\bar{u}$), D^+ ($c\bar{d}$), and F^+ ($c\bar{s}$), are the subject of this review. The family of mesons consisting of $c\bar{c}$ are known as the psions and their discovery in 1974 in conjunction with the discovery of charmed mesons in 1976 gave the first compelling evidence for the validity of the charm theory. (See, for example, Chinowsky 1977.)

We begin our review by summarizing the first experimental indications for the existence of charm as obtained from experiments in e^+e^- annihilation. This includes a brief discussion of the role of charm in the understanding of the ψ mesons, as well as the unraveling of the intricate structure present in the e^+e^- total hadronic cross section. Next we discuss the discovery of the D^0 and D^+, and detail those properties crucial to their identification as charmed particles. We then review the properties of the D^0 and D^+ learned through studies at the $\psi(3770)$ resonance. We summarize

Table 1   Quark quantum numbers

	u	d	s	c
Baryon number	1/3	1/3	1/3	1/3
Spin	1/2	1/2	1/2	1/2
Charge	+2/3	−1/3	−1/3	+2/3
Isospin	1/2	1/2	0	0
I_3	+1/2	−1/2	0	0
Strangeness	0	0	−1	0
Charm	0	0	0	1

the compelling evidence that indicates that this state decays nearly exclusively into $D\bar{D}$, which thus makes it particularly useful in establishing inclusive and exclusive D branching fractions.

Our discussion of branching fractions includes two particularly important D decay modes, $D^0 \to \pi^+\pi^-$ and $D^0 \to K^+K^-$. These processes are suppressed relative to $D^0 \to K^-\pi^+$ in the standard charm model, and thus serve as a critical test of that theory. This is followed by a discussion of the D semileptonic decay modes, which provide useful information on the D^0 and D^+ lifetimes.

Turning our attention to the data collected beyond the $\psi(3770)$, we discuss the properties and production mechanisms of the excited charmed mesons, the D^{*0} and the D^{*+} · D production just above the $\psi(3770)$ appears to be dominated by the three quasi-two-body processes, $e^+e^- \to D\bar{D}$, $D^*\bar{D}+\bar{D}^*D$, and $D^*\bar{D}^*$, in accordance with early theoretical predictions. The relative yield of each process, on the other hand, are somewhat surprising, and have led to considerable theoretical speculation. Finally, we summarize evidence for the existence of the F meson, which is as yet not on as solid a footing as the D^0, D^+ isodoublet.

There has been a great deal of theoretical and experimental work on charmed particles within the last decade. Because of the limitations of space and time, we confine our review to the experimental aspects of charmed meson production through e^+e^- annihilation. Hence we do not review the many important results on production of charmed mesons by neutrino, photon, and hadron beams, nor discuss the important results on charmed baryons. Our emphasis is entirely on experimental matters. The theoretical aspects of charm were recently reviewed in this series by Appelquist, Barnett & Lane (1978). When we found it necessary to reference theoretical works, we did so in the spirit of illustration and made no attempt to be exhaustive or judgmental.

2 THE e^+e^- ANNIHILATION PROCESS

2.1 QED and Nonresonant Annihilation

The electron-positron annihilation process has been studied using the colliding-beam technique since the early 1960s (Schwitters & Strauch 1976). The early motivation for such experiments was first the study of QED processes, such as $e^+e^- \to \mu^+\mu^-$, and later the study of $e^+e^- \to$ hadrons. Both processes are assumed to be dominated by the s-channel exchange of a virtual photon as illustrated in Figure 1. The cross section for the process $e^+e^- \to \mu^+\mu^-$ (neglecting the mass of the muon and radiative corrections) is $\sigma_{QED} = 4\pi\alpha^2/3s$ where $s = E_{cm}^2$. In analogy with Figure 1a we expect, as indicated schematically in Figure 1c, that the total hadronic cross section at high energies is

$$\sigma_{\text{had}} = \frac{4\pi\alpha^2}{3s} \sum Q_i^2$$

where Q_i are the contributing quark charges, and i ranges over those flavors with quark masses $< \frac{1}{2}E_{\text{cm}}$ and the three quark colors.

Below a center-of-mass energy of about 1.5 GeV, hadronic production by e^+e^- annihilation is dominated by the decays of vector mesons as shown in Figure 1b with sizable contributions from the ρ, ω, and φ. Many of the beautiful results obtained in this region are summarized by Perez-y-Jorba (1969).

Early data collected beyond the ρ, ω, and φ resonance region but below 3 GeV appeared to be consistent with the predictions of the naive point-like parton model, as shown in Figure 1c. The σ_{had} should scale in the limit of negligible quark masses (and small QCD corrections) as s^{-1}. The

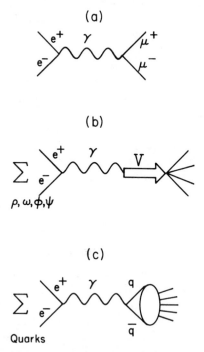

Figure 1 (a) The 1γ diagram for the QED process $e^+e^- \to \mu^+\mu^-$. Two types of 1γ contribution to $e^+e^- \to$ hadrons: (b) Hadronic decays of vector mesons. This should dominate when $s^{1/2}$ is near the mass of the vector mesons. (c) Hadronic contribution from the point-like production of quarks.

traditional way of expressing this scaling behavior evokes the famous ratio

$$R \equiv \frac{\sigma_{e^+e^- \to \text{hadrons}}}{\sigma_{\text{QED}}}.$$

As illustrated in Figure 1 the naive quark model prediction for this ratio is $R = \sum Q_i^2$.

Experimental measurements in the range $1.5 \lesssim E_{\text{cm}} \lesssim 3\,\text{GeV}$ taken prior to 1974 were consistent with values of R ranging from 2 to 3 (Richter 1974). Very recent data taken in this energy region by ADONE and DCI show consistency with $R = 2$ with considerably smaller error bars (summarized by Feldman 1978).

The value of $R = 2$ is precisely what one would expect in the original three-quark model (quark charges of $2/3$, $-1/3$, and $-1/3$) with the added color degree of freedom, whereby each quark is present in three colors.

As early as 1973, data accumulated at CEA indicated a rise in R to a value of 4.7 ± 1.1 nb at $s^{1/2} = 4\,\text{GeV}$ (Litke et al 1973, Tarnopolsky et al 1974). This rise in R was later corroborated by the SLAC-LBL collaboration working at SPEAR (Richter 1974). Although initial theoretical response was varied (Ellis 1974), one hypothesis for the rise in R was the opening up of new degrees of freedom such as the threshold of a new quark, or the production of a new heavy lepton (Perl 1980). We now believe there were indeed contributions to R from both effects.

2.2   The Discovery of the J/ψ

The J/ψ was simultaneously observed in p-Be collisions (Aubert 1974) and e^+e^- annihilations (Augustin et al 1974). In the hadronic production experiment the J appeared as a narrow enhancement in the electron pair invariant mass distribution in the reaction $p\,\text{Be} \to e^+e^-X$. This enhancement had a width of 20 MeV, which was consistent with experimental resolution. The ψ was observed in e^+e^- annihilations as an enhancement in the total hadronic and lepton pair cross section for center-of-mass energies near the ψ mass of 3.095 ± 0.002 GeV, with a measured width compatible with the e^+e^- experimental resolution of 2 MeV.

Because the ψ appeared as an s-channel resonance in e^+e^- annihilation, one could make a precision width measurement using $\int \sigma(E)\,dE$ (where E is the center-of-mass energy and σ is the total resonant cross section) and the measured muon pair branching ratio of $\sim 7\%$, even though the total width was much smaller than the center-of-mass resolution of a storage ring. This analysis yielded the remarkable result $\Gamma_\psi = 69 \pm 15$ keV. The properties of known vector mesons are summarized in Tables 2 and 2a. Although the ψ is over three times as massive as the conventional ρ, ω, and φ vector mesons, it is narrower by about two orders of magnitude.

Table 2 Resonance parameters for vector mesons[a]

State	Mass (MeV)	Total width Γ (MeV)	Partial width Γ_e (keV)	Branching fraction B_e	Reference
ρ	776 ± 3	155 ± 3	6.7 ± 0.8	$(4.3\pm0.5)\times10^{-5}$	b
ω	782.6 ± 0.3	10.1 ± 0.3	0.76 ± 0.17	$(7.6\pm1.7)\times10^{-5}$	b
φ	1019.6 ± 0.2	4.1 ± 0.2	1.31 ± 0.10	$(31\pm1)\times10^{-5}$	b
ψ	3095 ± 4	0.069 ± 0.015	4.8 ± 0.6	$(69\pm9)\times10^{-3}$	SLAC-LBL Mark I
ψ'	3684 ± 5	0.228 ± 0.056	2.1 ± 0.3	$(9.3\pm1.6)\times10^{-3}$	SLAC-LBL Mark I
ψ''	$\left\{\begin{array}{l}3772\pm6\\3770\pm6\\3764\pm5\end{array}\right.$	28 ± 5 / 24 ± 5 / 24 ± 5	0.35 ± 0.09 / 0.18 ± 0.06 / 0.28 ± 0.05	$(1.2\pm0.3)\times10^{-5}$ / $(0.7\pm0.2)\times10^{-5}$ / $(1.2\pm0.2)\times10^{-5}$	LGW / DELCO / Mark II
4.04[c]	4040 ± 10	52 ± 10	0.75 ± 0.10	$(1.4\pm0.4)\times10^{-5}$	DASP
4.16[c]	4159 ± 20	78 ± 10	0.77 ± 0.20	$(0.9\pm0.3)\times10^{-5}$	DASP
4.41	4414 ± 7	33 ± 10	0.44 ± 0.14	$(1.3\pm0.3)\times10^{-5}$	SLAC-LBL Mark I

[a] Other states have been reported between the φ and the ψ by experiments at Frascati and Orsay; we do not include them here.
[b] World averages compiled by the Particle Data Group (Bricman et al 1978).
[c] The SLAC-LBL and DELCO data do not separate this region into two states.

Table 2a New results on the Υ resonances

State	Mass (MeV)	Total width Γ (MeV)	Partial width Γ_e (keV)	Branching fraction B_μ	Reference
Υ(9.4)	9460 ± 10	~0.05	1.28 ± 0.27	$(2.1\pm1.4)\times10^{-2}$	DORIS[a]
	9433.1 ± 0.4[b]		1.15 ± 0.08	—	CLEO[c]
	9434.5 ± 0.4[b]				CUSB[d]
	9461.6 ± 0.5[e] }	$0.039\,^{+0.013}_{-0.008}$	1.31 ± 0.12	$(4.0\pm1.4)\times10^{-2}$	LENA[f,g]
	9463.0 ± 0.5[e] }		1.35 ± 0.11	$(2.9\pm1.3)\times10^{-2}$	DASP 2[g]
Υ'(10.0)	10015 ± 20		0.33 ± 0.14		DORIS[a]
	$M(Υ)+560.7\pm0.8$[h]		(0.23 ± 0.08)[i]		CLEO[j]
	9993.0 ± 1.0[b]		(0.39 ± 0.06)[i]		CUSB[d]
	$M(Υ)+553.6\pm1.7$[e]		(0.49 ± 0.09)[i]		LENA[g]
	$M(Υ)+557\pm2$[e]		(0.45 ± 0.05)[i]		DASP 2[g]
Υ''(10.3)	$M(Υ)+891.1\pm0.7$[k]		(0.31 ± 0.09)[i]		CLEO[j]
	10323.2 ± 0.7[b]		(0.32 ± 0.04)[i]		CUSB[d]
Υ'''(10.6)	$M(Υ)+1112\pm2$[l]	21.5 ± 5.7[m,n]	(0.21 ± 0.06)[i]		CLEO[c]
	$M(Υ)+1114\pm2$[l]	10.8 ± 0.9[n,o]	(0.25 ± 0.07)[i]		CUSB[p]

[a] Values for Υ are averages of PLUTO, DASP 2, and DESY-Heidelberg; and for Υ', DASP 2, and DESY-Heidelberg, as quoted by Meyer (1979).

[b] There is a ±30 MeV/c^2 systematic error including a 15 MeV/c^2 uncertainty due to the preliminary machine calibration at the Cornell Electron Storage Ring (CESR).

[c] Results from CESR quoted at the 1980 Experimental Meson Spectroscopy Conference at Brookhaven National Laboratory (Bebek 1980).

[d] Columbia University–Stony Brook experiment at CESR (Böhringer et al 1980).

[e] There is a ±10 MeV uncertainty in the DORIS ring energy calibration.

[f] LENA is a new NaI-Pb glass detector at DORIS.

[g] Results from DORIS quoted at the 1980 Experimental Meson Spectroscopy Conference at Brookhaven National Laboratory (Schröder 1980).

[h] In addition to ±3 MeV/c^2 systematic error.

[i] In units of $\Gamma_{ee}(Υ)$.

[j] CLEO experiment at CESR (Andrews et al 1980).

[k] In addition to a ±5 MeV/c^2 systematic error.

[l] In addition to a ±4 MeV/c^2 systematic error.

[m] The CESR beam spread has been unfolded from this width using a Breit-Wigner form for the resonance.

[n] This is a finite width resonance, and could well be the counterpart for the "B mesons" of what the ψ'' is for charmed mesons.

[o] The CESR beam spread has been folded from this width using a Gaussian form for the resonance.

[p] Results from CESR quoted at the 1980 Experimental Meson Spectroscopy Conference at Brookhaven National Laboratory (Lee-Franzini 1980, Finocchiaro 1980).

Because of the anomalous ψ width, it was historically important to establish that the ψ was indeed a vector meson as implied by its s-channel production in e^+e^- annihilation. This prejudice was partially borne out through the observations of psi-photon interference in the process $e^+e^- \rightarrow \mu^+\mu^-$, which established the ψ quantum numbers as $J^{PC}(\psi) = 1^{--}$ (Boyarski et al 1975). In addition, data on the photoproduction of the ψ (Knapp et al 1975, Camerini et al 1975, Gittleman et al 1975) demonstrated that the ψ-nucleon total cross section was ~ 1 mb—the same order of magnitude as for the conventional vector mesons.

A search for additional narrow vector mesons formed in e^+e^- annihilation found the $\psi(3684)$ or ψ' with a width of 228 keV, the $\psi(4400)$ with a width of 33 ± 10 MeV, and, considerably later, the $\psi(3770)$ with a width of 25 MeV.

As seen in Figure 2 these resonances occur in the region where R makes a transition from a value of ~ 2.5 to a plateau around 4 to 5.

2.2.1 THE OZI RULE AND THE ψ The discovery of the new narrow vector meson states encouraged the proponents of a new quark. In fact such states were being predicted by charm enthusiasts concurrently with the experimental observation of the ψ (Appelquist et al 1978). Within the charm picture the new vector mesons were assumed to be constructed from ^3S or ^3D states of $c\bar{c}$. Many such vector mesons can be constructed in the theory by placing the new quarks in various levels of radial excitation. The narrow width of the ψ could be explained as a manifestation of the phenomenological OZI rule (Okubo 1963, Zweig 1964, Iizuka 1966), which was previously proposed to explain the suppression of certain strong interaction processes that can only occur via "disconnected" quark diagrams.

A disconnected diagram allows at least one external particle to be isolated by making a cut that does not intersect any quark line. For example, Figure 3 demonstrates that $\varphi \rightarrow \pi^+\pi^-\pi^0$ proceeds via a disconnected or OZI-suppressed diagram, whereas $\varphi \rightarrow K^+K^-$ does not. Experimentally one does find that the decay $\varphi \rightarrow K\bar{K}$ is preferred over $\varphi \rightarrow \pi^+\pi^-\pi^0$, even though the former process has a much smaller phase space than the latter.

Using this rule one could explain the narrow widths of the ψ and ψ' by asserting that they had masses below the threshold for the pair production of the lowest-lying states (presumably D's) containing the new quark. If this were true, all decay diagrams for the ψ and ψ' would be disconnected and hence suppressed.

The dramatic increase in the width of the $\psi(4400)$ and, as determined much later, the $\psi(3770)$, would then represent the opening up of the $D\bar{D}$ decay channel. Hence one would expect the relation

$$\tfrac{1}{2}M_{\psi'} < M_D < \tfrac{1}{2}M_{\psi(3770)}$$

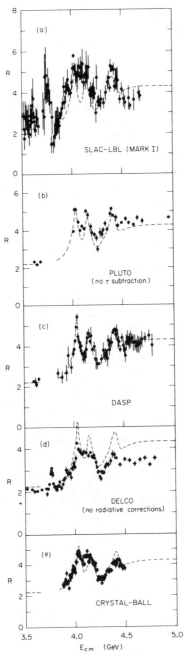

Figure 2 The observed values of *R* near charm threshold. All data are corrected for radiative effects and contamination from the τ, unless otherwise indicated. The hand-drawn curve, which follows the DASP points, has been applied to the other measurements in order to facilitate a comparison. Systematic errors, which range from 10 to 15%, are not included in the error bars. (Summary by Kirkby 1979.)

or $1.842 < M_D < 1.885$ GeV, with the upper bound coming through considerable hindsight.

2.2.2 PROPERTIES OF THE ψ If the psion family were comprised of $q\bar{q}$ states of a new quark, one would expect the presence of additional relatively narrow but strongly decaying states with $J^{PC} \neq 1^{--}$. These states would be constructed from p-wave or other excitations of the new quark. Several such additional states have been observed via radiative decays from the ψ'. The beautiful phenomenology associated with the many $c\bar{c}$ states below the charm threshold was reviewed by Chinowsky (1977) and hence is not discussed here. Suffice it to say that it is indeed difficult to account for the multitude of new narrow states produced in e^+e^- without accepting the presence of a new quark.

Additional evidence for this interpretation of the new mesons comes from a study of the ψ decay modes. We have previously noted that the OZI explanation for the narrow width of the ψ and ψ' implies the existence of a fourth quark. The large mass of the ψ suggests that this fourth quark is much heavier than the conventional three quarks, and hence if the new states consisted of quark-antiquark combinations of the new quark they must be isosinglet states. A study of pion multiplicities in the decays of the ψ demonstrated that it has odd G-parity and thus (in light of its odd C-parity) the ψ has even isospin (Jean-Marie et al 1976). Even isospin, coupled with the observation of a substantial $\psi \to p\bar{p}$ decay mode, proves that the ψ is an isosinglet as expected. Corroborating evidence comes from the observation of a $\Lambda\bar{\Lambda}$ decay mode (Peruzzi et al 1978).

To summarize, the observation and properties of the new mesons discovered in e^+e^- annihilation provide a vast body of data which is easily accommodated into the framework of a new, heavy quark. However, there is little to link this quark to the charm quark of the GIM model from a study of the ψ family alone. The definitive proof of the charm hypothesis comes from the observation of states of nonzero charm—the D^0 and D^+. We turn now to a brief discussion of the properties one would expect in the GIM model for these mesons.

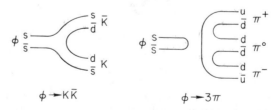

Figure 3   Illustration of OZI-allowed ($\varphi \to K\bar{K}$) and -forbidden ($\varphi \to \pi\pi\pi$) decays of the φ meson.

3 CHARM

3.1 *The GIM Model*

Several excellent and comprehensive reviews of the theoretical aspects of charm have been written since the seminal review of Gaillard, Lee & Rosner (1975). In this section we outline those aspects of the theory necessary for the understanding of the experimental material to follow.

As previously stated, the charmed quark was partly motivated to provide a means of cancelling first-order strangeness-changing neutral current effects such as $K_L \to \mu^+ \mu^-$ and a large weak $K_S - K_L$ mass difference. In Figure 4 we show a diagram for these processes within the three-quark theory. Although these diagrams appear to contribute to the second-order weak interaction, the loop integral enhances their strength, making them comparable to a first-order diagram. The factors appearing at the diagram vertices follow from the form of the precharm weak hadronic current:

$$J_h^\mu = \bar{u}\gamma^\mu(1 - \gamma^5)(d \cos \theta_c + s \sin \theta_c), \qquad 1.$$

where θ_c is the Cabibbo angle, introduced in 1963 to relate the strength of

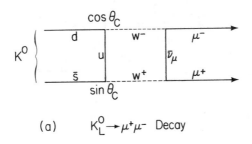

(a) $K_L^0 \to \mu^+ \mu^-$ Decay

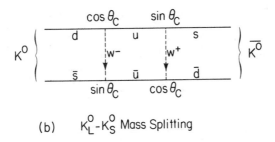

(b) K_L^0-K_S^0 Mass Splitting

Figure 4   Quark diagrams for the decay $K_L^0 \to \mu^+ \mu^-$ and the K_L^0/K_S^0 mass splitting with only u, d, and s quarks.

strangeness-conserving weak current to the strangeness-changing weak current. Present measurements indicate that the Cabibbo angle is small ($\theta_c = 13.2 \pm 0.5°$).

Figure 5 shows alternative diagrams for the processes employing the charmed quark in place of the up quark. These diagrams will tend to cancel the diagrams of Figure 4 provided the new vertex factors are as shown —that is, if Equation 1 is modified to

$$J_h^\mu = (\bar{u}\bar{c})\gamma^\mu(1-\gamma^5)\begin{pmatrix} \cos\theta_c & \sin\theta_c \\ -\sin\theta_c & \cos\theta_c \end{pmatrix}\begin{pmatrix} d \\ s \end{pmatrix}. \qquad 2.$$

Equation 2, employing two quark doublets and a unitary mixing matrix, is the hadronic current of the GIM model (Glashow et al 1970).

3.2 *Predicted Decays of Charmed Mesons*

Using Equation 2 and the bilinear weak Lagrangian $L = J^{\dagger\mu}J_\mu + J^\mu J_\mu^\dagger$, one obtains the results listed in Tables 3 and 4 for nonleptonic and semileptonic decays of charmed mesons. As illustrated in the tables, one expects a wide

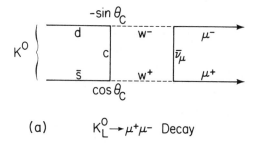

(a) $K_L^0 \rightarrow \mu^+\mu^-$ Decay

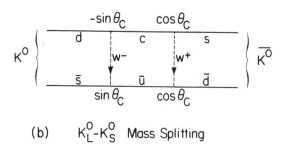

(b) K_L^0-K_S^0 Mass Splitting

Figure 5 Cancelling quark diagrams for the decay $K_L^0 \rightarrow \mu^+\mu^-$ and the K_L^0/K_S^0 mass splitting in the GIM model.

Table 3 Nonleptonic decays of charmed mesons in the GIM model

Cabibbo factors	Flavor structure	Selection rules	D decay examples	F decay examples
$\cos^4 \theta_c \cong 0.90$	$\bar{c}su\bar{d}$	$\Delta c = \Delta s$	$D \to K^- m\pi$	$F^+ \to m\pi, K^+K^-\pi$
$\sin^2 \theta_c \cos^2 \theta_c \cong 0.05$	$\bar{c}du\bar{d}$ and $\bar{c}su\bar{s}$	$\Delta s = 0$	$D \to m\pi, K^-K^+m\pi$	$F^+ \to K^+ m\pi$
$\sin^4 \theta_c \cong 0.0025$	$\bar{c}du\bar{s}$	$\Delta c = -\Delta s$	$D \to K^+ m\pi$	$F^+ \to K^+K^+m\pi$

Table 4 Semileptonic decays of charmed mesons

Cabibbo factors	Flavor structure	Selection rules	D examples	F^+ examples
$\cos^2 \theta_c \cong 0.95$	$\bar{c}s$	$\Delta c = \Delta s = \Delta q^a$	$D \to K^- e^+ \nu m\pi$	$F^+ \to m\pi e^+ \nu,$ $\to K^+K^- e^+ \nu$
$\sin^2 \theta_c \cong 0.05$	$\bar{c}d$	$\Delta c = \Delta q, \Delta s = 0$	$D \to m\pi e^+ \nu$	$F^+ \to Ke^+ \nu$

[a] The term Δq is the change in charge of the hadronic system

disparity in rates between Cabibbo-favored decays such as $D^0 \to K^- \pi^+$ or $D^0 \to K^- ev$, singly suppressed decays such as $D^0 \to \pi^+ \pi^-$ or $D^0 \to \pi ev$, and doubly suppressed decays such as $D^0 \to K^+ \pi^-$, owing to the smallness of the Cabibbo angle. We compare the predictions of Tables 3 and 4 to the data in Section 4.

3.3   Beyond Charm

Since the 1977 discovery of the Υ by Lederman and collaborators (Herb et al 1977), considerable indirect evidence has accumulated for the existence of another quark, the b quark, which is much more massive (~ 5 GeV) than the c quark. The simplest extension of the Weinberg-Salam-GIM model involves the introduction of a new quark doublet (t, b) where the b quark is the lighter quark with charge $-1/3$ responsible for the Υ, and the t quark is the heavier quark of charge $2/3$ whose presence is still speculation.

The model of Kobayashi & Maskawa (1973) incorporates this new doublet into the weak current by proposing that the u, c, t quarks mix with the d, s, b quarks via a general 3×3 unitary mixing matrix. Hence

$$J_h^\mu = (\bar{u}\bar{c}\bar{t})\gamma^\mu(1-\gamma^5)M \begin{pmatrix} d \\ s \\ b \end{pmatrix},$$

where

$$M = \begin{pmatrix} C_1 & S_1 C_3 & S_1 S_3 \\ -S_1 C_2 & C_1 C_2 C_3 - S_2 S_3 e^{i\delta} & C_1 C_2 S_3 + S_2 C_3 e^{i\delta} \\ S_1 S_2 & -C_1 S_2 C_3 - C_2 S_3 e^{i\delta} & -C_1 S_2 S_3 + C_2 C_3 e^{i\delta} \end{pmatrix}$$

and

$$C_i = \cos\theta_i, \quad S_i = \sin\theta_i, \quad i = 1, 2, 3.$$

Within this picture the single Cabibbo angle present in Equation 2 (now called θ_1) is augmented by two new angles θ_2 and θ_3 and a phase factor ($e^{i\delta}$). We note that the original GIM predictions for the weak transitions between the u, d, s, and c quarks are recovered in the limit $\theta_2, \theta_3 \to 0$.

Nonzero values for the new mixing angles will affect the weak decays of conventional particles as well as the decays of charmed particles. In particular, we see that the Cabibbo-suppressed processes $D^0 \to \pi^+ \pi^-$ and $D^0 \to K^+ K^-$ will no longer have identical mixing-angle factors as was the case in the GIM model. Within the sector of older phenomena, we see that the predictions for processes involving $u \rightleftarrows d$ transitions such as neutron beta decay still involve only the original Cabibbo angle θ_1. We can thus retain the value $\theta_1 = 13.2 \pm 0.5°$ as measured by comparing the rate of $n \to pe\bar{v}$ to that for $\mu \to ev\bar{v}$ (Nagels et al 1976).

The amplitude for u $\rightleftarrows$ s transitions, however, acquires a new mixing factor to become $\sin \theta_1 \cos \theta_3$. The success of the original Cabibbo model applied to processes such as $\Lambda \to pe\bar{\nu}$ and $K \to \pi e \nu$ thus limits the Kobayashi-Maskawa correction to $|\cos \theta_3| > 0.87$ (Schrock & Wang 1978). Less direct theoretical arguments based on the possible contributions of the t quark to the $K_L - K_S$ mass difference set a limit of $\theta_2 < 30°$ (Harari 1977). These limits, when applied to the D decays could cause deviations from the GIM predictions by factors of two, although the basic pattern of enhanced vs suppressed decays would be expected to hold.

4 D MESONS

4.1 Discovery of D Mesons

D mesons were first observed by the SLAC-LBL Mark I collaboration at SPEAR in 1976 using data collected at center-of-mass energies ranging from 3.9 to 4.6 GeV (Goldhaber et al 1976, Peruzzi et al 1976). The first D decay modes observed included $D^0(1863) \to K^-\pi^+$, $K^-\pi^+\pi^+\pi^-$ and $D^+(1868) \to K^-\pi^+\pi^+$. A substantial body of evidence soon accumulated linking the narrow enhancements in the $K^-\pi^+$, $K^-\pi^+\pi^+\pi^-$, and $K^-\pi^+\pi^+$ invariant mass distributions to the D^0, D^+ charmed isodoublet. We outline the evidence below.

4.1.1 EVIDENCE FOR ASSOCIATED PRODUCTION   Both the $D^0(1863)$ and $D^+(1868)$ are produced in final states containing a $D\bar{D}$ pair, as one would expect for particles containing a quantum number conserved by the electromagnetic interaction. This is evidenced by two observations: (a) No D's are observed in e^+e^- annihilation at either the ψ or ψ', although a substantial amount of Mark I data was collected at these resonances. The ψ' in particular is located just below $D\bar{D}$ threshold. (b) No evidence is seen in recoil mass spectra against the D^0 or D^+ system for events with recoil masses smaller than the D candidate mass of 1863 MeV.

4.1.2 EVIDENCE FOR WEAK HADRONIC DECAYS   Particles carrying a quantum number conserved in the strong or electromagnetic interaction must decay weakly. This is evidenced by five observations:

Narrow width   All reported sightings of the D^0 and D^+ into inclusive decay modes report an observed width that is consistent with experimental mass resolutions. The data with the best mass resolution set a limit $\Gamma_{D^0,D^+} < 2$ MeV.

Parity violation   The observation of parity violation in the decays $D^0 \to K^-\pi^+$ and $D^+ \to K^-\pi^+\pi^+$ is reminiscent of the $\theta-\tau$ problem for K decays of the 1950s, which led to the hypothesis of a parity-violating weak interaction. Because of the small mass difference between the $D^0(1863)$ and $D^+(1868)$ it is natural to assume that they are members of the same isodoublet and hence have the same spin and parity. The $D^0 \rightarrow K\pi$ decay final state must have a natural spin parity of $0^+, 1^-, 2^+, \ldots$. A study of the $D^+ \rightarrow K^-\pi^+\pi^+$ Dalitz plot (Wiss et al 1976) rules out D^+ final state spin-parity assignments of $J^P = 1^-$ and 2^+, while 0^+ is forbidden by angular momenta considerations for three pseudoscalars. Hence, neglecting possible higher spin assignments for the D system, one is left with a contradiction most naturally resolved by assuming that the D^0 and D^+ decay through the parity-violating weak decay.

An exotic final state   Because of the $\Delta s = \Delta c$ selection rule for the Cabibbo-favored hadronic decays of charmed particles, the D^+ must always decay hadronically into final states of positive charge and negative strangeness. Within the context of the conventional three-quark model such final states are labeled as exotic because they cannot be constructed from quark-antiquark pairs using the u, d, or s quarks. If the $K^-\pi^+\pi^+$ enhancement at 1863 MeV that we implicitly associate with the weak decay $D^+ \rightarrow K^-\pi^+\pi^+$ actually represented the strong decay of a noncharmed meson, that meson would be exotic and would be the first compelling observation of such a state. The observation of the $I_z = 3/2$ $K^-\pi^+\pi^+$ enhancement when combined with the nonobservation of an enhancement in $K^-\pi^+\pi^-$, the $I_z = -1/2$ brother, rules out such an interpretation, however.

Semileptonic decay   The observation of an appreciable semileptonic branching ratio, as discussed below, again suggests that D's do not decay strongly.

Evidence for a GIM pattern of decays   The Cabibbo-suppressed decay modes $D^0 \rightarrow K^-K^+$ and $\pi^+\pi^-$ have recently been observed in the SLAC-LBL Mark II detector at SPEAR. The dominant two-body decay mode, however, is $D^0 \rightarrow K^-\pi^+$. Hence the D^0 is observed to decay into both strange and nonstrange final states, which implies that the decay mechanisms do not conserve strangeness and are thus weak. This pattern of decay modes is characteristic of charm in the GIM model.

Much of the information presented above linking the early $(K\pi)^0$, $(K3\pi)^0$, and $(K2\pi)^+$ signals to the D^0D^+ charmed doublet was performed at the Mark I detector using data collected from 3.9–4.6 GeV with an emphasis on the 4.028-GeV resonance region. Although charmed mesons are copiously produced near 4.028 GeV, the considerable structure in the total cross section near this enhancement precludes a clean Breit-Wigner fit to determine the cross section beneath the peak. Such information would have been useful, for, as we discuss below, it provides a means of measuring the absolute branching fractions for the D decay modes.

4.2 Hadronic Decays of D Mesons

4.2.1 THE $\psi(3770)$ RESONANCE   After the Mark I detector at SPEAR stopped taking data, the $\psi(3770)$ or ψ'' resonance was discovered in the lead glass wall (LGW) and direct electron counter (DELCO) experiments (Rapidis et al 1977, Bacino et al 1978). Comparing the width of the ψ' ($\Gamma = 228$ keV) with that of the ψ'' ($\Gamma = 25$ MeV) we note that the effect of the OZI suppression at the ψ' is no longer present at the ψ'', since it lies above $D\bar{D}$ threshold. Furthermore the ψ'' lies below DD* threshold; it can only decay into D's via the process $D^0\bar{D}^0$ or D^+D^-.

The ψ'' has been studied extensively in the LGW and DELCO experiments and recently again in the SLAC-LBL Mark II experiments (Lüth 1979, Schindler et al 1980). Figure 6 shows the R distribution

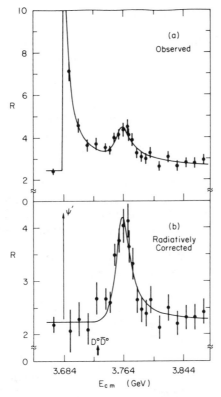

Figure 6   The value of $R \equiv \sigma_{had}/\sigma_{QED}$ in the vicinity of the ψ'' obtained by the Mark II collaboration (Schindler et al 1980); (a) before and (b) after radiative correction using the technique of Jackson & Scharre (1975). The curve is a fit of the data to Equation 3.

observed in the Mark II experiment, where R is the ratio of the observed hadronic cross section to the theoretical QED μ pair cross section σ_{QED}. The latter is obtained from calibration against observed Bhabha pairs. Figure 6a gives the R distribution where the $\tau^+\tau^-$ cross section has been subtracted. Figure 6b gives the R distribution after the radiative tails from the J/ψ and ψ' have been subtracted as well. The errors shown are statistical. The resonance is fitted to a p-wave Breit-Wigner expression (Barbaro-Galtieri 1968) with an energy-dependent total width $\Gamma_{tot}(E_{cm})$ that takes account of closeness to the different $D^0\bar{D}^0$ and D^+D^- thresholds. The explicit fitting function employed is

$$R(E_{cm}) = \frac{1}{\sigma_{QED}} \frac{3\pi}{M^2} \frac{\Gamma_{ee}\Gamma_{tot}(E_{cm})}{(E_{cm}-M)^2 + \frac{1}{4}\Gamma_{tot}^2(E_{cm})} \qquad 3.$$

and

$$\Gamma_{tot}(E_{cm}) \propto \frac{p_+^3}{1+(rp_+)^2} + \frac{p_0^3}{1+(rp_0)^2}$$

where p_+ (p_0) is the momentum of the pair-produced D^+ (D^0) and r the interaction radius. The quantities M (the resonance mass) and Γ_{ee} (the partial width to electrons) were additional free parameters. The fit is not sensitive to r, which was fixed at 2.5 Fermi. Table 5 summarizes the results of this fit for the various experiments. We note that the Mark II results are consistent with those of DELCO and the LGW except for a shift in the central mass that is 6–8 MeV lower than previous values. In addition the Mark II value for the width of the decay into e^+e^- of 276 ± 50 eV lies between the earlier reported values.

From theoretical arguments (Eichten et al 1975, Lane & Eichten 1976, Gottfried 1978) the ψ'' is believed to be a $^3D_1$ state of charmonium which is, however, mixed with the 2^3S_1 state, dominant in the ψ'. The relatively large Γ_{ee} value gives an estimate for this mixing angle of $20.3\pm2.8°$.

4.2.2 CHARMED-MESON BRANCHING RATIOS   Without knowledge of the total D production cross section it is difficult to measure the branching ratio

Table 5   Measurements of the $\psi(3770)$ resonance parameters

Experiment	Mass (MeV/c^2)	Γ_{tot} (MeV)	Γ_{ee} (eV)	ΔM[a] (MeV/c^2)
DELCO	3770 ± 6	24 ± 5	180 ± 60	86 ± 2
LGW	3772 ± 6	28 ± 5	345 ± 85	88 ± 3
Mark II	3764 ± 5	24 ± 5	276 ± 50	80 ± 2

[a] ΔM is the mass difference between the $\psi(3684)$ and $\psi(3770)$.

for a given exclusive final state. One can, however, readily measure the product $\sigma \cdot B$ by counting the number of events observed in a given channel and dividing by the acceptance and luminosity. Charm production at the ψ'' offers the considerable advantage that $\sigma(D\bar{D})$ can be determined if one is willing to assume two things: (a) The ψ'' is a state of definite isospin (0 or 1); this allows a prediction of the D^0/D^+ production ratio, namely $\sigma(D^0)/\sigma(D^+) \simeq p_0^3/p_+^3$ as expected for p-wave production. This reflects the D^0 and D^+ mass difference (1863.3 MeV and 1868.3 MeV respectively). (b) The ψ'' decays nearly entirely into $D\bar{D}$ ($\sim 99\%$). This is based on the $\Gamma_{tot}(\psi')$ to $\Gamma_{tot}(\psi'')$ ratio ($\sim 1/100$); i.e., the OZI-suppressed portion of the ψ'' decay width is of the same magnitude as the $\Gamma_{tot}(\psi')$.

Assumption (b) has been tested with limited statistical accuracy using events where both members of a $D\bar{D}$ pair are observed decaying into exclusive final states. The D branching ratio obtained by this technique can be used to compute the absolute D production cross section at the ψ'' (Schindler 1979).

Using these assumptions and their fit to the radiatively corrected ψ'' resonance shape, the LGW collaboration determined that the $D^0\bar{D}^0$ and D^+D^- cross sections averaged over their particular set of ψ'' running energies were 11.5 ± 2.5 nb and 9.1 ± 2.0 nb, respectively (Peruzzi et al 1977). Most of this running was done within 2 MeV of their nominal ψ'' mass. The Mark II collaboration, on the other hand, collected 49,000 hadronic events at an energy 7 MeV above their nominal ψ'' mass. They find that the total $D^0\bar{D}^0$ and D^+D^- cross sections for their running conditions are 8.0 ± 1.0 (± 1.2) nb and 6.0 ± 0.7 (± 1.0) nb, respectively (Schindler 1979), where the numbers in parentheses are the systematic error estimates.

The beam-constrained mass distributions obtained for several D^0 and D^+ final states are shown in Figure 7 for the LGW data and Figures 8–10 for the Mark II data. The beam-constrained mass M_b is calculated using the relationship

$$M_b = (E_b^2 - p^2)^{1/2},$$

where E_b is the storage-ring single-beam energy and p is the momentum of the D^0, D^+ candidate as determined by the magnetic detector. Such a technique implicitly assumes D^0's and D^+'s are pair produced at the ψ'' and hence have exactly half the total center-of-mass energy. In the spirit of this assumption, some background is eliminated by histograming only events with a detector-measured energy within 50 MeV of the beam energy. Use of the beam-constrained mass offers unparalleled resolution and background rejection.

Fits to the signals shown in the above figures have been used to determine $\sigma \cdot B$ and B for various decay modes. This information is summarized in Table 6.

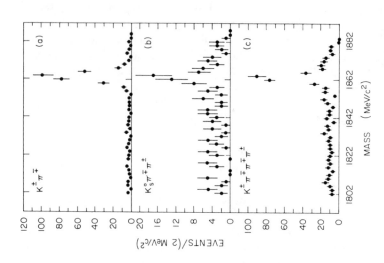

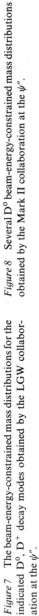

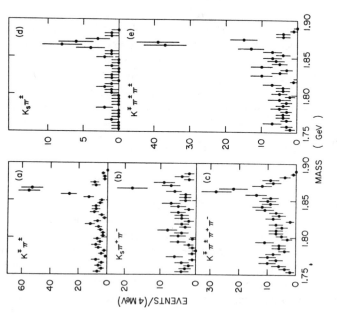

Figure 7   The beam-energy-constrained mass distributions for the indicated D^0, D^+ decay modes obtained by the LGW collaboration at the ψ''.

Figure 8   Several D^0 beam-energy-constrained mass distributions obtained by the Mark II collaboration at the ψ''.

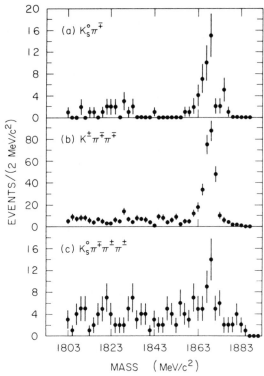

Figure 9 Several D^+ beam-energy-constrained mass distributions obtained by the Mark II collaboration at the ψ''.

Table 6 D meson branching ratios

Mode	LGW B (%)	Mark II B (%)
$K^-\pi^+$	2.2 ± 0.6	3.0 ± 0.6
$\bar{K}^0\pi^0$	—	2.2 ± 1.1
$\bar{K}^0\pi^+\pi^-$	4.0 ± 1.3	3.8 ± 1.2
$K^-\pi^+\pi^0$	12.0 ± 6.0	8.5 ± 3.2
$K^-\pi^+\pi^+\pi^-$	3.2 ± 1.1	8.5 ± 2.1
$\pi^+\pi^-$	—	0.09 ± 0.04
K^+K^-	—	0.31 ± 0.09
$\bar{K}^0\pi^+$	1.5 ± 0.6	2.3 ± 0.7
$K^-\pi^+\pi^+$	3.9 ± 1.0	6.3 ± 1.5
$\bar{K}^0\pi^+\pi^0$	—	12.9 ± 8.4
$\bar{K}^0\pi^+\pi^+\pi^-$	—	8.4 ± 3.5
$K^-\pi^+\pi^+\pi^+\pi^-$	—	$<4.1^a$
$\bar{K}^0K^+$	—	0.5 ± 0.27

^{a} 90% confidence limit.

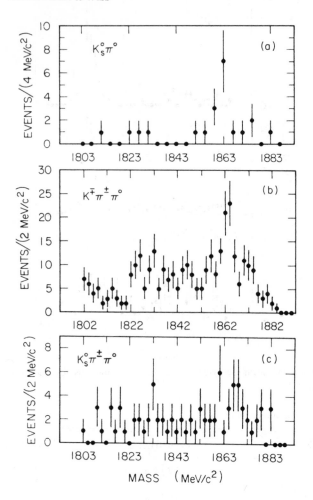

Figure 10   Several D beam-energy-constrained mass distributions for decays involving π^0's obtained by the Mark II collaboration at the ψ''.

4.2.3 CABIBBO-SUPPRESSED DECAY MODES   An intrinsic part of the GIM mechanism for charm is the prediction that in addition to the principal (Cabibbo-favored) D decay modes, which lead to K^- or $\bar{K}^0$ in the final states, there are also Cabibbo-suppressed modes producing zero strangeness final states. The Cabibbo-favored and -suppressed modes for D^0 two-particle final states are illustrated by the quark diagrams in Figure 11, where θ_A is the familiar Cabibbo angle and θ_B is a new angle that in the four-quark model is associated with the flavor mixing of charmed quarks. The

GIM assumption is that $\theta_A = \theta_B$. Experimentally one can independently measure these angles using

$$\tan^2 \theta_A = \frac{\Gamma(D^0 \to K^- K^+)}{\Gamma(D^0 \to K^- \pi^+)} \quad \text{and} \quad \tan^2 \theta_B = \frac{\Gamma(D^0 \to \pi^- \pi^+)}{\Gamma(D^0 \to K^- \pi^+)}.$$

The above expressions neglect the phase space corrections due to the K, π mass difference, which will raise the $\pi^+ \pi^-$ rate by 7% and lower the $K^+ K^-$ rate by 8%.

Figure 12 shows the Mark II $\pi^- \pi^+$, $K^- \pi^+$, and $K^- K^+$ invariant mass distributions for two-particle combinations with momentum within 30 MeV/c of the expected D pair momentum at the ψ''. Aside from the signals in the three channels at the D mass one notes kinematic reflections shifted by about ± 120 MeV/c^2 from the D mass due to $\pi \leftrightarrow K$ misidentifications. A fit to the data yields 235 ± 16 $K^\mp \pi^\pm$ events, 22 ± 5 $K^+ K^-$ events, and 9 ± 3.9 $\pi^+ \pi^-$ events. After correcting for the relative efficiencies one obtains

$$\frac{\Gamma(D^0 \to K^- K^+)}{\Gamma(D^0 \to K^- \pi^+)} = 0.113 \pm 0.03 \qquad\qquad 4.$$

and

$$\frac{\Gamma(D^0 \to \pi^- \pi^+)}{\Gamma(D^0 \to K^- \pi^+)} = 0.033 \pm 0.015, \qquad\qquad 5.$$

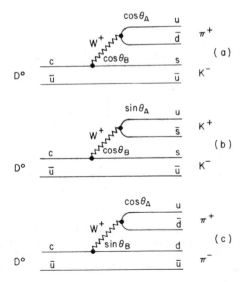

Figure 11   Quark diagrams for D^0 decays into two charged particles.

where the quoted errors include systematic effects. The results clearly demonstrate the existence of the Cabibbo-suppressed decay modes of roughly the expected magnitude, $\tan^2 \theta \simeq 0.05$, although the $\pi\pi$ ratio is lower by about one standard deviation and the KK ratio is higher by about two standard deviations.

We note that the discrepancy between Equation 4 and the pre-charm measurements of the Cabibbo angle cannot be explained by the presence of additional mixing angles in the Kobayashi-Maskawa model. Hence the discrepancy (if not statistical in origin) implies a violation of SU(3) invariance due to unknown dynamical effects. It is thus premature to use Equation 5 as a measure of the "new" Cabibbo angle (Abrams et al 1979).

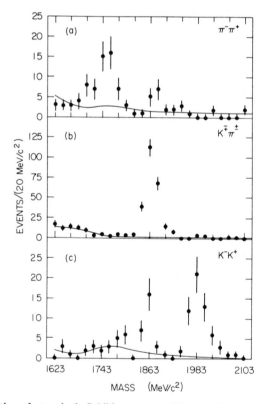

Figure 12 Evidence for two-body Cabibbo-suppressed decay modes obtained by the Mark II collaboration at the ψ''. Candidates are required to have momenta within 30 MeV of that expected for D pair production. Both (a) and (c) show prominent reflection peaks not centered at the D^0 mass because of π/K misidentification by the time-of-flight system. The solid curves are background estimates.

4.3   Semileptonic Decay Modes

The prompt leptons created in the semileptonic decay of charmed mesons have been used to estimate the rate for hadronic charm production, as well as to trigger detectors looking for the decay of charmed particles into exclusive hadronic final states. Because it is possible to estimate theoretically the D semileptonic width [i.e. $\Gamma(D \to Ke\nu)$], a measurement of the semileptonic branching ratio can be used to estimate the D lifetime. At the time of this writing, emulsion as well as precision bubble and streamer chamber lifetime studies are being performed. These experiments can observe D's traveling finite distances before decaying, directly measure their lifetimes, and thus check current theoretical ideas on D semileptonic decays (Voyvodic 1979).

Because D semileptonic decays produce a neutrino in the final state, their study in e^+e^- annihilation invariably requires the measurement of inclusive final state electron or muon rates at center-of-mass energies where charmed mesons are known to be copiously produced. Inclusive electrons rather than muons are generally studied since the charged lepton in $D \to \bar{K}e^+\nu$ or $D \to \bar{K}\mu^+\nu$ has a momentum spectrum that peaks near 500 MeV for D's produced near threshold. Hence the muons would be produced with momenta too low to be separated clearly from pions using conventional hadron filters.

4.3.1   THE AVERAGE SEMILEPTONIC BRANCHING RATIO   Several basic techniques for extracting the D semileptonic branching ratio have been discussed in the literature. The simplest method (experimentally) estimates the branching fraction $B(D \to eX) \equiv \Gamma(D \to eX)/\Gamma(D \to all)$ from the ratio of the total electron to the total charmed meson inclusive cross section:

$$B(D \to eX) = \frac{\sigma(e^+e^- \to e^{\pm} + \text{hadrons})}{\sigma(e^+e^- \to DX)}. \qquad 6.$$

This method, the first historically (Braunschweig et al 1976, Burmester et al 1976), continues to provide the best statistical information on the D semileptonic branching ratio. It suffers from inevitable systematic problems, however. In evaluating the numerator, one must be careful to exclude (or correct for) contributions from heavy lepton decays or electromagnetic processes. The most common way of minimizing this contamination is to demand that the electrons be accompanied by two or more charged hadrons. Since the τ heavy lepton is known to decay predominantly into final states containing leptons and single hadrons, the multihadronic background due to $e^+e^- \to \tau^+\tau^-$ production are expected to lie at about the 25% level (Perl 1980, Barbaro-Galtieri 1978) and can be subtracted using the measured τ branching fraction and computable production cross sections.

Several techniques, all subject to various systematic uncertainties, have been used to estimate the denominator of Equation 6. Extraction of $\sigma(e^+e^- \rightarrow DX)$ is relatively straightforward at the ψ'' since, as discussed earlier, it is natural to assume that this resonance decays exclusively into D's via

$$\psi'' \rightarrow (56 \pm 3)\% \ D^0\bar{D}^0 \rightarrow (44 \pm 3)\% \ D^+D^-. \qquad\qquad 7.$$

The fractions used in Equation 7 follow from the assumption of equal D^0, D^+ production corrected for p-wave threshold factors. Since it is impossible to separate the semileptonic decays of the D^0 from those of the D^+ without using tagged events or measuring the dielectron rate as well as the single electron rate, one in effect measures a weighted average of the D^0 and D^+ semileptonic branching ratios with weights given by Equation 7.

At center-of-mass energies above the ψ'' it is sometimes possible to extract $\sigma(e^+e^- \rightarrow DX)$ by counting hadronic D decays into exclusive final states such as $D^0 \rightarrow K^-\pi^+$, $D^+ \rightarrow K^-\pi^+\pi^+$ and dividing by the hadronic branching ratios measured for these states at the ψ''. When this is not possible because of statistical limitations, or in data predating the ψ'', the D inclusive cross section can be estimated at a given center-of-mass energy via

$$\sigma(e^+e^- \rightarrow DX) = (R - R_{old}) \times \sigma_{QED},$$

where R is the τ-lepton-corrected ratio of hadrons to μ pairs, and R_{old} is the value of R below the $\psi(3095)$, which presumably represents the cross-section contribution of the "old" u, d, and s quarks. Being cognizant of the possible contributions to both the numerator and denominator of Equation 6 from other charmed objects such as the F^+ and charmed baryons, some authors refer to the semileptonic branching ratios obtained through this technique as the average "charm" semileptonic branching ratio rather than the D semileptonic branching ratio.

Table 7 summarizes the average semileptonic branching ratios derived through measurements of the single electron rate by various groups. Figures 13–15 give the average semileptonic branching ratio vs E_{cm} for the DASP, LGW, and DELCO data (note there is a factor of two difference between the quantities plotted in Figure 15 and the other two). All data sets show a remarkable constancy in the value for the average branching ratio even though the relative contribution of the D^0, D^+, F^+, and Λ_c must be changing as a function of energy. At $E_{cm} = 4.028$ GeV, for example, measurements of exclusive final states show that $70 \pm 10\%$ of D's are neutral compared to the $56 \pm 3\%$ neutral-D fraction assumed at the ψ'' (Rapidis 1979).

Table 7 The branching ratio for $D \to e\nu X$

Experiment	E_{cm} (GeV)	Branching ratio (%)
DASP Wiik & Wolf 1978	$3.99 \to 4.08$	8.0 ± 2.0
LGW Feller et al 1978	ψ''	7.2 ± 2.8
DELCO Kirkby 1979	ψ''	8.0 ± 1.5
Mark II Lüth 1979	ψ''	9.8 ± 3.0
	Average	8.0 ± 1.1

4.3.2 THE INCLUSIVE ELECTRON MOMENTUM SPECTRUM   Aside from studying the rate for charm-associated inclusive electron production, it is interesting to study the momentum distribution. Figure 16 shows the inclusive momentum distribution obtained at the ψ'' by DELCO for events with ≥ 2 additional charmed particles. As the curves of the figure show, the momentum spectrum is consistent with the distribution expected for a mixture of $D \to Ke\nu$, $K\pi e\nu$, and $\pi e\nu$ semileptonic decays. The $Ke\nu$ and $K\pi e\nu$ contributions dominate over the Cabibbo-suppressed $\pi e\nu$ mode and

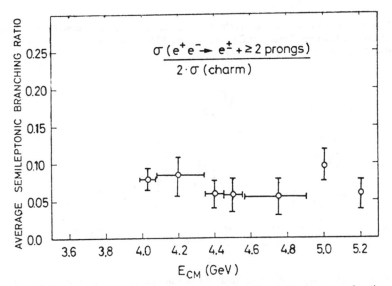

Figure 13 The average semileptonic branching ratio for charmed hadrons as a function of energy. The error bars are statistical only (from DASP data).

appear to be roughly equal. The exact ratio of Kπev to Kev depends sensitively on how much of the Kπ contribution comes from primary K*(890) formation.

4.3.3 TAGGED EVENTS   A second, potentially much cleaner, technique for extracting the D semileptonic branching ratio involves counting the number of events containing an electron recoiling against a "tagging"

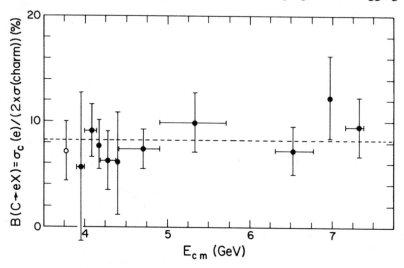

Figure 14   The branching fraction for charmed-particle decay into an electron plus additional particles as a function of energy (LGW collaboration) The dashed line indicates the average value of the ratio for $3.9 < E_{cm} < 7.4$ GeV.

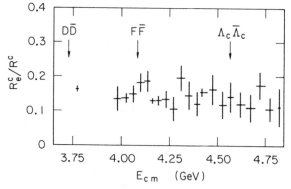

Figure 15   The ratio of R_e^c to the total charm hadronic cross section, R^c, measured by DELCO. The ratio R_e^c/R^c is equal to $2B_e(1-B_e)$, where the branching ratio $B_e = B$(charm $\rightarrow evX$) (Kirkby 1979).

$D^0 \to K^- \pi^+$, $K^- \pi^+ \pi^+ \pi^-$ or $D^+ \to K^- \pi^+ \pi^+$ candidate produced at the ψ''. Because the ψ'' cannot decay into final states containing a D^*, one is guaranteed that D^-'s are always produced against tagged D^+'s and $\bar{D}^0$'s are always produced against tagged D^0's. Hence the tagging technique allows one to measure separately the D^+ and D^0 semileptonic branching ratio. Owing to the smallness of tagging branching ratios, however, the number of tagged electron events is much smaller than the number of inclusive electron events, and such studies suffer from larger statistical errors.

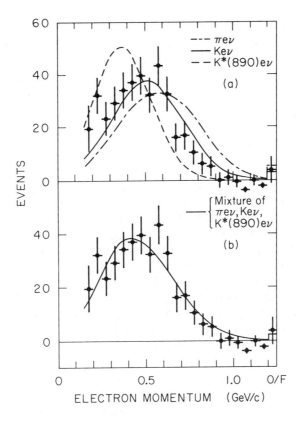

Figure 16 The electron momentum spectrum from D decays at the ψ'', measured by DELCO. The curves have been fitted to the data below 1 GeV/c and correspond to the following hypotheses:

(*a*) D $\to \pi e \nu$ (*dot-dashed curve*, $\chi^2/$dof = 80.9/16), D $\to$ Keν (*solid curve*, $\chi^2/$dof = 23.4/16), D $\to$ K*(890) eν (*dashed curve*, $\chi^2/$dof = 53.8/16).

(*b*) Contributions from D $\to$ Keν (55%), D $\to$ K*eν (39%), and D $\to \pi e \nu$ (6%) with $\chi^2/$dof = 11.2/15 (Kirkby 1979).

Table 8 summarizes the Mark II collaboration's information on the D^0 and D^+ semileptonic branching ratio (Lüth 1979). Wrong-sign e-tag events were used to compute the background level due to an $\sim 6\%$ π/e misidentification. We see from Table 8 that the D^0 and D^+ semileptonic branching ratios are unequal at about a two-standard-deviation level of significance.

We should also mention that the use of tagged events at the ψ'' has provided important information on the kaon content of D decays, and the D decay final state multiplicity distribution (Feller 1979, Schindler 1979, Lüth 1979, Vuillemin et al 1978).

4.3.4 TWO-ELECTRON FINAL STATES   The above result is corroborated by the DELCO collaboration's analysis of the two-electron vs one-electron rate at the ψ'' (Kirkby 1979). In the limit of perfect acceptance, the single (N_1) and double (N_2) electron event rates due to D decays are related to the neutral (B_0) and charged (B_+) semileptonic branching ratios by

$$N_1 = 2N_0 B_0(1 - B_0) + 2N_+ B_+(1 - B_+)$$
$$N_2 = N_0 B_0^2 + N_+ B_+^2,$$

8.

where N_0 and N_+ are the number of $D^0\bar{D}^0$ and D^+D^- events produced. One sees from Equation 8 that in the limit of small branching ratios a measurement of N_1 determines essentially a line in the B_0 vs B_+ plane whereas a measurement of N_2 determines an elliptical arc. One thus expects that a simultaneous measurement of N_2 and N_1 will lead to two ambiguous solutions for B_0 and B_+. Figure 17 indicates the experimental regions in the B_+ vs B_0 plane that are consistent to within one standard deviation with the data presented in Table 9. The data have been corrected for electron detection efficiency using the two extreme models that $D \to Kev$ (Figure 17a) or $D \to K^*ev$ (Figure 17b). Under either assumption it appears that $B_0 \gg B_+$ or $B_+ \gg B_0$. DELCO uses the K_S content in the two-electron events to distinguish between these two possibilities. Although both D^0's

Table 8   Semileptonic decays of D^+ and D^0, Mark II data at the ψ''

Decay mode	Tags (no.)	Electrons (no.)	Background	$B(\%)$
$D^+ \to e^+$		38	15 ± 1	
$\to e^-$	295 ± 18	4	3.9 ± 0.5	16.8 ± 6.4
$D^+ \to e^+$		36	19 ± 1	
$\to e^-$	480 ± 23	19	12 ± 1	5.5 ± 3.7

Table 9 The DELCO multiprong electron data sample at the ψ''

Event description	Event topology		
	1 electron	2 electrons	2 electrons + "V" (K_s^0)
Observed	1416	21	8
Background	692	4.6	1.8
Charm signal	724	16.4	6.2

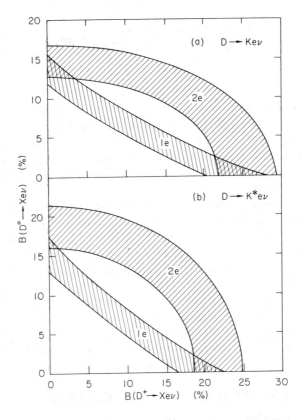

Figure 17 The allowed solutions for the D^0 and D^+ semileptonic branching ratios in the DELCO 1e and 2e multiprong data at the ψ''. The shaded regions, which correspond to $\pm 1\sigma$ limits, are plotted for two extreme assumptions of the detection efficiencies: (a) all $D \to Ke\nu$ and (b) all $D \to K^*e\nu$ (Kirkby 1979).

and D^+'s can decay into K_S, the D^0 must do so via the decay sequence $D^0 \to (K\pi)^- e^+ v$ where the isodoublet $K\pi$ system decays into a charged kaon 2/3 of the time ($I = 1/2$) and a neutral kaon only 1/3 of the time. The D^+, on the other hand, can produce K_S via both $D^+ \to K_S e^+ v$ and $K_S\pi^0 e^+ v$. In fact, a rather large fraction of the two-electron events (8 of 16.4) have a K_S, which suggests the solution $B_+ \gg B_0$. Combining the information on the single-electron and two-electron rate at the ψ'' with the K_S content of the two-electron events, DELCO finds the D^+ and D^0 semileptonic branching ratios to be $24 \pm 4\%$ and $< 5\%$ (95% confidence level), respectively, in agreement with the trend and values of the Mark II data of Table 8.

4.3.5 THE D^0, D^+ LIFETIMES   Because of the isosinglet character of the Lagrangian responsible for the Cabibbo-favored D semileptonic decay, one expects that $\Gamma(D^+ \to Km\pi e^+ v) = \Gamma(D^0 \to Km\pi e^+ v)$ for any number m of pions (see for example Pais & Treiman 1977). Hence, neglecting the Cabibbo-suppressed decay modes one has the relation $B_0\Gamma(D^0 \to \text{all}) = B_+\Gamma(D^+ \to \text{all}) = \Gamma(D \to KeX)$. In terms of lifetimes, one thus obtains $\tau(D^+)/\tau(D^0) > 4$ (95% confidence level) for the DELCO data, and $\tau(D^+)/\tau(D^0) = 3.1^{+4.1}_{-1.3}$ for the Mark II. One can estimate the order of magnitude of the D lifetime using the theoretical calculation for the D $\to Kev$ width of about 10^{11} sec^{-1} (see for example Fakirov & Stech 1978) and the ratio $\Gamma(D \to Kev)/\Gamma(D \to eX) = 45 \pm 24\%$ obtained through the DELCO fit to the ψ'' electron momentum spectrum. These results suggest that the D^+ lifetime is on the order of 10^{-12} sec while the D^0 is three to five times shorter lived.

Owing to the large phase space and number of hadronic decay modes available to both charmed mesons, it appears surprising that hadronic decays of the D° can be enhanced by at least a factor of three to five relative to those of the D^+, as suggested by the recent data. We note that the average charm semileptonic branching ratio reported by all groups appears to be remarkably constant as a function of center-of-mass energy. Within the limits of the present statistical and systematic uncertainties these measurements are not inconsistent with the observation of different lifetimes; however, one should eventually be able to measure significant variations in the semileptonic branching ratio in data taken at different center-of-mass energies.

5   THE EXCITED STATES OF CHARM

5.1   Observation of $D^{*+} \to \pi^+ D^0$

Structure present in the early D meson recoil spectra obtained by the SLAC-LBL Mark I collaboration suggested the presence of D*'s or heavier, new charmed mesons produced against the D in $e^+ e^-$ annihilation.

Direct evidence for the D*⁺ was obtained by this group (Feldman et al 1977) in data collected from 5 to 7.8 GeV, the energy limit of SPEAR. In an effort to observe the pion cascade process $D^{*+} \rightarrow D^0\pi^+$, $D^0 \rightarrow K^-\pi^+$ candidates with masses in the D region from 1.820 to 1.910 GeV and momenta exceeding 1.5 GeV were paired with extra pions of the appropriate charge. Figure 18 shows the resulting mass difference, $M(D^0\pi^+)$ $- M(D^0)$, distribution. Relatively large momenta D^0's were required in order that the pions produced in the process $D^{*+} \rightarrow \pi^+ D^0$ have sufficient momenta to be observed in their detector.

A clear, nearly background-free signal is seen in the $D^0\pi^+$, $\bar{D}^0\pi^-$ distribution at a mass difference of 145.3 ± 0.5 MeV. The narrow width of this peak sets the limit $\Gamma_{D^{*+}} < 2$ MeV/c^2 (90% confidence level). The slight enhancement at the same mass difference for $D^0\pi^- + \bar{D}^0\pi^+$ events can be explained by π/K misidentification by the time-of-flight system. The relative smallness of this peak compared to the peak of Figure 18a sets a

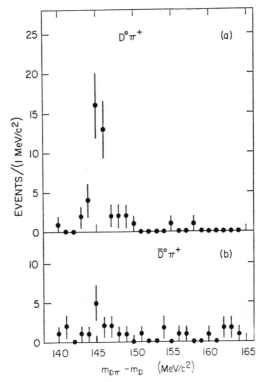

Figure 18 The $D^0\pi^+$ (a) and $\bar{D}^0\pi^+$ (b) mass difference distributions for $D^0 \rightarrow K^-\pi^+$ candidates lying within 45 MeV of the D^0 mass obtained by the Mark I collaboration. The slight signal in (b) is consistent with that expected from double time-of-flight misidentification. The charge conjugate reactions are included.

limit on the conjectured $D^0 - \bar{D}^0$ mixing process: less than 16% of produced D^0's mix into $\bar{D}^0$'s within the D^0 lifetime. Finally it was found that a substantial (i.e. $25 \pm 9\%$) fraction of D^0's produced in their data sample with momenta exceeding 1.5 GeV are from the D^{*+} pionic decay.

5.2   Evidence for the D^{*0}

A natural extrapolation of this D^{*+} observation is that there exists an excited state of the D^0, as well. Thus one would have an excited (D^{*0}, D^{*+}) isodoublet which might, for example, be a spin excitation of the (D^0, D^+)

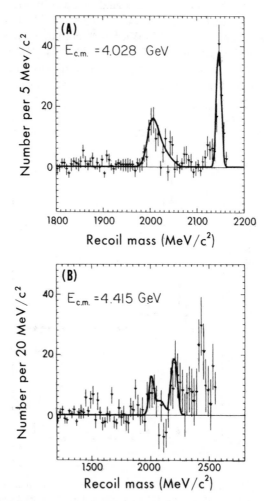

Figure 19   The D^0-subtracted recoil spectra obtained by the Mark I collaboration at two fixed energies. Please note the scale changes.

system. Figure 19a, which shows the recoil mass distribution against D^0 $\rightarrow K^- \pi^+$ candidates collected at a center-of-mass energy of 4.028 GeV, provides evidence for the existence of the D^{*0} (Goldhaber et al 1977). This distribution has background $K^{\mp} \pi^{\pm}$ combinations with masses straddling the D^0 signal subtracted out. The two peaks present in Figure 19b are attributed to the process $e^+ e^- \rightarrow D^* \bar{D}$ or $\bar{D}^* D$, which produces a peak near 2.01 GeV, and $e^+ e^- \rightarrow D^* \bar{D}^*$; $D^* \rightarrow \pi^0 D^0$, which produces a narrow reflection peak near 2.15 GeV. We note the relative smallness of the reaction $e^+ e^- \rightarrow D\bar{D}$, which would produce a peak near 1.863 GeV.

The narrowness of the reflection peak near 2.15 GeV must follow from the small Q value for the pionic cascade process as well as the improved recoil mass resolution for D^*'s produced nearly at threshold. This small Q value allows the radiative D^* decay process, $D^{*0} \rightarrow \gamma D^0$, to compete favorably with the pionic cascade, as we demonstrate later. Processes such as $e^+ e^- \rightarrow \bar{D}_1^0$, D^{*0}; $D^{*0} \rightarrow D_2^0 \pi^0$ where D_2^0 rather than D_1^0 is observed decaying into $K^- \pi^+$ give rise to the trailing tail of the peak near 2.15 GeV.

The solid curves superimposed on Figure 19a give the expected shapes for the $D^* \bar{D}$ and $D^* \bar{D}^*$ contributions to the D^0 recoil spectrum at E_{cm} = 4.028 GeV. These curves neglect any $D^* \rightarrow \gamma D$ contributions, however. Figure 19b shows the D^0 spectrum obtained for data collected at E_{cm} = 4.415 GeV overplotted with the expected shape of contributions from these same two processes.

Comparison of the second peak in Figure 19a and b shows that this peak moves and broadens in the manner expected for a kinematic reflection. The origin of the peak near 2.44 GeV in the E_{cm} = 4.415 GeV is as yet unclear—it may arise from multibody final states such as $D^{*0} \bar{D}^{*0} \pi^0$ or it may be possible evidence for a D^{**}.

5.3   The Masses and Branching Ratios of the D^*'s

In order to analyze quantitatively D production at E_{cm} = 4.028 GeV, a fit was performed to the joint D^0 and D^+ momenta spectra. Owing to the kinematic simplicity of the symmetric annihilation process, the D momentum can be trivially related to the value of the recoil mass against the D. The momentum variable does, however, offer an advantage because the D momentum resolution is a relatively insensitive function of momentum near threshold. The decay mode $D^{*+} \rightarrow \pi^+ D^0$ couples the charged and neutral D momentum spectra, thus necessitating a single fit to both.

Figure 20a illustrates eight contributions of the D^0 momentum spectrum at E_{cm} = 4.028 GeV in terms of the three basic processes:

(i) $e^+ e^- \rightarrow D\bar{D}$
(ii) $e^+ e^- \rightarrow D\bar{D}^* + \bar{D}D^*$
(iii) $e^+ e^- \rightarrow D^* \bar{D}^*$

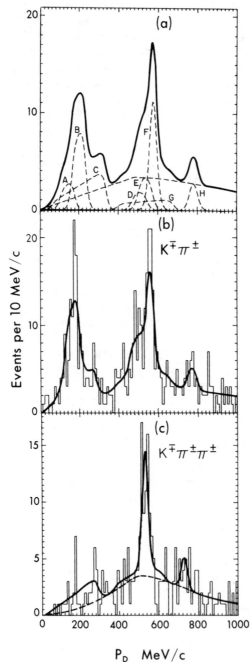

Figure 20   The D momentum spectrum at 4.03 GeV from Goldhaber et al (1977).

(*a*) Contribution to the expected D^0 momentum spectrum from:

A: $e^+e^- \rightarrow D^{*+}D^{*-}, D^{*+}$
$\rightarrow \pi^+ D^0$

B: $\rightarrow D^{*0}\bar{D}^{*0}, D^{*0} \rightarrow \pi^0 D^0$

C: $\rightarrow D^{*0}\bar{D}^{*0}, D^{*0} \rightarrow \gamma D^0$

D: $\rightarrow D^{*+}D^-, D^{*+} \rightarrow \pi^+ D^0$

E: $\rightarrow D^{*0}\bar{D}^0, D^{*0} \rightarrow \pi^0 D^0$

F: $\rightarrow \bar{D}^{*0}D^0, \text{direct } D^0$

G: $\rightarrow D^{*0}\bar{D}^0, D^{*0} \rightarrow \gamma D^0$

H: $\rightarrow D^0\bar{D}^0, \text{direct } D^0$

(*b*) $D^0 \rightarrow K^-\pi^+$ momentum spectrum, the curve is the result of the fit;

(*c*) $D^+ \rightarrow K^-\pi^+\pi^+$ momentum spectrum where the curve is the result of the fit and the dashed line is the background.

where D*'s decay ultimately into D's via the reactions:

(iv) $D^{*+} \to \pi^+ D^0$
(v) $D^{*+} \to \pi^0 D^+$
(vi) $D^{*+} \to \gamma D^+$
(vii) $D^{*0} \to \pi^0 D^0$
(viii) $D^{*0} \to \gamma D^0$.

The fit does indeed show that D production at this energy is overwhelmingly dominated by the two-body processes (i) through (iii). Less than 10% of the D^0's were found to arise from the three-body process $D^0 \bar{D}^0 \pi^0$. The positions and shapes of these contributions depend sensitively on the D* masses and $D^* - D$ mass differences. The relative areas of these contributions are functions of the rates for processes (i) through (iii) and the various D* branching ratios.

Table 10   Results from simultaneous fits to the D^0, D^+ momentum spectra at $E_{cm} = 4.028$ GeV (from Goldhaber et al 1977)

	Fit parameter	Normal fit	Isospin constrained fit	Estimated values
Masses in MeV/c^2	M_{D^0}	1864 (1.5)[a]	1862 (0.5)[a]	1863 ± 3[b]
	M_{D^+}	1874 (2.5)	1873 (2.0)	1874 ± 5
	$M_{D^{*0}}$	2006 (0.5)	2007 (0.5)	2006 ± 1.5
	$M_{D^{*+}}$	2009 (1.5)	2007 (0.5)	2008 ± 3
Branching ratios	$B(D^{*0} \to \gamma D^0)$	0.45 (0.08)	0.75 (0.05)	0.55 ± 0.15
	$B(D^{*+} \to \pi^+ D^0)$[c]	—	0.60 ± 0.15	—
	$\dfrac{B(D^+ \to K^- \pi^+ \pi^+)^c}{B(D^0 \to K^- \pi^+)}$	—	1.60 ± 0.60	—
D^0 source fractions	$D^0 \bar{D}^0$	0.05 (0.03)	0.05 (0.02)	0.05 ± 0.03
	$D^0 \bar{D}^{*0} + \bar{D}^0 D^{*0}$	0.42 (0.04)	0.34 (0.04)	0.38 ± 0.08
	$D^{*0} \bar{D}^{*0}$	0.47 (0.05)	0.32 (0.05)	0.40 ± 0.10
	$D^{*+} D^-$; $D^{*+} \to \pi^+ D^0$	0.03 (0.02)	0.09 (0.04)	0.06 ± 0.05
	$D^{*+} D^{*-}$; $D^{*+} \to \pi^+ D^0$	0.03 (0.03)	0.20 (0.07)	0.11 ± 0.10
D^+ source fractions	$D^+ D^-$	0.09 (0.05)	0.09 (0.05)	0.09 ± 0.05
	$D^{*+} D^- + D^{*-} D^+$	0.65 (0.07)	0.58 (0.06)	0.62 ± 0.09
	$D^{*+} D^{*-}$	0.26 (0.08)	0.33 (0.08)	0.29 ± 0.10

[a] Quantities in parentheses are typical statistical errors for a single fit.
[b] Errors quoted include estimated systematic uncertainty.
[c] These values can only be obtained under the assumptions of the isospin constrained fit. The quoted errors do not reflect possible breakdown of these assumptions.

Table 10, reprinted from Goldhaber et al (1977), summarizes the information obtained from two fits to the joint $D^0 \to K^- \pi^+$, and $D^+ \to K^- \pi^+ \pi^+$ momentum spectra. The detailed assumptions for both fits are described in this reference, but a few words are in order. Both fits embody the constraint $M_{D^{*+}} - M_{D^0} = 145.2$ MeV/c^2. The "normal" fit treats the two D^{*+} contributions to the D^0 spectrum, via process (iv), as independent parameters, whereas the "isospin-constrained" fit relates the number of D^0 arising from $D^{*+}D^{*-}$ production to that from $D^{*+}D^-$ production via a universal $D^{*+} \to \pi^+ D^0$ branching ratio and the assumption that, apart from p^3 threshold factors, the rates for the charged versions of processes (i) through (iii) equal the rates for the neutral versions.

Both fits match the experimental momentum spectra reasonably well. In addition, the D masses and ratio of D^+/D^0 branching fractions obtained in this fit agree well with the results of later work by the LGW and Mark II collaborations obtained at the ψ''. These results are compared in Table 11. Figure 21 summarizes the mass relationships between the D and D* systems.

We see from Table 10 that D^0 production at $E_{em} = 4.028$ GeV is dominated by nearly equal contributions from reactions (ii) and (iii). This is notable in light of the 16-MeV Q value for reaction (iii) compared to a Q of 159 MeV and 312 MeV for reactions (ii) and (i) respectively. One expects some enhancement of reaction (iii) due to the larger number of available final state spins, but the spin factor is considerably smaller than the enhancement implied by the data when corrected by the expected p^3 threshold factors. Various explanations for this enhancement have been offered, such as the possibility that the 4.028 resonance is a "molecular" state, or that the radial nodes of the $c\bar{c}$ wave function dictate the relative rates of reactions (i) to (iii) (see Appelquist et al 1978 for a discussion and references to the theoretical papers).

Table 11   Comparison of D results

	SLAC-LBL (Mark I)	LGW	Mark II
M_{D^+} (MeV/c^2)	1874 ± 5	1868.3 ± 0.9[a]	1868.4 ± 0.5[a]
M_{D^0} (MeV/c^2)	1863 ± 3	1863.3 ± 0.9[a]	1863.8 ± 0.5[a]
$\dfrac{B(D^+ \to K^- \pi^+ \pi^+)}{B(D^0 \to K^- \pi^+)}$	1.6 ± 0.6	$\dfrac{3.9 \pm 1}{1.8 \pm 0.5} = 2.2 \pm 0.8$	$\dfrac{6.3 \pm 1.5}{3.0 \pm 0.6} = 2.1 \pm 0.7$

[a] These masses are still dependent on the absolute SPEAR calibration. They are thus defined relative to the ψ mass. The mass difference $\Delta = M_{D^+} - M_{D^0}$ is known more precisely as some errors cancel. The values are: $\Delta = 5.0 \pm 0.8$ MeV/c^2 for the LGW and $\Delta = 4.7 \pm 0.3$ MeV/c^2 for the Mark II experiments respectively.

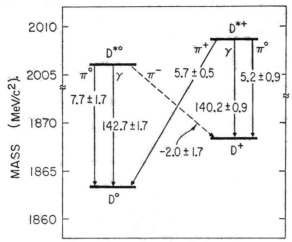

Figure 21   Mass level diagram giving the best current values for D* and D states. The arrows represent different decay modes of the D*; the numbers across the lines represent the Q for each decay expressed in MeV. The decay $D^{*0} \to D^+ \pi^-$ is kinematically forbidden. The masses are (D^0) 1863.3 ± 0.9, (D^+) 1868.3 ± 0.9, (D^{*0}) 2006.0 ± 1.5, and (D^{*+}) 2008.6 ± 1.0 MeV/c^2.

5.4   Spins of the D, D* Mesons

Because the D and D* are the two lightest charmed particles, it is a priori probable that the D is a pseudoscalar residing in the SU(4) multiplet of the pion, while the heavier D* is a vector residing in the SU(4) multiplet of the $\rho(770)$. The mass difference between particles in the vector and pseudo-scalar multiplets ($^3S_1$ and $^1S_1$) is presumably due to the hyperfine interaction between the quarks. Because of the large mass splitting within these SU(4) multiplets, and the presumably reduced strength of the quark hyperfine splitting for states containing a heavy, charmed quark, these conclusions may not necessarily hold. Present data are insufficient to establish the unique spin-parity assignments of the low-lying charmed mesons, although there is experimental information from SPEAR.

Some information is available from the study of the D* decay modes discussed earlier. Observation of the decay $D^* \to \pi D$ along with the reaction $e^+ e^- \to D\bar{D}^*$ demonstrates that the D and D* are not both spinless.[1] Evidence for the radiative decay $D^* \to \gamma D$ corroborates this conclusion.

The LGW collaboration (Peruzzi et al 1977) obtained information on the angular distribution (in θ) of the D momentum vector with respect to the $e^+ e^-$ annihilation axis for the reaction $e^+ e^- \to D\bar{D}$ obtained in data

[1] If the D and D* were both spinless they would require even relative parity to couple to a photon via $e^+ e^- \to DD^*$. However, then the decay $D^* \to \pi D$ would fail to conserve parity.

collected at the ψ''. Figure 22 shows the background-subtracted $\cos\theta$ distributions for $D^0(\bar{D}^0) \to K^{\mp}\pi^{\pm}$ and $D^{\pm} \to K^{\mp}\pi^{\pm}\pi^{\mp}$ events. Fits of these distributions to the form

$$\frac{dn}{d\cos\theta} \propto 1 + \alpha\cos^2\theta$$

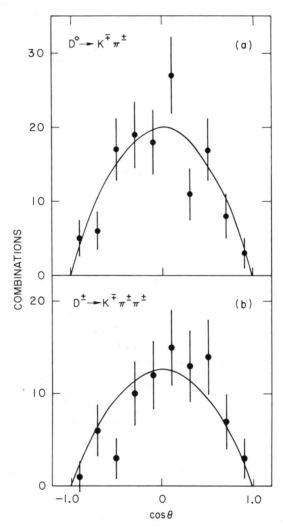

Figure 22 Distribution of events in the cosine of the angle between the incident e^+ beam and the D momentum for (a) $D^0(\bar{D}^0) \to K^{\mp}\pi^{\pm}$ and (b) $D^{\pm} \to K^{\mp}\pi^{\pm}\pi^{\pm}$, after background subtraction. The curves represent $\sin^2\theta$, the required distribution for the production of spinless D mesons.

yield the values $\alpha = -1.04 \pm 0.10$ for the $D^{\pm}$ and $\alpha = -1.00 \pm 0.09$ for the $D^0(\bar{D}^0)$, which are remarkably close to the value $\alpha = -1$ required for production of spinless particles. Their result suggests that the D is spinless as expected; however, two D's of higher spin could couple fortuitously to give a $\sin^2 \theta$ polar distribution as well.

Spin information on the D and D* is available from SLAC-LBL collaboration data on D's produced in e^+e^- annihilations at a center-of-mass energy of 4.028 GeV. At this energy D^0's are primarily produced via the processes $e^+e^- \rightarrow D^*\bar{D}^*$ and $D\bar{D}^*$. One can obtain a relatively pure sample of D's from either process by applying an appropriate cut on the measured recoil mass against the D^0.

The distribution for the angle between the D* momentum and the annihilation axis, obtained for $D^*\bar{D}^*$ production, was fit to the form

$$\frac{dn}{d \cos \theta} \propto 1 + \alpha \cos^2 \theta.$$

The measured value $\alpha = -0.3 \pm 0.3$ tends to rule out spinless D*'s at the two-standard-deviation level.

A study of the joint production and decay angular distribution for the process $e^+e^- \rightarrow \bar{D}^{*0}D^0$, $D^0 \rightarrow K^-\pi^+$ provides additional information on the charmed-meson spins. One can uniquely predict this distribution for the case of spin-0 D and spin-1 D* and vice versa if one assumes that the two mesons have even relative parity (as evidenced under these spin assignments by the observation of the reaction $D^* \rightarrow \pi D$). In particular there would be a considerable anisotropy in the $D^0 \rightarrow K\pi$ decay if a vector D^0 was produced against a pseudoscalar D*. Such an anisotropy is inconsistent with the SLAC-LBL data at about the three-standard-deviation level. Their data are fully consistent with the expected distribution for a pseudoscalar D and vector D* (Nguyen et al 1977).

In summary, all known data on charmed-meson spin is consistent with the expected vector character of the D* and pseudosclar character of the D. The two mesons cannot both be spinless, and the case of vector D's and pseudosclar D*'s is explicitly excluded. Nothing as of yet is known about the possibilities where the sum of the D and D* spins exceeds one.

6 THE F MESON

Little is known experimentally about the third charmed meson—the F^+. This isosinglet meson is constructed from a c and $\bar{s}$ quark and should decay predominantly into final states of zero strangeness according to the GIM model. Since the many possible multipion F^+ decay modes are expected to have huge backgrounds, investigators have tended to search for F^+

decaying into less common particles such as $K^+K^-\pi^+$, K^+K_S, or η and multipions.

At the time of this writing the only published observation of the F^+ comes from the DASP collaboration (Brandelik et al 1979), who looked for the process $e^+e^- \to F\bar{F}^*$ where the $F \to \pi\eta$ and $\bar{F}^* \to \gamma\bar{F}$ (the $F^* \to \pi F$ decay does not conserve isospin).

Figure 23 shows the $\gamma\gamma$ invariant mass distribution for hadronic events collected at the indicated center-of-mass energies. The most pronounced η signal occurs near $E_{cm} = 4.42$ (Siegrist et al 1976), which is at the location of a 33-MeV wide resonance in R. Requiring the presence of an additional soft γ ($E_\gamma < 140$ MeV) in the event, as would be the case for F^*F or F^*F^* production, appears to enhance the η signal relative to the background in the 4.42-GeV data (see Figure 24). These observations are thus suggestive of a substantial F^* contribution to η's produced at $E_{cm} = 4.42$ GeV.

In order to observe F^+'s decaying into a specific final state, events collected at 4.42 GeV were fitted to the hypothesis $e^+e^- \to FF^* \to \pi\eta F\gamma$ where the $\eta \to \gamma\gamma$ candidates were constrained to the precise η mass, and the $\pi\eta$ system was constrained to the mass of the missing F. A scatter plot of the $\pi\eta$ mass versus its recoil mass is shown in Figure 25 for events with an acceptable χ^2. There is a clustering of six events present in this plot that are F, F^* candidates, where $M_F = 2.04 \pm 0.01$ GeV/c^2 and $M_{F^*} = 2.15 \pm 0.04$ GeV/c^2. These six events also tend to fit the hypothesis $e^+e^- \to F^*\bar{F}^*$, where $M_F = 2.00 \pm 0.04$ and $M_{F^*} = 2.1 \pm 0.02$ GeV. Here the DASP collaboration take average values and quote $M(F) = 2.03 \pm 0.06$ GeV and $M(F^*) = 2.14 \pm 0.06$ GeV.

The six events imply $\sigma \cdot B(F^+ \to \eta\pi^+) = 0.41 \pm 0.18$ nb at the 4.42 resonance, a value close to the Mark II 95% confidence level upper limit of 0.26 nb at $E_{cm} = 4.42$ GeV and 0.33 nb at $E_{cm} = 4.16$ GeV. Clearly further work must be done to clarify the physics of the F.

ACKNOWLEDGMENTS

We want to thank Mrs. C. Frank-Dieterle for her help and meticulous care in preparing and compiling this manuscript.

This work was supported primarily by the US Department of Energy under Contracts No. W-7405-ENG-48 (Lawrence Berkeley Laboratory) and DE-AC02-76ER01195 (University of Illinois).

We wish to thank the Staff of the Aspen Center for Physics for the hospitality extended to us during the Summer of 1979 while this review was being prepared in part.

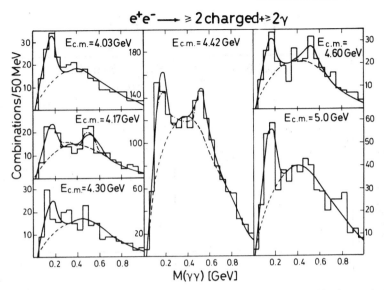

Figure 23 DASP data on inclusive η production. Note that the η signal is cleanest in the vicinity of the $\psi(4.42)$.

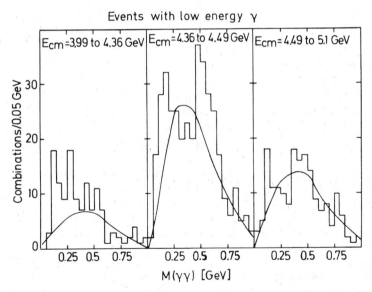

Figure 24 DASP data on η production for events with low energy γ rays. Note the η signal in the energy region near the $\psi(4.42)$ is improved if óne demands the presence of an additional soft γ ($E_\gamma < 140$ MeV). Such a γ could result from the process $F^* \rightarrow \gamma F$.

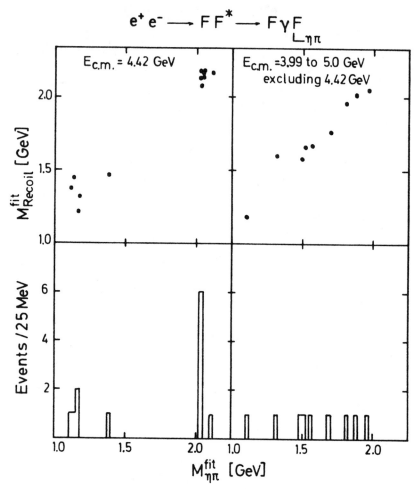

Figure 25   DASP data showing evidence for F production. A clustering of six events is seen in the $M_{\eta\pi}$ vs M_{Rec} scatter plot near $M_{\eta\pi} = 2.04$ GeV/c^2.

Literature Cited

Abrams, G. S. et al. 1979. *Phys. Rev. Lett.* 43:481
Andrews, D. et al. 1980. *Phys. Rev. Lett.* 44:1108
Appelquist, T., Barnett, R. M., Lane, K. 1978. *Ann. Rev. Nucl. Part. Sci.* 28:387
Aronson, S. H. et al. 1970. *Phys. Rev. Lett.* 25:1057
Aubert, J. J. 1974. *Phys. Rev. Lett.* 33:1404
Augustin, J.-E. et al. 1974. *Phys. Rev. Lett.* 33:1406

Bacino, W. et al. 1978. *Phys. Rev. Lett.* 40:671
Barbaro-Galtieri, A. 1968. *Advances in Particle Physics*, Vol. D, ed. R. Cool, R. Marshak, p. 193. New York: Interscience
Barbaro-Galtieri, A. 1978. *Production and Decay of Charm Particles in e^+e^- Collisions*, Lawrence Berkeley Lab., Rep. LBL-8537, LBL-9247 (1979)
Bebek, C. 1980. *Exp. Meson Spectrosc. Conf. EMS-80.* Brookhaven Natl. Lab.

Böhringer, T. et al. 1980. *Phys. Rev. Lett.* 44:1111

Boyarski, A. M. et al. 1975. *Phys. Rev. Lett.* 34:1357

Brandelik, R. et al. 1979. *Z. Phys. C* 1:233

Braunschweig, W. et al. 1976. *Phys. Lett. B* 63:471

Bricman, C. et al. 1978. *Phys. Lett. B* 75:1

Burmester, J. et al. 1976. *Phys. Lett. B* 64:369

Camerini, U. et al. 1975. *Phys. Rev. Lett.* 35:483

Carithers, W. A. et al. 1973. *Phys. Rev. Lett.* 31:1025

Chinowsky, W. 1977. *Ann. Rev. Nucl. Sci.* 27:393

Clark, A. R. et al. 1971. *Phys. Rev. Lett.* 26:1667

Eichten, E. et al. 1975. *Phys. Rev. Lett.* 34:369

Ellis, J. 1974. *Proc. XVII Int. Conf. High Energy Phys., London,* p. IV-20

Fakirov, D., Stech, B. 1978. *Nucl. Phys. B* 133:315

Feldman, G. J. et al. 1977. *Phys. Rev. Lett.* 38:1313

Feldman, G. J. 1978. *Proc. XIX Int. Conf. High Energy Physics, Tokyo,* p. 777

Feller, J. M. 1979. PhD thesis. Univ. Calif., Berkeley. *LBL-9017*

Feller, J. M. et al. 1978. *Phys. Rev. Lett.* 40:1677

Finocchiaro, G. 1980. See Bebek 1980

Gaillard, M. K., Lee, B. W., Rosner, J. 1975. *Rev. Mod. Phys.* 47:277

Gell-Mann, M. 1964. *Phys. Lett.* 8:214

Gittleman, B. et al. 1975. *Phys. Rev. Lett.* 35:1616

Glashow, S. L., Illiopoulos, J., Maiani, L. 1970. *Phys. Rev. D* 2:1285

Goldhaber, G. et al. 1976. *Phys. Rev. Lett.* 37:255

Goldhaber, G. et al. 1977. *Phys. Lett. B* 69:503

Gottfried, K. 1978. *Phys. Rev. Lett.* 40:598

Harari, H. 1977. *Proc. 1977 Summer Inst. on Part. Phys., Stanford Linear Accel. Cent. SLAC-204,* p. 1

Herb, S. W. et al. 1977. *Phys. Rev. Lett.* 39:252

Iizuka, J. 1966. *Suppl. Prog. Theor. Phys.* 37:21

Jackson, J. D., Scharre, D. L. 1975. *Nucl. Instrum. Methods* 128:13

Jean-Marie, B. et al. 1976. *Phys. Rev. Lett.* 36:291

Kirkby, J. 1979. In *Proc. 9th Int. Symp. Lepton and Photon Interactions at High Energies,* p. 107. Batavia, Ill.: Fermilab

Klems, J. H., Hildebrand, R. H., Stiening, R. 1970. *Phys. Rev. Lett.* 24:1086

Knapp, B. et al. 1975. *Phys. Rev. Lett.* 34:1040

Kobayashi, M., Maskawa, T. 1973. *Prog. Theor. Phys.* 49:652

Lane, K., Eichten, E. 1976. *Phys. Rev. Lett.* 37:477

Lee-Franzini, J. 1980. See Bebek 1980

Litke, A. et al. 1973. *Phys. Rev. Lett.* 30:1189

Lüth, V. 1979. In *Proc. 9th Int. Symp. Lepton and Photon Interactions at High Energies,* p. 78. Batavia, Ill.: Fermilab

Meyer, H. 1979. In *Proc. 9th Int. Symp. Lepton and Photon Interactions at High Energies,* p. 214. Batavia, Ill.: Fermilab

Nagels, M. M. et al. 1976. *Nucl. Phys. B* 109:1

Nguyen, H. K. et al. 1977. *Phys. Rev. Lett.* 39:262

Okubo, S. 1963. *Phys. Rev. Lett.* 5:165

Pais, A., Treiman, S. B. 1977. *Phys. Rev. D* 15:2529

Perez-y-Jorba, J. 1969. *Proc. IV Int. Symp. on Electron and Photon Interactions at High Energies, Liverpool,* p. 213

Perl, M. 1980. *Ann. Rev. Nucl. Part. Sci.* 30:299–335

Peruzzi, I. et al. 1976. *Phys. Rev. Lett.* 37:569

Peruzzi, I. et al. 1977. *Phys. Rev. Lett.* 39:1301

Peruzzi, I. et al. 1978. *Phys. Rev. D* 17:2901

Rapidis, P. A. et al. 1977. *Phys. Rev. Lett.* 39:526

Rapidis, P. A. 1979. PhD thesis. Stanford Univ. *SLAC-220*

Richter, B. 1974. *Proc. XVII Int. Conf. High Energy Phys., London,* p. IV-37

Schindler, R. H. 1979. PhD thesis. Stanford Univ. *SLAC-219*

Schindler, R. H. et al. 1980. *Phys. Rev. D.* In press

Schrock, R. E., Wang, L. L. 1978. *Phys. Rev. Lett.* 41:1692

Schröder, H. 1980. See Bebek 1980

Schwitters, R. F., Strauch, K. 1976. *Ann. Rev. Nucl. Sci.* 26:89

Siegrist, J. L. et al. 1976. *Phys. Rev. Lett.* 36:700

Tarnopolsky, G. et al. 1974. *Phys. Rev. Lett.* 32:432

Voyvodic, L. 1979. In *Proc. 9th Int. Symp. Lepton and Photon Interactions at High Energies,* p. 569. Batavia, Ill.: Fermilab

Vuillemin, V. et al. 1978. *Phys. Rev. Lett.* 41:1149

Wiik, B. H., Wolf, G. 1978. DESY-78/23 (Unpublished report); also Balian, R., Llewellyn-Smith, C. H., eds. 1977. *École d'été de physique théorique, 29th session, Les Houches, 1976.* Amsterdam: North-Holland

Wiss, J. E. et al. 1976. *Phys. Rev. Lett.* 37:1531

Zweig, G. 1964. *CERN-TH-401*

Ann. Rev. Nucl. Part. Sci. 1980. 30: 383–436

LARGE-SCALE SHELL-MODEL CALCULATIONS[1]   ×5621

J. B. McGrory[2]
Physics Division, Oak Ridge National Laboratory, Oak Ridge, Tennessee 37830; and Institute for Theoretical Physics, University of Frankfurt, Frankfurt, West Germany

B. H. Wildenthal[3]
Cyclotron Laboratory, Michigan State University, East Lansing, Michigan 48824

CONTENTS

[1] Authored by a contractor of the US government under contract W-7405-eng-26. Accordingly, the US government retains a nonexclusive, royalty-free license to publish or reproduce the published form of this contribution, or allow others to do so, for US government purposes.

[2] Research sponsored by the Division of Basic Energy Sciences, US Department of Energy, under contract W-7405-eng-26 with the Union Carbide Corporation.

[3] Research sponsored in part by US National Science Foundation, Grant PHY-7822696.

INTRODUCTION

Experimental studies in nuclear spectroscopy are now revealing with high precision an amazingly wide variety of properties of the low-lying energy levels of atomic nuclei. A given nuclear quantum state of A nucleons can be characterized not only by its energy and the quantum numbers for its total angular momentum (J), parity (π), and (approximate) isospin (T), but also by its static electromagnetic multipole moments (and even the radial distribution of these moments), the matrix elements by which it is connected via the electromagnetic and weak interaction operators to many other nuclear states of the same A-value, and the matrix elements by which it is connected to states of A-values differing from it by from one to at least four nucleons. The goal of a large-scale shell-model calculation is to create a theoretical structure by which the detailed features of these diverse phenomena can be comprehended as the consequences of the quantum-mechanical behavior of different ensembles of A particles, the particles themselves having simple, stable, and well-defined properties.

A rigorous, a priori theory of nuclear structure does not yet appear feasible, even if the requisite empirical data on the basic nuclear constituents and their interactions were fully and precisely known. Contemporary nuclear theory is therefore factored into a number of semi-autonomous components each of which treats a limited realm of observation at some specific level of detail and each of which is based upon a distinct ad hoc and more or less phenomenological set of assumptions. A shell-model calculation of the sort we discuss here will take as its province the complete gamut of spectroscopic features exhibited by the low-lying energy levels of a group of nuclei spanning up to several dozen contiguous A-values. It attempts to describe these features, subject to a few constraints, with the maximum possible quantitative accuracy.

The shell model assumes an intermediate mix, insofar as nuclear theory goes, of fundamental principles and phenomenological premises. It assumes neutrons and protons to be the constituents of atomic nuclei. Mesons and isobars, let alone quarks, are assumed to play no explicit role in nuclear structure, although allowance is made for their possible effects upon the apparent properties of the nucleons treated in the calculation. The constituent nucleons are assumed to exist in quantized single-particle orbits, and it is assumed that a finite (small) number of these single-particle states suffices to construct a basis for describing spectroscopic features. No further recourse is made to general assumptions about such issues as preferred clusters of nucleons, dominant excitation modes, or guiding macroscopic variables. Detailed phenomenology enters into this kind of theory in the assignment of the properties of the neutrons and protons explicitly considered in the calculation. These properties are expressed in the formal context of the model as the one- and two-particle matrix elements of the strong, electromagnetic, and weak interaction operators. To varying degrees, these matrix elements are treated as parameters in an attempt to achieve the closest resemblance between corresponding pairs of calculated and measured features.

The constraints upon this phenomenological parametrization of shell-model calculations come from requirements that the properties of the model nucleons be (a) consistent with the conclusions drawn from the properties of free and pairwise interacting neutrons and protons in conjunction with fundamental considerations of the many-body problem as applied to nuclei, and (b) internally consistent during the treatment of a significantly wide range of nuclei, many levels, and various types of phenomena. Large-scale shell-model calculations are, within these constraints, oriented strongly toward producing a comprehensive, detailed, and accurate accounting of the complete range of spectroscopic data experimentally exhibited by the nuclei of a given range of masses. This emphasis is to be contrasted to possible alternative approaches whose emphasis might be characterized variously as the achievement of economy and elegance of representation, of more rigorous adherence to first principles or of intuitive insights provided by utilization of classical analogies.

In this review we first sketch the general principles of the shell model and briefly describe the techniques by which large-scale calculations are carried out. (For an extensive and detailed discussion of modern shell-model techniques, see Brussaard & Glaudemans 1977.) We then address the relationships between the properties of nucleons as they are employed in the model and the properties they exhibit empirically in free space. The main body of the review presents results from representative calculations of nuclei in the $A = 17$–39 region. We hope from this presentation to illustrate the capabilities of this type of calculation to model the behavior of real nuclei. We follow this with a cursory overview of shell-model calculations as applied to other regions of the periodic table. Finally, we summarize the present status of large-scale shell-model calculations, and we speculate on possible future directions of the field.

PRINCIPLES AND TECHNIQUES OF SHELL-MODEL CALCULATIONS

General Remarks

The basic implicit assumptions of the nuclear shell model are that nuclear properties can be described in terms of neutron and proton coordinates and that in these coordinates the nuclear Hamiltonian contains at most two-body interactions. Thus, the nuclear Hamiltonian can be written in the shell model approximation as

$$\mathcal{H} = \sum_i T(r_i) + \sum_{i<j} V(r_i, r_j), \qquad\qquad 1.$$

where i and j label the nucleons of the nucleus. However, an exact solution of such a Hamiltonian cannot be obtained. The phenomenological shell model assumes that most of the effects of $V(r_i, r_j)$ can be absorbed into a central one-body potential $U(r_i)$ so that

$$\mathcal{H} \approx H_1(r_1) + H_{12}(r_1, r_2), \qquad\qquad 2.$$

where

$$H_1(r_i) = T(r_i) + U(r_i),$$

and

$$H_{12}(r_1, r_2) = V(r_1, r_2) - U(r_1),$$

is small enough to be treated in perturbation theory. The mechanics of shell-model calculations that follow from these assumptions involve

1. choice of H_1, the dominant central field;
2. calculation of the one-particle eigenstates of H_1 and the selection, from this set, of the orbits of the model;
3. construction, for a given number of nucleons and the chosen model orbits, of the multinucleon eigenstates of H_1;
4. specification of a residual two-body interaction of H_{12}; and
5. evaluation of the matrix elements of H_{12} between the multinucleon eigenstates of H_1 and calculation of the eigenvalues and eigenvectors of this matrix.

Many calculations assume that H_1 is a sum of a spherical harmonic oscillator potential, a spin-orbit $(l \cdot s)$ interaction, and a term proportional to l^2. The single-particle eigenstates $\rho(nlj)$ of a typical such interaction are shown in Figure 1 where the labels (n, l, j) specify the principal radial quantum number, the orbital angular momentum, and the total angular momentum. The simplest shell-model calculations, similar to those that first established the validity of the basic approximation (Mayer 1949), assume wave functions in which the nucleons fill the lowest available single-particle orbits, where this availability is determined by the Pauli principle. Thus from Figure 1, the ground-state wave function of ^{17}O would consist of the configuration $(0s_{1/2})^4$, $(0p_{3/2})^8$, $(0p_{1/2})^4$, $(0d_{5/2})^1$. The completely filled ("closed") shells are assumed to form an inert core with total angular

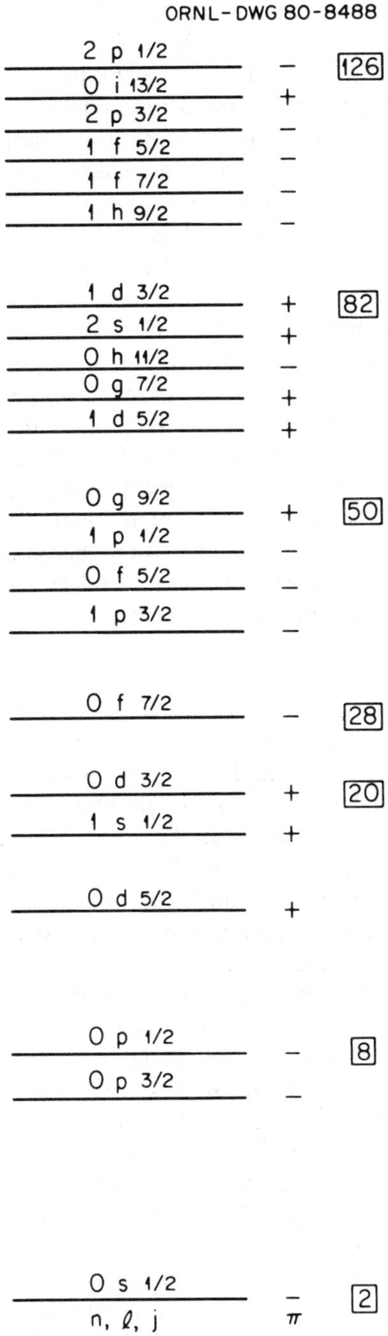

ORNL-DWG 80-8488

2 p 1/2	−	126
0 i 13/2	+	
2 p 3/2		
1 f 5/2	−	
1 f 7/2	−	
1 h 9/2	−	

1 d 3/2	+	82
2 s 1/2	+	
0 h 11/2	−	
0 g 7/2	+	
1 d 5/2	+	

0 g 9/2	+	50
1 p 1/2	−	
0 f 5/2	−	
1 p 3/2	−	

| 0 f 7/2 | − | 28 |

| 0 d 3/2 | + | 20 |
| 1 s 1/2 | + | |

| 0 d 5/2 | + | |

| 0 p 1/2 | − | 8 |
| 0 p 3/2 | − | |

| 0 s 1/2 | − | 2 |
| n, ℓ, j | π | |

Figure 1 Typical shell-model single-particle spectrum in a spherical harmonic oscillator potential with $l \cdot l$ and $l \cdot s$ terms: n, l, and j are the usual radial, orbital angular momentum, and total angular momentum quantum numbers; π labels the orbit parity. The numbers in the square boxes are the number of identical particles (neutrons only or protons only) to fill all orbits up to the orbit adjacent to the given box.

momentum $J = 0$, which generates the dominant central field H_1 [in the example of ^{17}O, $(0s_{1/2})^4$, $(0p_{3/2})^8$, $(0p_{1/2})^4$ constitutes the closed-shell "core" ^{16}O]. As nucleons are added to form ^{18}O, ^{19}O, ^{20}O, etc, they fill the $d_{5/2}$ orbit, and couple pairwise to $J = 0$, so that the properties of the model wave functions are essentially those of the last unpaired nucleon. In this simplest model, excited states are formed by elevating a nucleon to a higher energy orbit. Thus, the two excited states in ^{17}O should be $1/2^+$ and $3/2^+$, corresponding to the 17th nucleon occupying the orbits $\rho(1s_{1/2})$ and $\rho(0d_{3/2})$, respectively. If the lowest observed $J^\pi = 1/2^+$, $3/2^+$, and $5/2^+$ states in ^{17}O are identified with the $s_{1/2}$, $d_{3/2}$, and $d_{5/2}$ single-particle eigenstates of H_1, empirical estimates can be obtained for the energies of configurations with several particles excited from the $d_{5/2}$ to the other orbits. Many observed features in nuclei can be accounted for in this simple picture and this success established the basic validity of the shell-model approach. However, this scheme is inadequate to explain most of the details of nuclear spectroscopy. For this, a more detailed treatment of H_{12} is necessary, together with the freedom to admix many different multiparticle configurations into a given eigenstate. Shell-model calculations become more credible as techniques are found to work with larger and larger configuration spaces and to determine appropriate forms of H_{12}.

Computer Programs

Given the specifications of H_1 and H_{12}, and the identity of the valence orbits included in the active model space, there are a number of computer codes with which it is possible to construct the complete set of basis states, calculate the matrix elements of H_{12}, and diagonalize H_{12} in this basis. One widely used method of effecting this mechanical aspect of the calculation is incorporated in the Oak Ridge–Rochester program (French et al 1969) and another is incorporated in the Glasgow program (Whitehead et al 1977).

The distinctive ingredient of the Oak Ridge–Rochester program is that the multiparticle basis vectors are constructed so as to have good total angular momentum J and good isospin T. Within this basis for a state of N active model particles with specific J and T, matrix elements of a given shell-model Hamiltonian $H_1 + H_{12}$ are constructed and stored. Standard matrix diagonalization algorithms are then used to obtain the low-lying eigenstates and eigenvalues. The eigenvectors are then used with auxiliary programs to calculate matrix elements of other operators which yield predictions of strengths for other observables, i.e. the electric, B(EL), and magnetic, B(ML), multipole (L) transition probabilities and the singular or multiple nucleon transfer spectroscopic factors.

The Glasgow program takes a fundamentally different approach to determining the eigenvectors and eigenvalues of $\mathscr{H} = H_1 + H_{12}$. It does not work in a basis constructed to have good (J, T) but rather specifies the basis states by the occupation (either 0 or 1) of the orbital $(j_i m_i, \frac{1}{2}, m_t)$. Thus, the bases treated are significantly larger than those in the Oak Ridge–Rochester codes. (For a given number of model nucleons, the full matrix dimensions in this m-scheme basis are equal to the sum of the dimensions of all the different (J, T) matrices. Restrictions on the total M-value and symmetry considerations can, in practice, significantly reduce these m-scheme dimensions in special cases, however.) Within the large m-scheme basis, the Glasgow program uses the so-called Lanczos algorithm to obtain the eigenvalues and eigenvectors. In brief, the Lanczos method chooses an initial guess to an eigenvector, ψ_1, and then performs the following operations:

$$\mathcal{H}\psi_1 = \alpha_{11}\psi_1 + \alpha_{12}\psi_2 \qquad\qquad \psi_1 \perp \psi_2$$
$$\mathcal{H}\psi_2 = \alpha_{12}\psi_1 + \alpha_{22}\psi_2 + \alpha_{23}\psi_3 \qquad \psi_3 \perp \psi_2, \psi_3 \perp \psi_1 \qquad\qquad 3.$$
$$\mathcal{H}\psi_3 = \alpha_{23}\psi_2 + \alpha_{33}\psi_3 + \alpha_{34}\psi_4 \qquad \psi_4 \perp (\psi_1\psi_2\psi_3)$$

$\mathcal{H}$ is Hermitian, so $\alpha_{31} = \alpha_{13}$, but by construction α_{13} is 0. Thus, the generated matrix representation of $\mathcal{H}$ is tridiagonal. This operation can be continued. If, after each step in this procedure, the resulting tridiagonal matrix is diagonalized, it is found that the algebraically largest eigenvalues converge very rapidly to the exact largest eigenvalues in the full space. Experience in the sd shell has shown that less than 100 iterations are needed to get the lowest eigenvalues and eigenvectors for any size matrix. In any calculation, the actual matrix diagonalizations then involve relatively small matrices, with dimensions on the order of 100. The Glasgow codes are written so that no limitations of the size of the basis space are imposed by the available core size. The original Oak Ridge–Rochester code was limited in its dimensionality because the matrix diagonalization routine employed, based on the Givens-Householder algorithm, required the full matrix to be in the computer core at one time. Recent revisions permit diagonalization of the (J, T) Hamiltonian matrix with a Lanczos subroutine. A recent paper by Sau & Heyde (1979) indicates that, for a given matrix, the Lanczos algorithm generally is faster than the Givens-Householder algorithm.

No results of systematic comparisons of the running times of the Glasgow and Oak Ridge–Rochester codes have been published. Comparison of our experience with the latter code to published timings of the Glasgow code suggests that the Oak Ridge–Rochester code is more efficient for smaller ($< 500 \times 500$) matrices, while the Glasgow code is more efficient for larger matrices. In the range of economically feasible calculations, it appears that the timing differences are less than an order of magnitude. On an IBM-3033 or IBM-360/195, typical calculations in the middle of the sd shell (matrices $< 7000 \times 7000$) take ~ 3–10 hours.

The essential point to remember when considering different techniques for doing shell-model calculations is that, if alternate codes are free of technical errors, the identical physics output is obtained from the same physics input. That is, for a given Hamiltonian and model space, the same energies and wave functions (even if in different representations) are obtained.

Hamiltonians

Once the inert core and the set of active valance orbits, $\{j_1, j_2, \ldots, j_i\}$ are specified for a given shell-model calculation, H_1 and H_{12}, the one-body and two-body components of the Hamiltonian, must be specified. H_1 is usually assumed to be a diagonal one-body operator whose eigenvalues are the measured energies of the low-lying states in the one-particle nuclei of the model space whose properties are consistent with single-particle structure. Thus, in our example of the sd shell, the eigenvalues of H_1 are the observed energies of the lowest $J = 5/2^+$, $1/2^+$, and $3/2^+$ states in ^{17}O. The specifications of H_{12} are not so easily accomplished, since once configuration mixing of several orbits is allowed, the eigenvalues of even the two-particle systems of the model are, in general, complex functions of the individual terms of H_{12}.

It can be shown (e.g. French et al 1969) that any n-particle shell-model matrix element can be reduced to a linear combination of two-particle matrix elements

$$\langle \psi_A | H_{12} | \psi'_A \rangle = \sum_{\substack{i,j,k,l \\ J,T}} C_{AA'}^{ijkl,JT} \langle ijJT | H_{12} | klJT \rangle \qquad 4.$$

(with $i \leq j$, $k \leq l$, $i \leq k$, and $j \leq l$ when $i = k$) where i, j, k, l label single-particle orbits and $| ijJT \rangle$ is an antisymmetrized normalized two-particle state with total angular momenta J and isospin T. In the $d_{5/2} - s_{1/2} - d_{3/2}$ orbit space there are 63 such matrix elements. H_{12} for the sd shell is hence fully specified by these 63 numbers. There are two general methods by which these numbers are determined. We refer to them as "empirical" and "realistic." Empirical interactions are based in one fashion or another on parameters whose values are determined by requiring agreement between shell-model eigenvalues and measured level energies. Realistic interactions are constructed for a given shell-model space from known data on the free nucleon-nucleon force as described below. We describe here some of the ways these two methods have been used.

EMPIRICAL INTERACTIONS

Phenomenological potentials   In early shell-model calculations, matrix elements of H_{12} were calculated from phenomenological two-body potentials of the form

$$V(r) = \sum_{S,T} A^{S,T} f(a_{ST} r) P^{S,T} \qquad\qquad 5.$$

where S, T label the total intrinsic spin and isospin of the two interacting particles; $f(a_{ST} r)$ is a form factor for the interaction in the S, T channel and a_{ST} is the range of this interaction component; $P^{S,I}$ projects out the S, T channel; and $A^{S,T}$ is a strength parameter. This is a pure central force. More complicated spin dependences (tensor forces, two-body spin-orbit forces) can be added. The radial term $f(a_{ST} r)$ was usually assumed to have either a Gaussian or Yukawa form. The strength parameters were chosen initially so that $V(r)$ had the distinctive features of the free nucleon-nucleon force and then the strengths and ranges were varied to fit observed spectra. In order to evaluate matrix elements from such potentials, the size and shape of the radial wave functions, $\rho(nlj)$, must be specified. The usual approximation was to use harmonic oscillator wave functions for $\rho(nlj)$. The long-range tails of these wave functions are probably wrong, but the ranges of the potential are short enough so that this defect is usually not serious.

Phenomenological potentials are specified by four or more parameters. A very popular potential of this genre that is still used extensively is the "modified surface delta interaction" (MSDI) (Arvieu & Moszkowski 1966, Glaudemans et al 1967):

$$V_{\mathrm{MSDI}} = \sum_{T=0} A^T P^T \delta(\mathbf{r}_1 - \mathbf{r}_2) \delta(|\mathbf{r}_1| - |\mathbf{r}_N|) + \sum_{T=1} B^T P^T, \qquad 6.$$

where A^T and B^T are strengths and P^T the projection operator for the states with isospin T. This is an ordinary delta force interaction further constrained to act only at the nuclear radius r_N, plus an isospin-dependent monopole term. It contains four parameters to be determined from a fit to observed energy levels and, for such a simple formulation has been very effective, especially for systems of identical particles. Aside from the MSDI, few large shell-model calculations are now done with such phenomenological potentials.

Another method of potential parametrization is to make no explicit assumption about $V(r)$ except for the spin exchange character. The two-body matrix elements can then be reduced to linear combinations of radial integrals of $V(r)$ and these integrals treated as parameters (Cohen et al 1968). In this way, no assumption need be made about the single-particle wave functions.

Unconstrained empirical interaction   The logical extension of these empirical potentials is to treat the two-body matrix elements themselves as the parameters. For example, in the sd shell, one could try to fit all 63 two-body matrix elements to a set of experimental energy levels via the relationships of Equation 4. Such an approach avoids the rigidities inherent in the assumption of potentials. Again, no assumption about single-particle wave functions enters the energy level calculations. While any phenomenological parametrization of Hamiltonians raises some questions about the meaningfulness of the results obtained, the much larger number of parameters in unconstrained empirical Hamiltonians demands special consideration. Questions to be answered include the following:

What constitutes an adequate set of data to determine a given set of interaction parameters? For example, in the beginning of the sd shell, there are many known levels but few that significantly involve the $d_{3/2}$ orbit. Hence the energies of these levels have little dependence on matrix elements involving the $d_{3/2}$ orbit, and this aspect of the Hamiltonian is not well determined by these data alone.

What levels should be included in the fit? In any actual nucleus there are "intruder" states that cannot even approximately be described by the model. Some intruders are obvious, such as negative parity states in $A = 17$–39 nuclei in the context of an $(sd)^n$ model space. However, there may also be low-lying positive parity states dominated by complex core-excitation components, and these likewise cannot be described by sd shell configurations. To attempt to fit the energies of these states would introduce errors into the empirical H_{12}.

Does a set of best-fit parameters represent an absolute minimum in χ-squared space, or only a local relative minimum? How independent is the final solution for H_{12} from the initial assumptions?

There are no clear answers to these questions, and consistency, both internal and with respect to experimental data, plays the dominant role in establishing correct procedures. This approach to effective interactions obviously is not a fully predictive theory, since some agreement between theory and experiment is built in with the fitting procedure. However, for a given model space, this approach provides a standard by which to judge other interactions. It allows the study of how well the broad range of nuclear spectroscopic information in a given mass region can be described by a given shell-model space while minimizing the uncertainty of whether the Hamiltonian itself might be faulty. It is the central purpose of this review to answer this question, and thus we devote a large part of the paper to a discussion of shell-model results obtained with unconstrained empirical matrix elements for H_{12}.

REALISTIC INTERACTIONS   This term refers to matrix elements of H_{12} calculated for use in a specified shell-model space from the properties of the free nucleon-nucleon interaction. Calculations of such Hamiltonians represent an important step in the formulation of a fully microscopic theory of nuclear structure: accordingly there has been in the past decade considerable interest in the use of realistic interactions in shell-model calculations. In the remainder of this section we outline the procedures and some results of the calculation of realistic interactions. The nature in which these interactions are "effective" may thereby be made clearer and the present "state of the art" in this field summarized. After this summary, we briefly describe the relationships between large shell-model calculations and realistic interaction theory.

The basic problem can be outlined as follows. The exact eigenvectors and eigenvalues of the nucleus can, in principle, be obtained by solving the Schrödinger equation of the exact nuclear Hamiltonian

$$H\psi_i = E_i\psi_i. \qquad\qquad 7.$$

However, this is impossible in practice. The shell model attacks the problem of describing the nucleus by assuming that there is an effective Hamiltonian $\mathcal{H}$, acting in a small (model) subspace ϕ of the complete nuclear Hilbert space ψ, such that

$$\mathcal{H}\Phi_i = E_i\Phi_i. \qquad\qquad 8.$$

The goal is thus to calculate the effective operator $\mathcal{H}$ in a systematic way from the properties of free nucleons. An analogous problem exists for any other operator assumed to operate on the nuclear coordinates, since the operator $\mathcal{O}$ that acts upon the true nuclear state to yield a matrix element for $\langle\psi'|\mathcal{O}|\psi\rangle =$ OBS (observable) cannot, in general, be identical with the effective operator $\mathcal{O}'$ for operating on the model wave functions to yield $\langle\phi'|\mathcal{O}'|\phi\rangle =$ OBS. A vast literature exists on this topic. For more detailed reviews of the subject, see Barrett & Kirson (1973) and Barrett (1980).

It can be shown that if the total exact Hamiltonian has the form

$$H = H_0(r_i) + V(r_ir_j) = \sum T_i(r_i) + U(r_i) + \sum_{i<j} v(r_i,r_j) - U(r_i), \qquad 9.$$

then an energy-independent effective interaction, $\mathcal{V}$, in a truncated model space, has the form

$$\mathcal{V} = V - V\frac{Q}{H_0 - E_0}\mathcal{V}, \qquad\qquad 10.$$

where Q is the projection operator that projects onto all states outside the valence space of the model. In principle, $\mathscr{V}$ can have any particle rank from two-body to n-body even though V is only a two-body operator. The first problem to be dealt with is that in most potential parametrizations of the two-nucleon scattering and bound-state data, V has a very strong short-range repulsive component, in some cases an infinitely hard core. The short-range component is so singular that it cannot be treated in a perturbation theory. The solution to this problem is to replace V in Equation 10 by the Brueckner reaction matrix or G-matrix (see Day 1967), i.e.

$$\mathscr{V} = G - G \, \frac{Q}{H_0 - E_0} \, \mathscr{V}, \qquad \qquad 11.$$

where G is obtained from an exact solution of the problem of two particles scattering outside a closed core. There are now generally accepted techniques for solving the G-matrix equation to arbitrary accuracy. Given G, Equation 11 is used iteratively to generate a power series expansion for the effective interaction. The next problem is how to choose the central potential $U(r_i)$. There is no generally accepted prescription for this and harmonic oscillator energies and wave functions are typically used.

This leaves two major questions on convergence. One, does the effective interaction converge term by term in G? Two, for a given term in some power of G, does the intermediate-state expansion converge? It has been shown that for terms to second order in G, convergence is obtained by approximately 10 $\hbar\omega$ in excitation. For a large class of terms to third order in G, convergence appears to occur by roughly 6–8 $\hbar\omega$ of excitation. There are some terms at third order in G that do not converge to very high values of $\hbar\omega$. The question of convergence with respect to powers of G is more troublesome. For situations in which intruder states and model states overlap in energy, the series expansion for the effective operator will not converge (Shucan & Weidenmüller 1973). The best that can be hoped for in this case is asymptotic convergence so that limits can be put on the accuracy of the perturbation series.

In summary, the theory of effective shell-model interactions is incomplete. The question of how well the series converges is still essentially unanswered. There are other, nonperturbative, methods of calculating nuclear spectra in terms of the realistic interaction, and of these, the so-called e^S method (Kümmel et al 1978) seems the most accurate in principle. However, the first results with this method are not in satisfactory agreement with experiment.

Despite their inherent theoretical uncertainties as just outlined, realistic interactions have been used to calculate nuclear spectra in many different studies. The most widely used "realistic" two-body matrix elements are those constructed by Kuo & Brown (1966) in the sd and fp shell. They used the Hamada-Johnston nucleon-nucleon interaction for the free interaction and included in their calculation the bare Brueckner G-matrix plus one second-order diagram, the core-polarization diagram, in which a valence particle excites a core particle through $2\hbar\omega$. They calculated other second-order diagrams to $2\ \hbar\omega$ excitation, but found that the best agreement with the mass-18 spectrum could be obtained with only the bare term plus the core-polarization term. Kuo (1967) subsequently made some detailed refinements to the mass-18 calculations and found that best agreement was obtained when all second-order terms through $2\ \hbar\omega$ excitation were included in the calculations. The Kuo-Brown and Kuo interactions were used to calculate spectra in the mass region $A = 18$–22 (Halbert et al 1971) and $A = 34$–38 (Wildenthal et al 1971). Similar interactions were later constructed for nuclei near ^{40}Ca and ^{56}Ni (Kuo & Brown 1968) and ^{208}Pb (Kuo & Herling 1972). Results of shell-model calculations in all these regions with realistic interactions were equal to or better than results with phenomenological potential interactions.

A number of other calculations of the realistic effective interaction for two particles outside ^{16}O have been made (see Barrett & Kirson 1973, Barrett 1980). Some of the calculations are similar to Kuo-Brown calculations in philosophy but use different nucleon-nucleon interactions. The Sussex group developed a method to deduce the bare matrix elements directly from the nucleon-nucleon phase shift analysis, thus bypassing the need for a nucleon-nucleon potential. Higher-order terms in the reaction matrix elements have been studied, as have higher-order excitations in intermediate states. The net picture from all this effort remains blurred. For one thing, there are many intruder states in $A = 18$ nuclei, so there are ambiguities as to what data should be compared to theory. The calculated results have oscillated with the number of terms included in the realistic interaction calculation. For instance, in the Kuo-Brown calculation, the first-order core-polarization term made a large contribution to the $J = 0^+$ matrix elements and so had a strong beneficial effect on the ground-state energy of ^{18}O. But as successive improvements to the calculations were made, the higher-order corrections tended to cancel the core polarization and dilute the quality of agreement with experiment.

How have large shell-model calculations provided useful information about the realistic interactions? The most extensive shell-model calculations in which results with different effective two-body interactions were compared were the calculations by Halbert et al (1971) of the $A = 18$–22 nuclei, and by Wildenthal et al (1971) for the $A = 34$–38 nuclei. In various sections of those papers, shell-model results that used the Kuo and Kuo-Brown interactions and a number of more empirical interactions were compared. The qualitative features of the calculated properties of the "ground-state-band"-like states were well described by all effective interactions, including the realistic interactions. The various interactions yielded different predictions for the states outside the ground-state band, but no one interaction consistently yielded better agreement for all states. One consistent feature in these calculations is worth emphasizing. In many of the $A = 18$–22 nuclei there are sets of states that can be described as an "excited-state rotational band." The bandhead states of these bands consistently appear at too low an energy in the calculated spectrum vis-à-vis experiment. A particularly good example of this is in ^{25}Mg (Cole et al 1974). This is true for both the realistic interactions and the phenomenological potential interactions. The defect is apparently related to the centroids, $C_{jj'}$, of the diagonal matrix elements where

$$C_{jj'} = \frac{\sum_{J,T} (2J+1)(2T+1)\langle jj'JT | H | jj'JT \rangle}{\sum_{J,T} (2J+1)(2T+1)}. \qquad 12.$$

The $C_{jj'}$ are the average strengths of interactions between particles in orbits j and j'. The best available empirical interaction for sd shell calculations is the Chung-Wildenthal (C-W) interaction (Chung 1976). With this interaction, the positions of the excited-state bandheads are accurately reproduced. A principal difference between the Kuo interaction and the C-W interaction is that the centroids for the (5/2, 1/2) and (5/2, 3/2) interactions are less attractive by 200–300 keV. There is other evidence in the sd shell (Cortes & Zuker 1979) and the fp shell (Pasquini & Zuker 1977) that such shifts are called for. The effects of differences in centroids are rather small for calculations involving only a few active particles. This might suggest that the empirical shifts are an approximation to terms in the effective interaction that have three-body and higher particle ranks. Note that the equations for effective interactions derived from a microscopic theory include particle rank higher than two. Three- and four-particle forces become important only when large numbers of valence particles are present. There have been very few calculations of effective three-body forces for shell-model calculations. Goldhammer et al (Norton & Goldhammer 1971) studied the potential effects of three-body forces in the p shell, and the Glasgow group (Cole et al 1975a) studied possible effects of three-body forces on binding energies. Cortes & Zuker (1979) recently studied the effects of three-body and four-body monopole interactions in the sd shell. They present evidence that it may be possible to describe all sd shell nuclei for $A = 18$–38 with one interaction, which contains three-body and four-body effective interactions. Thus, there is a suggestion in large shell-model calculations that the shell-model effective interactions have non-negligible terms with particle rank greater than 2. Considering the uncertainty and complexity of the calculations of the two-body terms, it is unlikely that there will be calculations of these many-body operators in a perturbation approach in the near future.

In summary, the best first guess for any large shell-model calculation is certainly a calculated "realistic" interaction. Many large shell-model calculations suggest that such interactions have a consistent defect in that the average attraction between different shell-model orbits is too strong. Whether this defect arises in the assumption that the effective residual interaction is strictly two-body or whether it represents a defect in the calculation of the two-body interaction remains an open question.

LARGE-SCALE SHELL-MODEL CALCULATIONS IN THE sd SHELL

General Considerations

RATIONALE FOR CHOOSING THE sd SHELL   The mass region $A = 17$–39 offers an ideal realm in which to explore the possibilities of accounting for experimentally observed features of nuclear structure with shell-model calculations.

First, as outlined above, a reasonable and easily defined candidate for a model, one which comprises a complete major oscillator shell, exists for these nuclei. The shell gaps between the 0p and 0d, 1s orbits at $A = 16$ and between the 0d, 1s and 0f, 1p orbits at $A = 40$ are sufficiently large that, at least for the $A = 19$–37 region, the basis defined by the single-particle quantum states $0d_{5/2}$, $1s_{1/2}$, and $0d_{3/2}$ should incorporate the essential degrees of freedom necessary to model the typical low-lying positive-parity state. (Of course, an implicit if not explicit goal of any shell-model calculation is to establish in detail the quantitative accuracy of qualitative assertions about the adequacy of the assumed model space such as the preceding one.)

Second, in shell-model calculations of many-particle systems it should be possible to use the complete set of many-particle basis states allowed by the single-particle orbits that define the chosen model for as wide a range of nuclei as possible. At most, 6957 basis states are required to describe a state of given NJT in the complete sd shell space. This dimensionality happens to correspond approximately to what current computers can practically manage in the context of the Oak Ridge–Rochester programs and for the analogous (larger) dimensionalities of the Glasgow programs. Hence, calculations for all nuclei in the $A = 17$–39 region can be, and in fact have been, carried out in the complete $d_{5/2} - s_{1/2} - d_{3/2}$ basis space. The unique position of the sd shell in regard to dimensionalities can be appreciated by noting that the equivalent dimensionality for the 0p shell is 2×10^1 and for the 0f, 1p shell is 10^7.

Lastly, given the advantages in principle and in practice of choosing the sd shell region as a laboratory for large-scale shell-model studies, it is a happy circumstance that the nuclei encompassed in this region exhibit a wide variety of structural features and that the experimental exploration of these features is well developed (Ajzenberg-Selove 1978, Endt & van der Leun 1979). Thus, when we investigate the capability of large-scale shell-model techniques to account accurately and comprehensively for observed features of nuclear structure, we are confined to examples from the sd shell but we are constrained neither by a lack of diversity in the phenomena to be reproduced nor by a lack of empirical information about these phenomena. It is also rewarding that various other formulations of nuclear structure theory, both microscopic and macroscopic collective, are typically applied to sd shell nuclei. This makes it possible to study the relative merits of a wide variety of different approaches (Harvey 1968, Ripka 1968, Cusson & Lee 1972, Malik & Scholz 1967).

COMMENTS ON SPECIFIC CALCULATIONS   Motivated by the above considerations, we illustrate our review of large-scale shell-model calculations with examples taken from the sd shell. We focus upon the detailed features of two representative nuclei, ^{23}Na and ^{33}S, in addition to surveying systematic trends and average values. This juxtaposition of the particular details of a selected nucleus or two against the backdrop of average trends reflects our opinion that large-scale shell-model calculations should aim simultaneously for high accuracy in specific systems and for comprehensive, consistent reproductions of features as a function of A and Z. The significance of detailed agreement between calculation and experiment for selected levels of a particular nucleus is greatly augmented if comparable agreement is obtained in a consistent fashion for many other levels of neighboring nuclei. Likewise, the significance of a model's ability to account for the observations of one type of phenomena is enhanced if it can simultaneously account for other unrelated types of phenomena.

The calculations presented and discussed here, for ^{23}Na and ^{33}S in particular as well as for shell-wide surveys, are excerpted from a study of all sd shell nuclei; this study uniformly employed the complete $d_{5/2} - s_{1/2} - d_{3/2}$ basis space and either one or the other of two similarly derived unconstrained empirical Hamiltonians (Chung 1976, Wildenthal 1977). The Hamiltonian used for ^{23}Na (and all other systems in the $A = 17$–28 region) was obtained by iteratively adjusting a starting Hamiltonian composed of realistic two-body matrix elements calculated (Kuo 1967) for

$A = 18$ to obtain an rms best fit to 200 experimental level energies in $A = 18$–24 nuclei. The one-body energies were based on data from ^{17}O (Ajzenberg-Selove 1977). The Hamiltonian used for ^{33}S (and all other $A = 28$–39 systems) was similarly obtained by adjusting realistic two-body matrix elements calculated (Wildenthal et al 1971) for $A = 38$ to obtain an rms fit to 140 experimental level energies in $A = 32$–38 nuclei. The single-particle energies were based on data from $A = 39$ (Endt & van der Leun 1979).

The two-body matrix elements and single-particle energies for each of these two Hamiltonians were constrained to be A independent. This means that exactly the same 66 numbers which specify the one plus two-body Hamiltonian for the $d_{5/2} - s_{1/2} - d_{3/2}$ shell model were used for every nucleus from $A = 17$ through $A = 28$ and a different fixed set of 66 numbers used for $A = 28$ through $A = 39$. Ideally, the "true" effective Hamiltonian matrix elements would exhibit regular dependences upon nuclear size or A. There is strong empirical evidence (Chung 1976) that one A-independent two-body Hamiltonian cannot give good results for the entire $A = 17$–39 region. The use of two different A-independent Hamiltonians can be thought of as employing a very crude mass dependence, the dependence being a step function at $A = 28$. [The two different empirical Hamiltonians predict very similar structure for the $A = 28$ region. Thus, the A dependence of the Hamiltonian(s) might better be characterized more as a change in slope at $A = 28$ than as a step function.]

These particular two sets of one- and two-body Hamiltonian matrix elements are one stage in an evolving study of the optimal empirical Hamiltonian for complete space sd shell-model calculations. Their antecedents are the "K + 12fp" Hamiltonian described by Halbert et al (1971) and the Preedom-Wildenthal interaction (Preedom & Wildenthal 1972). The Preedom-Wildenthal and the present $A = 17$–24 Chung-Wildenthal interactions have been used extensively in calculations for sd shell nuclei that employ both the Glasgow and Oak Ridge–Rochester techniques (Preedom & Wildenthal 1972, Robertson & Wildenthal 1973, Wildenthal 1977, 1978, Wildenthal & Chung 1979, Cole et al 1975b,c,d,e, 1976, Kelvin et al 1977, Chung et al 1979, Brown et al 1978, 1980a,b,c).

Relations Between Measured and Calculated Energies

Comparison of the eigenvalues E^{NJTv} obtained from the diagonalizations of a model Hamiltonian with the energies measured for the levels with the matching quantum numbers in the corresponding actual nucleus constitutes the first stage in evaluating almost any theoretical study of nuclear structure. The comparison of theoretical to experimental energies can take many forms. Emphasis can be placed on broad-scaled aspects and directed at the qualitative features of the low-lying spectra of a cluster of neighboring nuclei, the trends of ground-state binding energies or energy differences between the ground and first excited states as functions of A, at level densities as functions of A, Z and/or excitation energy, or, simply, at the total average deviation between calculated and measured values. Alternatively, the focus can be shifted to emphasize individual systems, in which case the details of spin sequences and energy gaps and state-to-state correspondences become paramount. As we have emphasized, large-scale shell-model calculations can address all of these aspects simultaneously.

SYSTEMATIC FEATURES OF ENERGY SPECTRA AS A FUNCTION OF MASS   The rich variety of features exhibited in the energy level spectra of sd shell nuclei is illustrated in Figures 2a–d by an overview of the nuclei $A = 22$–33. These spectra illustrate the degree to which the shell-model calculations qualitatively reproduce experimental observations on individual energy spacings, ordering of spin sequences, and level densities. The doubly even (even A, even Z) nuclei ^{22}Ne and ^{24}Mg bracketing ^{23}Na (Figures 2a and b) exhibit a sequence and spacing of $J^{\pi} = 0^{+}, 2^{+}$, and 4^{+} levels that are characteristic of the spectrum of the rotational excitations of permanently deformed ellipsoids, namely $E_{J} = kJ(J+1)$. The levels in ^{22}Ne and ^{24}Mg, which by their energies (and by their electric quadrupole properties as well) can be characterized as rotational states of the ground-state shape, are noted with heavy lines in the figures. On the other hand, the spectrum of ^{24}Ne (Figure 2b), differing from ^{24}Mg by the change of two protons into neutrons, is suggestive of the 0-, 1-, and 2-phonon spectrum of a vibrating system. Likewise, the two doubly even nuclei ^{32}S and ^{34}S, which bracket ^{33}S, also have spectra suggestive of vibration (Figure 2c). These states are also indicated by heavy lines. The region in between ^{32}S and ^{24}Mg, centered around ^{28}Si, has qualitative structural characteristics that are neither clearly rotational nor vibrational. These characteristics, however, are within the compass of the same shell-model approach. The predictions of both empirical Hamiltonians are compared to the experimental spectrum for ^{28}Si in Figure 2d. Even though this nucleus is four mass units removed from the regions of data used in fixing the Hamiltonians, the essential observed features of ^{28}Si are unmistakable in the calculated spectra. This is an impressive illustration of the high degree to which all systems in the sd shell are correlated.

The conclusions to be drawn from Figures 2a–d are as follows.

1. Different qualitative structures and excitation mechanisms are suggested by the observed spectra of sd shell nuclei.
2. These features sometimes change rapidly as the numbers A, Z change.

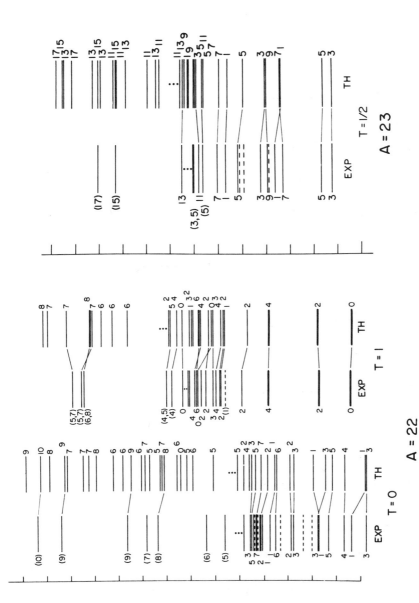

Figure 2a Calculated and experimentally measured level energies in $A = 22, 23$. Heavy lines indicate members of ground-state bands. Dashed lines indicate negative-parity intruder levels in the experimental spectra. Presumed counterparts in the experimental and theoretical spectra are connected. The excitation energy scales are marked in MeV.

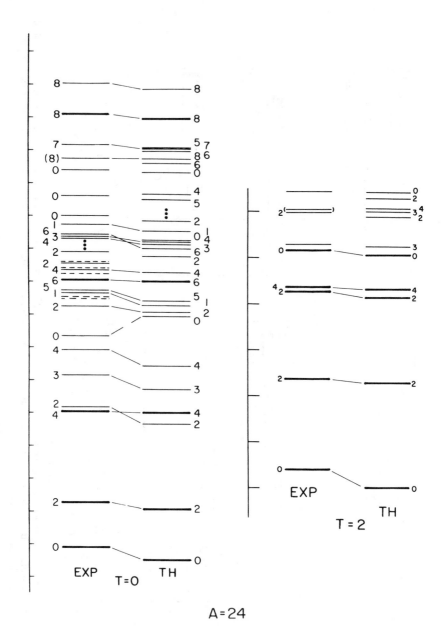

Figure 2b Calculated and experimentally measured level energies in $A = 24$. Lines and scales are as indicated in Figure 2a.

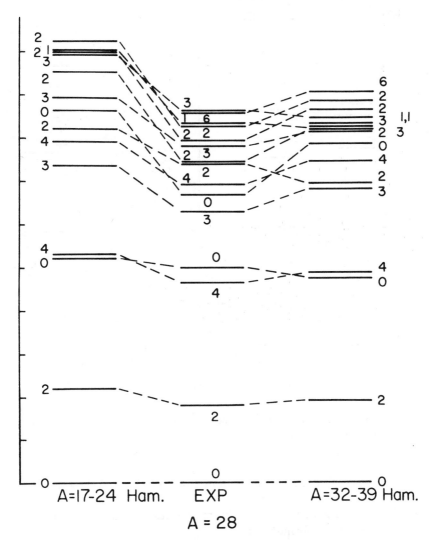

Figure 2c   Calculated and experimentally measured level energies in $A = 28$. Lines and scales are as indicated in Figure *2a*.

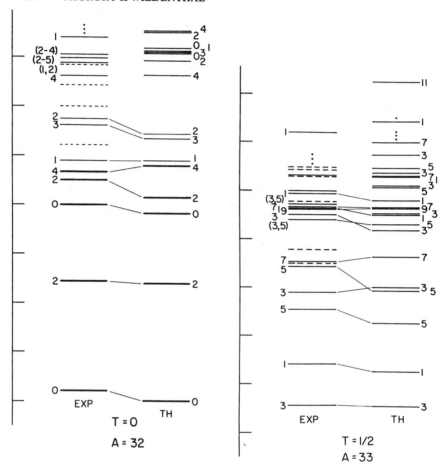

Figure 2d Calculated and experimentally measured level energies in $A = 32, 33$. Lines and scales are as indicated in Figure 2a.

3. Large-scale shell-model calculations can reproduce all these qualitatively different features correctly.
4. These different qualitative structure features emerge from internally consistent model formulations.

To elaborate upon point 4, we emphasize that spectra of completely different character are obtained from the shell-model calculations not by altering the available degrees of freedom, not by postulating a different intrinsic deformation or change in mode of excitation for the different nuclei, and not by changing from nucleus to nucleus the parameters that characterize the nature of the constituent nucleons of the model. Rather, the changes emerge unforced from the diagonalization of the same constant-valued model Hamiltonians for systems with different numbers of neutrons and protons.

DETAILS OF ENERGY SPECTRA OF ^{23}Na   A magnified view of the comparison between calculated and measured (Endt & van der Leun 1979) energy levels, focused on ^{23}Na, is provided in Figure 3. A discussion of the various aspects of the comparison illustrates some of the current issues in the area of experimental nuclear spectroscopy, issues that concern identification of levels and their spin-parity assignments, and the relationship between this area and shell model theory. Obviously, a comparison of calculated to experimental energy levels would be straightforward if all experimental

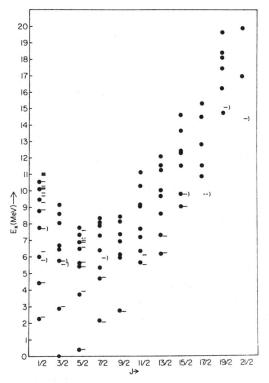

Figure 3   Calculated and experimentally measured levels of $A = 23$, $T = 1/2$ (^{23}Na). The solid circles indicate calculated excitation energies, the dashes energy levels with firmly assigned J^π, and dashes with right-hand parentheses levels with only tentative or contradictory assignments of J^π. Only eight calculated levels of each J^π value are shown and experimental levels above 6 MeV are shown only in special cases.

levels were identified, labeled with quantum numbers, and, ultimately, further characterized as to their internal structure. This situation obtains for the first 5 MeV of excitation in ^{23}Na, as is typical for odd-A nuclei in the sd shell. Within the first 5 MeV of excitation, there is a one-to-one correspondence between calculated and observed levels (nine of each) and close agreement in the values of their energies.

Above 5-MeV excitation the density of calculated levels increases rather abruptly from two to ten per MeV. At the same time, the density of experimentally observed levels also sharply increases. This is accompanied by a sharp increase in unassigned spin-parity values for these levels; it also becomes probable that some existing levels are as yet undetected experimentally. This ambiguity and incompleteness in experimental identifications and assignments of levels severely crimps our pursuit of the calculation to higher excitation energies, but further knowledge can sometimes be extracted for particular sets of J^{π} values.

Levels with large J-values can be isolated experimentally and identified against a background of lower spin states because of their different decay properties. Their densities at a given excitation energy are lower than those levels with lower J-values. We see from Figure 3 that states of $J^{\pi} = 11/2$, 13/2, and 15/2 are experimentally identified in the region of 5–9-MeV excitation and that the correspondence between these energies and the calculated values is as accurate as for those levels in the first 5 MeV of excitation. States of still higher spin are both calculated and tentatively identified, but the issue of agreement between the two sets is cloudy because of the experimental ambiguities that commence with the identity of the 9.81-MeV level. This state is identified in different experiments (Kekelis et al 1977, Evers et al 1976) as having $J^{\pi} = 17/2^{+}$ and $15/2^{+}$. The $15/2^{+}$ value is in excellent agreement with the second calculated $15/2^{+}$ energy and lower by 1 MeV than the energy of the lowest calculated $17/2^{+}$. Arguments similar to those upon which the $17/2^{+}$ assignment is based lead to assignments of $19/2^{+}$ and $21/2^{+}$ for states at 15.1 and 14.4 MeV, respectively. The $21/2^{+}$ assignment, as can be seen, is in striking disagreement with the calculated values. Final evaluation of the present theoretical results is thus partly dependent upon resolution of these experimental issues.

Exceptionally complete experimental evidence is available for states in ^{23}Na with $J^{\pi} = 1/2^{+}$ and $5/2^{+}$. We see from Figure 3 that good one-to-one correspondence between experiment and calculation persists at least up through the sixth state of each spin, corresponding to about 9 MeV for $J^{\pi} = 1/2$ and about 8 MeV for $J^{\pi} = 5/2$. Above these excitation energies, the experimental density abruptly becomes much larger than those calculated. This is the sort of breakdown to be expected for a theory, such as the shell model, which is based upon a restriction to selected "lowest energy" degrees of freedom. At high excitation energies in the actual systems the excluded degrees of freedom (single-nucleon orbits in the context of the shell-model formulation) are able to compete with the higher energy combinations of the model-allowed modes, and the observed densities then grow beyond the calculated ones. It appears that for all odd-A nuclei in the middle of the sd shell, intruder states ("intruder" in the sense that they arise from excitations excluded from the model space) begin to occur at 6–7-MeV excitation energy or at about the sixth or seventh level of a given J^π value in the model spectrum. Of course, negative-parity states, by definition intruders, begin to appear at about half this excitation energy.

The converse sort of disagreement between shell-model predictions of level densities and observation, in which levels are predicted at too low an energy or with too great a density, has no such natural origin and explanation. This sort of disagreement must result from the employment of an incorrect model or incorrect specification of the parameters of the model.

Relationships Between Nuclear States of Differing A-values

Nuclear reactions initiated with beams of projectiles in the energy range 5–100 MeV per nucleon in which a cluster of nucleons is added to or subtracted from a target (ground) state, commonly referred to as stripping and pickup reactions, respectively, tend to populate strongly only a few states in the spectra of the residual nuclei (Macfarlane & French 1960). This selectivity suggests that such experiments can provide insight into how states of a given A-value are related to states in neighboring systems of both larger and smaller A. Such information can then ultimately reveal much about how the multinucleon systems of a given model space are constructed by the successive addition of nucleons, both singly and in clusters, to the model core (^{16}O in our present example) or conversely by subtraction from the nucleus corresponding to the fully occupied model space (^{40}Ca in our present example). The shell model is particularly well suited for the description of these processes since the building blocks of the model (single-nucleon states) are the entities that are added or subtracted in the reaction processes.

GENERAL COMMENTS ON THE ANALYSIS OF PARTICLE TRANSFER REACTIONS WITH SHELL-MODEL WAVE FUNCTIONS   The relationships between states, which are qualitatively suggested by the experimental cross sections of transfer reactions, can be made quantitative via analysis with a theoretical description of the reaction process. Such analyses are typically carried out with the distorted-wave Born approximation (DWBA) (Austern 1970). The predictions of the DWBA theory typically yield angular distribution shapes (plots of the differential cross section as a function of angle) that are characteristic of the quantum of orbital angular momentum transferred by the cluster; the overall normalization of these shapes is derived from the assumption that the heavier mass state has unit overlap with the wave function formed by coupling the transferred cluster to the lighter mass state.

The comparison of calculated and experimentally observed angular distribution shapes provides a spectroscopic tool by which spins and parities of states can be assigned, but this aspect of transfer reactions does not concern us here, since we implicitly assume that the J^π values of both the final and target states are known. Rather, we are concerned with the ratios of the magnitudes of the measured and calculated differential cross sections, so that we have quantitative measures of what are strong, medium, and weak transitions. We assume that the experimental data for a given transition can be characterized in terms of a strength for each value of orbital angular momentum that can be transferred between the initial and the final state, and we assume that the predictions of DWBA have been used to "reduce" these strengths.

The reduced strength for nucleon transfer between a pair of states is proportional to the square of the matrix element of the operator for the creation or destruction of the proper number (and type) of nucleons. In terms of the shell-model wave functions Ψ^{NJT} and the nucleon creation operator $a_j^\dagger$, the reduced cross section for the transfer of a nucleon in the single-particle orbit nj can be expressed in terms of the matrix element

$$\langle \psi^{N+1,J,T} \, ||| \, a_j^\dagger \, ||| \, \psi^{N,J',T'} \rangle \qquad\qquad 13.$$

[The triple bars indicate a matrix element reduced in both J and T space. See Brussaard & Glaudemans (1977).] These matrix elements are labeled by the j-values of the single-nucleon states. Transitions of different l or j between a given pair of states are incoherent and so each matrix element in Expression 13 can, in principle, be related to a separate experimental quantity. The matrix elements (Expression 13) are large to the degree that (a) it is possible to add a nucleon in the orbit (nlj) to $\psi^{NJ'T}$, and (b) $\psi^{N+1,J,T}$ resembles this "target-plus-single-nucleon" state. The first aspect provides (via Pauli principle arguments) a measure of the occupancy of the particular orbit (nlj) in the initial state. The second aspect is particular to the chosen final state. A sum over all final states thus removes this second aspect and leaves only the first, yielding "sum rule" estimates of the occupancy of the various orbits (Macfarlane & French 1960).

For two-nucleon transfer processes the relevant shell-model matrix elements are

$$\langle \psi^{N+2,J,T} ||| (a_j^\dagger a_{j'}^\dagger)_{\Delta J;\Delta T} ||| \psi^{N,J',T'} \rangle \qquad\qquad 14.$$

The values of ΔJ and ΔT with which these matrix elements are labeled are those of the two-particle states allowed in the model space. For the $d_{5/2} - s_{1/2} - d_{3/2}$ space these are $\Delta J - \Delta T = 0\text{-}1$, $1\text{-}0$, $2\text{-}1$, $3\text{-}0$, $4\text{-}1$, and $5\text{-}0$. The number of different pairs j and j' contributing to a particular ΔJ–ΔT term equals the dimensionality of that state in the two-particle system.

The essential differences between one-nucleon transfer and two-, three-, or four-nucleon transfer are that (a) unlike one-nucleon transfer, several matrix elements can contribute coherently to a transition of a given total orbital angular momentum L in multinucleon transfer, thus giving rise to interference effects, and (b) the shell-model matrix elements for multinucleon transfer, e.g. those in Expression 14, must be projected upon the wave function of the relevant transferred cluster in order to obtain predictions for observed transfer strengths.

This second condition is a consequence of the fact that what is measured experimentally is not just the probability that state ψ^{NJT} is reached from $\psi^{N'J'T'}$ by the creation of the appropriate number of nucleons, but rather the probability that this transition takes place through the formation of an intermediate state, that of the transferred cluster. This cluster is presumed to be either the deuteron or the dinucleon $T = 1$ state for two-nucleon transfer, the triton (or mirror ^{3}He nucleus) for three-nucleon transfer, and the alpha particle for four-nucleon transfer.

Three- and four-nucleon transfers are treated analogously to two-nucleon transfer. For transitions involving only sd shell orbits, the relevant shell-model matrix elements are, respectively

$$\langle \psi^{N+3,J,T} ||| (a_j^\dagger a_{j'}^\dagger a_{j''}^\dagger)_{\Delta J = 1/2 - 13/2; \Delta T = 1/2} ||| \psi^{NJ'T'} \rangle \qquad\qquad 15.$$

and

$$\langle \psi^{N+4,J,T} ||| (a_j^\dagger a_{j'}^\dagger a_{j''}^\dagger a_{j'''}^\dagger)_{\Delta J = 0,2,4,6,8; \Delta T = 0} ||| \psi^{N,T'T'} \rangle. \qquad\qquad 16.$$

As in two-nucleon transfer, there is a correspondence between the number and dimensionality of the $\Delta J - \Delta T$ matrix elements for three- and four-particle transfer and the states of the three- and four-particle nuclei of the model space.

In order to impose the condition that multinucleon transfer pass through a cluster state, the three- and four-nucleon spectroscopic amplitudes are conventionally overlapped with the three- and four-nucleon eigenfunctions of the SU(3) operator, these solutions corresponding to the maximum spatial symmetry (the minimum relative internal motion) obtainable in the model space. These "scalar products" yield numbers that are compared to the reduced strength extracted from the analyses of experimental data under the assumption of a cluster transfer in DWBA. Two-nucleon transfer could be treated in the same approximation but, instead the various terms are conventionally treated individually.

CALCULATIONS OF SINGLE-NUCLEON TRANSFER REACTIONS   The predictions for single-nucleon transfer matrix elements for transitions to the states of ^{23}Na are shown in comparison to experimental information (Endt & van der Leun 1979) in Figure 4 in a format that displays the energy positions as well as the magnitudes for both pickup and stripping matrix elements. Cross sections are, of course, proportional to the squares of the transition matrix elements. We have chosen to display the matrix elements unsquared, that is, the square roots of the transition intensities, so as to provide a better view of the smaller values. Remember, however, that the smaller experimental values in the transfer data have large uncertainties because of residual components of the observed cross sections that may not be well described by the conventional DWBA theory.

The results shown in Figure 4 clearly demonstrate how single-nucleon transfer matrix elements provide specific and unambiguous characterizations of selected states in terms of the identity of the transferred nucleons

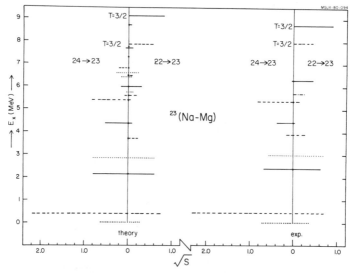

Figure 4   Calculated and experimentally based reduced strengths (shown as the square roots of the so-called spectroscopic factors) for transitions leading to states of ^{23}Na and ^{23}Mg. Transitions characterized by $l = 2, j = 3/2$ transfer are indicated by dotted lines, $l = 2, j = 5/2$ transfers by dashed lines, and $l = 0, j = 1/2$ transfers are indicated by solid lines.

and the magnitudes of the matrix elements. These characterizations relate the states in the residual nucleus, in this example the $A = 23$ system, to the ground-state wave functions of the target nuclei, in these examples $A = 22$ and 24. It is also clear from Figure 4 that there is very close agreement between theory and experiment both in the overall pattern of the spectra and in individual fine details. Surveys of the data of the sd shell confirm that this close agreement is a consistent feature of these systems (Cole et al 1976, Wildenthal & Chung 1980).

A convenient frame of reference for such a discussion of single-nucleon matrix elements is provided by noting that the stripping matrix elements have a maximum value of unity and the pickup matrix elements a maximum value of $\langle j \rangle^{1/2}$ where $\langle j \rangle$ is the maximum possible occupation of the orbit j in the target nucleus. The matrix elements presented in Figure 4 show that the $3/2^+$ ground state of ^{23}Na does not have a large overlap via $a^\dagger_{d_{3/2}}$ with the ground state of either ^{22}Ne or ^{24}Mg. This suggests that the structure of this state is dominated by $d_{5/2}$ configurations. This is the structure of the model wave function. The second $3/2^+$ state has larger matrix elements for both stripping and pickup. The model wave functions indicate that the $d_{3/2}$ admixture in ^{24}Mg is essentially exhausted by these two transitions but that about half the available strength for adding a $d_{3/2}$ nucleon to ^{22}Ne lies at energies higher than those experimentally or theoretically studied.

The $5/2^+$ (first excited) state of ^{23}Na has large matrix elements connecting it via $a^\dagger_{d_{5/2}}$ to both ^{22}Ne and ^{24}Mg. This state closely resembles the idealized state to be obtained by either adding or subtracting a $d_{5/2}$ particle to the respective target states. Remaining detectable strength for $d_{5/2}$ transitions to states with $T = 1/2$ symmetry is distributed over the second and third $5/2^+$ states in a distinct pattern; the second state having almost no connection to ^{24}Mg but a measurable connection to ^{22}Ne, and the third having quite an appreciable pickup matrix element but a very small stripping matrix element.

Analogous information on states in the mass-33 region can be deduced from a similar analysis of the single-particle strengths for nuclei in that mass region. The only significant discrepancy is in the $d_{5/2}$ pickup reactions. There the shell model shows the strength concentrated in one state, while experiment indicates the strength is spread over two states.

In summary, those features of the experimental spectra of energy levels that are revealed by single-nucleon transfer reactions, namely the angular momentum of the transferred nucleons, the relevant reduced strengths of the transitions, and the energy distribution of these strengths, are reproduced very accurately by the present shell-model calculations. For the limited class of states whose spins are equal to the spins of the single-particle orbits of the model and whose matrix elements corresponding to single-nucleon transfer are a significant fraction of the maximum allowed values, this sort of information is the most basic and least ambiguous that exists for the characterization of nuclear levels. This type of characterization is limited because it is available only for a limited range of J-values, and because the "allowed-but-weak" transitions only exclude a certain type of structure. The experimental status of this subject is surprisingly incomplete (Endt 1977), particularly as regards the A dependence of the reduced transition strengths and the strengths for the transfer of one kind of l-value relative to the strengths for another.

CALCULATIONS OF TWO-NUCLEON TRANSFER REACTIONS   Two-nucleon transfer data are, as mentioned, conventionally treated by generating the full theoretical differential cross-section profile and extracting its ratio to that of the corresponding experimental state. Agreement between calculated and theoretical matrix elements is then expressed in terms of these cross-section ratios at the expense of emphasizing the ratio of either experimental or theoretical strength to the intrinsic cluster transfer cross sections. The results obtained by analyzing the $^{21}\text{Ne}(^3\text{He,p})^{23}\text{Na}$ reaction (Fortune et al 1978a), presented in Table 1, are typical for such reactions in the sd shell and analysis with DWBA and the present family of wave functions in that the average deviation between theoretical and experimental matrix elements is about 25%.

For odd-A residual nuclei such as our present examples, the two-nucleon transfer can populate some states with a combination of L-values, and hence the model predictions can be tested not only on the basis of their overall magnitudes but also on the ratios predicted for the different possible

Table 1   Ratios of predicted and measured strengths $N = \sigma_{\text{exp}}/\sigma_{\text{th}}$, expressed as $\sqrt{N}$, for the population of the states of $A = 23$, $T = 1/2$ (^{23}Na and ^{23}Mg) by the two-nucleon transfer reactions $^{21}\text{Ne}(^3\text{He,p})^{23}\text{Na}$ and $^{25}\text{Mg}(\text{p,t})^{23}\text{Mg}$

State	$\sqrt{N}(^3\text{He,p})$	E_{exp} (MeV)	State	$\sqrt{N}(\text{p,t})$	E_{exp} (MeV)
$(1/2)_1$	1.37	2.39	$(5/2)_1$	1.07	0.45
$(1/2)_2$	1.03	4.43	$(5/2)_2$	0.95	3.86
$(3/2)_1$	0.79	0.00	$(5/2)_3$	1.00	5.29
$(3/2)_2$	1.17	2.98	$(5/2)_4$	1.30	5.56
			$(5/2)_5$	0.87	6.57
$(7/2)_1$	0.74	2.08	$(5/2)_6$	1.80	6.90
$(7/2)_2$	0.92	4.28	$(5/2)_7$	0.72	6.98
$(9/2)_1$	0.77	2.70	$(5/2)_8$	2.70	7.58
			$(5/2)_{T=3/2}$	0.95	7.79
$(5/2)_1$	0.99	0.44			
$(5/2)_2$	1.03	3.91			

L contributions. This latter test is made by comparing just the shapes of the observed and calculated profiles of differential cross sections. A recent high precision study (Nann & Wildenthal 1980) of ^{25}Mg$(5/2_1^+)$(p,t)^{23}Mg$(5/2_v^+)$ transitions, which can proceed via transfers of $L = 0, 2,$ and 4, has yielded angular distributions for nine (eight $T = 1/2$ and one $T = 3/2$) states of $J^\pi = 5/2^+$. (^{23}Mg is the mirror nucleus of ^{23}Na, and hence their wave functions should be essentially identical in these excitation energies.) The observed transitions have strikingly different shapes, indicative of different states having different ratios of matrix elements for $L = 0, 2,$ and 4 transfer. A comparison between observed and calculated shapes (see Figure 5 and Table 1) shows remarkably good agreement in this aspect of the structure of these states, as well as in the aspect of overall intensities.

The data for two-nucleon transfer perhaps do not offer as reliable a view into nuclear structure as do those of single-nucleon transfer because the typical cross sections are lower, and hence various multistep processes may contribute relatively more to the experimentally measured strengths. A variety of evidence suggests this sort of problem (King et al 1976, Nann & Wildenthal 1979). As in the case of single-nucleon transfer, there is a need to establish better the A dependence of experimental strengths. Only preliminary work on this topic has so far been accomplished (Fortune et al 1979a). In spite of these problems, the shell-model analyses of these data offer unique insights into certain properties of nuclear states. Existing analyses suggest that the shell-model wave functions incorporate the essential features of these properties even if uncertainty remains as to quantitative accuracies.

CALCULATIONS OF FOUR-NUCLEON TRANSFER REACTIONS   The states of ^{23}Na have also been studied experimentally (Fortune et al 1978b, Chung et al 1979, Eswaren et al 1979) with the four-nucleon transfer reaction ^{19}F(^{6}Li,d)^{23}Na. This "alpha transfer" reaction on a $J = 1/2^+$ target should populate states in ^{23}Na with $J = L \pm 1/2$, which have structures resembling alpha-like clusters with the appropriate orbital angular momentum L relative to the ^{19}F ground state. The relative values of the calculated matrix elements for such a connection are compared to experimentally based values in Figure 6. The profile of the matrix elements for the states below 6-MeV excitation is very different from that of either one-nucleon or two-nucleon stripping. With two possible exceptions, the agreement between theory and experiment is quite close. The lowest two states, $J^\pi = 3/2^+$ and $5/2^+$ are populated with moderate strength by $L = 2$, and the second $3/2^+$ state has a comparable magnitude. The predicted strength of the second $5/2^+$ state is larger than the probable measured value, although incomplete resolution of the data leaves this point unsettled. The lowest $9/2^+$ and the

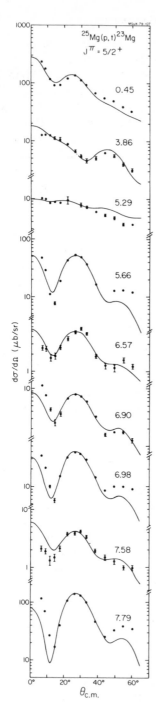

Figure 5 Theoretical angular distributions calculated with the distorted wave Born approximation from shell-model transfer amplitudes compared to the corresponding measured differential cross sections for the ^{25}Mg(p,t)^{23}Mg reaction.

two lowest $7/2^+$ states are observed to be populated via $L = 4$ and the calculated intensities agree well with the measured values, both relative to each other and to the values for the $L = 2$ transitions. Likewise, the $L = 0$ transitions to the first and second $1/2^+$ states show agreement in strength relative both to each other and to the $L = 2$ and $L = 4$ transitions. Finally, the model predicts significant $L = 6$ strength to the lowest $11/2^+$ and $13/2^+$ states. The experimental result for the $11/2^+$ state is in qualitative agreement with the prediction, and the $13/2^+$ state has not been measured directly.

Recent studies (Chung et al 1979, Anantaraman et al 1980) of the systematics of reduced alpha transfer strengths as a function of A will help to tie together the various existing studies of individual nuclei in the sd shell. Results indicate that within the $A = 20$–36 range, the shell-model predictions successfully explain the experimentally measured trends, but that for the ^{36}Ar–^{40}Ca connection in particular, the sd shell-model wave functions do not predict nearly enough strength. Studies with wave functions incorporating heuristic admixtures of fp shell configurations show that admixtures of less than 10% can easily make up the observed discrepancies between experiment and pure sd shell predictions.

In summary, many studies of the sort briefly described here suggest that the wave functions generated in large-scale shell-model calculations which

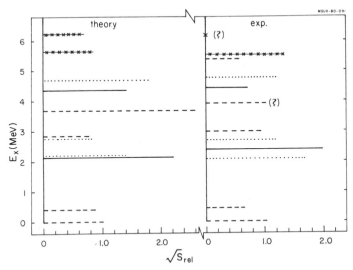

Figure 6 Comparison of the calculated and experimentally based reduced strengths (displayed as the square roots of the spectroscopic factors) for the reaction ^{19}F(^{6}Li,d)^{23}Na. The various L transfers are indicated by solid lines ($L = 0$), dashed lines ($L = 2$), dotted lines ($L = 4$), and cross-hatched lines ($L = 6$).

employ empirical Hamiltonians can, in the typical case, successfully predict the strengths for nucleon transfer reactions extracted from experimental data to within the uncertainties in these numbers. These uncertainties arise in some degree from inadequate data, to a larger extent from uncertainties in the correct formulation of the DWBA parameters, and even more to the inadequacy of the DWBA for the description of many transitions, particularly those of multi-nucleon transfer reactions.

Nuclear Moments, Electromagnetic, and Weak Interaction Transition Rates

In this section we describe how predictions for electromagnetic moments and transition rates and weak interaction decay rates are calculated from shell-model wave functions. We also discuss the assumptions underlying a comparison of these calculations to experimental reduced strengths. We then discuss what the calculated values of such observables reveal about the characteristics of these wave functions and present some comparisons between model results and experimental measurements to illustrate how well large-scale shell-model calculations account for these sorts of phenomena.

GENERAL COMMENTS   The model predictions for all moments and transition rates are based upon matrix elements of the form

$$\langle \psi^{NJT} ||| 0p^{(\Delta J)} ||| \psi^{NJ'T'} \rangle$$

$$= \sum_{j,j',\Delta T} \langle \psi^{NJT} ||| (a_j^\dagger a_{j'})_{\Delta J;\Delta T} ||| \psi^{NJ'T'} \rangle \langle \rho_j ||| 0p^{(\Delta J;\Delta T)} ||| \rho_{j'} \rangle. \qquad 17.$$

The "one-body density" matrix elements

$$\langle \psi^{NJT} ||| (a_j^\dagger a_{j'})_{\Delta J;\Delta T} ||| \psi^{NJ'T'} \rangle \qquad\qquad 18.$$

yield the probabilities that the final state can be formed by annihilating a nucleon in orbit j' from the initial state and then creating its replacement in orbit j, the pair being coupled to rank $\Delta J; \Delta T$. These matrix elements, which correspond to the nucleon transfer matrix elements (Expressions 13–16), embody the entire information content of the wave functions, and hence the Hamiltonian, which is relevant to these processes. The "single-particle" matrix elements (spme)

$$\langle \rho_j ||| 0p^{(\Delta J;\Delta T)} ||| \rho_{j'} \rangle \qquad\qquad 19.$$

express the response of the model nucleons to the operator which corresponds to the observed phenomena and, in effect, weight the probability that the initial- and final-state wave functions are connected via the $(a_j^\dagger a_{j'})$ path by the probability that the pertinent operator can effect a single-nucleon transition from orbit j' to orbit j. We note that the relative phases, as well as the magnitudes of the spme, are important in the determination of how the individual one-body density matrix elements combine to yield a total matrix element. The assumption underlying Equation 17 is that observed transitions between states ψ and ψ' are "one-body" in nature, namely that they correspond to the change in the nlj quantum number of at most one nucleon between any pair of basis vectors.

As just noted, the values of the one-body densities are uniquely defined by the choice of the orbits of the shell-model space and the matrix elements of the model Hamiltonian. The values of the single-particle matrix elements, however, are subject to a variety of interpretations. The simplest and most naively appealing of these interpretations is that the single-particle wave functions ρ_j are those of neutrons and protons with intrinsic properties equal to those measured for these particles in free space, and with spatial properties consistent with the measured size of the relevant nuclei. However, if the model matrix elements (Equation 17) are to correspond to the action in real nuclei of the same physical operator, the parameters of this operator in the context of the model must be altered ("renormalized") in ways quite analogous to those used to construct "realistic" residual interactions, as discussed above. (For a review of effective one-body operators, see Barrett & Kirson 1973.) In the limit of the impulse approximation (consideration restricted to one-body operators of the form in Equation 17), the effects of such renormalizations can be simply incorporated as changes in the values of the spme of Expression 19. Different physical operators appear to require differing degrees of re-normalization from the free nucleon values of their spme (Halbert et al 1971). The measured electric quadrupole moments of ^{17}O and ^{17}F imply that the electric charges of the model nucleons in the sd shell model we discuss here must be increased by about 0.4e each to explain E2 effects with the model wave functions. On the other hand, the magnetic dipole moments of these same $A = 17$ states are consistent in the model formulation with the free nucleon values of the M1 spme.

STATIC MAGNETIC DIPOLE AND ELECTRIC QUADRUPOLE MOMENTS   The magnetic dipole moments of the $J^\pi = 3/2^+$ ground states of ^{23}Na and ^{33}S are, respectively, $+2.22$ and $+0.64$ nuclear magnetons (n.m.). The ^{33}S value is not too dissimilar from the value ($+1.15$ n.m.) of the moment of a free neutron in an $l = 2, j = 3/2$ orbit. This is consistent with the properties of this state as viewed with the single-nucleon stripping and pickup reactions and further supports its qualitative interpretation as a neutron in the $d_{3/2}$ orbit coupled to the ^{32}S ground state. The ^{23}Na value, on the other hand, is incompatible with that ($+0.12$ n.m.) of an $l = 2, j = 3/2$ proton. We recall that the single-nucleon transfer properties of this ^{23}Na state suggested that its wave function bore little resemblance to a $d_{3/2}$ nucleon coupled to the neighboring $J = 0$ ground state of ^{22}Ne.

The static electric quadrupole moments of these same states in ^{23}Na and ^{22}S are, respectively, $+10.8$ and -6.4 e fm^2. The ^{33}S value can be compared to the value (0.0) of the moment of a free neutron in a $d_{3/2}$ orbit and to the value (-1.3 e fm^2) obtained when the same neutron has its charge renormalized ("effective charge") from 0 to $+0.35$e. The ^{23}Na value can be compared to the value (-4.6 e fm^2) of a free proton in the $d_{3/2}$ orbit a and the value (-6.2 e fm^2) obtained when the proton has its charge renormalized from 1.0 to 1.35e. Once again, the observed properties of the ^{33}S ground state are qualitatively similar to those of a $d_{3/2}$ neutron plus ^{32}S core, while the ^{23}Na properties strongly contradict an analogous interpretation. The positive quadrupole moment of ^{23}Na indicates that the charge distribution of this state is prolate ("cigar-shaped") rather than the oblate ("pancake-shaped,") distribution obtained from a wave function corresponding to a single particle orbiting a spherical core.

The experimental values of the magnetic dipole and electric quadrupole moments of ^{23}Na and ^{33}S ground states are compared to the results of the large-scale shell-model calculations in Table 2. It can be inferred from Table 2 that the ^{33}S model wave function incorporates corrections to the basic "single-particle-like" structure of this state, corrections that reduce the dipole moment value by a factor of two and increase the quadrupole moment value by a factor of three from the single-particle values, thus bringing both calculated values into approximately 10–15% agreement with the observed values. For ^{23}Na the shell model results are completely different from the single-particle predictions and reproduce the observed values with, again, 10–15% accuracy. The M1 operator used in the calculations is unrenormalized, while the E2 operator is renormalized by

Table 2 Experimental and calculated values for the static magnetic dipole (units of nuclear magnetons) and electric quadrupole (units of e fm^2) moments of the ^{23}Na and ^{33}S ground states

State	Single-particle		Shell-model		Experiment	
	M1	E2[a]	M1	E2[a]	M1	E2
^{23}Na $(3/2^+)_1$	$+0.12$	-6.2	$+2.05$	$+10.3$	$+2.22$	$+10.8 \pm 0.8$
^{33}S $(3/2^+)_1$	$+1.15$	-1.3	$+0.58$	-7.0	$+0.64$	-6.4 ± 1.0

[a] Calculated with $e_p = 1.35e$, $e_n = 0.35e$.

adding charges of 0.35e to both the neutron and proton. These simple parametrizations of these operators suffice to account for the M1 and E2 moments over the complete $A = 17$–39 region with an average accuracy equivalent to that shown for $A = 23$ and 33 (Wildenthal & Chung 1979, Brown et al 1980a, Wildenthal et al 1980).

MAGNETIC DIPOLE AND ELECTRIC QUADRUPOLE TRANSITIONS   The values of electromagnetic transition probabilities between pairs of nuclear states reveal aspects of how their structures are related to each other. Large matrix elements for electric quadrupole transitions identify pairs of states whose intrinsic macroscopic charge distributions are similar in shape. Large magnetic dipole matrix elements suggest pairs of states whose spatial symmetry is the same, the M1 operator not being able to alter the orbital angular momentum of single-nucleon states.

Values of the E2 matrix elements that reflect the connections between the first $3/2^+$, $5/2^+$, and $1/2^+$ states of ^{23}Na and other low-lying states in ^{23}Na are shown in Figure 7. Theoretical values, calculated with the same effective charges used in obtaining the quadrupole moments, are shown for all

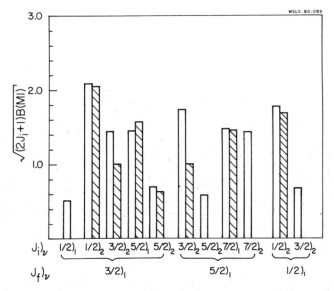

Figure 7 Comparison of magnetic dipole matrix elements extracted from experimentally measured electromagnetic decay rates with shell-model predictions based on the free-nucleon form of the M1 operator. Theoretical results are indicated for decays to the lowest $3/2^+$, $5/2^+$, and $1/2^+$ states in ^{23}Na from other low-lying states by the open rectangular bars. Existing experimental data for these same transitions are shown, without indications for the typical 10–20% experimental uncertainties, by the hatched rectangular bars.

allowed transitions, but the corresponding experimental values are not available in all cases. We note that the value of the ^{23}Na matrix element for a $d_{5/2} \rightarrow d_{5/2}$ transition of a particle with charge 1.35e is 11 e fm^2. Hence the connections via the E2 operator between some pairs of model states are significantly enhanced over the single-particle value, while others are suppressed. The pattern of these connections is "nucleus-specific" in the sense that the ^{23}Na pattern is quite distinct from, say, the ^{33}S pattern. Comparison of the existing experimental values in ^{23}Na to these predictions suggests that the model wave functions incorporate coherent many-body effects by which some matrix elements are enhanced as well as those by which the suppression of other transitions are suppressed. A survey (Schwalm et al 1977) of accurately measured (mostly large) $B(E2)$ values in the mass range $A = 20$–37 has established that the level of accuracy noted

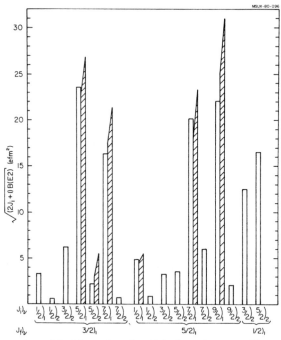

Figure 8   Comparison of electric quadrupole matrix elements extracted from experimentally measured electromagnetic decay rates with shell-model predictions based on a neutron effective charge of 0.35e and a proton effective charge of 1.35e. Theoretical results are indicated for decays to the lowest $3/2^+$, $5/2^+$, and $1/2^+$ states in ^{23}Na from other low-lying states by the open rectangular bars. Existing experimental data for these same transitions are shown by the hatched rectangular bars, with the experimental uncertainties indicated by the heights of the triangular areas on the tops of these bars.

here for ^{23}Na is typical for transitions between the lowest-lying nuclear states over the entire sd shell.

Values of the M1 matrix elements between the same pairs of states in ^{23}Na considered in the case of the E2 operator are shown in Figure 8. These matrix elements differ from those of the E2 operator in that they are, overall, small relative to the single-particle values (6.8 n.m. for a $5/2 \to 5/2$ transition). This means that the typical M1 transition is significantly hindered, either by the absence of allowed $a_j^\dagger a_{j'}$ paths or by cancellations between different paths. Consequently, these matrix elements are less characteristic of the complete wave function than are the large E2 matrix elements. In addition, as can be seen, there is less variation in magnitude from transition to transition than in the case of the E2 operator, so that qualitative characterizations of transitions on the basis of their strength is less clear cut. As is obvious from Figure 8, the model predictions not only account for the observed characteristic suppression of the M1 transition matrix elements in ^{23}Na, but also reproduce with fair accuracy the detailed variations from transition to transition of these values. The theoretical values shown are based on the free nucleon M1 spme. A renormalization of this operator may possibly improve agreement between the shell-model predictions and observation just as it did for static dipole moments (Wildenthal & Chung 1979). A particular class of M1 transitions, those connecting ground states to states at excitation energies of 5–15 MeV which can be reached by $\Delta T = 1$ transitions, uniquely characterize some states by virtue of large reduced M1 strengths (Fagg 1975). The present wave functions successfully predict these "giant M1" phenomena (Wildenthal & Chung 1980).

BETA TRANSITIONS   The form of the operator assumed to effect Gamow-Teller beta decay, $\sigma\tau$, is similar to, but simpler than, that of the $\sigma + 1$ operator assumed to effect magnetic dipole decay. A systematic study of Gamow-Teller beta decay in the sd shell (Brown et al 1978) established that the present shell-model wave functions account for the strong variation in relative decay strength as a function of A but failed, by up to 30%, to account for the absolute intensities when the free nucleon spme were used. This suggests that a renormalization of the Gamow-Teller operator may be needed for such model calculations.

OTHER ELECTROMAGNETIC PHENOMENA   Experimental measurements to determine matrix elements of the higher multipoles of the electric and magnetic operators must typically be carried out via electron scattering. Within the sd shell model, moments and transition matrix elements for M3, E4, and M5 operators can be calculated and compared with data from elastic and inelastic electron scattering experiments. Measurements on ^{23}Na have not yet been carried out. The very few data on M3 decay transitions and the more abundant but still sketchy data on E4 scattering transitions are accounted for with the same detailed and systematic accuracy that has been demonstrated for the E2, M1, and Gamow-Teller data (Brown et al 1980b,c). The electron scattering data may enable us to extend tests of the shell model's predictive power to include the question of radial distribution of the transition matrix elements or, equivalently, the momentum transfer dependence of the transition strengths. Results to date of such studies (Hammerstein et al 1973, Singhal et al 1977, 1979) show that the dominant features of these data are adequately reproduced by the present wave functions. Discrepancies at the larger q-values for some transitions may possibly arise from wave function inadequacies within the context of sd shell configurations, to the neglect of relatively important extra-sd shell configurations, to the details of the radial dependence of the single-particle wave functions, or to non-nucleonic effects.

Conclusions

The family of model wave functions for nuclei in the mass region $A = 17$–39 discussed in this section are constructed out of the complete set of multiparticle states available in the $d_{5/2} - s_{1/2} - d_{3/2}$ single-particle space. We have used unconstrained empirical residual interactions, which we believe to be the best available interactions for these shell-model calculations, and we have considered a broad variety of observables. We find that the model wave functions, when combined with empirically and theoretically sound effective one-body operators, yield predictions that agree with experimental measurement at the level of 10–20% across the board. When we consider the large number of states for which this conclusion holds, the internal consistency of the model formulation that ties them all together, and the widely different properties sampled by the different operators, we are driven to the conclusion that the theoretical wave functions constitute effective models for the actual states. Current experimental spectroscopy does not appear to yield any general class of phenomena that clearly falls outside the compass of this model formulation. Rather, the issues remaining to be solved appear to be those of detailed accuracy, namely how can the inadequacies of the present wave functions be quantitatively measured and what type of correction to or expansion of the present model will most efficiently yield improved agreement with experiment. One step in this latter direction is the recent efforts of the Glasgow group (R. R. Whitehead, personal communication) to include a more accurate treatment of Coulomb energy effects.

SHELL-MODEL CALCULATIONS IN OTHER MASS REGIONS

Nuclei in the 0p Shell

Other than the members of the 1s, 0d shell, there is only one group of nuclei for which calculations can now be carried out uniformly for all states in the complete space of the leading shell-model configurations. This is the group lying between ^{4}He and ^{16}O, corresponding to the 0p shell. The largest dimensionalities of the $0p_{3/2} - 0p_{1/2}$ space are $\sim 10^3$ times smaller than those of the 1s, 0d shell, and hence 0p shell calculations are much less demanding of shell-model techniques. For this reason, both historically and at present, most new shell-model developments, whether in the interpretation of experimental phenomena or in the exploration of the theoretical foundations, have taken place in the context of 0p shell calculations. The structure of these nuclei exhibit a wide range of multiparticle phenomena, but the limited numbers of orbits and active particles does not allow 0p shell systems to develop all of the diversity exhibited by 1s, 0d shell nuclei.

Early studies of 0p shell nuclei with empirical interactions culminated in the work of Cohen & Kurath (1965), who thoroughly explored possible variations of energy level data and constraints on one- and two-body matrix elements in determining model Hamiltonians. A series of papers (Cohen & Kurath 1970, Kurath 1973, Kurath & Millener 1975) presented calculated results for comparison to data from most of the standard spectroscopic probes. The predictions of the Cohen-Kurath wave functions have been very successful in many tests against experimental measurements. Alternative formulations of 0p shell Hamiltonians yield wave functions either not very different from the Cohen-Kurath values or not appreciably in better overall agreement with experiment.

One type of alternative 0p shell calculation used realistic two-body matrix elements (Halbert et al 1966, Hauge & Maripuu 1973). In these calculations the nuclear size parameter (which affects the values calculated for realistic two-body matrix elements) and/or the spin-orbit splitting of energies of the $p_{3/2} - p_{1/2}$ orbits are adjusted for each mass value independently to yield best agreement between calculated and observed energy level values. With this type of empirical optimization, agreement in local energies as well as for other observables, is comparable to that obtained with empirical interactions.

The importance of three-body forces has been investigated by 0p shell calculations (Norton & Goldhammer 1971) in which either empirical or realistic two-body terms are employed. The strong dependence of three-body contributions upon the number of active nucleons ensures that the incorporation of these extra degrees of freedom into the fixing of the Hamiltonian has significant effects. The determination of the degree to which this type of correction to the usual shell-model formulation is necessary remains subtle, however.

Current investigations of 0p shell structure emphasize the incorporation of configurations ($0s_{1/2}$ and $1s, 0d$) from outside the p shell space into the dominant 0p structure (Millener & Kurath 1975, Jaeger & Kirchbach 1977, Teeters & Kurath 1977). Calculations in which the model space includes orbits from more than one major oscillator shell are afflicted (except in a few special cases) by the admixture of spurious states, corresponding to the motion of the center of mass of the model nucleons, into the family of normal states that characterize relative motion among the model nucleons. These spurious states can be exactly separated from the normal states if the complete space of n $\hbar\omega$ configurations is treated. The great enlargement of the model space attendant to such exact treatment of spurious center-of-mass motion is such that the 0p shell with 1 $\hbar\omega$ excitation is the only general calculation presently feasible.

Nuclei in the 0f, 1p Shell

A third popular region for shell-model calculations is the 0f, 1p shell. ^{40}Ca is the heaviest doubly closed shell nucleus where neutrons and protons occupy the same single-particle orbits (i.e. $0f_{7/2}$, $1p_{3/2}$, $0f_{5/2}$, and $1p_{1/2}$). The dimensions of calculations in (fp)^5 spaces are on the order of 10^3, and beyond $n = 5$ complete space calculations are essentially impractical. The $f_{7/2}$ level has the largest j-value in the space and, in the spectrum of ^{41}Ca (or ^{41}Sc), it is more than 2 MeV in energy below the other orbits. The pairing force is largest for large j orbits and the effective nucleon-nucleon potential has a strong pairing force component. Because of these factors, a pure $(f_{7/2})^n$ model might be expected to be particularly good for this mass region; this has proven to be so. Many nuclei can be described by the $(f_{7/2})^n$, $n = 0$–16 model, and there are only eight two-body matrix elements in the model space, so there are many more levels than parameters in such calculations. The major early $(f_{7/2})^n$ calculations were by Talmi & Unna (1960), Ginocchio & French (1963), and McCullen et al (1964). These calculations were successful in describing many levels in these nuclei near ^{40}Ca. The calculations were recently repeated by Kutschera et al (1978) using the most recent information on the relevant two-body matrix elements.

Among the $(f_{7/2})^n$ nuclei are the Ca isotopes with $A = 41$–50. It is practical to calculate the Ca isotopes in more complete spaces because "identical particle" (maximum-isospin) spaces are much smaller than spaces with $T < T_{max}$. Thus, there have been a number of calculations of the Ca isotopes involving first $(f_{7/2}, p_{3/2})$ spaces (Raz & Soga 1965, Engeland &

Osnes 1966, Federman & Talmi 1966), and then calculations (McGrory et al 1970) in the full (fp)^n space with some restrictions on the number of particles excited to the $p_{3/2}$ and $f_{5/2}$ spaces. With each increase in the space, the number of parameters increased, and with each increase a few more low-lying levels were accounted for by the model. In the (fp)^n calculations a "realistic" interaction was used. Even in the full (fp)^n spaces there were obvious intruder states at low energies (~ 2 MeV). In ^{42}Ca and ^{44}Ca low-lying 0^+ states are experimentally observed but have no good theoretical analogs. In the (fp)^n calculations clear evidence was found that the monopole part of the two-body interaction was too strong, as discussed above.

Complete (fp)^n calculations for nonidentical particles are reported for $A = 42$–44 nuclei. The first such calculations (McGrory 1973) used realistic interactions. These calculations were recently redone (Motoba & Itonaga 1979) with particular emphasis on α transfer strengths and the question of the "rotational" nature of ^{44}Ti. The ground-state band spectrum of ^{44}Ti is distinctly less characteristic of a rotational band than is the sd shell analog of ^{44}Ti, ^{20}Ne. The SU(3) purity of ^{44}Ti depends, of course, on the residual interaction, but the largest SU(3) component in any ^{44}Ti ground-state calculation (Motoba & Itonaga 1979) is $\sim 40\%$. The comparable number in ^{20}Ne is 80%. Relatively large E2 effective charges are needed in the (fp)$^{2-4}$ calculations, and there are, if anything, more intruder states in the ^{40}Ca core calculations than in the ^{16}O core calculations. Cline (1977) studied the quadrupole moments of nuclei near ^{40}Ca and found that the signs of the quadrupole moments are opposite to those predicted by spherical shell models. All this indicates that core-excitation effects are critical near the ^{40}Ca closed shell and, as discussed below, are not yet accounted for in any large-basis spherical shell model.

Nuclei in the A = 90 Region

Another popular "good closed-shell" region is the $A = 90$ region, where ^{88}Sr ($Z = 38$, $N = 50$) is the closed core. The active proton orbits are the $p_{1/2}$ and $g_{9/2}$ orbits. The lowest neutron orbits are the $d_{5/2}$, $s_{1/2}$, $d_{3/2}$, and $g_{7/2}$ orbits. The inclusion of a relatively high spin orbit such as the $g_{9/2}$ orbit leads to a rather rich spectroscopy of high spin states. There have been a number of calculations in this mass region involving empirical and realistic interactions (Talmi & Unna 1960, Auerbach & Talmi 1965, Bhatt & Ball 1965, Vervier 1966, Cohen et al 1969, Ball et al 1972). More recently, Gloeckner & Serduke (1974) redid the pure proton $(p_{1/2}, g_{9/2})^n$ calculation, and they included in their "fit" certain E2 and M4 transition rates. The inclusion of these transitions in the search introduced a dependence on the phase of the two-body matrix elements that was not present in energy level fits. Serduke et al (1976) studied $N = 49$ nuclei and found evidence for an isotopic-spin nonconserving component of the effective residual interaction, but they point out that the nonconservation could be a truncation effect and not necessarily a fundamental property of the nucleon-nucleon force.

A recent calculation of particular interest in this mass region involves the sudden appearance of a ground-state rotational band in the Zr isotopes around $A = 100$. Outside the $N = 50$ core, one of the valence neutron orbits is the $g_{7/2}$ orbit. Federman & Pittell (1979) show that the sudden onset of deformation in the Zr isotopes essentially coincides with the case where all neutron orbits up to the $g_{9/2}$ are filled. As the neutrons populate the $g_{7/2}$ orbit, the strong n-p interaction between the $g_{7/2}$ neutron and $g_{9/2}$ proton produce the observed deformation.

Nuclei in the ^{208}Pb Region

The prevalent consensus is that ^{208}Pb is the best shell-model closed core. There have been numerous calculations of the two-particle systems around ^{208}Pb (i.e. ^{208}Pb, ^{210}Po, ^{208}Bi, etc) (see McGrory & Kuo 1975 for references). The largest systematic calculations (McGrory & Kuo 1975) in the region are for systems of from 2 to 4 protons (^{210}Po, ^{211}At, ^{212}Rn), 2 to 4 neutrons (^{210,211,212}Pb), and 2 to 4 neutron holes (^{206,205,204}Pb), wherein the residual interactions were constructed by Kuo & Herling (1972). The calculations account for much of the systematic data on energy levels, transfer cross sections for one- and two-particle transfers, and electromagnetic transitions. The calculations involve single-particle orbits with $j = 11/2, j = 13/2,$ and $j = 15/2$ so there are states with very high spins in the model space. Thus, the shell model should be very useful in studying high spin phenomena around ^{208}Pb. The Stockholm group (Lindén et al 1978) have already made such studies. They use, as much as possible, empirical residual interaction matrix elements deduced directly from "two-particle" spectra. Their calculated spectra agree with experiment for high spin states in some three- and four-particle systems to within 5–10 keV in many cases.

Shell-Model Calculations with a ^{56}Ni Core

The shell-model calculations in the p shell, the sd shell, and the fp shell described previously assume cores that are closed in both LS and j-j coupling. There are a number of calculations in the region around ^{56}Ni in which ^{56}Ni is assumed to be a closed core in j-j coupling, i.e. the ground state is ^{40}Ca $+ (f_{7/2})^{16}$, and the valence active orbits are the $f_{5/2}$, $p_{3/2}$, and $p_{1/2}$. The dimensions in this model space are identical to the sd shell calculations described above. In ^{57}Ni the $p_{1/2}$ orbit, the highest single-particle orbit, is at ~ 1.1 MeV, so the three orbits are very close in energy.

One might expect strong configuration mixing. One could also expect core-excitation effects to be strong here, since the gap between the $f_{7/2}$ orbit and the $p_{3/2}$, $p_{1/2}$, and $f_{5/2}$ orbits is probably only 3–4 MeV. The most extensively studied nuclei are the Ni isotopes. The Ni isotopes have been treated as vibrational nuclei in the collective model, so the shell-model calculations possible for the Ni isotopes are a good vehicle for studying the structure of vibrational wave functions. The ^{56}Ni core shell-model calculations (Auerbach 1967, Cohen et al 1967, Hsu 1967) in general are quite successful in treating all the known Ni isotopes. To do so, however, the quadrupole effective neutron charge must be very large ($\sim 2.0e$). The measured magnetic moments in the Ni isotopes cannot be described in the ^{56}Ni core calculations (Hass et al 1978). The M1 renormalizations due to $f_{7/2} - f_{5/2}$ neutron excitations are too large and too mass dependent to be absorbed in one effective operator. The shell-model calculations reproduce all the so-called vibrational characteristics of the Ni isotopes. However, the seniority of the "two-phonon" states is predominantly two, while true two-phonon states would have seniority four; this casts serious doubt on the validity of the vibrational model here. The structure of the Cu and Zn isotopes have been calculated (Koops & Glaudemans 1977, van Hienen et al 1976) in a $(p_{3/2}, p_{1/2}, f_{7/2})^{4-56}$ space. The calculations are in reasonably good agreement with experiment, but not as quantitatively successful as are the (sd)^n calculations. There have been a number of successful calculations (Horie & Ogawa 1973, Morrison et al 1975, Bendjaballah et al 1977, McGrory & Raman 1979, Vennink & Glaudemans 1980) of the Fe isotopes where protons are restricted to the $f_{7/2}$ shell and neutrons are in the same orbits as in the Ni calculations discussed above.

The large E2 effective charges and the strong state dependence of the M1 operator suggest that the $(f_{7/2})^{16}$ approximation of ^{56}Ni is not a good one. There have been a series of calculations (Wong & Davies 1968, Pittel 1970, Oberlechner & Richert 1972) of the ground states of ^{56}Ni, where various $2p-2h$ and $4p-4h$ states are included in the model space. Most of the calculations use realistic interactions and treat the $f_{7/2} - p_{3/2}$ splitting as a parameter. There is little consistency among the various calculations. Pasquini & Zuker (1977) pointed out the importance of the monopole interactions (see Equation 12) between the $f_{7/2}$ particles and the $p_{3/2}$, $p_{1/2}$, and $f_{5/2}$ orbits. The average interaction of a particle in orbit j_1 with the closed-shell configuration $(f_{7/2})^{16}$, is 16 times the $7/2 = j_1$ monopole interaction. These interactions act as mass-dependent, single-particle energies. Because of the large factor, 16, a small error in the monopole can have a large impact on the effective single-particle energy of the orbit j; this in turn can have a large effect on the zero-order position of np–nh states. As discussed above, there is strong evidence that the realistic energy monopoles are too strong. By reducing the strength of the monopole, the ground-state admixture of the $(f_{7/2})^{16}$ configuration is enhanced. This suggests that with a more accurate effective interaction, the $(f_{7/2})^{16}$ assumption for ^{56}Ni may appear more reasonable, but there is no conclusive calculation on this point to date.

This still leaves the question of the strong renormalizations. In the sd shell a $(d_{5/2})^{12}$ approximation to the ground state of ^{28}Si would be analogous to the $(f_{7/2})^{16}$ assumption for ^{56}Ni. The exact diagonalization of the Si isotopes in the sd shell is practical. It would be enlightening to compare such calculations to calculations of the Si isotopes with a ^{28}Si $(d_{5/2})^{12}$ core to see explicitly the origin of the large renormalizations around j-j-coupled semiclosed shells.

Shell-Model Calculations of "Closed-Shell" Nuclei

We have presented considerable evidence that in all mass regions studied to date, there exist sets of low-lying discrete states, which are well described by conceptually simple configuration-mixing shell-model calculations. All these models assume an inert core. But for all the "cores" discussed here, there is experimental evidence of a rich spectroscopy of excited states that must be intruders. Good agreement between theory and experiment can be obtained in many cases only by ignoring some low-lying intruder states. For instance, in the sd shell nuclei, the $(sd)^n$ model can only describe positive-parity states. But there are many observed low-lying negative-parity "intruder" states in nuclei around ^{16}O, e.g. there is a $J = 1/2^-$ state almost degenerate with the ^{19}F ground state. In addition, there are often intruder states that cannot be identified as such a priori. They are usually identified as intruders only after many efforts to fit them into the shell-model calculations have failed.

Several calculations of "closed-shell" nuclei and nuclei near closed shells attempt to provide a shell-model description of intruder states. The most successful of these involve ^{16}O and nuclei in that mass vicinity. These calculations describe the properties of many levels in nuclei in the mass-16 region. But perhaps more important is the pedagogical insight they provide to the structure of shell-model calculations. In these calculations, the model core was assumed to be ^{12}C. The active orbits were the $p_{1/2}$, $d_{5/2}$, and $s_{1/2}$ orbits. Since the p orbit and the sd orbits have opposite parity, the model includes states of both parities. In the first calculations in this model (Zuker et al 1968) the single-particle energies and two-body matrix elements were more or less hand adjusted to give an optimal description of the ^{16}O spectrum. The resulting single-particle energies are close to the energies of lowest $1/2^-$, $5/2^+$, and $1/2^+$ states observed in ^{13}C, the "single-particle nucleus" in the model. The interactions determined by fitting ^{16}O were subsequently used (McGrory & Wildenthal 1973) to calculate the structure of nuclei with $A = 18$–20. In addition, using data from $A = 13$–20 nuclei, a computer search on all the matrix elements in the calculation (30 two-body matrix elements plus 3 single-particle energies) was made (Reehal & Wildenthal 1973) to find a "best-fit" interaction. All the calculations did a remarkable job of describing most of the features of low-lying states in these $A = 13$–20 nuclei. The three distinctive experimental features of ^{16}O that are qualitatively reproduced by the calculation are

1. a large gap between the ground state and all excited states,
2. a rotational band of states starting with the 0^+ at 6.06 MeV which are primarily 4p – 4h states, and
3. a rotational band of 2p – 2h states starting with a $J = 0^+$, $T = 0^+$ state around 12 MeV.

As a representative sample, the observed spectra for the $A = 19$ systems are shown in Figure 9, as are the calculated $A = 19$ spectra. The most distinctive feature here is the existence of a band of states with negative parity that are essentially degenerate with the positive-parity ground-state band. In all the calculations the relative binding energies of all the nuclei are qualitatively reproduced.

The results in these calculations were obtained after much parameter adjusting, so they are hardly "first principles" calculations. Nevertheless, they are a dramatic example of the ability of the shell model to describe simultaneously states usually referred to as spherical states and deformed states where the final parameters of the model appear to be quite reasonable.

Beyond this qualitatively satisfactory result, these "open-shell" calculations provide significant insight into the structure of shell-model calculations. In the "open-shell" calculations, 60% of the ground-state wave function of ^{16}O is in the closed-shell configuration $(p_{1/2})^4$. In the usual shell-model assumption for the sd shell, this admixture is assumed to be 100%. Similarly, the ground-state wave functions in the $A = 17$–20 mass regions have very large (30–50%) admixtures of core-excited states. But the calculated spectra for these nuclei still reproduce the same data that was reproduced by the shell-model calculations in which an inert ^{16}O core was assumed. For instance, the $J = 5/2^+$ ground state of ^{17}O has 35–40% "core-excited" configurations in the open-shell models. However, the calculated spectroscopic factor for the $^{16}O(d,p)^{17}O$ ground-state transition is approximately equal to unity, the value that results when ^{17}O is described as a $d_{5/2}$ particle outside a closed ^{16}O core. This reflects the fact that the calculation of the spectroscopic factor involves an overlap function, and in both the closed-shell ^{16}O and open-shell ^{16}O models, the structure of the ground state of ^{17}O is one $d_{5/2}$ particle coupled to the ground state of ^{16}O. For the calculation of most overlaps, the structure of the core is irrelevant, since one integrates over all the core coordinates. Zuker (1969) analyzed the wave functions of many of the states in these nuclei in the open-shell model and found that many states could be described as sd shell particles coupled to the ground-state wave function of ^{16}O in the open-shell model. These results are analogous to the well-known pseudonium calculations of Lawson & Soper (1967). They first performed calculations in an $(f_{7/2}, d_{3/2})^n$ model space with a given interaction. The results of this two-shell calculation were used as experimental data to be "fit" into a pure $(f_{7/2})^n$ model. The $(f_{7/2})^n$ model reproduced most of the "data" on low-lying states. The shell-model calculations could hide a great deal of two-shell configur-

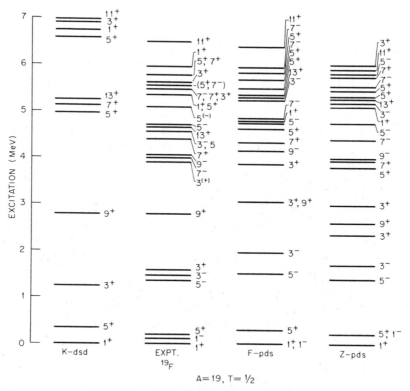

$A = 19$, $T = 1/2$

Figure 9 Calculated and experimentally observed spectra of ^{19}F. Spectra labeled K-dsd are calculations in the $(sd)^n$ space with closed ^{16}O core. Spectra labeled F-pds and Z-pds represent calculations of ^{19}F in a model space where ^{12}C is a closed core, and the active valence orbits are the $0p_{1/2}$, $0d_{5/2}$, and $1s_{3/2}$, so that the ^{19}F calculation involves seven active particles. Different effective residual interactions were used to obtain the two calculated spectra.

ation mixing. The pseudonium results were held by many to be a condemnation of the shell model. In fact, the pseudonium results and the broken ^{16}O core calculations are necessary corollaries to shell-model theories. There is no question of finding an exact solution of the many-body problem. If it were not possible to ignore many effects of core excitation, there would be little hope of doing any systematic studies of nuclear structure in a microscopic model. Such calculations issue a not-to-be-ignored caveat as to the interpretation of shell-model results, and they also provide a basis for optimism in performing shell-model calculations.

The success of the shell-model calculations in the $A = 16$ region with an open ^{16}O core led naturally to a similar attempt around the ^{40}Ca region. As is the case for nuclei with $A \approx 16$, there are many "intruder" states in the nuclei around $A = 40$. The qualitative n-particle n-hole (np-nh) structure of many states in the ^{40}Ca region has been determined. Much of the structure can be interpreted with phenomenological wave functions first suggested for this region by Gerace & Green (1967); these wave functions are expressed as linear combinations of spherical states and np-nh states constructed from a deformed basis. Data on one-, two-, and four-particle transfer can be accounted for by such models (Fortune et al 1979b and references therein). Recently, such wave functions were used to describe E0 transition data successfully (Fortune & Dieperink 1979). There have been several attempts to perform spherical shell-model calculations of these same states in a model analogous to the ^{12}C core model for ^{16}O. All the published calculations performed in a completely spherical basis include configurations of the type $(d_{3/2}^{n_1}, f_{7/2}^{n_2})$. The earliest of these were by Federman & Pittel (1969) and by Zuker (1971). The calculations in these spaces apparently accounted for the ground-state binding energies and excitation energies of a number of $J = 0^+$ and $J = 2^+$ states, and in some cases $B(E2)$ values for transitions between these low spin states. However, subsequent detailed studies of two- and four-particle transfer data indicated these two-shell wave functions were inadequate to describe the observed strengths in these transfers. A good example of this is a study of the $^{42}Ca(p,t)^{40}Ca$ reaction by Seth et al (1974). To our knowledge, no calculation in a $(d_{3/2}^{n_1}, f_{7/2}^{n_2})$ space accounts for all transfer and decay data to anywhere near the accuracy found for the analogous calculations in the $A = 16$ region. The fundamental problem is easily seen. The existence of low-lying intruder states is qualitatively accounted for by models where the underlying nuclear fields were strongly deformed (Gerace & Green 1967). In the sd shell the deformed single-particle states are strongly mixed linear combinations of s and d orbitals. In the ^{12}C core calculations the $d_{5/2}$ and $s_{1/2}$ orbits are included explicitly in the model space, and thus the model space has the ingredients to describe some deformation properties. In the

^{40}Ca region the deformed states in the fp shell are strong fp admixtures. A pure $f_{7/2}$ model cannot produce strong deformations. At a minimum, the shell-model space needed to describe the "intruder" states around ^{40}Ca should include $(d_{3/2}^{n_1}, f_{7/2}^{n_2}, p_{3/2}^{n_3})$ configurations, and most likely the $s_{1/2}$ hole orbit will also play an important role. Systematic shell-model calculations in such model spaces will probably remain impractical in the near future.

Shell-Model Calculations with No Core

Finally, there have been a few attempts to perform no-core shell-model calculations. Irvine et al (1976) studied a number of nuclei in the 0p shell in a model space where all orbits from the 0s to the (0f, 1p) shell were active and a variety of configurations with various degrees of excitation included. The spaces chosen were big enough that states with different center-of-mass motion appeared to be well separated. The residual interactions used were essentially realistic interactions constructed from the Reid nucleon-nucleon potential, where short-range effects were included through variational correlation functions. The calculations are being repeated with improved residual interactions. Ceuleneer et al (1976) reported a calculation of ^{6}Li in a model space where all configurations within 4 $\hbar\omega$ of the lowest (0s^4, p^2) configurations were included (i.e. effects of orbits through the 2p, 1f, 0h shell). They use the Sussex realistic interaction for the p shell. The calculated energy levels were within 1–2 MeV of the observed levels. Significant effects of 4 $\hbar\omega$ excitations were found, a measure of the slowness of convergence of the "realistic" calculations. Such no-core calculations of very light systems are useful as a measure of the ultimate capabilities of the shell model and of the convergence rate of such calculations. Again, technical capabilities preclude such calculations much beyond the 0p shell.

SUMMARY

In the last two decades the simultaneous rapid development of the techniques of shell-model calculations and of computer technology have made it possible to perform shell-model calculations in large multiparticle, multiorbit bases with reasonable economy. In this review, we have tried to illustrate the quantitative accuracy with which large-basis shell-model calculations of energy levels, particle-transfer strengths, and electromagnetic moments and transition rates reproduce observed nuclear data. We have used the sd shell as an example because it covers a wide range of nuclear species, the nuclei therein display many interesting systematic features that are often described in terms of collective variables, and because the same, simply defined, "complete" shell model can be applied throughout the entire shell. For the types of data we have discussed, there are no striking discrepancies for either the systematic features viewed over the entire mass region, or for extensive details in any one nucleus. We have pointed to similar calculations in other nuclear mass regions. Taken collectively, the results of these calculations suggest the shell model in reasonably complete spaces is a useful model in all mass regions.

What does the future hold in this field? Improvements in shell-model techniques may make for more efficient shell-model calculations, and there will certainly be larger and faster computers available. But shell-model space dimensions increase much more rapidly than will any hoped-for technical improvement, and it is not likely that any more "complete" shells will become as accessible as the sd shell. Thus, future efforts in the area of the shell model will be in two general areas: (a) searches for effective truncation schemes, and (b) application of existing shell-model capabilities to new types of observables.

There are, of course, a diversity of approximations to shell-model calculations, i.e. various derivatives of Hartree-Fock calculations (see Goodman 1979), group-theoretical truncation schemes such as SU(3) (Harvey 1968) and pseudo-SU(3) models (Arima et al 1977, Strottman 1977), and hybrid models involving particle core and cluster core coupling. One approach, a weak-coupling scheme (Arima & Hamamoto 1971), has been recognized for some time but has not been extensively pursued. In general terms, this involves solving for the eigenstates of the neutron and proton Hamiltonians exactly, generating a truncated neutron-proton space by directly coupling the low-lying eigenstates in the separate neutron and proton spaces, and diagonalizing the neutron-proton Hamiltonian in this truncated space. This approach has been studied in the region of the iron isotopes (Horie & Ogawa 1973, Paar 1972) and near the ^{208}Pb core (Lindén et al 1978, Strottman 1979a) with considerable success. These are regions where the neutrons and protons are filling inequivalent orbits, so that the neutron-proton interaction is relatively weak. The scheme is surprisingly successful (McGrory 1980) in the sd shell where the neutron-proton interaction is strong. Such an approach is philosophically very close to the interacting boson model (Arima & Iachello 1978), which has enjoyed great success recently.

Large shell-model calculations will play an essential role in such frontiers of nuclear physics as intermediate and high energy electron scattering, meson physics, and heavy ion physics. A major part of the electron scattering program will center around the precise determination of charge densities and transition densities. Only very preliminary efforts have been made to calculate the radial dependence of the effective transition operators, which can be extracted from electron scattering experiments. Nuclear structure studies with pion probes will demand the best possible nuclear wave functions. There is already evidence (Strottman 1979b) that the calculated strengths of π-induced charge exchange reactions (π^-, π^0) are very sensitive to nuclear wave functions. A major part of the low energy heavy ion physics program has involved, and will continue to involve, nuclear spectroscopy of high spin states and of low spin states of nuclei far from stability.

The ability of the nuclear shell model to give a consistent quantitative accounting of almost all aspects of nuclear structure over the entire mass range is an outstanding feature of modern nuclear physics. The model will remain a vital tool as the search for new facets of nuclear physics and a deeper understanding of the nucleus as a many-particle quantum mechanical system moves forward.

Literature Cited

Ajzenberg-Selove, F. 1977. *Nucl. Phys. A* 281:1–148

Ajzenberg-Selove, F. 1978. *Nucl. Phys. A* 300:1–224

Anantaraman, N., Fulbright, N. W., Stwertka, P. M. 1980. *Phys. Lett.* In press

Arima, A., Hamamoto, I. 1971. *Ann. Rev. Nucl. Sci.* 21:55–92

Arima, A., Iachello, F. 1978. *Ann. Phys. (NY)* 111:201–38

Arima, A., Sebe, T., Inoue, T., Akiyama, Y., Shirai, K. 1977. *Proc. Int. Conf. on Nucl. Struct., Tokyo,* p. 269. Tokyo: Int. Acad. Print. 883 pp.

Arvieu, R., Moszkowski, S. A. 1966. *Phys. Rev.* 145:830–36

Auerbach, N., Talmi, I. 1965. *Nucl. Phys.* 65:458–80

Auerbach, N. 1967. *Phys. Rev.* 163:1203–18

Austern, N. 1970. *Direct Nuclear Reaction Theories,* New York: Wiley. 390 pp.

Ball, J. B., McGrory, J. B., Larson, J. S. 1972. *Phys. Lett. B* 41:581–84

Barrett, B. R., Kirson, M. W. 1973. *Adv. Nucl. Phys.* 6:219–82

Barrett, B. R. 1980. *The (p, n) Reaction and the Nucleon-Nucleon Force,* Telluride, Col., March 29–31, 1979, ed. C. D. Goodman, pp. 57–63. New York: Plenum

Bendjaballah, N., Delaunay, J., Jaffrin, A., Nomura, T., Ogawa, K. 1977. *Nucl. Phys. A* 284:513–33

Bhatt, K. H., Ball, J. B. 1965. *Nucl. Phys.* 63:286–304

Brown, B. A., Chung, W., Wildenthal, B. H. 1978. *Phys. Rev. Lett.* 40:1631–34

Brown, B. A., Chung, W., Wildenthal, B. H. 1980a. *Phys. Rev. C.* In press

Brown, B. A., Chung, W., Wildenthal, B. H. 1980b. *Phys. Rev. C.* In press

Brown, B. A., Massen, S. E., Chung, W.,

Wildenthal, B. N., Shibata, T. A. 1980c. *Phys. Rev. C.* In press

Brussaard, P. J., Glaudemans, P. W. M. 1977. *Shell Model Applications in Nuclear Spectroscopy.* Amsterdam: North-Holland. 452 pp.

Ceuleneer, R., Erculisse, M., Gilles, M. 1976. *Phys. Lett. B* 65:101–2

Chung, W. 1976. *Empirical renormalizations of shell model Hamiltonians and magnetic dipole moments of sd-shell nuclei.* PhD thesis. Mich. State Univ., East Lansing. 135 pp.

Chung, W., Fortune, H. T., Wildenthal, B. H. 1979. *Phys. Rev. C* 19:530–32

Cline, D. 1977. *Physics of Medium Light Nuclei,* pp. 89–119. Bologna, Italy: Editrice Compositori. 544 pp.

Cohen, S., Kurath, D. 1965. *Nucl. Phys.* 73:1–41

Cohen, S., Kurath, D. 1970. *Nucl. Phys. A* 141:145–57

Cohen, S., Lawson, R. D., Macfarlane, M. H., Pandya, S. P., Soga, M. 1967. *Phys. Rev.* 160:903–15

Cohen, S., Lawson, R. D., Pandya, S. P. 1968. *Nucl. Phys. A* 114:541–50

Cohen, S., Lawson, R. D., Macfarlane, M. H., Soga, M. 1969. *Phys. Lett.* 10:195–98

Cole, B. J., Watt, A., Whitehead, R. R. 1974. *Phys. Lett. B* 49:133–36

Cole, B. J., Watt, A., Whitehead, R. R. 1975a. *Phys. Lett. B* 57:9–12

Cole, B. J., Watt, A., Whitehead, R. R. 1975b. *J. Phys. G* 1:17–34

Cole, B. J., Watt, A., Whitehead, R. R. 1975c. *J. Phys. G* 2:213–29

Cole, B. J., Watt, A., Whitehead, R. R. 1975d. *J. Phys. G* 3:303–23

Cole, B. J., Watt, A., Whitehead, R. R. 1975e. *J. Phys. G* 9:935–55

Cole, B. J., Watt, A., Whitehead, R. R. 1976. *J. Phys. G* 2:501–18

Cortes, A., Zuker, A. P. 1979. *Phys. Lett. B* 84:25–30

Cusson, R. Y., Lee, H. C. 1972. *Nucl. Phys. A* 211:429–62

Day, B. D. 1967. *Rev. Mod. Phys.* 39:719–70

Endt, P. M. 1977. *Atom. Data Nucl. Data Tables* 19:23–61

Endt, P. M., van der Leun, C. 1979. *Nucl. Phys. A* 310:1–751

Engeland, T., Osnes, E. 1966. *Phys. Lett.* 20:424–27

Eswaren, M. A., Gove, H. E., Cook, R. E., Draayer, J. P. 1979. *Nucl. Phys. A* 332:525–32

Evers, D., Denhofer, G., Assmann, W., Harasim, A., Konrad, P., Ley, C., Rudolph, K., Sperr, P. 1976. *Z. Phys. A* 280:287–307

Fagg, L. W. 1975. *Rev. Mod. Phys.* 47:683–711

Federman, P., Pittel, S. 1969. *Nucl. Phys. A* 139:108–12

Federman, P., Pittel, S. 1979. *Phys. Rev. C* 20:820–29

Federman, P., Talmi, I. 1966. *Phys. Lett.* 22:469–72

Fortune, H. T., Powers, J. R., Middleton, R., Nann, H., Wildenthal, B. H. 1978a. *Phys. Rev. C* 18:1–8

Fortune, H. T., Powers, J. R., Middleton, R., Bethge, K., Pilt, A. A. 1978b. *Phys. Rev. C* 18:255–64

Fortune, H. T., Bland, L., Middleton, R., Chung, W., Wildenthal, B. H. 1979a. *Phys. Lett. B* 87:29–31

Fortune, H. T., Al-Jadie, M. N. I., Betts, R. R., Bishop, J. N., Middleton, R. 1979b. *Phys. Rev. C* 19:756–64

Fortune, H. T., Dieperink, A. E. L. 1979. *Phys. Rev. C* 19:1112–13

French, J. B., Halbert, E. C., McGrory, J. B., Wong, S. S. M. 1969. *Adv. Nucl. Phys.* 3:193–256

Gerace, W. J., Green, A. M. 1967. *Nucl. Phys. A* 93:110–32

Ginocchio, J. N., French, J. B. 1963. *Phys. Lett.* 7:137–39

Glaudemans, P. W. M., Brussaard, P. J., Wildenthal, B. H. 1967. *Nucl. Phys. A* 102:593–601

Gloeckner, D. H., Serduke, F. J. 1974. *Nucl. Phys. A* 220:477–92

Goodman, A. L. 1979. *Adv. Nucl. Phys.* 11:263–366

Halbert, E. C., Kim, Y. E., Kuo, T. T. S. 1966. *Phys. Lett.* 20:657–61

Halbert, E. C., McGrory, J. B., Wildenthal, B. H., Pandya, S. P. 1971. *Adv. Nucl. Phys.* 4:316–443

Hammerstein, G. R., Larson, D., Wildenthal, B. H. 1973. *Phys. Lett. B* 39:176–78

Harvey, M. 1968. *Adv. Nucl. Phys.* 1:67–180

Hass, M., Benczer-Koller, N., Brennan, J. M., King, H. T., Goode, P. 1978. *Phys. Rev. C* 17:997–1001

Hauge, P. S., Maripuu, S. 1973. *Phys. Rev. C* 8:1609–20

Horie, H., Ogawa, K. 1973. *Nucl. Phys. A* 216:407–28

Hsu, L. 1967. *Nucl. Phys. A* 96:624–40

Irvine, J. M., Mani, G. S., Pucknell, V. F., Vallieres, M., Yazici, F. 1976. *Ann. Phys. NY* 102:129–55

Jaeger, H. V., Kirchbach, M. 1977. *Nucl. Phys. A* 291:52–62

Kekelis, G. J., Lumpkin, A. H., Kemper, K. W., Fox, J. D. 1977. *Phys. Rev. C* 15:664–85

Kelvin, D., Watt, A., Whitehead, R. R. 1977. *J. Plasma Phys.* 3:1539–58

King, C. H., Shakabudden, M. A. M., Wildenthal, B. H. 1976. *Nucl. Phys. A* 270:399–412

Koops, J. E., Glaudemans, P. W. M. 1977. *Z. Phys. A* 280:181–209

Kümmel, H., Luhrman, K. H., Zabolitsky, J. G. 1978. *Phys. Rep.* 38:1–63

Kuo, T. T. S., Brown, G. E. 1966. *Nucl. Phys.* 85:40–86

Kuo, T. T. S. 1967. *Nucl. Phys. A* 90:199–208

Kuo, T. T. S., Brown, G. E. 1968. *Nucl. Phys. A* 114:241–79

Kuo, T. T. S., Herling, G. 1972. *Nucl. Phys. A* 181:113–31

Kurath, D. 1973. *Nucl. Phys. C* 7:1390–95

Kurath, D., Millener, D. J. 1975. *Nucl. Phys. A* 238:269–86

Kutschera, W., Brown, B. A., Ogawa, K. 1978. *Rev. Nuovo Cimento* 1:1–116

Lawson, R. D., Soper, J. M. 1967. *Proc. Int. Nucl. Phys. Conf.*, ed. R. L. Becker, C. D. Goodman, P. H. Stelson, A. Zucker, pp. 511–25. New York: Academic. 1121 pp.

Lindén, C. G., Bergström, I., Blomqvist, J., Roulet, C. 1978. *Z. Phys. A* 284:217–31

Macfarlane, M. H., French, J. B. 1960. *Rev. Mod. Phys.* 32:567–91

Malik, F. B., Scholz, W. 1967. *Nuclear Structure*, ed. A. Hasoain, pp. 34–50. Amsterdam: North-Holland. 360 pp.

Mayer, M. G. 1949. *Phys. Rev.* 75:1969–82

McCullen, J. D., Bayman, B. F., Zamick, L. 1964. *Phys. Rev. B* 134:515–38

McGrory, J. B., Wildenthal, B. H., Halbert, E. C. 1970. *Phys. Rev. C* 2:186–212

McGrory, J. B. 1973. *Phys. Rev. C* 8:693–710

McGrory, J. B., Wildenthal, B. H. 1973. *Phys. Rev. C* 7:974–93

McGrory, J. B., Kuo, T. T. S. 1975. *Nucl. Phys. A* 247:283–316

McGrory, J. B., Raman, S. 1979. *Phys. Rev. C* 20:830–41

McGrory, J. B. 1980. Private communication and to be published

Millener, D. J., Kurath, D. 1975. *Nucl. Phys. A* 255:315–38

Morrison, I., Smith, R., Amos, K. 1975. *Nucl. Phys. A* 244:189–204

Motoba, T., Itonaga, K. 1979. *Prog. Theor. Phys. Suppl.* 65:136–62

Nann, H., Wildenthal, B. H. 1979. *Phys. Rev. C* 19:2146–54

Nann, H., Wildenthal, B. H. 1980. *Phys. Rev. C.* In press

Norton, J. L., Goldhammer, P. 1971. *Nucl. Phys. A* 165:33–55

Oberlechner, G., Richert, J. 1972. *Nucl. Phys. A* 191:577–95

Paar, V. 1972. *Nucl. Phys. A* 185:544–52

Pasquini, E., Zuker, A. P. 1977. *Physics of Medium-Light Nuclei,* ed. P. Blasi, R. A. Ricci, pp. 62–78. Bologna, Italy: Editrice Compositori. 544 pp.

Pittel, S. 1970. *Phys. Lett. B* 33:158–60

Preedom, B. M., Wildenthal, B. H. 1972. *Phys. Rev. C* 6:1633–44

Raz, B. J., Soga, M. 1965. *Phys. Rev. Lett.* 24:924–28

Reehal, B. S., Wildenthal, B. H. 1973. *Part. Nucl.* 6:5–30

Ripka, G. 1968. *Adv. Nucl. Phys.* 1:183–260

Robertson, R. G. H., Wildenthal, B. H. 1973. *Phys. Rev. C* 8:241–46

Sau, J., Heyde, K. 1979. *J. Phys. G* 5:1643–50

Schwalm, D., Warburton, E. K., Olness, J. W. 1977. *Nucl. Phys. A* 293:425–80

Serduke, F. J., Lawson, R. D., Gloeckner, D. H. 1976. *Nucl. Phys. A* 256:45–86

Seth, K. K., Saha, A., Benenson, W., Lanford, W. A., Nann, N., Wildenthal, B. H. 1974. *Phys. Rev. Lett.* 33:233–36

Shucan, T. H., Weidenmüller, H. A. 1973. *Ann. Phys. NY* 76:483–509

Singhal, R. P., Kelvin, D., Knight, E. A., Watt, A., Whitehead, R. R. 1979. *Nucl. Phys. A* 323:91–108

Singhal, R. P., Knight, E. A., Macauley, M. W. S., Kelvin, D., Watt, A., Whitehead, R. R. 1977. *Phys. Lett. B* 68:133–35

Strottman, D. 1977. *Nucl. Phys. A* 279:45–52

Strottman, D. 1979a. *Phys. Rev. C* 20:1150–54

Strottman, D. 1979b. *LASL Preprint LA-UR-79-307*

Talmi, I., Unna, I. 1960. *Ann. Rev. Nucl. Sci.* 10:353–408

Teeters, W. D., Kurath, D. 1977. *Nucl. Phys. A* 283:1–11

van Hienen, J. F. A., Chung, W., Wildenthal, B. H. 1976. *Nucl. Phys. A* 269:159–88

Vennink, R., Glaudemans, P. W. M. 1980. *Utrecht Univ. Preprint, Holland*

Vervier, J. 1966. *Nucl. Phys.* 75:17–78

Whitehead, R. R., Watt, A., Cole, B. J., Morrison, I. 1977. *Adv. Nucl. Phys.* 9:123–75

Wildenthal, B. H., Halbert, E. C., McGrory, J. B., Kuo, T. T. S. 1971. *Phys. Rev. C* 4:1266–1314

Wildenthal, B. H. 1977. *Elementary Modes of Nuclear Excitations,* ed. R. Broglia, A. Bohr, pp. 383–462. Amsterdam: North-Holland. 545 pp.

Wildenthal, B. H. 1978. *Nucleonika* 33:459–506

Wildenthal, B. H., Chung, W. 1979. *Mesons in Nuclei,* ed. M. Rho, D. H. Wilkinson, pp 723–53. Amsterdam: North-Holland. 1155 pp.

Wildenthal, B. H., Chung, W. 1980. See Barrett 1980

Wildenthal, B. H., Chung, W., Schwalm, D. 1980. *Phys. Rev. C.* In press

Wong, S. S. M., Davies, W. G. 1968. *Phys. Lett. B* 28:77–78

Zuker, A. P., Buck, B., McGrory, J. B. 1968. *Phys. Rev. Lett.* 21:39–43

Zuker, A. P. 1969. *Phys. Rev. Lett.* 23:983–87

Zuker, A. P. 1971. *The Structure of $f_{7/2}$-Nuclei,* ed. R. A. Ricci, pp. 95–110. Bologna, Italy: Editrice Compositori. 536 pp.

Ann. Rev. Nucl. Part. Sci. 1980. 30 : 437–73

ULTRASENSITIVE MASS SPECTROMETRY WITH ACCELERATORS

×5622

A. E. Litherland

Department of Physics, University of Toronto, Toronto, Ontario, Canada M5S 1A7

CONTENTS

INTRODUCTION

This review describes how ion counting and other techniques that originate in nuclear and atomic physics have been used recently to determine the abundance of natural radioisotopes by mass spectrometry instead of by the direct observation of the radioactive decay by low level counting (1). These techniques increase the sensitivity of mass spectrometry by at least a factor of a thousand, and have already led to the study of isotope ratios below 10^{-15} and to the use of the term ultrasensitive mass spectrometry. Also, as accelerators of various types (2) are an important part of the present mass

437

0163-8998/80/1201-0437$01.00

spectrometric apparatus, the title of this review follows immediately. Ultrasensitive mass spectrometry is not confined to the study of radioisotopes; stable isotopes can also be rare due to their low natural abundance and to geochemical processes. The techniques to be described are applicable to both rare stable isotopes and radioactive isotopes.

Since the first mass spectrometers were developed, great improvements have been made in the precision with which isotope ratios can be measured (3) and in their ability to separate ions of atoms or molecules with very similar masses (4). In addition the chemical properties of different atoms can be exploited (5) to increase the sensitivity to certain atoms, such as those with low ionization potentials by using thermal ionization, or to those that are chemically inert like the noble gases. It is also possible, for example, to use chemical reactions at near atmospheric pressure in the gas phase to select preferentially certain atoms or molecules. The atmospheric pressure chemical ionization sources for mass spectrometers make sensitivities approaching 1 part in 10^{12} (1 ppt) possible for some molecules in air (6). The high selectivity of lasers can also be added to mass spectrometry (7) to increase the sensitivity to particular elements and isotopes. The developments described in this review represent other ways of greatly increasing the sensitivity of mass spectrometers by using techniques developed mainly for use in nuclear physics.

The basic ideas of this rapidly growing subject are the discrimination between atomic species at the ion source of a mass spectrometer by the use of negative ions, the destruction of interfering molecular species, and finally the counting and, if possible, identification of the individual accelerated ions. These ideas and their practical realization are discussed in later sections.

The increased sensitivity of the mass spectrometric methods described above is presently being exploited in efforts to develop new methods of radioisotope dating with particular emphasis on the radiocarbon dating of milligram carbon samples (8), to develop new methods of stable isotope dating with ion microprobes (9–11), and to assay rare stable elements in materials and minerals especially with ion microprobes (10).

A complete description of the historical processes that led to the present activity in ultrasensitive mass spectrometry is beyond the scope of this review. Several general accounts (12–14) make contributions toward such a description. Current general descriptions (15, 16) tend to stress the discovery of a new ^{14}C dating method because of the importance to archaeology. The emphasis on radiocarbon dating is somewhat premature because there is an enormous difference between detecting ^{14}C atoms at natural levels with existing equipment, which is easy using negative ions and tandem accelerators as described in Section 3.1, and measuring the

$^{12}C:^{13}C:^{14}C$ isotope ratios with the necessary precision for dating. In contrast, the assay of the radioactive isotope ^{10}Be, for example, does not compete with an accurate well-developed method and so the precision now needed for ^{10}Be assay is not as great as for radiocarbon dating. Exploratory measurements of great interest are being carried out on radioisotopes such as ^{10}Be, ^{36}Cl etc. These and other examples not requiring the precision of radiocarbon dating are also discussed in Section 3.

In spite of the problems of isotope ratio measurement (Section 3), the great sensitivity of the new techniques has been demonstrated again and again for different isotopes. When added to the other new techniques being developed (6, 7) it is clear that we will eventually have new analytical instruments of very high sensitivity and, if the past development of mass spectrometry is followed, of steadily increasing precision.

1.1 Limitations of the Direct Observation of Radioactive Decay

The observation of the radioactive decay of a single atom is possible; consequently, with efficient apparatus for the detection of the decay particles and a radioactive species with a half-life of seconds or minutes, it is possible to detect all or nearly all of a small number of radioactive atoms in the presence of a large number of nonradioactive atoms. This technique is now highly developed and, for example, the chemistry of only a few atoms of the new element rutherfordium, $Z = 104$, has been studied (17a). However, as the half-life increases the time taken to carry out an experiment with a small number of radioactive atoms naturally increases; for half-lives of, say, 10^6 years efficient detection of the radioactive decay products becomes impossible unless the experiment can be continued for 10^6 years. Therefore, studies of long-lived radioactivities invariably use very large numbers of atoms and the apparatus detects the decay of only a small fraction of the total during the experiment. In this situation the mass-spectrometric detection sensitivity surpasses by far the sensitivity of radioactive counting methods (17b).

An important example is the study of the ^{14}C (half-life 5730 years), generated in the atmosphere by cosmic rays, in connection with radiocarbon dating (1, 18). The observed beta-ray counting rate from one gram of contemporary carbon of biological origin is about 15 per minute per gram. However this low counting rate is supported by the presence of 6.5×10^{10} atoms of ^{14}C in the one-gram sample. Clearly if the ^{14}C atoms could be counted efficiently by mass spectrometry it would be possible to determine the ^{14}C content of very small quantities of carbon. This has now been accomplished for milligram carbon samples even though the ratio $^{14}C/^{12}C$ is near or below 10^{-12} (see below).

1.2   Limitations of Conventional Mass Spectrometry

The detection of radioactive atoms, such as ^{14}C, by conventional mass spectrometry has been discussed for some time, (1, 19a,b). Ion sources capable of producing over 1.6 μA of C^+ or C^- ions exist and this implies that beams of 10^{13} ions per second are available. As the ratio of natural ^{14}C to ^{12}C for contemporary carbon is about 10^{-12} (1), the ^{12}C ions would be accompanied by ten ^{14}C ions per second. This rate compares favorably with the 15 beta rays emitted per minute per gram of contemporary carbon. In addition 1.6 μA of carbon ions is equivalent to only 0.7 μg of carbon per hour. Even with an efficiency of 1% for converting atoms in the carbon sample to carbon ions, this represents 70 μg of carbon generating a total of 36,000 ^{14}C ions. The 70 μg of carbon would emit 36,000 beta rays in 65 years, which dramatically illustrates the importance of trying to apply mass spectrometry to the study of natural radiocarbon (1) in particular and long-lived radioactive atoms in general.

The detection of rare stable or long-lived radioactive atoms, such as ^{14}C, by conventional mass spectrometry is complicated by the presence of much more abundant molecules and atoms, such as $^{12}CH_2$ and ^{14}N, with nearly the same mass. Their masses can be compared with the mass of ^{14}C by evaluating the average masses divided by the difference in masses $(M/\Delta M)$, which are 1134 and 84,000 respectively.

Mass spectrometers that have separated ^{40}Ca and ^{40}Ar $(M/\Delta M = 193,500)$ have been made (4) but to achieve this separation a narrow object slit a few micrometers wide was used. In addition, highly monoenergetic ions were necessary. These requirements create a dilemma in trying to apply conventional mass spectrometry to radioisotope detection. The very high mass resolution required to eliminate molecular and atomic interferences can only be achieved at the expense of low efficiency for the already rare atoms. The alternative of building a very large and expensive magnetic spectrometer does not seem attractive. Fortunately, using the techniques described in later sections, it is easy to separate many rare atoms, such as ^{14}C, efficiently with the help of conventional magnetic and electrostatic spectrometers. The removal of molecular interferences also promises to solve one of the major problems in using ion microprobes (9–11).

2   ULTRASENSITIVE MASS SPECTROMETRY

2.1   Stable Isotopes and the Elimination
of Molecular Interferences

It is a remarkable fact that the only element that has no stable isotope with a unique mass number is indium. This is because of the very long half-lives (over 10^{14} years) of ^{113}Cd and ^{115}In. As a result, only indium cannot be uniquely identified by observing one of the stable isotopes.

The assay of stable elements by mass spectrometry, down to and well below parts per billion of a sample, should be possible with relatively low mass resolution spectrometry if molecular interferences are eliminated and especially if negative ions are used instead of positive ions (19a,b).

There are many solutions to the molecular interference problem other than the use of high resolution. For example, Tyrell et al (20) reported that they could discriminate against molecules by a factor of ten by using pulse height analysis of the ion pulses from their secondary emission detector (21). Shimizu et al (22) used the fact that the high energy tail of the kinetic energy distribution of the secondary ions following the sputtering process is depleted in molecules relative to atoms, often by several orders of magnitude. Molecules can also be destroyed selectively by using lasers, as proposed by Hurst et al (7). However, at present the most efficient way seems to be the exploitation of the fact that multiply charged molecules break up rapidly ($\ll 1~\mu s$). Any process involving the removal of more than two electrons from a molecule can therefore be used to eliminate them from the mass spectrum. Several stable or metastable doubly charged molecules are known (23, 24), but no triply charged molecules are known at present.

At least two methods can be used to produce highly charged atoms and molecules. The first requires the use of an ion source that generates significant quantities of multiply charged ions (25) as discussed briefly in Sections 3.1.1, 3.1.9, and 3.2. The second requires the acceleration of singly or doubly charged ions to a velocity such that on the average a total of three electrons (26) are removed from the atom or molecule after passage through a thin foil or appropriate thickness of gas. For light elements this requires ion energies of about 3 MeV, and for heavy elements ion energies of about 6 MeV, to reach the optimum efficiency as shown in Figure 1. This simple highly efficient procedure (27) is the basis of much of the work to be discussed later. Atoms with more than three electrons removed can also be used.

The generation of ion energies of 3–6 MeV requires quite small accelerators. If positive ion sources are used, a small electrostatic accelerator (2) is adequate. However, these accelerators have the disadvantage of requiring the ion source or the detector system to be at high voltage.

In tandem accelerators (2) the change from negative to positive ions takes place in the central positive electrode. The subsequent acceleration of the positive ions increases the energy of the ions still further. In this case both ion source and ion beam analysis systems are outside the high voltage electrode.

The use of multiply charged atoms in mass spectrometry, to eliminate molecular interferences, introduces another set of ambiguities, which fortunately are rare in conventional mass spectrometry. Ion sources that generate +3 ions also generate several other charge states; in a system employing charge exchange of negative ions of a few MeV, all charge states below about 10 are present. Only the charge state +3 (or higher) is required for analysis and the other charge states must be eliminated. Consequently both mass and charge spectrometry are required.

2.2   Mass and Charge Spectrometry

The electric and magnetic fields, used for the analysis of ions, provide only information about the two quantities E/q and M/q where E, M, and q are the energy, mass, and charge of the ion. There are four ways in which the quantities E/q and M/q may be determined (28):

1. magnetic selection $(\mathbf{B}\rho)^2 = 2(M/q)(E/q)$

2. electrostatic selection $\mathbf{E}\rho = 2(E/q)$

3. cyclotron selection $1/f = \dfrac{2\pi}{\mathbf{B}}(M/q)$

4. velocity selection $v^2 = 2(E/q)(q/M)$

where $\mathbf{B}$ is the magnetic field, ρ the radius of the ion path, $\mathbf{E}$ the electric field,

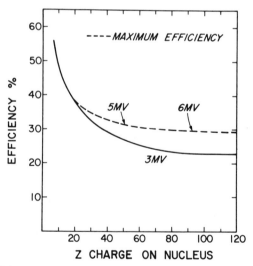

Figure 1   The efficiency for generating the +3 charge state of an ion, by charge exchange in argon gas at 3-MeV energy, is shown as a function of the charge on the nucleus Z. The solid curve (26) is for singly charged ions that have been accelerated by a potential difference of 3 MV. The maximum efficiency, shown dashed, requires a larger ion energy (voltage) for larger Z.

f the cyclotron frequency, and *v* the ion velocity. Low resolution measurements that separate neighboring isotopes produce as their final output only the quantities M/q and E/q and, as in this approximation M can be regarded as an integer, ambiguities can arise if M and q have common factors. It is for this reason that some flexibility in the choice of q is desirable. If it is possible to measure the energy of the ion also, then it is possible to determine q and so determine the mass from the ratio M/q. The requirements for energy resolution of a heavy ion detector (29) are modest, if charge states as low as $+3$ are used, because the energies from the (E/q, M/q) selection will then be in the ratios $3:2:1$ if the mass M is a multiple of 3. If the charge $+2$ or $+1$ ions are very intense compared with the charge $+3$ ion, the addition of a charge exchange to, for example, $+4$ by passing the ions through a foil or gas and then analyzing the ions with an electric (30) or magnetic field can eliminate all but the wanted ion with an efficiency near 50%. This procedure also eliminates the need for the heavy ion detector to measure the ion energy. See Sections 3.8 and 4.2 for further comments on this procedure.

The use of energy, mass, and charge signatures, at energies such that charge state $+3$ or higher is dominant, is the basis for the ultrasensitive mass spectrometry of all stable isotopes except for small traces of indium in either cadmium or tin. An example is discussed in Section 3.8.

2.3   *Radioactive Isotopes and the Separation of Isobars*

Radioactive isotopes ultimately decay to stable isotopes and in the case of beta decay the mass difference between a stable and radioactive isotope pair is often so small that they can be regarded as having the same mass at low mass resolution. The pair of such isotopes are usually referred to as isobars. The separation of isobars with high efficiency and the elimination of molecules are the keys to the ultrasensitive mass spectrometry of radioisotopes such as ^{14}C and ^{14}N without using extremely high mass resolution. In theory any difference between the isobars can be exploited to separate isobars. However, even though they are different chemical elements, the separation by conventional chemical methods is very difficult because the unwanted element is often more than 10^9 times as abundant as the wanted element.

A practical general method for discriminating between isobars does not exist. However, if atoms can be completely stripped of all their electrons after all molecules are eliminated, then the ratio M/q uniquely discriminates between isobars. The procedure was used recently by Raisbeck et al (32) in experiments that successfully separated the isobars ^{26}Al and ^{26}Mg. For these elements this procedure requires hundreds of MeV energy and very large heavy ion accelerators (2), so it is used only in cases where isobar separation is important and is impossible by other methods. In addition it is applicable only when the rare isobar has larger nuclear charge than the abundant isobar: Otherwise the small amount of the remaining abundant isobar with a single electron attached would cause confusion. A lower energy accelerator with a storage ring for identifying similar fully stripped ions has also been proposed (33).

The most dramatic method of separating some isobars lies in the exploitation of the properties of negative ions. In the case of ^{14}C and ^{14}N there is fortuitously a complete separation due to the stability of C$^-$ (binding energy 1.25 eV) and instability of N$^-$ (unbound). However, in general each isobaric pair must be investigated carefully to find out empirically how suitable the negative ion properties are for isobar separation. In this connection it is well worth remembering that the degree of separation required is often enormous. In the case of ^{14}C and ^{14}N it is at least ten orders of magnitude. Also, although a negative ion ground state may be unstable, there may exist metastable excited states with lifetimes long enough for the ions to transit the relevant portions of the mass spectrometer. An example is Be$^-$. In the case of ^{14}N$^-$ the metastable states appear (34) to have lifetimes short enough to ensure that they do not interfere with the detection of ^{14}C. The energy level diagram for N$^-$ is shown in Figure 2 (35, 36a,b) to illustrate the complexity of the situation.

The separation of isobars can also be accomplished by exploiting the fact that, after acceleration to a suitable energy, light isobars can be separated in

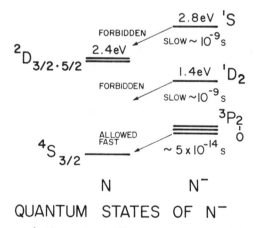

Figure 2   The expected quantum states of nitrogen negative ions and their LS coupling assignments are shown in relation to those of the neutral nitrogen atom. All negative ion states are unbound. The decay of the $^1S_0$ and $^1D_2$ excited states is forbidden in LS coupling but not in intermediate coupling (36).

some cases by either range or rate of energy loss (37). Ions such as ^{14}C and ^{14}N have ranges in solids or gases that differ by over 20% at energies above 14 MeV. Although it is difficult in the ^{14}C–^{14}N case, the range technique has been used to detect ^{14}C in the presence of huge fluxes of ^{14}N (38a–c) and it has become the widely used procedure for separating ^{10}B and ^{10}Be (38a–c, 39a,b). ^{10}Be and ^{10}B both form negative ions so that the range technique is at present the only way of separating them. In this case the ranges differ by over 28% at energies above 10 MeV.

Differences in the rate of energy loss (dE/dx) in gases can also be used to separate light isobars from one another. A heavy ion counter suitable for making such measurements has been described (40). The measurement of dE/dx has proved to be a useful additional constraint in some cases and

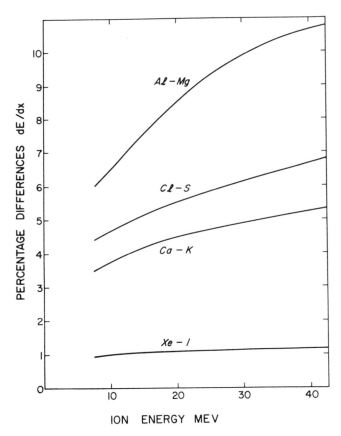

Figure 3   The differences between the rate of energy loss (dE/dx) values for various isobars is shown as a percentage and as a function of the ion energy (42).

essential in the separation of ^{36}Cl and ^{36}S (41), both of which form negative ions readily. This is discussed in Section 3.4.

The separation of isobars by range or by rate of energy loss becomes progressively more difficult as the charge on the nucleus increases (42) as shown in Figure 3. It is quite likely that it will be difficult to separate ($\Delta Z = 1$) isobars beyond Ca ($Z = 20$) by range or dE/dx techniques especially if the intensity ratios are large.

Other, as yet untried, techniques for separating isobars are the use of resonance ionization spectroscopy with lasers (7), variations on the technique of atmospheric pressure chemical ionization (6), which exploits selective chemical reactions in the ion source, and the use of autoionizing states that are electron number specific.

The use of autoionizing states to separate for example ^{10}Be and ^{10}B ions (K. H. Chang, personal communication) is based upon the fact that the $^4P_{5/2}$ state of the three electrons of Be$^+$ decays spontaneously to Be^{+2} plus an electron. B$^+$ has no such autoionizing state. The autoionizing states of the three-electron system have been studied extensively (43).

3 APPLICATIONS OF ULTRASENSITIVE MASS SPECTROMETRY

3.1   Carbon-14 Detection and Dating with Tandem Accelerators

3.1.1 THE DEVELOPMENT OF CARBON-14 DETECTION USING NEGATIVE IONS   The possibility of separating ^{14}C from ^{14}N using negative ions has been known since the early 1960s when it was found that negative ions of nitrogen atoms could not be used to accelerate nitrogen in tandem accelerators whereas carbon atoms could be used. However, the first attempts in the early 1970s to separate ^{14}C from ^{14}N atoms for dating purposes, using negative ions and conventional mass spectroscopy, were carried out (1, 19, 45) using the ^{14}C^{15}N$^-$ molecular ion. At that time it was strongly suspected experimentally and theoretically (35, 36) that both N$^-$ and N$_2^-$ were unstable so that there would be no interference from ^{14}N^{15}N$^-$. Negative atomic ions were not used because of their lower yield from the particular ion source used. It is ironical that the principal ion interfering with ^{14}C^{15}N$^-$, which prevented success at the time, was an atomic ion ^{29}Si$^-$. The separation of ^{29}Si$^-$ and ^{14}C^{15}N$^-$ could, in principle, be carried out by the mass spectrometry of the dissociated ^{14}C^{15}N$^-$ molecular ion into, for example, ^{14}C$^+$ plus ^{15}N albeit at reduced efficiency. However, also in the early 1970s, ion sources were developed (46–48) that gave very large C$^-$ ion currents from solid samples by sputtering those samples, usually in the form of graphite, with 20-keV Cs^+ ions. These new negative ion sputter sources renewed interest in ^{14}C dating with negative ions and tandem accelerators as early as 1974. This interest led after much discussion to the first exploratory experiments in the spring of 1977. Also in 1976 a proposal was made (19, 49) to combine laser enrichment of the ^{14}C in a sample followed by conventional mass spectrometry with positive ions and finally a conversion to negative ions. The use of isotope enrichment combined with ultrasensitive mass spectrometry could be of importance in the future and is discussed in Section 3.1.4. In parallel with the discussions on the use of tandem accelerators, experiments being carried out were leading to the proposal (37) to use positive ions and a cyclotron for radioisotope dating.

^{14}C ions at natural concentrations ($^{14}C/^{12}C \approx 10^{-12}$) were first detected during a study of the degree of instability of the N^- ion (34), a study originated to assist in the design of the negative ion source for a proposed small ^{14}C-dating tandem machine. The investigators were surprised to find that the N^- ions were so unstable that ^{14}C ions from a natural sample could be detected readily. A limit approaching 10^{-15} for $^{14}C/^{12}C$ was found in the very first measurements. This fortuitous property of the N^- ion is illustrated in Figure 2, which shows that theoretically not only is the ground state unstable but the isomeric states also probably have short lifetimes because of deviations from pure LS coupling (36a,b). On the whole the conventional experimental work on the degree of instability of N^- states is still contradictory (35, 50, 51). There is, however, general agreement that the N^- states are too unstable under the conditions used to interfere with ^{14}C detection at natural concentrations (34, 52a,b, 53).

It is worth noting that the negative ion sputter source (48), used primarily because of the high yield of C^- (tens of microamps) and the high efficiency ($\sim 10\%$) (54) for turning atoms into useful negative ions, also discriminates strongly against weakly bound negative ions, as expected from the local thermal equilibrium model and the Saha-Eggert equation (9). This feature combined with the rapid spontaneous breakup of N^- into N plus an electron, makes negative ions from sputter ion sources very suitable for discriminating between the ^{14}C and ^{14}N isobars. Further studies of the N^- ion would, however, be desirable.

Two other·key experiments, relevant to the design of small ^{14}C-dating machines based upon tandem accelerators, have been reported (24, 55). The first experiment was on the destruction of the mass-14 molecules. The CH_2^{+2} molecule is metastable (24) with a lifetime of about 10 μs, which indicates that charge $+3$ ions must be used to ensure that all molecules are destroyed. No CH_2^{+3} molecules have been observed. The second experiment (55) showed that $^{14}C^{+3}$ ions at natural abundances could be detected at 12 MeV, which is near the optimum for $+3$ ions and a tandem accelerator.

As a result of the work described above, three small 3-MV tandem accelerators have been designed (56) and are now being built to measure ^{14}C in natural samples (see also Section 4.1). These machines are expected to play an important role in the further development of ^{14}C dating by atom counting because it is so difficult to carry out such a development with unmodified existing nuclear physics machines, which are available for research and development only a small fraction of the time. However, the general principles of ^{14}C measurements with these new machines have been established with existing machines.

The separation of ^{14}C and ^{14}N by conventional negative ion mass spectrometry (19) has been considered several times in recent years. Hall & Hedges (49) proposed a charge exchange from $+1$ to -1 to eliminate ^{14}N. The existence of $^{12}CH_2^-$ and $^{13}CH^-$ ions led to discussions on the possibility of a charge exchange from $+2$ to -1 to eliminate both ^{14}N and the molecules. These discussions were followed by measurements (24) at Rochester that showed the $^{12}CH_2^{+2}$ ion to be metastable. Middleton (57) then proposed to exploit the charge exchange from $+3$ to -1. At present there is little research on these possibilities although once the 3-MV tandem accelerators, being built specifically for ^{14}C dating, are in operation they will be ideal to check the nature of the mass-14 ions produced in this way. It is possible, for example, that the conversion of CH_2^{+2} to CH_2^- by the capture of three electrons after many collisions with other atoms will result in the destruction of all the CH_2^{+2} interfering molecules.

3.1.2 SOME EXPERIENCE IN RADIOCARBON DATING   Research and development on ^{14}C detection and dating using existing tandem accelerators is continuing at many laboratories (52a,b, 58–63). The apparatus used at Rochester University for such experiments (63) and for ultrasensitive mass spectrometry in general is shown schematically in Figure 4. The negative ion sputter source produces up to 10 μA of $^{12}C^-$ ions at 20 keV; after the ions have been magnetically analyzed they are accelerated and focused onto the aperture AP1. The Faraday cup FC1 can be used to monitor the ion source current. The negative carbon ions are then accelerated to 8 MV and five electrons removed to produce C^{+4} ions. These are then accelerated further and focused onto the next aperture AP2 and analyzed by two magnets that specify the product $(M/q) \times (E/q)$. The electrostatic analyzer then specifies (E/q) and the heavy ion counter (40) specifies the energy E as well as dE/dx, the rate of energy loss. The time-of-flight detector is used mainly for heavy element work (Sections 3.6 and 3.8).

Some results on ^{14}C detection taken with the heavy ion counter (40), before the use of the electrostatic analyzer, are shown in Figure 5. The clear separation of the residual ^{14}N ions from the ^{14}C ions is shown. The ^{12}C and ^{13}C ions, which have a different E/q, are eliminated when the electrostatic analyzer is used. They are generated from the ^{12}CH$_2^-$ and ^{13}CH$^-$ beams that are injected with the ^{14}C$^-$ beam and destroyed by the removal of several electrons. A second charge exchange of the positively charged ^{12}C and ^{13}C ions, in the residual gas of the high voltage accelerating tubes of the tandem accelerator, generates a continuum of ion energies and some of these ions are then transmitted by the magnetic analysis. The residual ^{14}N is generated mostly from negative molecular ions injected into the tandem due to the poor mass resolution ($M/\Delta M \approx 25$) of the 30° inflection magnet. The ^{14}N positive ions then undergo additional charge exchanges in the residual gas of the accelerating tubes. The ^{14}N ions are indistinguishable from the ^{14}C ions except by the final dE/dx measurement, but the nitrogen-containing molecules can be eliminated before they enter the accelerator by employing an electrostatic analyzer and a high resolution injection magnet system. ^{14}N ions, generated from the molecules during voltage fluctuations of the tandem accelerator, can be eliminated also by better voltage control (56).

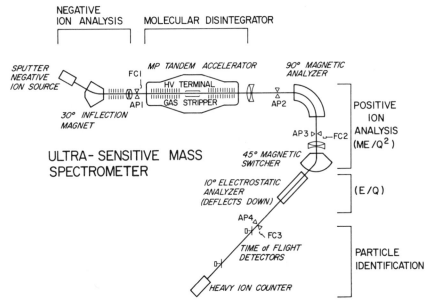

Figure 4 The ion beam transport system of the Rochester University tandem accelerator used as an ultrasensitive mass spectrometer is shown schematically. Ion-beam-defining apertures are designated by AP and Faraday cups for ion beam current measurements are designated FC.

The ^{14}C counting rate can be obtained with the system shown in Figure 4 and compared with the injected $^{12}C^-$ ion current in FC1. The ratio of these numbers is simply related to the $^{14}C/^{12}C$ ratio, which in turn can be related to the age of the sample by the exponential radioactive decay law. The results of some 1977 measurements (24, 64) are shown in Figure 6. The geological samples had been dated previously (M. Rubin, personal communication) and the graph shows, within the errors, good agreement between the two methods. Measurements can also be made by comparing the $^{13}C^{+4}$ current at FC2, the $^{12}C^{+4}$ current at FC2 but attenuated by a known amount near FC1 before entering the tandem, and the $^{14}C^{+4}$ ion counting rate. Later individual measurements with an accuracy approaching $\pm 1\%$ have been made (65), but experience demonstrates that dating with the existing apparatus is much more difficult than simple detection of the ^{14}C ions. This was not unexpected because existing tandem accelerators and their peripheral equipment were not designed to measure isotope ratios to the needed accuracy (better than $\pm 1\%$). However, work on ^{14}C dating continues because the experience is of great value in designing machines specifically for ^{14}C dating.

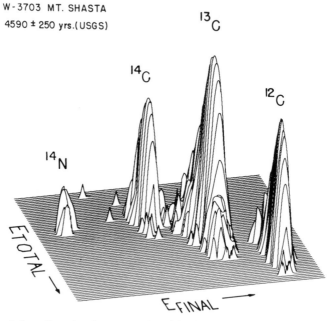

Figure 5   A three-dimensional representation of data from the heavy ion counter (40) used at Rochester. The vertical axis is the logarithm of the number of counts. E TOTAL is the total ion energy measured in the counter and E FINAL is the energy measured in the final section of the counter. E FINAL is related to the ion range and the rate of energy loss dE/dX.

Since the measurements shown in Figure 6 were made, many of the components of the tandem accelerator at Rochester have been upgraded and a redesigned ion source system is to be installed in early 1980. In addition many ultrasensitive measurements on other radioisotopes have been made at Rochester and elsewhere, as well as on stable isotopes (Sections 3.3 to 3.8).

3.1.3 SAMPLE PREPARATION   Preparing the solid carbon in a form suitable for the negative ion sputter source required much developmental work (61, 62, H. W. Lee and R. P. Beukens, personal communication). This problem now appears to have satisfactory solutions as suitable graphite-like material yielding high C^- currents has been made from acetylene gas. At present the study of ion sources that use gases, such as CO_2 which might

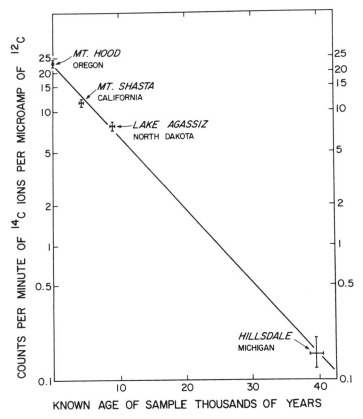

Figure 6   The results of dating milligram samples (64) by atom counting compared to the ages measured by beta-decay counting. The observed ^{14}C counting rate is compared with the $^{12}C^-$ current at the ion source.

appear to be simpler or more convenient, has been inhibited by the existence of a large memory effect (60). The sputter negative ion source is expected to be much superior to the gas source in this respect because the local environment of each sample in a sputter ion source is different and the material sputtered during the analysis of a sample can be prevented from reaching the samples waiting for analysis. To date no clear evidence for a memory effect has been observed with sputter ion sources. Ion sources operating with flowing gases are also not as efficient in their use of carbon as sputter sources that use solid material.

3.1.4 BACKGROUND AND CONTAMINATION   A persistent background of ^{14}C ions, near 10^{-3} of the contemporary level (equivalent to about 50,000-year-old material), is observed when graphite is used in negative ion sputter sources (53). No ^{14}C is expected to be present in graphite. This background is usually attributed to contamination by contemporary carbon during sample preparation, or to small amounts of contamination after insertion of the samples into the ion source. Residual hydrocarbons containing ^{14}C in the vacuum system would be expected to crack on the surface of a graphite sample because of the bombardment of the cesium sputtering beam. This level of "contamination" is difficult to study with existing machines; such studies are time consuming because the observed ratio $^{14}C/^{12}C$ is near 10^{-15}. The incentive to make such studies is not great because new machines, which will have superior ion sources and vacuum systems and which have not been used previously for nuclear reaction studies, are being built (56) to make ^{14}C measurements. Section 4.1 contains additional discussion.

Contamination from contemporary ^{14}C during the preparation of the samples for analysis is a problem facing all radiocarbon-dating laboratories. However, it is worth noting that, because there are no monoenergetic ^{14}C ions in the cosmic ray or radioactivity background, the atom-counting method suffers only from contamination problems. Consequently contamination from whatever source will be one of the main problems in extending the time range of radiocarbon-dating measurements using atom counting. The use of ^{14}C enrichment (19, 61a,b) may be an attractive option to alleviate this problem. An enrichment of the ^{14}C by a factor of eight could increase the time range of dating by as much as three half-lives if the same final level of contamination is assumed.

3.1.5 ION TRANSMISSION PROBLEMS   The transmission of the ^{14}C ions through the electron stripping canal at the center of the tandem accelerator is variable, reflecting partly the problem of adjusting the magnetic field of the ion source injection system and the ion beam steering controls for the low fluxes of ^{14}C ions. Although existing tandem accelerators were designed for high transmission (approaching 100%) through the electron stripping canal in the high voltage electrode, it was naturally assumed that rapid adjustments of the focusing and the steering controls would be used to maximize the transmission of the ion currents. This procedure is straightforward for ion beams over one nanoamp and consequently the electron stripping canal was made as small in cross section as possible to minimize gas flow into the accelerator while transmitting nearly 100% of the ion beam. Unfortunately this means that the transmission through the electron stripping canal depends critically on the adjustments of the focusing and steering controls. With fluxes of rare atoms of 10 per second it would take several minutes to determine the transmission to a statistical accuracy of $\pm 1\%$ for one adjustment of the focusing and steering controls, so clearly other methods must be used.

In the case of ^{10}Be and ^{36}Cl the accompanying beams of ^{10}B and ^{36}S can be used effectively as pilot beams to optimize the transmission through the tandem and beam transport system, as their fluxes are usually much larger than those of radioisotopes. However, in other cases, such as ^{14}C, there is no accompanying beam so that other less suitable ion beams must be used. These ion beams do not have the same M/q and E/q and so at the typical tandem accelerator, where the beam transport system is a mixture of electric and magnetic elements, the adjustment of the beam transport system can be a lengthy and somewhat uncertain process. The use of graphite enriched in ^{14}C can expedite such adjustments, but of course care must be taken to avoid contamination of the unknown samples. ^{7}Li$_2^-$ ion beams from the injection of the mass-14 ^{7}Li$_2^-$ beam can be used in principle, if the mass difference of $M/\Delta M \approx 500$ between ^{7}Li$_2^-$ and ^{14}C$^-$ is taken into account. The E/q values are the same for the final ^{7}Li^{+2} and ^{14}C^{+4} ions, and the M/q values differ by only $\frac{1}{2}\%$.

The machines being built (56) for radiocarbon studies have been designed to simplify adjustment of the beam transport system by avoiding a mixture of electric and magnetic elements, and in addition they have a larger diameter electron stripping canal to make the adjustments less critical and to ensure that variations in ion source emittance during a measurement do not cause variations in transmission (66).

3.1.6 ISOTOPE RATIO MEASUREMENTS   Measurements of the ^{12}C : ^{13}C : ^{14}C ratios are necessary in order to estimate a date for carbon that left the biosphere on the death of a plant or animal (18). The ^{12}C : ^{13}C ratio is needed to correct for the isotope fractionation (18) due to the slight dependence of chemical reaction rates on mass. At present these ratios are obtained by measuring in sequence the ^{12}C and ^{13}C ion currents and the ^{14}C counting rate in a heavy ion detector. This necessitates cycling the magnetic elements, before and after the accelerator, under computer control (65). Such an arrangement has been made to work successfully and during periods of apparatus stability isotope ratios with external and internal errors better than $\pm 1\%$ can be obtained. This is the accuracy that must be reached consistently before dating by atom counting of carbon from the past 10,000 years can be taken seriously. However, for dating samples older than 10,000 years before the present, a lower accuracy suffices and between 10,000 years and possibly 50,000 years useful dating measurements can be made on very small (milligram) samples. For example, recently a few-mg piece of mammoth muscle was dated (63, 65) to be 27,000 $\pm$ 1,000 years old.

There are many solutions to the problem of the measurement of the carbon isotope ratios. One simple solution is to inject the isotopes sequentially and rapidly by varying the voltage on the injection magnet box (56). This will allow more rapid cycling of the isotopes and, by varying the period of the injection of the different ions, the average $^{12}C^-$ current injected into the accelerator can be reduced. This is the procedure to be used for the new ^{14}C machines under construction. Another possible system uses one of the many designs for achromatic magnet systems (67). These complex systems could make possible the attenuation of the ^{12}C-beam by an appropriate amount and the simultaneous measurement of the $^{12}C : ^{13}C : ^{14}C$ ratios.

3.1.7 ISOTOPE FRACTIONATION PROBLEMS   Isotope fractionation by the measuring apparatus is a problem that has to be taken as seriously in ultrasensitive mass spectrometry as in conventional mass spectrometry (3) if comparable accuracies are to be obtained. An additional form of isotope fractionation can be observed in mass spectrometry involving destruction of molecules at a few MeV. The electron stripping cross sections are velocity dependent (26) and since the ions are stripped at the same energy the probability of generating a given charge state is not the same for all isotopes. This effect can change the measured isotope ratios by many percent but, as actual measurements are usually made with respect to some standard, the problem is not as serious as it first appears. However, there are velocities or ion energies where the effect is a minimum as can be seen from the calculated probabilities of charge state $+3$ for ^{14}C and ^{12}C ions in Figure 7. Clearly at a negative ion energy of about 2.68 MeV, according to these calculations, the correction will be zero for $^{14}C : ^{12}C$ but will be finite for $^{13}C : ^{12}C$. There will be an excess of about 2% of ^{13}C ions according to Figure 7. As charge state probabilities (26) are in general not well enough known, it will be necessary to check such calculations in due course. The figure also shows that the energy of the negative ions must not be allowed to vary before stripping the electrons if accurate comparisons are to be made. A potentially more serious problem may be encountered in isotope fractionation at the sputter ion source. Some measurements (68, 69) show that the isotope ratios will vary with time as well as with angle of emission from the surface being sputtered. It will obviously be necessary to study such effects in detail as the accuracy with which the isotope ratios can be measured is improved.

3.1.8 NEGATIVE ION PROBLEMS AND OPPORTUNITIES   More subtle isotope fractionation effects due to the use of the less familiar negative ions are worth mentioning. The carbon negative ion also exists in isomeric excited states (70) with very low electron binding energies of 35 meV. These isomeric states have large cross sections (71) for destruction at low velocity (< 100 keV). If such cross sections are velocity dependent, as expected, then

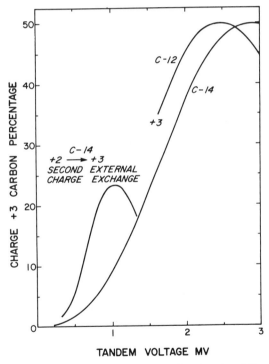

Figure 7   The calculated (26) percentage of charge +3 carbon ions following stripping of electrons from a C^- ion is shown as a function of the voltage on the central electrode of a tandem accelerator (2). Also shown is the calculated percentage for generation of +2 carbon ions in the central electrode followed by a second charge exchange to +3 outside the tandem accelerator.

isotope fractionation prior to injection into the accelerator may have to be considered as a result of differential loss. Fortunately, as discussed in Section 3.1.1, such weakly bound negative ions are expected to be rare from negative ion sputter sources although no measurements have been made to check this point. The isomeric states, however, dominate the C^- ions from positive ion sources that are followed by charge exchange at 20 keV or higher to form negative ions (70). This has been demonstrated by the electric dissociation (70, 71) of the metastable ions as shown in Figure 8. He^-, C^-, Si^- and O^- are shown for comparison. He^- is readily dissociated by an electric field as are the isomeric states of C^- and Si^- that are close to the electron binding energy. O^- has no nearly unbound state and cannot be

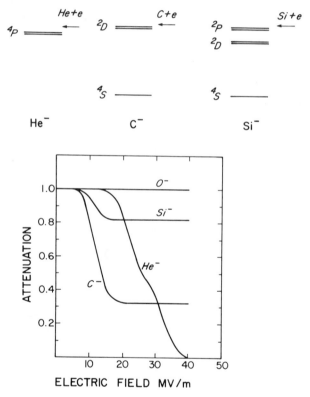

Figure 8   The energy level diagrams of the negative ions of He, C, and Si are shown to illustrate that excited states of these negative ions occur near their dissociation energies (shown by horizontal arrows) of 70 meV, 1.27 eV, and 1.3 eV respectively. The LS assignments of the quantum states are given also. The attenuation of the negative ion beams due to dissociation of the ions in excited states nearest the electron binding energies is shown as a function of the electric field (71). The negative ion of oxygen has no excited states.

dissociated using practical electric fields. The electric dissociation of C^- (Figure 8) indicates that about 70% of the C^- ion beam was in the isomeric states following charge exchange of C^+ ions in magnesium vapor. It is worth noting that the presence of dissociatable isomeric states is also isobar specific (33) and that, for example, $^{32}Si^-$ could in principle be separated from $^{32}S^-$ by electric fields because of the absence of isomeric states of $^{32}S^-$. This approach will need much research but may be applicable to some heavier isobaric pairs.

3.1.9 POSITIVE ION PROBLEMS   Isotope fractionation effects, in addition to those at the ion source (68), can also take place when positive ions are used. It is usually forgotten that the electrons attached to a positive ion are not necessarily in their ground state (72). Not only are long-lived excitations of a single electron to states with high principal quantum numbers (Rydberg states) possible, but so are complicated excitations involving several electrons. Sometimes these excitations can be long lived and decay by autoionization (43). The series of 3-electron atoms, including C^{+3}, are well known to have such states, which decay spontaneously by the emission of an electron. C^{+3} has a mean life of about 10^{-7} seconds for the $^4P_{5/2}$ autoionizing state (43), which can be formed by charge exchange in a gas. Clearly more $^{12}C^{+3}$ ions in the $^4P_{5/2}$ state will survive to the ion detector than $^{14}C^{+3}$ because of the higher velocity. This will generate a small additional isotope fractionation. In the unfamiliar field of ultrasensitive mass spectrometry it is inevitable that many careful tests of the procedures to be adopted will be necessary as the accuracy with which isotope ratios are measured improves. It is stimulating to be reminded that measurements in conventional mass spectrometry can reach accuracies of $\pm 0.01\%$ (74) for Nd-Sm dating.

3.2   Carbon-14 Detection and Dating with Cyclotrons

Following experiments to search for integrally charged quarks (75), Muller proposed that radioisotope dating and in particular radiocarbon dating with a cyclotron (37) might be possible. This suggestion led to the extensive work of Raisbeck et al on ^{10}Be (Sections 3.3 and 3.5) and to the work in which the idea of mass spectrometry at high and very high energies is being explored. This subject was reviewed recently (76).

The use of a cyclotron solves the problems of ultrasensitive mass spectrometry in different ways from the use of electrostatic accelerators. First, the molecular ambiguities are eliminated by extracting highly charged positive ions from an ion source. This is then followed by the resonant acceleration, dependent only upon the M/q of the ions, to a final energy decided by the size of the cyclotron. An important point made by

Muller (37) is that not only is the mass resolution of the cyclotron quite high, as masses differing by $\Delta M \approx M/20,000$ can be resolved as described by Muller et al (38), but the mass resolution function is more favorable than the usual Gaussian resolution function owing to the resonant nature of the acceleration. This feature makes the cyclotron a superior $M/\Delta M \approx 20,000$ mass spectrometer that without further modification appears to be a very useful device for ultrasensitive mass spectrometry. Unfortunately the separation of most isobars is not possible even with the high mass resolution of the cyclotron and this poses a serious problem because the stable isobar is always more abundant than the radioactive isobar. A partial solution to this problem was proposed by Muller (37) and was shown to work satisfactorily for the detection of ^{14}C and to work very well for ^{10}Be and ^{3}H. In the separation of ^{14}C from ^{14}N it was shown that greater range in matter of the ^{14}C could be used to ensure that the ^{14}N was absorbed completely. The ^{14}C could be detected and counted even though the ^{14}N flux was 10^{10} times that of the ^{14}C flux. If, however, the more abundant stable isobar has longer range than the radioactive isobar, as in ^{26}Al and ^{26}Mg, then this solution is inadequate.

The use of a cyclotron for attempts at radiocarbon dating has, so far, had mixed success (38a–c, 77) and at present modifications to the ion source system are being carried out (38a–c, R. A. Muller, personal communication).

A basic problem with the use of existing cyclotrons as part of ultrasensitive mass spectrometers is the problem of isotope ratio measurement. The difficulties of the rapid cycling of a cyclotron radio frequency and adjusting the magnetic field to tune the various isotopes through the cyclotron, with unvarying transmission efficiency, seem to be formidable. For comparison measurements at the level of near $\pm 10\%$ accuracy, the cyclotron may possibly be used simply as the detector of the rare isotope and the abundant isotope can be measured separately as reported by Raisbeck et al (39a,b). In many instances, discussed in Section 3.3, this procedure can be quite acceptable. Unfortunately, the radiocarbon dating of archaeological samples (18) requires at least $\pm 1\%$ accuracy in the isotope ratio measurements and the complexity of getting the ion beam into and out of the cyclotron may make such an accuracy difficult to obtain.

3.3 Beryllium-10 Analysis

The study of ^{10}Be, produced from the high energy proton spallation of carbon, by mass spectrometry was carried out many years ago (78). However, the molecule $^{9}BeH^{+}$ proved to be troublesome and precluded completely the mass spectrometry of ^{10}Be generated by cosmic rays in the atmosphere. The use of the cyclotron as part of a mass spectrometer together with the elimination of the interfering isobar ^{10}B by the range technique (37, 38a–c) has proved to be of great value in the detection of ^{10}Be at natural levels, and many exploratory measurements have been made. In the early work (39a,b) tests of the method with known samples showed that the ratios of $^{10}Be/^9Be$ from 10^{-8} to 10^{-10} could be studied quite easily and that the ^{10}Be in a sample containing as few as 10^7 ^{10}Be atoms could be detected. This is some three orders of magnitude lower than can be achieved by the most sensitive counting techniques, which demonstrates the power of the method. A sample containing 10^7 ^{10}Be atoms would emit only five beta rays a year, far below the background level in a beta-ray counter.

The early tests of Raisbeck et al (39a,b) were followed by a series of exploratory measurements including a detection of the ^{10}Be in ice from Antarctica (79), the search for ^{10}Be variations in marine sediments (80), and the study of the residence time of ^{10}Be and its concentration in the ocean surface layer (81).

^{10}Be measurements are now being carried out in many laboratories and tandem accelerators are widely used (82–85). The measurement of ^{10}Be in manganese nodules has been reported (82) using a tandem accelerator to accelerate $^{10}Be^{16}O^-$ ions and subsequent absorption of the ^{10}B ions by the range technique. A comparison of the results (82) obtained by both the accelerator measurements and the beta decay measurements is shown in Figure 9. ^{10}Be is very easy to observe at low concentrations; ratios as low as 7×10^{-15} for $^{10}Be/^9Be$ have been reported (83) using a tandem accelerator. In a series of experiments using ^{10}Be generated by the $^9Be(n,\gamma)^{10}Be$ reaction, a wide range of concentrations was studied and a linearity of better than 1% observed over three orders of magnitude in concentration variation. These measurements were made both by varying the appropriate magnetic elements manually and later with the computer-controlled system. These results were taken during a period of good apparatus stability and they demonstrate what can be achieved with an existing tandem accelerator and peripheral equipment. In addition, as discussed in Section 3.1.5, the presence of the easily absorbed ^{10}B beam makes accurate adjustment of the ion beam transport system much easier than for ^{14}C.

The reproducibility achieved with the cyclotron as an element of a mass spectrometer is near $\pm 10\%$ but as pointed out by Raisbeck (personal communication) this is more than adequate for the present exploratory phases of work that has not been possible before.

3.4 Chlorine-36 Analysis

The radioisotope ^{36}Cl (half-life 308,000 years), is created in the atmosphere by cosmic rays and it is expected to be significant in hydrogeological studies, especially in selecting nuclear-waste storage sites (86). This cosmic-ray-generated isotope has now been detected in ground water with the help of a tandem accelerator and atom identification techniques (41). Exploratory tests using a cyclotron have been reported (87).

^{36}Cl has to be distinguished from the isobars ^{36}Ar and ^{36}S as well as from any molecules such as $^{12}C_3$, $^{18}O_2$ etc. Fortunately Ar^- is quite unstable and ^{36}S is a very rare isotope (0.014% of sulfur) although sulfur forms negative ions quite readily. The mass number of ^{36}Cl is divisible by 2, 3, 4, 6, etc so that ideally charge state $+5$ or $+7$ must be analyzed; otherwise many fragments of molecules with identical E/q or M/q could be transmitted through the low resolution magnetic and electric analyzers and could pose a counting rate problem in the critical final detector (40).

Preliminary measurements on ^{36}Cl detection at Rochester (88) indicated the importance of an electrostatic analyzer. After such an analyzer was added to the system (Figure 4) ^{36}Cl at natural abundances was detected (41). The early measurements (88) also demonstrated the size of the memory effect. When CCl_4 vapor containing ^{36}Cl was used in a duoplasmatron ion source it generated $^{36}Cl^-$ beams for tens of hours even after CCl_4

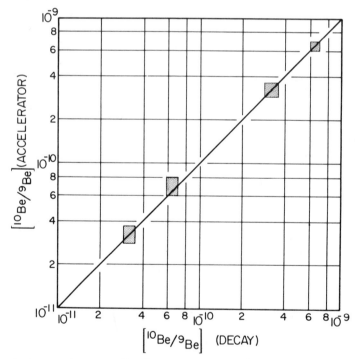

Figure 9 A comparison of the $^{10}Be/^9Be$ ratios determined by beta-ray counting (DECAY) and atom counting (ACCELERATOR).

containing no ^{36}Cl was used in the ion source. The solid AgCl samples used in the negative ion sputter source showed no evidence of a memory effect.

Figures 10*a* and 10*b* show the output of the heavy ion counter for a sample of chlorine from Lake Ontario and for chlorine from reagent grade AgCl (41). The lower limit reached in the latter case was ^{36}Cl/Cl $\approx (3 \pm 1) \times 10^{-15}$.

In these experiments great care had to be taken to keep the sulfur

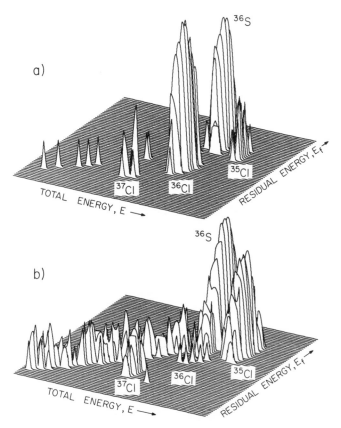

Figure 10a   A three-dimensional representation, similar to that shown in Figure 5, of data (41) from the heavy ion counter at Rochester. The ions are observed from an AgCl sample from Lake Ontario when the apparatus was adjusted to transmit only mass-36 charge-7 ions. The ^{35}Cl and ^{37}Cl peaks are part of the background and the ^{36}S peak is due to the residual sulfur in the purified sample.

Figure 10b   The ^{36}Cl^{+7} ions from a zone-refined reagent-grade AgCl sample are shown together with the background ions. The few counts shown for ^{36}Cl represent a ^{36}Cl/Cl ratio of $(3 \pm 1) \times 10^{-15}$.

contamination low because of the presence of the ^{36}S ions. Zone-refined AgCl was used for the results shown in Figure 10*b*. In later measurements on ^{36}Cl in Antarctic meteorites (89) the lower limit reached was 7×10^{-16} for ^{36}Cl/Cl when improved chemical preparation techniques were used to lower the level of the sulfur in the AgCl samples. The quoted accuracy of the ^{36}Cl/Cl ratios was between $\pm 5\%$ and $\pm 10\%$, which, as in the case of the ^{10}Be/Be ratios, is considered to be quite satisfactory for exploratory work where no previous measurements exist.

3.5 *Aluminum-26 Analysis*

^{26}Al is also created by cosmic rays in the atmosphere by the spallation of argon. However, it is produced at a yield lower by about a factor of 100 than

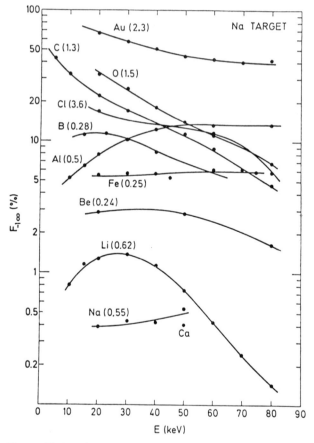

Figure 11 The equilibrium fractions for negative ions produced by charge exchange of positive ions in sodium vapor (91) are shown as a function of ion energy for various elements. The electron binding energy is also shown for each element.

^{10}Be so that it is considerably more difficult to study (1). Also the abundant stable isobar ^{26}Mg has a longer range than the ^{26}Al in matter so that the isobars are probably impossible to separate by using positive ions unless very high energies are used. At very high energies (Section 2.3) a significant fraction of the atoms can be created as bare nuclei and the ^{26}Mg^{+12} and ^{26}Al^{+13} can be easily distinguished.

This procedure was first successfully used by Raisbeck et al (32) who accelerated ^{26}Al^{+9} and ^{26}Mg^{+9} to 200 MeV and then analyzed the ions magnetically after stripping off all their electrons in a foil. A ratio of ^{26}Al/Al of 10^{-11} was measured and it was estimated that 10^9 atoms of ^{26}Al could be measured under the conditions of the experiment. This is the number expected to be present in about 40 grams of ocean sediment (32).

The radioisotope ^{26}Al has also been detected at lower energies using tandem accelerators (84, 90). As in the case of the pair of isobars, ^{14}C and ^{14}N, the negative ion of ^{26}Al is stable and the ^{26}Mg$^-$ ion appears to be either metastable with a short half-life or too fragile to survive the electric and magnetic fields of the apparatus. A ratio of $(3.8 \pm 0.7) \times 10^{-12}$ for ^{26}Al/Al was measured with a sensitivity for ^{26}Al/Al of about 10^{-14} reached (90). The total efficiency in this case was near 10^{-4}, for atoms of ^{26}Al from the sample detected in the final detector divided by the number of ^{26}Al atoms in the sample. By sputtering ^{26}Al followed by a charge exchange to ^{26}Al$^-$ in potassium vapor at 25 keV (91) as shown in Figure 11 it should be possible to reach a total efficiency of over 1% for the detection of ^{26}Al atoms in a sample. Such a high efficiency brings one step closer the possibility of exploiting a chronology independent of cosmic-ray fluctuation by measuring the ^{10}Be/^{26}Al ratio (92) in small samples with atom counting.

3.6 Iodine-129 Analysis

The radioisotope ^{129}I (half-life 1.6×10^7 years) is created in meteorites by cosmic rays. It is also a product of the spontaneous fission of ^{238}U and the neutron-induced fission of ^{235}U. ^{129}I can be studied quite readily by neutron activation analysis (93) because the large neutron capture cross section of 30 barns turns ^{129}I into ^{130}I, which has a half-life of 12 hours. ^{129}I has also been detected by conventional negative ion mass spectrometry (94) with a ^{129}I/^{127}I ratio of less than 10^{-7}. These measurements were limited by the presence of mass-129 negative molecular ions indistinguishable from ^{129}I$^-$.

Iodine forms negative ions very readily and as ^{129}Xe$^-$ is expected to be unstable, or metastable with a very short half-life, the separation of these isobars can be accomplished at the ion source and the mass-129 molecules eliminated with certainty by removing three or more electrons from them.

A successful exploratory study of ^{129}I by atom counting (95) has been made recently at Rochester using the tandem accelerator as a molecular disintegrator. Unfortunately the mass resolution at the ion source $(M/\Delta M \approx 25)$ was quite insufficient to resolve ^{129}I and the stable isotope ^{127}I. The usual heavy ion detectors, silicon counters or ionization chambers, were also unable to resolve the ^{129}I ions from the residual ^{127}I ions since this requires a counter resolution of 1% or better. Also the resolution of the electrostatic analyzer shown in Figure 4 was inadequate. Consequently, a time-of-flight apparatus (30) with a one-nanosecond timing resolution over a 2.2-meter flight path was used. Use of milligram samples of AgI in the negative ion sputter source reached a sensitivity for ^{129}I/^{127}I of less than 10^{-12} in about half an hour (95). This was sufficient to

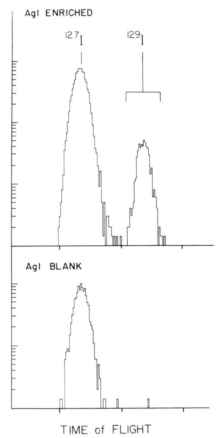

Figure 12   The time-of-flight spectrum for the ions emerging from the electrostatic analyzer of Figure 4. Silver iodide enriched to ^{129}I/^{127}I $\approx 10^{-11}$ was used for the top spectrum and no ^{129}I should be present in the bottom spectrum. The logarithm of the ion counts is shown at the left.

observe the ^{129}I in two meteorite samples. The separation of ^{129}I and ^{127}I achieved in one series of experiments is illustrated in Figure 12.

The sensitivity achieved was about two orders of magnitude better than the best previous measurements (96) and in principle the sensitivity for the atom counting procedure can be increased further by at least three orders of magnitude to approximately the same sensitivity as for ^{36}Cl/Cl. Surface ionization negative ion sources (97) promise to be very useful for very high sensitivity measurements of the halogens such as ^{129}I and ^{36}Cl.

3.7   Analysis of Other Radioisotopes

Both ^{3}H (37) and ^{32}Si (84) have been detected by atom counting. ^{3}H (half-life of 12 years) has a very low natural abundance, and an isotope enrichment of 500 times was used to increase the ^{3}H to a level at which it could be detected. The dating of water samples by atom counting of the ^{3}He from the decay of ^{3}H is, however, already a well-established and accurate technique (98).

^{32}Si (half-life about 400 years) was created (84) by nuclear reactions in pure silicon and was detected at the level of $10^{-8} > {}^{32}$Si$/^{28}$Si $> 10^{-14}$. The method used involved the measurement of the difference in energy between ^{32}Si and ^{32}S after they traversed an aluminum foil stack. A large magnetic spectrograph was used to measure the energy difference. The general problem of detecting naturally occurring ^{32}Si is made more difficult by the fact that both ^{32}Si and ^{32}S form negative ions. However, the range of ^{32}Si is greater than ^{32}S by about 14% at 64 MeV so that range separation (37) may be useful in this case (see also Section 3.1.8).

3.8   Stable Isotope Analysis

As discussed in Sections 1.1 and 2.1, the elimination of molecules from mass spectra should increase greatly the sensitivity for the detection of trace quantities of stable elements because all elements, except indium, have at least one unique isotope. The detection and measurement of trace quantities of elements with high sensitivity, better than one ppb, can have many applications in the area of materials research and the study of minerals (10). Some of the techniques developed for the study of radioisotopes can be of great value in the study of stable elements; the early work on the element platinum (J. C. Rucklidge et al, personal communication) illustrates this point.

The apparatus used for the study of the platinum isotopes was the same as that used to study ^{129}I and is shown in Figure 4. The final heavy ion counter in this case was a silicon surface barrier counter. Platinum makes negative ions quite readily because of the high (2 eV) electron binding energy. However, because of the high magnetic rigidity of the platinum ions the voltage on the central electrode of the tandem was lowered to 3.3 MV and 19.8-MeV Pt^{+5} ions were analyzed. The samples used in the negative ion sputter source were pressed powder samples from South Africa (3740 ppb) and Sudbury, Canada (59 ppb). Their compositions had been analyzed previously by neutron activation analysis (99). A three-dimensional plot of the $^{194}Pt^{+5}$ ion energy, time of flight, and logarithm of the total ion counts is shown in Figure 13, which is for the 59-ppb platinum sample from Sudbury. The 3740-ppb sample gave similar results with an appropriately larger yield of platinum ions. The total ion counts of $^{195}Pt^{+5}$ were similar as expected from the natural isotope abundances. Also ^{197}Au ions were readily observed at the appropriate magnet settings. The sensitivity of the particular arrangement used was at least 1 ppb and further exploratory work is in progress.

These preliminary data demonstrate, as one might expect from the radioisotope work, that the ultrasensitive mass spectrometry of stable isotopes is possible and promising. The data also demonstrate the problems that will be encountered in such work. Other ions from both samples were observed with the same time of flight but with different energies. These are the ions expected from the breakup of molecules with masses near 194 that are from more abundant elements in the samples and that are injected

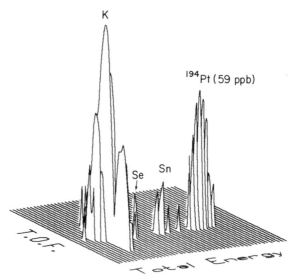

Figure 13   A three-dimensional representation, similar to that given in Figure 5, of the data from the time-of-flight apparatus and a final solid state heavy ion counter, which measures the total energy. Rock powder containing 59 ppb of platinum was used in the ion source of the apparatus shown in Figure 4.

into the tandem accelerator because of the low resolution of the inflection magnet. The moderate resolution ($E/\Delta E \approx 200$) of the $10°$ electrostatic analyzer at high energies ensures that molecular fragments with $M/q = 38.8 \pm 0.2$ will be transmitted to the detector. The tin ions are probably a mixture of $^{116}Sn^{+3}$ and $^{117}Sn^{+3}$ with values of M/q of 38.67 and 39 respectively; the $^{39}K^{+1}$ ion also has M/q of 39.

These measurements demonstrate that an inflection magnet for the low energy negative ions (Figure 4) capable of resolving individual masses near 194 is highly desirable. Then it may only be necessary to select the charge state of the ion at high energy in order to have very high sensitivity. A simple ion-energy-measuring detector would also remove any ambiguities due to molecular fragments. The addition of a second charge exchange at high energies followed by a second electrostatic analysis could eliminate the need for an energy-sensitive detector. The possible presence of hydride ions requires that the optimum electrostatic analyzer should have a resolution $E/\Delta E \approx 1000$. The required accuracy of the rare element abundance measurement can be achieved by using the standard isotope dilution method (3).

4   CONCLUSIONS

4.1   *Present Situation*

Modifications are being made to existing nuclear physics facilities so that programs on ultrasensitive mass spectrometry can be carried out. For example, an external ion source for the Berkeley 88-inch cyclotron is being added (38a–c, R. A. Muller personal communication) and this is expected to solve many of the problems that frustrated the early attempts to measure isotope ratios.

A different set of frustrating problems have been encountered with tandem accelerators as described in Section 3, and two strategies are being followed to remove or alleviate these problems.

The first approach involves the installation of machines specifically for ultrasensitive mass spectrometry (56). The subject has been considered important enough for three small systems to be funded at the University of Arizona at Tucson, Oxford University in England, and the University of Toronto in Canada respectively. The first two will be devoted mainly to the development of radiocarbon dating by atom counting. The project at the University of Toronto is for ultrasensitive mass spectrometry in general and for the development of a new generation of ion microprobes for geology and materials research. A plan of the layout of the components of the machine for Toronto is shown in Figure 14. The accelerator, known as a 3-MV tandetron, is about five times shorter than the accelerator shown in Figure 4.

The accelerator size was chosen to maximize the efficiency for the detection of ^{14}C ions. If an efficiency lower by a factor of two could be tolerated, an accelerator one third the size of that shown in Figure 14 could be used. In this latter case a second charge exchange would be necessary after the accelerator, as shown in Figure 7. The operation of the devices shown in Figure 14 has been described in some detail (56).

The second approach is being adopted at laboratories such as that at the University of Rochester where extensive modifications are planned to improve the existing apparatus at the nuclear structure laboratory for ultrasensitive mass spectrometry. The much larger accelerator at Rochester will be able to study those aspects of ultrasensitive mass spectrometry that require higher energy. It should be easier, for example, to separate isobars above ^{36}Cl-^{36}S by dE/dx measurements than at the laboratories containing a smaller accelerator.

4.2 Future Applications

The variety of the techniques to improve the sensitivity of mass spectrometry described earlier will undoubtably increase, and better methods for discriminating between elements and isobars will be found. However, even without significant improvements, many new programs on the detection of radioisotopes and stable isotopes are being started or being considered.

Work on the search for superheavy elements has already been carried out using a tandem accelerator (100), and quark searches using electrostatic

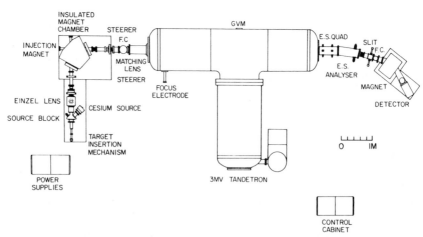

Figure 14 A plan of one version of the ultrasensitive mass spectrometers being built in various laboratories. The scale in meters is shown. Faraday cups are designated FC and the electrostatic elements at the exit of the 3-MV tandem or molecular disintegrator are labeled ES. The generating voltmeter for the measurement of the high voltage is designated GVM.

accelerators (101, 102) and a cyclotron (75) have been reported. These searches will undoubtedly continue with ever increasing degrees of sophistication using the many improvements possible in the systems.

One improvement related to the system used to search for nonintegrally charged quarks or possibly quarks attached to nucleons, or nucleons absorbed in quarks (103), employs an all electric analysis system (30) to avoid the tedious, time-consuming search of the mass spectrum. The all-electric system acts as a filter for unusually charged objects and works on the principle that charge exchange of atoms between a pair of electrostatic analyzers specifies the ratio of the charges on the atoms. So, for example, an object $X^{+1/3}$ changing to $X^{-2/3}$ by the capture of one electron would simulate the capture of three electrons by a singly ionized atom. Doubly charged negative ions are presently unknown: if they exist they are expected to be very fragile because of coulomb repulsion. Another charge exchange at somewhat higher energies would also be necessary in order to break up any interfering molecules. Finally, a charge exchange of the type $X^{+1/3}$ changing to $X^{+7/3}$ would simulate the loss of six electrons even though only two electrons were removed. The mass spectrum from such a charge filter could be measured with the help of time-of-flight and energy-sensitive detectors. Experiments of this type will be most valuable because, even if they produce no fractionally charged events, they will undoubtedly advance the techniques of ultrasensitive mass spectrometry.

Magnetic analyzers can also be used in the search for anomalous mass nuclei although such experiments are time consuming. An experiment on the search for anomalously heavy isotopes of oxygen was reported recently (104). Although unsuccessful, a useful lower limit on the existence of such objects was assigned.

Rare radioisotopes are continually being created by solar neutrinos as well as by cosmic rays and there has been interest recently on detecting such radioisotopes. The creation of ^{205}Pb from ^{205}Tl has been considered because it would be sensitive to the lowest energy solar neutrinos (105). Because the ratio $^{205}Pb/^{205}Tl$ at equilibrium is near 10^{-19}, the problem is a formidable one requiring the use of thallium ore that contains less than 1 ppm lead. Unfortunately both lead and thallium make negative ions so the selection of isobars cannot be carried out in the same way as ^{14}C and ^{14}N. A search (K. H. Chang, personal communication) for a suitable pair for negative ion selection suggests the pair ^{137}La, ^{137}Ba. Much preparatory research will be necessary for that type of experiment.

The double beta decay of tellurium and selenium isotopes has been observed with the help of mass spectrometry (106) by exploiting the fact that xenon and krypton isotopes are rare. Similar experiments using negative ion selection are possible (H. E. Gove, personal communication) provided careful selection is made of the material for study.

Undoubtedly the search for superheavy elements will continue to be modeled on the study of the platinum isotopes described in Section 3.8. Element 111 is expected to be eka-gold and to form stable negative ions (17). In this case only the destruction of interfering molecules and overall apparatus sensitivity is a concern.

The problems in using cyclotrons as precise ultrasensitive mass spectrometers seem formidable compared with those of tandem accelerators, as discussed earlier in Section 3.2. However, there are several important cases where only the positive ion cyclotron or very high energy heavy ion machine approach seems feasible for the detection of some long-lived radioactive nuclei. Atoms such as ^{41}Ca and the heavy noble gases, which do not form negative ions readily, present a problem for acceleration in tandem accelerators. The importance of ^{81}Kr for studies in hydrology (86) and ^{41}Ca for archaeological studies (107) have been stressed recently. In both cases the acceleration to very high energy now seems to be the only feasible method of detecting them and discriminating against their isobars ^{81}Br and ^{41}K. However, the possibility of using molecules such as $^{41}CaF^+$ to discriminate against the unstable $^{41}KF^+$ should not be forgotten (5).

There are many applications of ultrasensitive mass spectrometry to problems using long-lived radioisotopes as tracers. Two examples among many are the study of human metabolism using ^{14}C as a tracer (108) and the study of nervous-system problems using ^{26}Al as a tracer (109). These and other applications of negative ions, molecular interference elimination, and a much broader range of energy utilization than conventional mass spectrometry are likely to have a considerable effect on trace element and isotope analysis as the subject of ultrasensitive mass spectrometry matures.

ACKNOWLEDGMENTS

I would like to thank K. H. Purser, H. E. Gove, and my colleagues at Toronto, Rochester, and elsewhere for many valuable discussions that have helped to clarify the problems of ultrasensitive mass spectrometry. This work was supported in part by a grant from the Natural Sciences and Engineering Research Council, Ottawa.

Literature Cited

1. Oeschger, H., Wahlen, M. 1975. *Ann. Rev. Nucl. Sci.* 25:423–63 (see particularly Sect. 3.3)
2. Grunder, H. A., Selph, F. B. 1977. *Ann. Rev. Nucl. Sci.* 27:353–92
3. De Bievre, P. 1978. In *Advances in Mass Spectrometry*, ed. N. R. Daly, 7A:395–447. London: Inst. Petroleum, Heyden & Son. 816 pp.
4. Fukumoto, S., Matsuo, T., Matsuda, H. 1968. *J. Phys. Soc. Jpn.* 25:946–50
5. Ravn, H. L. 1979. *Phys. Rep.* 54:201–59
6. Karasek, F. W. 1978. *Ind. Res. Dev.* 20 (12):86–90

7. Hurst, G. S., Payne, M. G., Kramer, S. D., Young, J. P. 1979. *Rev. Mod. Phys.* 51:767–819
8. Gove, H. E., ed. 1978. *Proc. 1st Conf. on Radiocarbon Dating with Accelerators.* Univ. Rochester. 401 pp.
9a. Anderson, C. A., Hinthorn, J. R. 1972. *Science* 75:853–60
9b. Anderson, C. A., Hinthorn, J. R. 1973. *Anal. Chem.* 45:1421–38
10a. Purser, K. H., Litherland, A. E., Rucklidge, J. C. 1979. *Surface Interface Anal.* 1:12–19
10b. Benninghoven, A., Evans, C. A. Jr., Powell, R. A., Shimizu, R., Storms, H. A., eds. 1979. *Secondary Ion Mass Spectrometry (SIMS II).* New York: Springer. 298 pp.
11. Hinton, R. W., Long, J. V. P. 1979. *Earth Planet. Sci. Lett.* 45:309–25
12. Rothenberg, M. 1977. *Phys. Today* 30 (12):17–19
13a. Muller, R. A. 1979. *Phys. Today* 32(2): 23–30
13b. Litherland, A. E., Gove, H. E., Purser, K. H. 1980. *Phys. Today* 33(1):13–15
14. Bennett, C. L. 1979. *Am. Sci.* 67:450–57
15. Berger, R. 1979. *J. Arch. Sci.* 6:101–3
16. Banning, E. B., Pavlish, L. A. 1979. *Antiquity* 53:226–28
17a. Keller, O. L. Jr., Seaborg, G. T. 1977. *Ann. Rev. Nucl. Sci.* 27:139–66
17b. Hintenberger, H. 1962. *Ann. Rev. Nucl. Sci.* 12:435–506 (see particularly Figure 10)
18. Fleming, S. 1976. *Dating in Archaeology*, pp. 56–85. London: Dent. 272 pp.
19a. Anbar, M. 1978. See Ref. 8, pp. 152–55
19b. Wilson, H. W. 1979. In *Radio Carbon Dating, Proc. 9th Int. Conf. Los Angeles/La Jolla 1976*, ed. R. Berger, H. E. Suess, pp. 238–45. Berkeley: Univ. Calif. Press. 800 pp.
20. Tyrell, A. C., Ridley, R. G., Daly, N. R. 1968. *Int. J. Mass Spectrosc. Ion Phys.* 1:69–73
21. Daly, N. R. 1960. *Rev. Sci. Instrum.* 31:264–67
22. Shimizu, N., Semet, M. P., Allegre, C. J. 1978. *Geochim. Cosmochim. Acta* 42:1321–34
23. Bowie, J. H., Stapleton, B. J. 1976. *J. Am. Chem. Soc.* 98:6480–83
24. Litherland, A. E. 1978. See Ref. 8, pp. 70–113
25. Bock, R. 1974. In *Nuclear Spectroscopy and Reactions, Part A*, ed. J. Cerny, pp. 89–91. New York/London: Academic. 518 pp.
26. Wittkower, A. B., Betz, H. D. 1973. *Atomic Data* 5:113–66
27. Purser, K. H. 1977. *US Patent No. 4,037,100.*
28. Purser, K. H., Litherland, A. E., Gove, H. E. 1979. *Nucl. Instrum. Methods* 162: 637–56
29. Bromley, D. A., ed. 1979. Detectors in nuclear science. *Nucl. Instrum. Methods* Vol. 162. 737 pp.
30. Kilius, L. R. 1980. *Ultrasensitive mass spectrometry.* PhD thesis. Univ. Toronto, Canada
31. Deleted in proof
32. Raisbeck, G. M., Yiou, F., Stephan, C. 1979. *J. Phys. Lett.* 40:241–43
33. Kilius, L. R., Litherland, A. E. 1978. *Rep. Workshop on Dating Old Ground Water*, ed. S. N. Davis, pp. 62–79. Univ. Ariz., Tucson. 138 pp.
34. Purser, K. H., Liebert, R. B., Litherland, A. E., Beukens, R. P., Gove, H. E., Bennett, C. L., Clover, M. R., Sondheim, W. E. 1977. *Rev. Phys. Appl.* 12:1487–92
35. Fogel, Ya. M., Kozlov, V. F., Kalmykov, A. A. 1959. *Sov. Phys. JETP* 9:963–64
36a. Burke, P. G., Berrington, K. A., Le Dourneuf, M., Vo Ky Lan. 1974. *J. Phys. B* 7:L531
36b. Condon, E. U., Shortley, G. H. 1935. *Theory of Atomic Spectra*, pp. 282–83, Cambridge Univ. Press. 441 pp.
37. Muller, R. A. 1977. *Science* 196:489–94
38a. Muller, R. A., Stephenson, E. J., Mast, T. S. 1978. *Science* 201:347–48
38b. Mast, T. S. 1978. See Ref. 8, pp. 239–44
38c. Stephenson, E. J. 1978. See Ref. 8, pp. 187–95
39a. Raisbeck, G. M., Yiou, F., Fruneau, M., Loiseaux, J. M. 1978. *Science* 202:215–17
39b. Raisbeck, G. M. et al. 1978. See Ref. 8, pp. 38–46
40. Shapira, D., Devries, R. M., Fulbright, H. W., Toke, J., Clover, M. R. 1975. *Nucl. Instrum. Methods* 129:123–30
41. Elmore, D., Fulton, B. R., Clover, M. R., Marsden, J. R., Gove, H. E., Naylor, H., Purser, K. H., Kilius, L. R., Beukens, R. P., Litherland, A. E. 1979. *Nature* 277:22–25 and errata 246
42. Northcliffe, L. C., Schilling, R. F. 1970. *Nucl. Data Tables* 7:233–463
43. Dmitriev, I. S., Vinogradova, L. I., Nikolaev, V. S., Popov, B. M. 1966. *Sov. Phys. Lett.* 3:20–23
44. Deleted in proof
45. Schnitzer, R., Aberth, W. A., Brown, H. L., Anbar, M. 1974. *Proc. 22nd Ann. Conf. Am. Soc. Mass Spectrosc., Philadelphia, Pa.*, pp. 64–69
46. Hortig, G., Mokler, P., Mueller, M. 1968. *Z. Phys.* 210:312–13
47. Purser, K. H. 1973. *Proc. Int. Conf. on Technology of Electrostatic Accelerators*, ed. T. W. Aitken, N. R. S. Tait, pp. 392–400

48. Middleton, R. 1974. *Nucl. Instrum. Methods* 122:35–43; See also Ref. 8, pp. 196–219
49. Hall, E. T., Hedges, R. E. M. 1977. Presented at Intl. Symp. Archaeometry and Archaeological Prospection, Philadelphia, Pa.
50. Hird, B., Ali, S. P. 1978. *Phys. Rev. Lett.* 41:540–42
51. Hiraoka, H., Nesbet, R. K., Welsh, L. W. Jr. 1977. *Phys. Rev. Lett.* 89:130–33
52a. Nelson, D. E., Korteling, R. G., Stott, W. R. 1977. *Science* 198:507–8
52b. Nelson, D. E. et al. 1978. See Ref. 8, pp. 47–69
53. Bennett, C. L., Beukens, R. P., Clover, M. R., Gove, H. E., Liebert, R. B., Litherland, A. E., Purser, K. H., Sondheim, W. E. 1977. *Science* 198:508–9
54. Brand, K. 1977. *Nucl. Instrum. Methods* 141:519–20
55. Doucas, G., Garman, E. F., Hyder, H. R. M., Sinclair, D., Hedges, R. E. M., White, N. R. 1978. *Nature* 276:253–55
56. Purser, K. H., Hanley, P. R. 1978. See Ref. 8, pp. 165–86
57. Middleton, R. 1978. See Ref. 8, pp. 157–64
58. Andrews, H. R., Ball, G. C., Brown, R. M., Davies, W. G., Imahori, Y., Milton, D. C. 1980. *Proc. 10th Int. Radiocarbon Conf. Radiocarbon.* In press
59. Farwell, G. W., Schaad, T. P., Schmidt, F. H., Tsang, M. Y. B., Grootes, P. M., Stuiver, M. 1980. *Proc. 10th Int. Radiocarbon Conf. Radiocarbon.* In press
60. Shea, J. H., Conlon, T. W., Asher, J., Reed, P. M. 1980. *Proc. 10th Int. Radiocarbon Conf. Radiocarbon.* In press
61a. White, N. R., Wand, J., Hedges, R. E. M. 1980. *Proc. 10th Int. Radiocarbon Conf. Radiocarbon.* In press
61b. Hedges, R. E. M., Moore, C. B. 1978. *Nature* 276:255–57
62. Grootes, P. M., Stuiver, M., Farwell, G. W., Schad, T. P., Schmidt, F. H. 1980. *Proc. 10th Int. Radiocarbon Conf. Radiocarbon.* In press
63. Gove, H. E., Elmore, D., Ferraro, R., Beukens, R. P., Chang, K. H., Kilius, L. R., Lee, H. W., Litherland, A. E., Purser, K. H., Rubin, M. 1980. *Proc. 10th Int. Radiocarbon Conf. Radiocarbon.* In press; Also *Univ. Rochester Rep. UR-NSRL-196*
64. Bennett, C. L., Beukens, R. P., Clover, M. R., Gove, H. E., Kilius, L. R., Litherland, A. E., Purser, K. H. 1978. *Science* 201:345–47
65. Gove, H. E., Elmore, D., Ferraro, R., Beukens, R. P., Chang, K. H., Kilius,

L. R., Lee, H. W., Litherland, A. E., Purser, K. H. 1980. *Proc. 4th Int. Conf. Ion Beam Analysis. Nucl. Instrum. Methods* 168:425–33
66. Chapman, K. R. 1979. *Proc. 1978. Symp. Northeast. Accel. Personnel, Oak Ridge, Tenn. Oak Ridge Natl. Lab. Rep. CONF-781051*, pp. 80–81
67. Livingood, J. J. 1969. *The Optics of Dipole Magnets.* New York/London: Academic. 261 pp.
68. Russell, W. A., Papanastassiou, D. A., Tombrello, T. A. 1978. *Geochim. Cosmochim. Acta* 42:1075–90
69. Tombrello, T. A. 1979. *Proc. 10th Lunar Planet. Sci. Conf. Pt. 3*, pp. 1233–35
70. Oparin, V. A., Il'in, R. N., Serenkov, I. T., Solov'ev, E. S., Fedorenko, N. V. 1971. *Sov. Phys. Lett.* 13:249–52
71a. Il'in, R. N., Serenkov, I. T., Oparin, V. A. 1975. *9th Int. Conf. Physics Electronic Atomic Collisions, Seattle, Wash.*, pp. 39–40
71b. Oparin, V. A., Il'in, R. N., Serenkov, I. T., Solov'ev, E. S., Federenko, N. V. 1969. *7th Int. Conf. Physics Electronic Atomic Collisions, Amsterdam*, pp. 796–99
72. Betz, H. D. 1972. *Rev. Mod. Phys.* 44:465–539 (see particularly Sect. VI and Fig. 6.2)
73. Deleted in proof
74. Hamilton, P. J., O'Nions, R. K., Evensen, N. H. 1978. *Earth Planet. Sci.* 36:263–68
75. Muller, R. A., Alvarez, L. W., Holley, W. R., Stephenson, E. J. 1977. *Science* 196:521–23
76. Mast, T. S., Muller, R. A. 1980. *Nucl. Sci. Appl.* In press
77. Stephenson, E. J., Mast, T. S., Muller, R. A. 1979. *Nucl. Instrum. Methods* 158:571–77
78. Fontes, P., Perron, C., Lestringuez, J., Yiou, F., Bernas, R. 1971. *Nucl. Phys. A* 165:405–14
79. Raisbeck, G. M., Yiou, F., Fruneau, M., Lieuvin, J. M., Loiseaux, J. M. 1979. *Nature* 275:731–33
80. Raisbeck, G. M., Yiou, F., Fruneau, M., Loiseaux, J. M., Lieuvin, M., Ravel, J. C., Hays, J. D. 1979. *Geophys. Res. Lett.* 6:717–19
81. Raisbeck, G. M., Yiou, F., Fruneau, M., Loiseaux, J. M., Lieuvin, M. 1979. *Earth Planet. Sci. Lett.* 43:237–40
82. Turekian, K. K., Cochran, J. K., Krishnaswami, S., Lanford, W. A., Parker, P. D., Bauer, K. A. 1979. *Geophys. Res. Lett.* 6:417–18
83. Kilius, L. R., Beukens, R. P., Chang, K. H., Lee, H. W., Litherland, A. E., Elmore, D., Ferraro, R., Gove, H. E., Purser, K. H. 1980. *Nucl. Instrum. Methods* 171:355–60

84. Kutschera, W., Henning, W., Paul, M., Stephenson, E. J., Yntema, J. L. 1980. *Proc. 10th Int. Radiocarbon Conf. Radiocarbon.* In press

85. Nelson, D. E., Korteling, R., Southon, J., Nowikow, I., Hammaren, E., Burke, D. G., McKay, J. W. 1979. *Ann. Prog. Rep. McMaster Accel. Lab., Hamilton, Ontario,* pp. 93–96

86. Davis, S. N., ed. 1978. *Rep. Workshop on Dating Old Ground Water.* Univ. Ariz., Tucson. 138 pp.

87. Mast, T. S. 1978. See Ref. 8, pp. 239–44

88. Naylor, H., Elmore, D., Clover, M. R., Kilius, L. R., Beukens, R. P., Fulton, B. R., Gove, H. E., Litherland, A. E., Purser, K. H. 1978. See Ref. 8, pp. 360–71

89. Nishiizumi, K., Arnold, J. R., Elmore, D., Ferraro, R. D., Gove, H. E., Finkel, R. C., Beukens, R. P., Chang, K. H., Kilius, L. R. 1979. *Earth Planet. Sci. Lett.* 45: 285–92

90. Kilius, L. R., Beukens, R. P., Chang, K. H., Lee, H. W., Litherland, A. E., Elmore, D., Ferraro, R., Gove, H. E. 1979. *Nature* 282: 488–89

91. Heinemeier, J., Hvelplund, P. 1978. *Nucl. Instrum. Methods* 148: 425–29

92. Lal, D. 1962. *J. Oceanogr. Soc. Jpn., 20th Anniv. Iss.,* pp. 600–14

93. Edwards, R. R. 1962. *Science* 137: 851–53

94. McHugh, J. A., Sheffield, J. C. 1965. *Anal. Chem.* 37: 1099–1101

95. Kilius, L. R., Beukens, R. P., Chang, K. H., Lee, H., Litherland, A. E., Elmore, D., Gove, H. E., Finkel, R. C. 1979. *Bull. Am. Phys. Soc.* 24: 1186

96. Rook, A. L., Suddueth, J. E., Becker, D. A. 1975. *Anal. Chem.* 47: 1557–61

97. Rachidi, I., Monte, J., Pelletier, J., Pomot, C., Rinchet, F. 1976. *Appl. Phys. Lett.* 28: 292–94

98. Clarke, W. B., Jenkins, W. J., Top, Z. 1976. *Int. J. Appl. Radiat. Isotopes* 27: 515–22

99. Hoffman, E. L. 1978. *Platinum group elements and gold content of some nickel sulphide ores.* PhD thesis. Univ. Toronto, Canada

100. Schwartzschild, A. Z., Thieberger, P., Cumming, J. B. 1977. *Bull. Am. Phys. Soc.* 22: 94

101. Schiffer, J. P., Renner, T. R., Gemmell, D. S., Mooring, F. P. 1978. *Phys. Rev. D* 17: 2241–44

102. Boyd, R. N., Elmore, D., Melissonos, A. C., Sugarbaker, E. 1978. *Phys. Rev. Lett.* 40: 216–20

103. De Rujula, A., Giles, R. C., Yaffe, R. L. 1978. *Phys. Rev. D* 17: 285–301

104. Middleton, R., Zurmuhle, R. W., Klein, J., Kollarits, R. W. 1979. *Phys. Rev. Lett.* 43: 429–31

105. Freedman, M. S., Stevens, C. M., Horwitz, E. P., Fuchs, L. H., Lerner, J. L., Goodman, L. S., Childs, W. J., Hessler, J. 1976. *Science* 193: 117–19

106. Bryman, D., Picciotto, C. 1978. *Rev. Mod. Phys.* 50: 11–21

107. Raisbeck, G. M., Yiou, F. 1979. *Nature* 277: 842–43

108. Keilson, J., Waterhouse, M. D. 1978. See Ref. 8, pp. 391–97

109. Crapper, D. R., DeBoni, U. 1980. In *Experimental and Clinical Neurotoxicology,* Chap. 22, ed. P. S. Spencer, H. H. Schaumburg. Baltimore, Md: Williams & Wilkens. In press

Ann. Rev. Nucl. Part. Sci. 1980. 30 : 475–542

PARTICLE COLLISIONS ABOVE 10 TeV AS SEEN BY COSMIC RAYS

✻5623

T. K. Gaisser[1]

Bartol Research Foundation of The Franklin Institute, University of Delaware, Newark, Delaware 19711

G. B. Yodh

National Science Foundation, Washington, DC 20550 and Department of Physics, University of Maryland,[2] College Park, Maryland 20742

CONTENTS

[1] Supported in part by the US Department of Energy.
[2] Permanent address.

475

0163-8998/80/1201-0475$01.00

1   INTRODUCTION

In particle physics the primary attraction of cosmic rays is their high energy. Our main goal here is to discuss the extent to which cosmic ray experiments have provided in the past, and can provide in the future, information about particle interactions beyond accelerator energies, despite the problem of low flux. We also want to indicate how these same experiments bear on the fundamental astrophysical questions of origin, acceleration, and propagation of cosmic rays. The relation of the cosmic ray experiments to experiments at the new generation of colliding-beam machines will be a major theme of the paper. We will, of course, describe how novel cosmic ray phenomena might be manifest in machine experiments, but we will also describe how new machines may help solve the astrophysical questions listed above.

1.1   *Summary and Plan of the Paper*

To facilitate the use of this review we include here a self-contained overview of the main points of the paper, organized to follow the chapter headings. Generally the topics are arranged in order of increasing energy, decreasing flux, and increasing detector size. We avoid a rigid adherence to this scheme, however, in order to give a coherent account of such subjects as new particle production and energy dependence of cross sections.

1.2   *Fluxes and Techniques*

The spectrum of relativistic cosmic ray nuclei incident at the top of the atmosphere has been measured by a variety of techniques up to 10^{20} eV. (See Hillas 1975 and 1979a for reviews.) Its dominant feature is the rapid decrease of flux with increasing energy, as illustrated in Figure 1. Of outstanding interest are the shoulder and change of slope of the spectrum between 10^{14} and 10^{16} eV.

In the energy range of interest from 10^{13} eV to 10^{20} eV the beam flux decreases by nearly 14 orders of magnitude. It is immediately clear, therefore, that different energy regimes within this broad span require different experimental techniques. Detectors range from small emulsion chambers, with an acceptance of 0.1 m² sr, to giant air shower arrays with areas measured in square kilometers. Intermediate in size and energy range are large calorimeters and emulsion chambers.

1.3 Collisions at 10–20 TeV

The emulsion chamber technique yields the most detailed information about properties of hadronic interactions in this energy range because it allows detection and energy estimates of single charged hadrons as well as

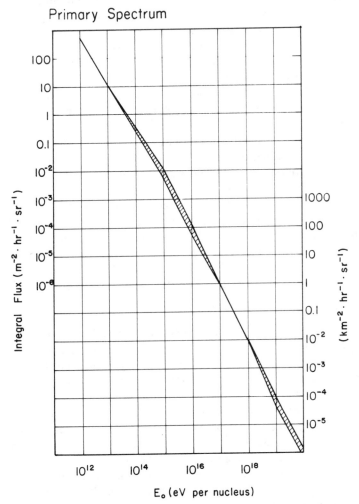

Figure 1 The flux of all primary cosmic rays at the top of the atmosphere, shown as an integral spectrum. Experimental uncertainties are represented by cross-hatching.

of neutral pions. There is limited information from these experiments about inclusive cross sections and multiplicity distributions. Events with fifty or more secondaries are relatively common. Inclusive cross sections are consistent in the fragmentation region with scaling from lower energies, but the average multiplicity appears to be somewhat higher than usually predicted by extrapolations from lower energy.

1.4   *New Particle Production*

Evidence for production of hadrons with new quantum numbers (flavors) by cosmic rays implies large cross sections (of the order of 1 mb per nucleon) at 10–100 TeV, not inconsistent with extrapolation of recent accelerator results on charm production. These experiments and the implications of such large cross sections for possible observation of new flavors in other cosmic ray experiments will be discussed. We find that, although it may be possible to observe production of new hadrons heavier than charmed particles in deep underground muon experiments, the anomalous events previously reported (Krishnaswamy et al 1976) most likely require other processes for their explanation.[3] Similarly, it seems virtually impossible to interpret the Tien-Shan observation of a threshold for anomalously slow cascade absorption in lead (Yakovlev et al 1979) in terms of production of new flavors.

Finally we mention that the presence of new long-lived ($\tau > 10^{-7}$ s) heavy objects is suggested by experiments that study simultaneously the distributions of energies and delay times of hadrons near air shower cores (Goodman et al 1979a). These experiments are also sensitive to primary composition from 10^{13} to 10^{15} eV, and suggest that the fraction of heavy nuclei in the beam increases significantly in this range.

1.5   *Collisions at 50–200 TeV*

To obtain significant numbers of events above 20 TeV it is necessary to use large, ground-based detectors, such as emulsion chambers, exposed for long periods. These have poorer resolution and typically detect only electromagnetic cascades with visible energies above a threshold of 0.2–2 TeV, depending on the detector. In one case (the Brazil-Japan collaboration experiment at Mt. Chacaltaya), more than 50 events in this energy range occurred in the target layer of the detector and so are relatively clean. Monte Carlo simulations, which incorporate the experimental biases, indicate that many features of the data can be understood as extrapolation of behavior familiar from accelerator energies, though it is apparent that a large amount of hard scattering (high p_T) and high multiplicity are required (Ellsworth et al 1980).

---

[3] They could, for example, be due to decay or interaction of new kinds of particles.

1.6   *Events Around 500 TeV and Above*

By extending the target to include a portion of the atmosphere over a large emulsion chamber it is possible to observe events up to above 10^{15} eV. These are really air shower cores. The famous Centauro events (Bellandi Filho et al 1979) are in this category. These are events with an anomalously small fraction of their energy in the electromagnetic (π^0) component as compared to the hadronic ($\pi^\pm$ and nucleon) component. We describe how these air-jets (A-jets) are selected, discuss event rates and compare the results of different experiments with each other and with results of simulations. Combining all experiments, it appears that the distribution of events in a scatter plot of electromagnetic vs hadronic energy indeed shows a significant excess in the region of small electromagnetic and large hadronic energy. This excess continues to defy a conventional explanation. We discuss briefly some unconventional explanations that have been suggested.

1.7   *Extensive Air Shower Interactions up to Millions of TeV*

The subject of extensive air showers (EAS) has been reviewed recently (Gaisser et al 1978), so we emphasize here some aspects not included in that review. These include especially (*a*) recent work on longitudinal development of air showers, including observations of atmospheric scintillation light with Fly's Eye (Cassiday et al 1979) and Cherenkov light using fast-timing detectors (Hammond et al 1978, Khristiansen 1979), and (*b*) multiple energetic muons in deep underground experiments. The fundamental problem with interpretation of EAS data arises from the fact that the primaries are not observed directly. Thus, for example, a shower of a given energy that develops rapidly could be a heavy nucleus with small energy per nucleon (i.e. a superposition of many low energy, rapidly developing subshowers) or a proton shower in which the high energy interactions are characterized by a high degree of energy sharing among many secondaries. (Recall that in the conventional picture one or two secondaries carry away most of the energy.) This example illustrates the necessity of determining simultaneously both the nature of the primary beam and the characteristics of high energy interactions above 10^{14} eV from shower properties. We will review the extent to which this has been done in past experiments and outline how the new experiments such as Fly's Eye can be expected to lead to significant progress.

1.8   *Energy Dependence of Cross Sections*

The original suggestion by Yodh et al (1972) that hadronic cross sections increase with energy was based on measurements of cosmic ray attenuation by the atmosphere. Measurement of the ratio of the real to the imaginary part of the forward proton-proton scattering amplitude, together with use of dispersion relations (Amaldi et al 1977), confirm that this increase continues up to 50 TeV, roughly the present limit of the cosmic ray attenuation method. We will summarize parametrizations which show that if the increase continues at this rate the p-p total cross section will reach 100 mb by 10^{17}–10^{18} eV, a prediction that can be checked by Fly's Eye.

1.9   Outlook

This is an interesting time for interaction between particle physics and cosmic rays. On the one hand, there are evidently new phenomena such as Centauro events, which should appear in experiments at the new $\bar{p}p$ colliding-beam facilities at CERN and FNAL (Cline & Rubbia 1979, Cline 1979a) and at the pp collider at BNL (Hahn et al 1977, White 1979) if they are features of hadronic interactions around 500 TeV lab energy rather than the signal of exotic components of the cosmic ray beam. On the other hand, studies of hadronic interactions at these machines should clarify the general features of particle production sufficiently so that significant progress can be made in the indirect determination of primary composition around 10^{15} eV from air shower measurements. In this way experiments with new machines could make a contribution to cosmic ray astrophysics in the particularly interesting range around 10^{15} eV (recall Figure 1).

New air shower investigations, including the measurements of air Cherenkov light, the Fly's Eye experiment, and the experiments at the Akeno array in Japan (Kamata 1979), are already in progress. These promise significant advances in our determination of behavior of high energy interactions and of primary composition up to 10^{18} eV and beyond.

2   FLUXES AND TECHNIQUES

The flux of particles with energies above 10^{14} eV is about 1 m^{-2} hr^{-1} (Figure 1). This energy therefore represents a practical dividing line between two broad classes of experiment: (a) those in which the primary particle and its interaction can be observed directly at the top of the atmosphere; and (b) those in which the low flux is overcome by using a detector of large area exposed at the Earth's surface for extended periods. In the latter case it is impossible to identify the charge of the primary nucleus directly. In this situation one must seek to determine simultaneously the properties of the interactions governing the development of the observed cascade and the nature of the primary.

We illustrate the relationship between flux and technique schematically in Table 1a. The numbers are meant only to be semiquantitative; for simplicity, overlaps between energy regions of adjacent techniques are not shown. The energy axis is labeled both in total lab energy of the incident nucleus (above the line) and in total center-of-mass energy of a nucleon-nucleon system, $s^{1/2}$ (below the line). The primary flux is shown below the energy axis and various experimental techniques and their parameters above it. The new $\bar{p}p$ and pp colliding-beam experiments (Table 1b) are planned to explore the range 540 GeV $\leq s^{1/2} \leq$ 2000 GeV, which is equivalent to $E_{\text{lab}} \approx 2 \times (10^{14}-10^{15})$ eV, overlapping the small air shower range.

Table 1a Cosmic ray experimental techniques

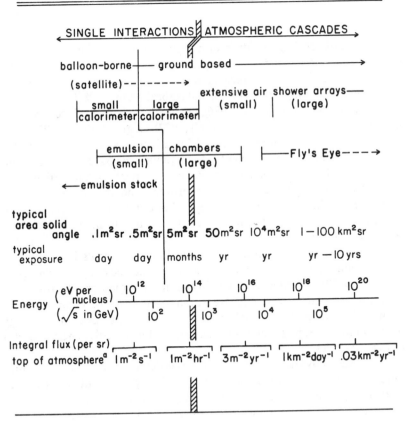

^a All particles

Table 1b New machines[a]

Machine		Center-of-mass energy (GeV)	Equivalent lab energy (GeV)	Planned completion
Colliding beams	CERN p̄p	540	1.5×10^5	1981
	ISABELLE pp (Brookhaven National Lab)	800	3.4×10^5	1986
	FNAL p̄p	2,000	2.1×10^6	1983
Possible future colliders	Super-ISABELLE	4,000	8.5×10^6	
	Future Fermilab	10,000	5.3×10^7	?
	Very Big Accelerator (VBA)	20,000	2.1×10^8	

[a] From L. W. Jones (1979a).

2.1 Primary Spectrum

At present, direct determination of nuclear composition of the primaries extends to about 4×10^{12} eV per nucleus, well below the 10^{14}-eV dividing line mentioned above. These measurements were made with balloon-borne instruments by Ryan et al (1972), Balasubrahmanyan & Ormes (1973), Juliusson (1974), Schmidt et al (1976), and Orth et al (1978). Recent reviews may be found in the papers of Caldwell (1977), Ormes & Freier (1978), and Simon et al (1980). The first column of Table 2 summarizes these results for the major elemental components. In a pioneering satellite experiment Grigorov et al (1971) measured the total energy spectrum (all particles) out to 10^{13} eV. The separation between protons and heavy nuclei in this experiment is, however, unreliable beyond about 4×10^{12} eV per nucleon because of problems with backscattered neutrons from the calorimeter (Ellsworth et al 1977).

At higher energies balloon-borne emulsion experiments have detected nuclei including protons, α-particles, and heavier nuclei up to energies of the order of 10^{15} eV. McCusker (1975) summarized these results from early exposures of emulsion stacks. He emphasized that the composition is known directly to be mixed up to this energy. Statistics and energy determination are, however, insufficient to establish relative energy spectra at these energies.

Experiments by the University of Chicago and the Chicago-Maryland-Goddard Space Flight Center groups in the Space Shuttle will extend statistically significant direct measurements of nuclear composition to about 5×10^{13} eV. At higher energies, knowledge of the spectrum and composition will continue to depend on various indirect methods. These include (a) secondary fluxes of muons, hadrons, and photons in the atmosphere; (b) charge ratio of muons, which reflects the charge ratio of incident nucleons and hence the composition, since all incoming neutrons may be assumed bound; (c) distributions of time delays of energetic hadrons behind air shower fronts; and (d) various other indirect air shower observations, such as depth of shower maximum. We will return to the first two points in the discussion below of cosmic rays in the atmosphere, and to some details of the situation with high energy air showers in Section 7.

Meanwhile, we can form the following consistent picture of composition throughout the range 10^{12}–10^{15} eV. We first note that the Grigorov all-particle spectrum coincides with the sum of the separately measured components below 10^{13} eV per nucleus and with air shower measurements around 10^{15} eV (Hillas 1979a). Between these energies we use measurements from the Maryland group.

Goodman et al (1979b) compared observations of delayed hadrons in small showers with Monte Carlo simulations to determine the composition necessary to account for the rather large flux of low energy hadrons near shower cores with significant delays behind the shower front. Such delayed particles can arise from nucleons released high in the atmosphere from heavy nuclei, which then cascade through the atmosphere losing energy and falling behind the shower front. For simplicity, nuclei of the He, CNO, and light heavy (LH) $(10 \leq Z \leq 16)$ charge groups were constrained to have

Table 2   Primary composition and spectra[a]

	Direct measurements		Power law approximations[c]	
Species	E^{b} (GeV/n)	$\dfrac{dn}{dE}$ $(m^2 sr\ s\ GeV/n)^{-1}$	Energy range[c] (GeV/n, approx.)	$\dfrac{dn}{dE} = KE^{-\gamma}$
Proton	2000	$(1.5 \pm 0.4) \times 10^{-5}$	10^4–10^5	$K = 1.50 \times 10^4$ $\gamma = 2.71 \pm 0.06$
Alpha	300	$(8 \pm 2) \times 10^{-5}$	5×10^3–5×10^4	$K = 533$ $\gamma = 2.71$
C,N,O	250	$(1.2 \pm 0.22) \times 10^{-5}$	2×10^3–2×10^4	$K = 36.2$ $\gamma = 2.71$
LH$(10 \leq Z \leq 16)$	75	$(1.2 \pm 0.2) \times 10^{-4}$	2×10^3–2×10^4	$K = 11.4$ $\gamma = 2.71$
Fe $(Z \geq 25)$	63	$(6.7 \pm 1.6) \times 10^{-5}$	2×10^3–2×10^4	$K = 1.3$ $\gamma = 2.36 \pm 0.06$

[a] Fluxes are stated as number of nuclei per GeV per nucleon per $m^2 \cdot s \cdot sr$.
[b] Energy up to which measurement errors are less than 25%.
[c] The experiment of Goodman et al (1979b) is sensitive to each species over this range, but the fits describe experimental data down to 25 GeV/n. Here E is in GeV/n.

energy spectra proportional to protons, as at low energies. The two components thus defined were each described by a magnitude and an exponent of the energy dependence that were allowed to vary to fit the data. The best fits for these parameters for all the components are given in the second column of Table 2.

In obtaining these results, each component was also constrained to pass through the highest energy direct measurement with a statistical error of $\leq 25\%$. Because the experiment is sensitive to showers in a certain size range at 2900 m altitude, the results for protons and for Fe refer to somewhat different ranges of primary energy per nucleus, as noted in Table 2. It is noteworthy that the slopes obtained in this way are consistent with those measured directly at low energy. It is therefore reasonable to use the forms shown in Table 2 to describe the composition and energy spectra over the entire range from 20 GeV per nucleon up to the highest energies probed in the delayed-particle experiment. For Fe this corresponds to 10^{15} eV per nucleus. The picture thus obtained is displayed in Figure 2, where we show integral spectra from 10^{12} to 10^{16} eV per nucleus. Consistency of this picture can be checked by adding the separate components and finding a sum in agreement with the all-particle spectrum measured by Grigorov. The consistency of this primary spectrum and composition with secondary fluxes in the atmosphere is discussed in the next section.

With the composition shown in Figure 2, the primaries are about 60% iron at 10^{15} eV per nucleus. It is just above this energy where the bend in the primary spectrum shown in Figure 1 occurs. It is therefore clear that the iron spectrum cannot continue much beyond 10^{15} eV with the flat slope of 2.36. It must steepen significantly in this region.

The region around 10^{15} eV is clearly of interest because of the shoulder and change of slope of the energy spectrum here. Several authors have discussed mechanisms that might lead to such a feature: the possible existence of a cutoff in magnetic rigidity above which cosmic rays leak out of the galaxy was discussed many years ago by Peters (1961). Such a cutoff would affect different nuclei at different total energies, with light nuclei escaping at lower energies. This would lead to an enrichment in heavy nuclei at high energy. The difference in slope for Fe is a hint that the situation is more complicated, however, perhaps reflecting a different source or production mechanism for the heavy nuclei.

If hadronic scaling is a valid concept up to 10^{17}–10^{18} eV then it can be inferred from air shower data that the primary beam is mixed, with significant amounts of heavy nuclei. A picture thus seems to be emerging in which the primary beam becomes enriched in heavy nuclei until it is predominantly Fe nuclei around 10^{15} eV. At higher energies the iron

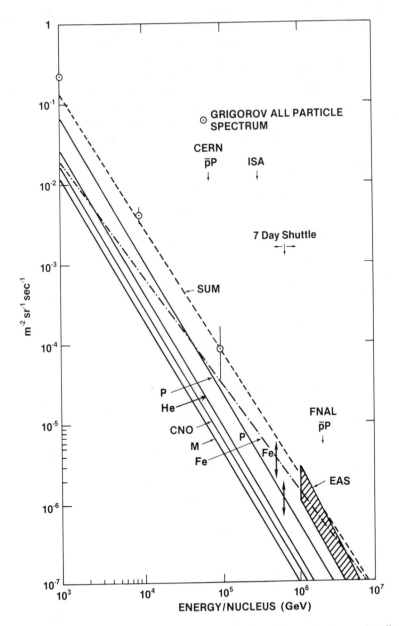

Figure 2   The separate components of the integral primary cosmic ray spectrum, as described in the text and Table 2. Label M corresponds to an LH group.

spectrum must steepen and the composition may become more like that at low energies. The determination of the primary composition above 10^{15} eV is a major goal of several current air shower experiments.

2.2   Cosmic Rays in the Atmosphere

Cosmic rays in the atmosphere are of interest both as beam particles for experiments at the surface of the Earth and for the inferences that can be made from them about the primary beam and about their interactions with nuclei of the atmosphere. (See Hayakawa 1969 for an account of this subject.) In either case we must solve the same set of cascade equations to describe the development of the haronic shower.

2.2.1 CASCADE EQUATIONS   In simplified form we may write the relevant coupled equations for the hadronic cascade as

$$\frac{dN(E, y)}{dy} = -\frac{N(E, y)}{\lambda_N(E)} + \int_E^\infty \frac{N(E', y)F_{NN}(E, E')}{\lambda_N(E')}\frac{dE'}{E} \qquad 1.$$

and

$$\frac{d\Pi(E, y)}{dy} = -\frac{\Pi(E, y)}{\lambda_\pi(E)} - \frac{\varepsilon_\pi \Pi(E, y)}{Ey \cos\theta} + \int_E^\infty \frac{\Pi(E', y)F_{\pi\pi}(E, E')}{\lambda_\pi(E')}\frac{dE'}{E}$$

$$+ \int_E^\infty \frac{N(E', y)F_{N\pi}(E, E')}{\lambda_N(E')}\frac{dE'}{E}, \qquad 2.$$

where N stands for nucleon and Π for charged pion (π^0's always decay to two photons before interacting and thus contribute only to the electromagnetic cascade). The quantities $N(E, y)\,dE$ and $\Pi(E, y)\,dE$ give the number of nucleons and charged pions respectively with energies between E and $E + dE$ at atmospheric depth y (in g cm^{-2}) measured along a direction with angle θ with respect to the vertical (i.e. zenith angle θ). $F_{ab}(E, E')\,dE/E$ is the number of particles of type "b" with energy between E and $E + dE$ produced on average in the collision between particle "a" with energy $E' > E$ and a nucleus of the atmosphere. The quantities $\lambda_i(E)$ are the energy-dependent interaction lengths in air for particles of type "i," given by

$$\lambda_i(\text{g cm}^{-2}) = \frac{2.4 \times 10^4}{\sigma_{i\text{-air}}^{\text{inel}}(\text{mb})}, \qquad 3.$$

where $\sigma_{i\text{-air}}^{\text{inel}}$ is the inelastic cross section for collisions between particles of type "i" and an average "air" nucleus (mass number $A \simeq 14.5$).

The second loss term in Equation 2 accounts for pion decay, $\pi \to \mu + \nu$, written in the approximation of an exponential atmosphere. The relative probability of decay (as compared to interaction) decreases with increasing pion energy (time dilatation) and increases with increasing altitude and decreasing depth y (as the atmosphere becomes less dense). The secant θ factor reflects the enhanced relative probability of decay for inclined particles, which spend more time in the less dense upper atmosphere. The quantity $\varepsilon_\pi = h_0 m_\pi c / \tau_\pi \approx 130$ GeV for pions. The corresponding quantity for kaons is about 1000 GeV.

The first term on the right-hand side of Equations 1 and 2 gives the flux lost via interactions, and the integrals account for particle production. In this simplified form production of nucleons by pions has been neglected, and other species, including kaons and antinucleons, have been neglected altogether. These may be added in a straightforward way, leading to more coupled equations with more terms.

2.2.2 BOUNDARY CONDITIONS   Two sets of boundary conditions for Equations 1 and 2 correspond to two classes of cosmic ray experiments. (a) The condition $N(E, 0) = N_0(E) =$ differential primary spectrum of nucleons (whether bound or not and measured in energy per nucleon) leads to a solution that is the differential flux of all hadrons as a function of atmospheric depth—the uncorrelated spectrum. (b) The condition $N(E, 0) = \delta(E - E_0)$ leads to a solution that is the flux of hadrons in an air shower. (The boundary condition must be changed appropriately in case the incident particle is a nucleus rather than a proton.)

2.2.3 SCALING   Analogous equations govern the development of electromagnetic cascades initiated by electrons and photons. In that case λ_i are characteristic lengths for pair production or bremsstrahlung, and F_{ij} represent the distributions of secondary momenta for these processes. In the electromagnetic case $F(E, E') = F(E/E')$ and $\lambda =$ constant. This scaling property simplifies solution of the cascade equations, which were first treated systematically and in the context of air showers by Rossi & Greisen (1941). Explicit solutions may be found for boundary conditions of type (a) when the incident spectrum is approximated by a power law,

$$N(E, 0) = \text{const} \times E^{-\gamma}. \qquad\qquad 4.$$

For the δ-function boundary condition, analytic solutions may still be found approximately by Mellin–Laplace transforms. These solutions are used to determine energies of electromagnetic cascades in emulsion chambers and to calculate the electromagnetic component of air showers given the π^0 (and hence photon) production spectrum.

The present conventional picture of strong interactions involves a form of hadronic scaling in which the quantities in Equations 1 and 2 obey

$$F_{ab}(E, E') \rightarrow F_{ab}(E/E') = F_{ab}(x_{lab}).$$ 5.

The relevant limit is that in which energies of both incident and produced hadrons are much larger than the masses involved. Note that in the absence of characteristic masses Equation 5 follows from dimensional analysis since F is defined to be a dimensionless function. The cascade equations are written in terms of lab energies, whereas the more fundamental expression of various scaling assumptions is in the center-of-mass system (cms). Three distinct forms of scaling have been distinguished: (a) limiting fragmentation (Benecke et al 1969), (b) Feynman scaling (Feynman 1969), and (c) radial scaling (Yen 1974). Limiting fragmentation applies only to secondaries that carry a nonvanishing fraction of available momentum in the center of mass in the limit $s^{1/2} \rightarrow \infty$. It follows both from (b) and (c), which differ from each other only in the central region, i.e. for secondaries that are slow in the cms frame. The quantities F_{ab} are related to the usual Lorentz invariant inclusive cross sections by

$$F_{ab} = \frac{1}{\sigma^{inel}} \int E \frac{d\sigma}{d^3p} \frac{dp_\parallel}{dE} d^2\mathbf{p}_T$$ 6.

The invariant inclusive cross section in terms of cms scaling variables is

$$E \frac{d\sigma}{d^3p} = \mathbf{E}^* \frac{d\sigma}{d^3\mathbf{p}^*} = f(x, p_T, s),$$

where E^*, $\mathbf{p}^*$ are the cms energy and momentum of a secondary particle. The Feynman scaling variable is $x = 2p_\parallel^* s^{-1/2}$, where $p_\parallel^*$ is the component of $\mathbf{p}^*$ along the incident direction. For radial scaling, $x = x_R = 2E^* s^{-1/2}$. Scaling implies possible dependence on one of the x's and $p_T = p_T^*$, but not on s. Radial scaling is in better agreement with accelerator data (Taylor et al 1976), but the difference is important only for relatively low energy particles, for example in a measurement of the multiplicity. For $x_R \gg 2\mu_T s^{-1/2}$, where $\mu_T = (p_T^2 + m^2)^{1/2}$, we find $x_R \cong x_{lab}$.

The general idea of scaling is in the spirit of the parton model, in which constituent quarks and gluons carry the momentum of the incident hadron with a well-defined distribution and then radiate or recombine to form hadrons after colliding with some target. Attempts to make specific parton models of hadronic production at small transverse momentum, however, suffer from lack of a theoretical foundation: there is no hard scattering that can be singled out to justify use of an impulse approximation. Nevertheless, considerable progress has been made in understanding particle ratios in terms of a simple quark-parton interpretation (see, for example, Duke &

Taylor 1978, Das & Hwa 1977, Ochs 1977, Van Hove & Pokorski 1975). It therefore seems appropriate to use scaling to summarize accelerator data on hadronic interactions and to formulate the following question: how different are interactions at $s^{1/2} = 10^3$ or 10^4 GeV from those below 100 GeV?

2.2.4 HADRONS IN THE ATMOSPHERE   The idea of hadronic scaling is quite old. If the kernels of the coupled integral equations 1, 2, etc have the scaling property shown in Equation 5, then for an incident spectrum described by a power, the equations separate and the solutions have the form

$$N(E, y) = N(E, 0)G_N(y) \qquad \text{7.}$$

and

$$\pi(E, y) = N(E, 0)G_\pi(y),$$

and similarly for other species. Thus, if scaling is valid (including constant cross sections), the energy spectra at any atmospheric depth have the same power dependence as the primary spectrum. This was recognized by Heitler & Janossy (1949), who, like Feynman (1969), motivated the scaling form for pion production by the analogy with bremsstrahlung. Generally speaking, measurements of cosmic ray hadrons in the atmosphere give energy spectra consistent with the primary spectrum at low energies and somewhat steeper above several TeV.

Since particle production presumably depends on energy per nucleon, the spectrum of secondaries in the atmosphere is sensitive to the relative flux of different nuclei in the all-particle spectrum. This sensitivity, together with the difficulty of measuring the uncorrelated flux of hadrons at high energy, prevents the secondary spectrum from being a definitive test of the scaling extrapolation. It will be useful, however, to estimate the fluxes of nucleons and charged pions as a function of depth in the atmosphere to have an idea of the beam composition for various experiments. Solution of Equations 2 and 3 gives

$$G_N(y) = e^{-y/\Lambda_N} \qquad \text{8a.}$$

and

$$G_\pi(y) = Z_{N\pi} \frac{\Lambda_\pi \Lambda_N}{\lambda_N(\Lambda_\pi - \Lambda_N)} e^{-y/\Lambda_\pi} \left[1 - e^{-y\left(\frac{\Lambda_\pi - \Lambda_N}{\Lambda_\pi \Lambda_N}\right)} \right], \qquad \text{8b.}$$

where the attenuation length for particles of type "a" is given by $\Lambda_a = \lambda_a(1 - Z_{aa})^{-1}$. The quantities

$$Z_{ab} = \int_0^1 dx \, x^{\gamma - 2} F_{ab}(x) \qquad \text{9.}$$

result from folding the steep primary energy spectrum (which favors low energy primaries) with the secondary spectra of particles produced in single interactions (which generally favor high energy primaries). This reflects a fundamental bias of cosmic ray experiments that appears time and again in various forms: because of the steep spectrum, observed particles of a given energy are more likely to be relatively high energy secondaries from low energy primaries than would be true for an unbiased sample of interactions at fixed energy. Thus, in particular, secondary fluxes in the atmosphere are sensitive primarily to the forward fragmentation region because of the factor $x^{\gamma-2} \cong x^{0.7}$ in Equation 9. Another example will be relevant in Sections 5 and 6 where we discuss experiments sensitive only to photons, electrons, and positrons. In that case, events with an unusually large fraction of the energy going into secondary π^0's will be preferentially selected.

The most comprehensive accelerator data on production of nucleons, mesons, and antinucleons in pp and πp collisions is in the range 100–400 GeV/c beam momentum from Fermilab (e.g. Johnson et al 1978, Cutts et al 1978, and Eisenberg et al 1978). Data on pp collisions from ISR extend up to the equivalent of nearly 2000 GeV/c. Accelerator data have been parametrized by Hillas (1979b) for cosmic ray cascade applications. His parametrizations give $Z_{NN} \approx 0.37$, $Z_{N\pi} = 0.081$, $Z_{NK} = 0.010$, and $Z_{\pi\pi} = 0.28$ for $\gamma = 2.71$, where N and π refer to sums and/or averages of appropriate charge states. This gives a pion-to-nucleon ratio of one at 500 g cm^{-2}. The present precision and coverage of the required kinematic regions are such that Z_{NN} and $Z_{\pi\pi}$ are known to about ± 0.04.

The sensitivity of Z's to spectral index differs for leading and produced particles. For a spectral index of 2.36 (a relatively flat iron spectrum) one finds that $Z_{NN}(\gamma = 2.36) = 1.1 \times Z_{NN}(\gamma = 2.75)$ and $Z_{N\pi}(\gamma = 2.36) = 1.6 \times Z_{N\pi}(\gamma = 2.7)$ using data referred to above. Nuclear target effects may reduce this ratio somewhat (Hillas 1979b), but we expect fluxes of pions and nucleons to be comparable at high mountain elevations (~ 5 km).

Hillas finds no evidence for violation of radial scaling over the accelerator energy range, within experimental uncertainties, provided low energy resonant effects are accounted for. This does not exclude the possibility that small violations are present which could become large at higher energies (see, for example, Wdowczyk & Wolfendale 1979). Indeed a major goal of much of the work described in this review is to test hadronic scaling over as great an energy range as possible. Anisovich & Shekhter (1978) suggested, within the context of the parton picture, that hadron collisions become more inelastic at energies high enough so that clouds of partons surrounding different constituent quarks overlap each other. They argue that this leads to a significant decrease in fast secondaries around 10^6 GeV.

2.2.5 MUONS   Fluxes of muons are the most directly measured of the secondary cosmic rays. The muon spectrum is calculated from the pion and kaon flux by multiplying the meson flux at depth y by the decay probability (as in the second term of Equation 2) and then integrating over depth. At low energies ($E < 100$ GeV) nearly all mesons decay rather than interact, and the muon energy spectrum has the same power as the primary spectrum. At high energies ($E > 1000$ GeV) the time dilatation factor, $1/E$ in Equation 2, leads to the result that the muon spectrum is steeper by one power of E than the primary spectrum.

The relevant primary flux is the flux of all nucleons. We use the parametrizations of Table 2 to find an expression for this flux at the top of the atmosphere. It is approximated by

$$\frac{dN}{dE} = 1.8 \times 10^4 \, E^{-2.71} + 73 \, E^{-2.36} \qquad\qquad 10.$$

for $25 \lesssim E \lesssim 5 \times 10^5$ GeV. The units are nucleons per ($\mathrm{m^2 \cdot s \cdot sr \cdot GeV/n}$). This spectrum can be represented by a single effective spectral index of 2.65 between 1 and 1000 TeV. The spectrum of Equation 10 agrees very well with the all-nucleon flux derived by Hillas (1979a) from the muon spectrum of Amineva et al (1973). (Note that it is possible to have more than one composition that simultaneously agrees with the all-particle spectrum and with the all-nucleon spectrum. Hillas has a different composition from that described here, but the same all-nucleon and all-particle spectra.)

A more recent summary of vertical muon fluxes (Allkofer et al 1979) is compared in Figure 3 with calculations (T. K. Gaisser and T. Stanev, private communication) based on two different primary spectra, (*a*) the one described in Table 2 and (*b*) a low energy spectrum extrapolated from the composition at 25 GeV/n (with a constant slope of 2.75). The latter gives a somewhat lower all-nucleon spectrum than Equation 10 and hence a somewhat lower muon spectrum. The experimental measurements are inadequate to distinguish between the two different compositions.

2.2.6 MUON CHARGE RATIO   The charge ratio of high energy muons has been measured up to muon energies of the order of 10 TeV, in the vertical and horizontal directions (Allkofer 1979, Jokisch et al 1979, Muraki et al 1979, and references therein). The positive excess reflects the proton-to-neutron ratio in primary cosmic rays at energies about an order of magnitude higher than those of the muons. It is, however, very difficult to make unambiguous statements about the neutron fraction as the relation between observed μ^+/μ^- ratio and primary n/p ratio depends sensitively upon the details of the inclusive single-particle cross sections. Not only are there few data on $\pi^\pm$ and $K^\pm$ production from light nuclei in the forward fragmentation region, but what data exist for pp, np, and πp collisions are not precise enough to make clearcut inferences. Claims have been made that vary from 25% neutron fraction down to only 9% neutron fraction in primary cosmic rays at 10 to 50 TeV per nucleon (Adair et al 1974, 1977, Jokisch et al 1979). We examine the problem a little more carefully.

One can relate (Adair et al 1974, Hoffman 1975) the muon charge ratio to ratios of moments of particle production cross sections

$$\frac{\mu^+}{\mu^-} = \frac{Z_{\pi^+}}{Z_{\pi^-}} \frac{\left(1 + \frac{Z_{K^+}}{Z_{\pi^+}} R\right)}{\left(1 + \frac{Z_{K^-}}{Z_{\pi^-}} R\right)}.$$

The ratio R is related to probabilities of decay versus interaction for kaons as compared to pions. R depends on energy and varies by a factor of the order of 3 to 4 in going from 100 GeV to asymptotic energies (Ramana

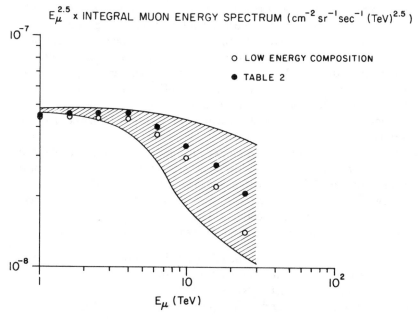

Figure 3   Integral muon spectra, times $E_\mu^{2.5}$, calculated using a scaling model (Ellsworth et al 1979a) with increasing cross section, are compared to a world survey of the data (*shaded region*) for two assumed compositions.

Murthy 1977). Energy variation of μ^+/μ^- depends on the difference between the ratios Z_{K^+}/Z_{π^+} as compared to Z_{K^-}/Z_{π^-}. If these are equal there can be no energy variation of μ^+/μ^-. If on the other hand $Z_{K^+}/Z_{\pi^+} \approx 2(Z_{K^-}/Z_{\pi^-})$ and $Z_{K^+}/Z_{\pi^+} \approx 0.1$, then one gets an energy variation of about 9%. If the low energy value of μ^+/μ^- is 1.3, then the asymptotic value should go to 1.42.

The actual values for the three ratios Z_{π^+}/Z_{π^-}, Z_{K^+}/Z_{π^+}, and Z_{K^-}/Z_{π^-} are not very well determined. The ranges of values, including uncertainties in inclusive cross sections and spectral indices are

$$1.6 < Z_{\pi^+}/Z_{\pi^-} < 2.0$$

$$0.12 < Z_{K^+}/Z_{\pi^+} < 0.13$$

$$0.065 < Z_{K^-}/Z_{\pi^-} < 0.09.$$

Furthermore, for cosmic ray applications proton-nucleus collisions must be considered, which will change the values somewhat.

In summary, if one fixes the μ^+/μ^- ratio at 300 GeV to the experimental value of 1.3 and calculates the energy variation to find its value at 3000 GeV, one finds that in order to obtain a constant value of μ^+/μ^- ratio over the whole range, the neutron fraction in primary cosmic rays must increase above a TeV. It is, however, currently impossible to quantitatively predict the neutron fraction because we lack precise data on inclusive cross sections for hadron-air scattering and because the measured charge ratio for $E_\mu \gtrsim 3$ TeV (Allkofer 1979) is uncertain. The composition given in Table 2 is consistent with the observations on muon charge ratio, giving a primary n/p ratio that ranges from 0.09 at 100 to 0.12 at 10^4 GeV/n.

3   COLLISIONS AT 10–20 TeV

3.1   *The Small Emulsion Chamber Technique*

The most detailed data on interactions with energies around 20 TeV come from exposures of small emulsion chambers in balloons near the top of the atmosphere. Niu and colleagues have described the structure of the chambers, as well as the scanning and measuring of tracks, in an elegant review paper (Fuchi et al 1979a). A typical chamber has two sections: a target layer consisting of lucite plates interleaved with thin layers of nuclear emulsion films, and an analyzer layer consisting of a sandwich of lead plates, nuclear emulsion, and x-ray film. Its overall size is $\sim (25 \times 20) \times 25$ cm, with a mass of 50 kg. A characteristic feature of these assemblies is that the particles traverse the sandwiches nearly perpendicular to their planes for most accepted events.

The lucite plates are one millimeter thick, so that in the target section the track of a charged particle is visible every one millimeter along its path. Because the target is a low Z material, electromagnetic cascading is minimal in this part of the detector and therefore does not obscure charged tracks from the vertex. The analyzer layer is about 10 radiation lengths thick, so that in this lower half of the chamber γ rays and electrons can be detected by their electromagnetic cascades. These can be found for $E > 30$ GeV, and the energy can be determined by fitting to standard electromagnetic cascade curves (Nishimura 1964, 1967). This is quite different from the situation in a pure emulsion stack, where most γ rays cannot be measured in this way.

Charged particle tracks can be found without bias provided their angle of divergence from the event axis is not too large ($< 15°$ or $p_{\parallel} > 1.5$ GeV/c for a typical transverse momentum). A crude estimate of the total energy of the charged particles can be made from the relation for relativistic particles, $E \cong p_T/\tan \theta$ by writing $\sum E_i \approx \sum \langle p_T \rangle /\tan \theta_i$. Here p_T is the transverse component of the momentum and $\langle p_T \rangle \cong 400$ MeV/c for pions. In addition, the precision with which the chambers are constructed allows energies of charged particles (up to 2 TeV in favorable cases) to be estimated by the relative scattering method, which relates the divergence of parallel tracks from each other to energy on the basis of multiple Coulomb scattering.

Events are located by scanning the x-ray films in the lower chamber for dark spots. Events with more than 400–600 GeV of energy in the electromagnetic component are found in this way. Since all neutral pions will have decayed into photon pairs, any event with at least this much energy shared among neutral pions should be found. The event is then traced back into the target layer until the point of interaction is found. Although the actual interaction vertex is normally in the lucite (which comprises some 90% of the target material), and is therefore not visible, the track of the incident particle can be found if it is charged, and its charge can be determined.

3.2 Multiplicity of Secondaries

Systematic scans of events around 20 TeV induced by singly charged particles have been reported by Sato et al (1976) and Fuchi et al (1979b). At balloon altitudes, most singly charged particles are protons, so it is safe to assume these are proton-induced interactions. Altogether there are 38 interactions in the plastic of which 22 have estimated $E_0 > 6$ TeV (average energy $= 20 \pm 5$ TeV, which corresponds to a total center-of-mass energy of about 200 GeV). The multiplicity distribution of charged secondaries for the high energy group has a mean of 35 with a standard deviation of about 20.

A number of corrections must be made to compare these values to those for pp collisions. One must correct for (*a*) effects of nuclear target, (*b*) bias against detection of low energy secondaries, (*c*) bias against events with < 500 GeV in neutral pions. Effects (*b*) and (*c*) combine to make a strong bias against detection of low multiplicity events with the secondaries in the backward center-of-mass hemisphere (known as target dissociation). Fortunately the chambers have been exposed extensively to accelerator proton beams in charmed particle searches and for calibration purposes. The nuclear target effect has been studied in emulsion chambers by Fumuro et al (1977). Sato et al (1976) estimate the bias against target dissociation to require a 10% reduction in the multiplicity. Fuchi et al (1979b) find a

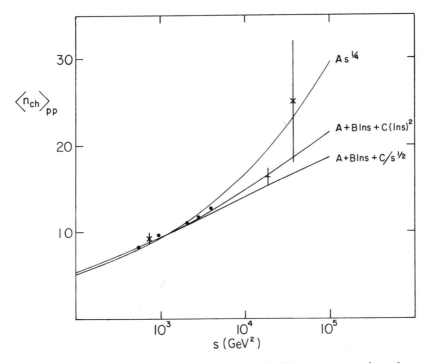

Figure 4 Cosmic ray data on average charged multiplicity are compared to three extrapolations of the accelerator data. Accelerator points (dots) are from Thomé et al (1977). The point shown as "X" at $s = 800$ GeV^2 was obtained in an exposure at FNAL of the same type of emulsion chamber (Fuchi et al 1979b) with which the highest cosmic ray result was obtained. Comparison with other accelerator results serves to calibrate the emulsion chamber technique. The cross is from Chaudhary & Malhotra (1975).

corrected mean charged multiplicity of 9.3 ± 0.7 in an emulsion chamber experiment in a 400-GeV/c proton beam. Their corresponding corrected result for the high energy portion of their events with $E_0 > 20$ TeV is 25 ± 7. The measured value of n_{ch} in pp collisions at 400 GeV/c is 9 (Albini et al 1976). Since this is consistent with the corrected value found by the emulsion chamber, it seems reasonable to use the value quoted by Fuchi et al at 20 TeV. It is consistent with the corrected value of 22 ± 3 for the high energy events of Sato et al (1976) where the quoted error is statistical only. The larger error of Fuchi et al (1979b) presumably includes an estimate of systematic effects, such as those arising from the steep spectrum.

These results are compared with older cosmic ray measurements (Chaudhary & Malhotra 1975) and with accelerator data (Thomé et al 1977) in Figure 4. Also shown are three extrapolations of accelerator data. An average multiplicity proportional to $s^{1/4}$ is characteristic of thermodynamic models (Fermi 1951, Landau 1953). The logarithmic forms are characteristic of the conventional scaling picture. In their review of the older emulsion stack data, Chaudhary & Malhotra (1975) selected events with one or at most two heavy prongs (fragments of the target nucleus) in order to examine only those interactions in the emulsion that are with proton targets or that involve a peripheral interaction with only one nucleon of a target nucleus. Also, the range of interaction energies is 2–45 TeV. Because of the steep spectrum, the median energy is lower than the mean. For both these reasons the point at 10 TeV (which corresponds to $s = 2 \times 10^4$ GeV^2) is a conservative estimate of the multiplicity. On the other hand, the threshold energy for detection of events in the emulsion chamber together with the steep spectrum could give a net selection effect in favor of high multiplicity. A detailed simulation would be necessary to quantify these effects, however. We conclude that Figure 4 indicates a multiplicity somewhat higher than conventional extrapolations to 20 TeV or $s = 4 \times 10^4$ GeV^2 (Albini et al 1976, Thomé et al 1977, Giacomelli 1978).

3.3   Angular Distributions

Fuchi et al (1979b) have also reported results for angular distributions, corrected for nuclear target effects. The pseudo-rapidity density at $x = 0$ obtained at cosmic ray energies is 3.9 ± 0.6 at a mean center-of-mass energy of 190 GeV. Pseudo-rapidity is $dn/d \log \eta$, where $\eta \doteq \tan(\theta_i/2)$. This value is significantly larger than that obtained by Thomé et al (1977) at ISR. The ISR values range from 1.4 to 1.9 in the range 23.6 GeV $< s^{1/2} <$ 62.8 GeV. Fuchi et al obtain a result similar to the ISR values from an emulsion chamber experiment in a proton beam at 400 GeV/c. Some increase is expected at the cosmic ray energy from radial scaling (Ellsworth 1979);

however, it is not clear whether this can accommodate such a large effect.

Because it is difficult to determine interaction energies event by event, it is not possible to make a very significant test of hadronic scaling with the few events around 20 TeV studied in emulsion chambers. Analyses that have been made show that the data are not inconsistent with scaling in the fragmentation region (Sato et al 1976, Gaisser 1977).

4 NEW PARTICLE PRODUCTION

4.1 *Emulsion Chamber Studies*

The most fruitful use of emulsion chambers has been in searches for new particles with lifetimes in the range 10^{-14}–10^{-12} s. Particles with lifetimes in this range and Lorentz factors (E/m) of 100–1000 have typical track lengths of millimeters to centimeters and are therefore nicely matched to the size of the small emulsion chamber. The first candidate for a charmed particle was reported by Niu et al in 1971 and soon after interpreted within the GIM (Glashow et al 1970) charm scheme by Hayashi et al (1972).

Evidence for production of new particles in cosmic ray interactions around 10–20 TeV has been reviewed recently by Niu (1979) and brought up to date by Fuchi et al (1979c). At present, seven events appear to have visible pairs of particles with lifetimes in the range 10^{-14}–10^{-12} s and masses of more than a GeV, as summarized in Table 3. Other events with only one visible candidate for an "X" particle exist, but backgrounds can be significant in this case. Candidates with visible pairs of new particles are

Table 3 Candidates for associated production of heavy flavors

Event[a]	E_0 (TeV)	n_{ch}	Comment
6B-23	10	70	Original event of Niu et al (1971); one of the X particles has 40% of visible energy, but multiplicity is very high, thus difficult to interpret as dissociation.
BEC-II[b]	18	27	Central production
11C-34	20	70	Complex, possible cascade
T-star	20	36	$X_1^0 \to \pi^0 x^0 + X_2^0 \to \pi^0 x^0$ $E_1 > 1.6\,\text{TeV}\ E_2 > 2.4\,\text{TeV}$ possibly diffractive production
ST-2	25	51	Possible cascade
6a-19L	20	20	
BEC-7e-31	25	56	Recently reported by Fuchi et al (1979c); central production

[a] See Niu (1979) for event identification and further information on decay modes.
[b] Sugimoto et al (1975).

very unlikely to be background, typically less than one chance in 10^{-4} (Gaisser & Halzen 1976).

Because the identity of charged particles and the sign of their charge are not determined and because neutral decay products such as K^0 and neutron are not seen, it is difficult to use the data to determine branching ratios and to identify the quantum numbers of the X particles. It was, however, pointed out quite early (Hoshino et al 1975) that the lifetime of the neutral candidates is about one fifth that of the charged candidates. In retrospect, the recent observation of this effect in other experiments at accelerator energies (Prentice 1979, Kirkby 1979) further confirms the emulsion chamber results. Assuming a mass of 2 GeV for the X particles, Niu (1979) estimates $0.7–0.8 \times 10^{-12}$ s for the overall lifetimes and $0.3–0.5 \times 10^{-12}$ s and $1–2 \times 10^{-12}$ s for neutral and charged particles respectively.

The cross section for production of X particles, though again difficult to estimate quantitatively because of scanning biases, is clearly large. Niu (1979) has estimated that the production rate is one event per 20–40 observed interactions at 20 TeV. This corresponds to 0.3–0.6 mb per nucleon if $\sigma_{charm} \propto A^{1.0}$ or 0.75–1.5 mb per nucleon if $\sigma_{charm} \propto A^{2/3}$, as could be the case for diffractively produced particles (Ringland & Wachsmuth 1980). The average multiplicity of charged secondaries produced in the X-particle events is 47 ± 8; this is somewhat higher than the overall average multiplicity for proton-plastic interactions in the same energy range, which is 36 ± 10 (Fuchi et al 1979b). Evidently this would indicate that the massive particles are produced predominantly in central rather than in peripheral collisions (Gaisser et al 1974). One must note, however, the puzzling fact that in the original Niu et al event (1971) one of the produced X particles had apparently 40% of the observed energy (10 TeV), which suggests diffractive production, but the charged multiplicity of the event was 70. Even if we bear in mind that a leading neutron would have escaped detection, this seems hard to interpret.

Finally, we note that two of the events in Table 3 apparently show cascading characteristic either of a higher charm quantum number or of production of higher flavors that decay by cascading through charm. Especially in view of these possibilities, the cross section of ~1 mb per nucleon is consistent with the observed rate of charm production in hadronic interactions at accelerator energies, as illustrated in Figure 5.

Because of the rapidly decreasing flux of cosmic rays the emulsion chamber technique is not practical at higher energies [although a single 100-TeV event observed in such a chamber by Ogata et al (1979) may involve production of neutral particles with lifetimes around 10^{-13} s that show cascading in their decays]. Sawayanagi (1979) reports a search

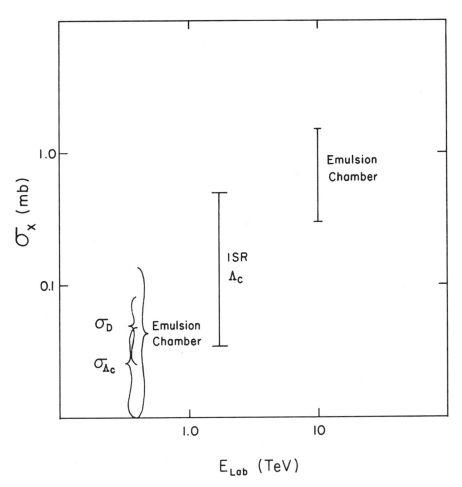

Figure 5 Comparison of emulsion chamber cross sections (Fuchi et al 1979a,c,d) for production of pairs of new particles with data on production of charmed hadrons. The emulsion chamber data as plotted here include a factor of 2.5 uncertainty corresponding to a range from 2/3 to 1 in the A dependence of σ_x. The estimates of σ_D and σ_{Λ_c} at 400 GeV/c are taken from the self-consistent estimates of beam dump and other measurements made by Ringland & Wachsmuth (1980). The range of values for σ_{Λ_c} at ISR is taken from the summary made by the same authors.

undertaken among events found in the large (40 m^2) two-storied emulsion chamber on Mt. Chacaltaya in Bolivia. Evidence here is less direct, consisting of diverging tracks in the lower chamber that appear to originate at vertices in the 1.5-m air gap between the two stories of the chamber. These candidates are found in families of secondary particles in the lower chamber that originate from hadronic interactions in the upper chamber. The decay path lengths of 10–100 cm correspond to lifetimes of order 10^{-13}–10^{-12} s for typical energies of one to several TeV. The interaction energies here are of the order of 100 TeV, but the cross-section estimate is very difficult because of selection effects. Several candidates were found out of only 12 events scanned in this way, but the events selected for scanning were chosen to have high multiplicity. The cross section for these events also appears to be quite large.

4.2   Calorimeter Studies

4.2.1 DELAYED PARTICLES   A traditional way to look for long-lived, massive particles is to look for particles delayed with respect to an air shower front (Damgard et al 1965, Chatterjee et al 1965; see Goodman et al 1979a for a review of the subject). The time delay of a relativistic particle is about $1666/\gamma^2$ ns per km of path length. A shower typically begins 15–20 km above sea level. The delay acquired by a particular particle depends both on its energy at production and on its interaction history, but 1 km (corresponding roughly to one interaction length at mountain altitude) can be taken as a characteristic length. The full width of the arrival time distribution of normal hadrons with $E > 30$ GeV in air showers is about 10 ns (Goodman et al 1979b). Thus particles with lifetimes longer than 10^{-7} s and mass greater than several GeV will show up if they interact in the calorimeter.

The University of Maryland group observed three events with delays greater than 25 ns and energies of $\gtrsim 45$ GeV as shown in Figure 6b, corresponding to one event per 3000 showers and hence to a flux of (4.3 $\pm$ 1.3) $\times 10^{-11}$ cm^{-2} sr^{-1} s^{-1} (Goodman et al 1979a). The experimental setup consisted of a 4-m^2 iron calorimeter 985 g cm^{-2} thick with sensitive layers of liquid scintillator and spark chambers with a shielded plastic scintillator below two interaction lengths of iron to measure time delay and pulse height. The timing and shower density were measured with several scintillation counters. Monte Carlo simulations of atmospheric cascades generated by primary nuclei, including the detector response, appear to rule out various conventional explanations such as locally produced nuclear fragments and nucleons from the original interaction. The simulations show that the observed showers are due primarily to cosmic rays with 10–100 TeV per nucleon primary energy (see Goodman et al 1979c for a detailed discussion of the Monte Carlo). A simulation that included a cross section expected for production of new heavy flavors did produce events of the type observed. With an inclusive cross section of the form $\exp(-7x) \times \exp(-6p_T)$ and an energy dependence given by Isgur & Wolfram (1979), the observed rate corresponds approximately to 50 μb at 100 TeV. [This estimate could be somewhat increased if a flatter p_T distribution were assumed, as is probably appropriate for production of a massive particle (Halzen 1977).] We know, however, that the delayed events cannot be due to the known new flavors, charm and bottom, because their lifetimes are too short (Rosner 1979, Niu 1979, Prentice 1979, Kirkby 1979).

The delayed events involve the interaction of hadrons in the calorimeter. The massive object responsible for the delay need not itself be a hadron if it has an appropriately long lifetime ($\sim 10^{-7}$ s) and contains hadrons among its decay products. The events cannot be due to conventional weak intermediate bosons, however, because their lifetimes are expected to be much too short (Li & Paschos 1971, Quigg 1977). Evidently a new kind of particle would be required to explain these events. Several other experiments have seen such events at rates not inconsistent with the Maryland experiment (Goodman et al 1979a). Of particular interest is the new, mountain-level experiment of the Indian group (Bhat et al 1979). This experiment includes a large multiplate cloud chamber to obtain visual information about the events. A new experiment with more than an order of

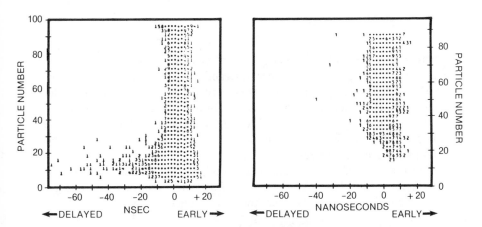

Figure 6   Distributions of pulse heights vs time delay behind the shower front for hadrons in air showers. The two groups of data correspond to different trigger conditions: (*left*) small pulse heights included and (*right*) small pulse heights vetoed. The tail of low energy delayed particles in the left diagram requires a large contribution from heavy primaries, as discussed in Section 2. The three energetic delayed hadrons are clearly visible in the right diagram (Goodman et al 1979a).

magnitude higher sensitivity is being constructed at sea level by the Maryland group. An intercomparison of observations at two different levels may shed light on the characteristics of such particles.

4.2.2 HADRON ATTENUATION   At the last several cosmic ray conferences, results reported from the very large calorimeter at Tien Shan (Yakovlev et al 1979 and references therein) show that the attenuation length of hadrons in lead increases significantly around 50–100 TeV. The calorimeter has an area of 36 m^2 and a total thickness of 850 g cm^{-2} of lead (about 4.5 mean free paths). It consists of 15 layers of lead ranging from 2.5 to 10 cm in thickness separated by crossed layers of ionization chambers. The resulting overall acceptance of the detector is 18 m^2 sr.

The attenuation length is obtained from a plot for each event of the amount of energy contained in each layer vs its depth in the calorimeter. At high energies (above 100 TeV) most of the events are air shower cores. At lower energies a significant subset of events are unaccompanied hadrons interacting in the calorimeter. Because of the high density of lead the ratio of the radiation length to the nuclear mean free path is quite high (about 30), so that the incident electromagnetic component is absorbed in the first few layers. After 280 g cm^{-2} the visible electromagnetic component is assumed to be in equilibrium with the hadronic core. Thus for a single hadronic component characterized by an interaction length in lead, λ, the rate at which energy is deposited is given by

$$\frac{dE}{dx} = \frac{k_\gamma E(x)}{\lambda},$$

where $E(x)$ is the hadronic energy present at depth x and k_γ is the mean fraction of energy deposited per interaction in electromagnetic form (mostly via neutral pions). In a normal cascade, energy deposition in the calorimeter is exected to be dominated by pions with $k_\gamma \cong 1/3$, corresponding to an attenuation length in lead $\Lambda \cong 3\lambda \cong 700$ g cm^{-2}. This is the value found experimentally up to about 50 TeV. Above 100 TeV, however, the attenuation length is about 1100 g cm^{-2}.

This is a large effect. If it is due to a second, new component produced in interactions above 100 TeV that carries a fraction F of the hadronic energy, then (neglecting coupling between the two channels)

$$E(x) = E(0)[(1-F)e^{-x/700} + F e^{-x/\Lambda_2}].$$

If we use the ratio $E(700)/E(300)$ to define an equivalent overall attenuation $\Lambda \cong 1100$ g cm^{-2}, we find $0.25 < F < 1$ and correspondingly $\infty > \Lambda_2 > 1100$. It has been suggested (Aseikin et al 1975, Cline 1979b) that the effect could be due to production of charm. In view of the size of the effect, we find this very unlikely, because it would require at least 25% of all hadronic energy to be in this component. Even the most optimistic estimates based on diffractive production of charmed baryons in pp collisions are much less than this (recall Figure 5).

Other clues to the nature of the effect (called production of a "long-flying component") are that the ratio of electromagnetic to hadronic energy in the shower cores increases at about the same total hadronic energy and that the anomalous attenuation is confined to the central region of the shower core, with the attenuation observed to be normal in the periphery of the events. The latter point confirms that there is an energy threshold for the effect. The ratio of electromagnetic to hadronic energy is defined as the energy deposited in the first three layers divided by the total hadronic energy (layers 4 through 15 plus a correction for energy lost out the bottom of the calorimeter). We note (L. W. Jones, private communication) that the increase in the electromagnetic component is opposite to the Centauro effect, discussed in Section 6 below. The long-flying component could be due to production of unstable particles (including leptons) if by chance they had decay modes and lifetimes appropriate to the calorimeter. Again, the production cross section and energy fraction of the component would have to be large. It would be useful to simulate cascades on the computer to study systematic effects in this detector and thus to eliminate the possibility of energy-dependent biases.

4.3 *Underground Experiments*

Several studies of high energy interactions using particle telescopes placed deep underground have been carried out. Some have operated in various arrangements for as long as ten years, such as the Kolar Gold Field detectors in India (Krishnaswamy et al 1971, 1976, 1977, 1979a,b,c) and the Case-Wits-Irvine detectors in South Africa (Crouch et al 1978). Others have come into operation recently, for example the Baksan telescope in the Caucasus, USSR (Alexeyev et al 1979) and the Homestake Mine detector, South Dakota, USA (Deakyne et al 1978, 1979). These detectors, in general, are capable of (a) multiple track recording and reconstruction; (b) determination of the direction of the primary giving rise to the event; and (c) particle identification, i.e. whether a given track or tracks imply a muon, a hadron, or an electromagnetic shower (e or γ). They are in general relatively large, either in acceptance or in their "mass" or both, and are located at great depths under rock (greater than 850 hg cm^{-2} or about >0.2 TeV equivalent muon energy is needed to penetrate the depth). A variety of detector elements have been used, including crossed flash-tube arrays, proportional chambers, and scintillator counters, generally interspersed with absorbing material to form a crude calorimeter as well as a directional hodoscope.

In what follows we discuss some interesting events collected in the Kolar

Gold Field (KGF) experiment. The more recent experiments (Homestake and Baksan) have yet to report on their "unusual" events, if any. We summarize KGF observations on (a) possible observation of new particles in multitrack events, (b) anomalous cascades deep underground, and (c) unusual double-core events.

4.3.1 MULTITRACK EVENTS   New investigations to study further the multitrack special events observed in 1975 (Krishnaswamy et al 1976, 1977) have been carried out using a modified detector at depth of 3375 hg cm^{-2} with stereoscopic capabilities (Krishnaswamy et al 1979a). In a run of 5.16 × 10^8 s, three such special events were seen along with 10,390 single muons, 307 multiple parallel muons and 1271 cascades, and 2 anomalous showers (described below). The main characteristic of these events is multitrack, large opening-angle configurations with one or more of the following features: (a) clearly separated tracks originating from a vertex in air, (b) multiple penetrating tracks (muons?) originating from more than one vertex, and (c) multiple cascades of penetrating tracks with large opening angles. These three events combined with the previous six examples give nine candidates for unusual events suggestive of production and decay of a new particle with $\tau \geq 10^{-9}$ s. At least five are from very large zenith angles, and so cannot be due to some unusual muon interaction.

At the 7000 hg cm^{-2} depth, three such special events were seen out of total of 17 events with $\theta > 50°$ and were therefore assumed to be induced by neutrinos (Krishnaswamy et al 1977).

Accelerator searches for this type of event were negative (Benvenuti et al 1975, Faissner et al 1976), which led to the conclusion (Sarma & Wolfenstein 1976) that they are not produced by ν_μ or $\bar{\nu}_\mu$ with energies of 10–100 GeV.

4.3.2 ANOMALOUS CASCADES   Anomalous cascades are steeply inclined showers traversing detectors placed at depths of 3375 and 7000 hg cm^{-2} (Krishnaswamy et al 1976, 1977, 1979b). The telescopes have vertical detector planes of crossed proportional counters, flash tubes, and/or scintillators. At 3375 g cm^{-2}, in a total exposure of 1.68×10^9 m^2 s and an angular range from 30° to greater than 90° (upwards), four events have been observed. The visible energy of these events exceeds several hundred GeV. The observed spectrum of bursts, when extrapolated to the energy range of these anomalous events gives a flux at least an order of magnitude less than that implied by these four events. Two of the four events were obtained prior to 1977 and two were observed recently (1979) with improved apparatus in about one third as much running time as the first run.

At the greater depth of 7000 hg cm^{-2}, in 1.8×10^9 m^2 s, two events at

$\theta > 45°$ were observed with more than 1000 particles in each, again corresponding to several hundred GeV. At this depth, in the same running time only ten other showers were seen, and they had energies less than 30 GeV.

The flux of these events seems to be depth independent and much greater than that which could be predicted from muon interactions or from ordinary neutrino interactions.

4.3.3 UNUSUAL DOUBLE-CORE EVENTS   A new KGF detector at a depth of 1840 hg cm^{-2} (Krishnaswamy et al 1979c) consisting of a 3-layer calorimeter of 26 m^2 with crossed proportional-counter hodoscopes was put into operation recently. The shallow calorimeter has 1 inch of Pb above the first layer, 1 inch of iron above the second, and 1/4 inch of iron above the bottom layer.

In the initial few months of operation two energetic double-core, high p_T events were detected. Analysis on one showed that it had two "jets" with energies greater than 30 GeV (conservatively) and with an opening angle of 26° coming from a common vertex in rock. Hence the minimum transverse momentum is about 7 GeV/c. Here the frequency is too high to be accounted for by v interactions. Given the flux of muons at this depth, if these events were produced by muon interactions in the rock, then the cross section would have to be very large (in view of the high p_T), $\sim 10^{-31}$ cm^2, which corresponds to several percent of the total interaction rate of muons.

4.3.4 DISCUSSION   The KGF events, which are not yet understood, are not contradicted by other underground experiments. The large detectors at Homestake and Baksan may give us further data on such events, although their spatial resolution is not as fine as that of the KGF experiments. The new proton decay detectors (*Physics Today*, Jan. 1980, p. 17) may be instrumented to give a resolution of the order of KGF with much larger collection volumes. Although such cosmic-ray-induced events would be nuisance background in the search for proton decay, they could prove to be interesting in themselves.

5   COLLISIONS AT 50–200 TeV

Three groups are currently using large ($\gtrsim 50$ m^2) ground-based emulsion chambers to study interactions above 100 TeV, where the flux is too low for the small balloon-borne chambers (Pamir collaboration, Budilov et al 1977 and Bayburina et al 1979a,b; Mt. Fuji collaboration, Akashi et al 1979a,b,c; Brazil-Japan collaboration, Lattes et al 1971 and Bellandi Filho et al 1979).

A somewhat smaller chamber (~ 10 m^2) has been in use on Mt. Kanbala in China (Ren et al 1979). Of these, the Brazil-Japan group has especially designed its emulsion chamber to be able to study local interactions in the detector as well as products of interactions in the overlying atmosphere. They have studied over 50 local events with more than 20 TeV visible in photons, corresponding to a mean interaction energy in the range 100–200 TeV. The systematics of these events is the subject of this section; the higher energy atmospheric cascades will be discussed in Section 6. We begin, however, with a description of the detector and of the morphology of air-jets (A-jets) and carbon-jets (C-jets), which is relevant to both topics.

5.1   Large Emulsion Chambers

The basic element of the large emulsion chamber is a sandwich of lead plates (each about 1 or 2 cm thick) and x-ray films. This device (called a Γ block by the Pamir collaboration) causes any incident energetic photon or electron to develop an electromagnetic cascade in the lead, which is sampled photographically by the x-ray films. The energy in the electromagnetic cascade can be measured by relating the spot darkness in successive layers of film to standard electromagnetic cascade theory (Ohta 1971, Akashi et al 1964). The darkness is related to the number of electrons by exposing the film to a 650-MeV/c electron beam and hence related to electromagnetic cascade theory (Nishimura 1964, 1967). The calibration is

Table 4   Large mountain emulsion chambers

	Chacaltaya (550 g cm^{-2})	Mt. Fuji (650 g cm^{-2})	Pamir (600 g cm^{-2})
Type of chamber	Two-story with air gap and target layer of pitch. The lead/x-ray sandwiches also include some layers of nuclear emulsion	Lead/x-ray sandwich	Alternating layers of lead/x-ray sandwich and carbon layers. Each of 4 carbon layers is $\sim 0.4\lambda_{int}$
Detector thickness	$\sim 1.5\lambda_{int}$	$\sim 1\lambda_{int}$	$\sim 2\lambda_{int}$
Exposure (m^2 year)	150	65	90 (complete scan) 210 (scan for Centauro)
Selection criteria	$n_\gamma \geq 4$	$n_\gamma \geq 4$	$n_\gamma \geq 3$ $r < 15$ cm
Jet threshold (TeV)	~ 2	~ 2	2–4
Number of families with visible energy ≥ 100 TeV	50	15	30 (complete scan) 100 (scan for Centauro)

also checked more directly by using chambers that also contain layers of nuclear emulsion in which the number of tracks (electrons and positrons) at a sequence of depths along the shower axis can be counted. Typically the minimum energy required for detection of a cascade with high efficiency during a scan of an x-ray film is 2–4 TeV, and the uncertainty in the energy estimate is about $\pm 15\%$.

5.1.1 COMPARISON OF DETECTORS   Of primary interest here are the chambers that are thick enough to allow an appreciable fraction of the incident hadrons to interact high enough in the chamber so that their energies can also be estimated. Table 4 summarizes the properties of the thick chambers used by the three groups. Hadrons are defined as jets beginning more than 4–6 radiation lengths deep in the chamber. (The γ background among hadrons, and vice versa, can be removed statistically in a straightforward way.) The Pamir and Chacaltaya chambers both have one or more layers of carbon, each about 20–30 cm thick (~ 0.4 radiation lengths and ~ 0.6 interaction lengths); this enhances the detector's capability to discriminate hadrons from photons and electrons. Unlike the small emulsion chambers, however, the light material here does not contain layers of nuclear emulsion. Tracks of singly charged particles are therefore not observable in the large chambers, and hadron energies can only be estimated (with large fluctuations) from their secondary electromagnetic cascades, which arise mainly from decay of neutral pions produced by their interactions in the chamber. Hadrons that do not interact in the chamber, or that deposit less than the threshold energy of ~ 2 TeV, are not seen at all.

5.1.2 CLASSIFICATION OF JETS   Each dark spot cascade is called a "jet." A jet or a family of jets can arise from a variety of sources. These are enumerated in Table 5 and shown pictorially in Figures 7a and 7b superimposed on a Brazil-Japan chamber with an air gap.

Table 5   Morphology of jets and jet-families

Source	Observed phenomenon	Special distinguishing features
Single e or γ	Single γ jet	Characteristic EM Cascade
Single $\pi^{\pm}$, N	Pb-jet	Delayed start, long nuclear EM cascade
Single $\pi^{\pm}$, N	C-jet	No jet in upper chamber, a jet family in lower chamber points to absorber
Several γ or e, $\pi^{\pm}$ and N	A-jet (γ-family)	One or more interactions in air making a space-associated jet family
Several γ or e	A-jet	An atmospheric EM cascade from photon or electron

5.1.3 C-JETS AND A-JETS   The essential feature of the Chacaltaya emulsion chamber that makes it suitable for studying local interactions in the detector (and hence minimizing the complicating effects of atmospheric cascading that contaminate A-jets) is the 1.5-m air gap between the upper and lower chambers (see Figure 7a). A typical atmospheric interaction occurs 0.5–1 km above the detector, so typical secondaries with $E = 5$–10 TeV and $p_T = 500$ MeV will have separations of some 5 cm in the detector. Correspondingly, typical secondaries with several hundred or a thousand GeV from interactions of hadrons in the carbon target will have separations of a few millimeters. The intrinsic size of the individual jets in the lead-

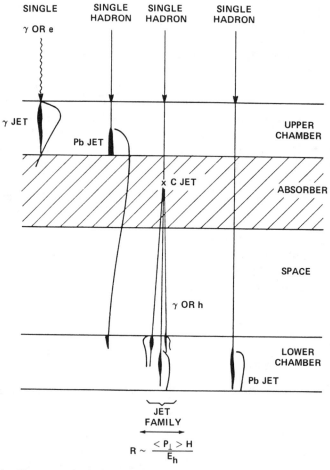

Figure 7a   Signatures of a single photon or electron and of single hadrons in the Chacaltaya emulsion chamber (Lattes et al 1971).

photosensitive chamber (Γ block) is 10–100 μm. There is thus a well-defined hierarchy of structure recognizable in the lower chamber of the Chacaltaya detector: A-jets, which contain C-jets, which in turn contain individual electromagnetic cascades of two types: γ-jets from photons produced in the carbon interaction and Pb-jets from those hadrons produced in the carbon interaction that happen to interact in the lower chamber. Cascades in the same family are virtually parallel, and the angular resolution and flux are such that the chance for confusing products of two different interactions is small (a few percent).

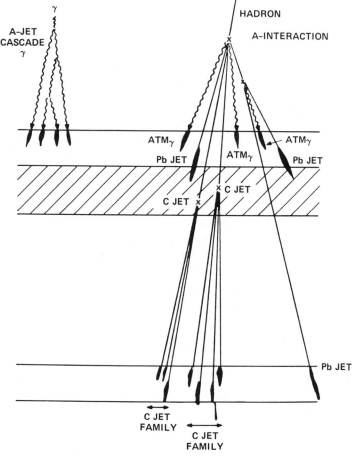

Figure 7b   Illustration of A-jet families produced by photon and hadron interactions in the atmosphere. Note that the hadronic A-jet family contains C-jets.

5.1.4 EVENT RATES   An important check on the consistency of data
obtained with the large emulsion chambers is to see that the rates for
various classes of events are sensible, given the exposure factor (area
× solid angle × exposure time) for each detector. We note first that results
from the three groups on the absolute rates of single hadrons are internally
consistent (Bayburina et al 1979a,b). Second and more important, the
hadron fluxes measured by large, deep, total energy calorimeters (Aseiken
et al 1975, Siohan et al 1978) are also in agreement with the emulsion
chamber results. For comparison, the deep calorimeter fluxes must be
converted to the higher elevation and corrected for the fact that they absorb
nearly the full hadron energy, whereas the emulsion chamber detects only
the fraction k_γ in the electromagnetic component. With these corrections,
emulsion chamber fluxes are still higher by a factor of five, which is to be
expected since the flux reported from the calorimeter measurements is that
of single, unaccompanied hadrons.

The flux measured in the emulsion chamber experiments is a complicated
combination of single hadrons and hadron families. In cases where jets from
an interaction above the chamber are very close together, they may be
counted together in the measured flux of hadrons. The ideal determination
of the all-hadron flux would require a detector of very fine spatial and
energy resolution that could count separately jets belonging to families as
well as isolated ones. It would then be straightforward to compare observed
fluxes with the calculations outlined in Section 2.2.4. In practice, this ideal
measurement cannot be made. What is measured must, however, lie
between two limits, bounded above by the all-hadron flux (Equation 8) and
below by the flux of uninteracted primary nucleons.

Referring to the extrapolations of the energy-dependent cross sections in
Section 8, we can estimate these bounds as follows. Over a limited energy
region from $\sim 10^3$ to $\sim 10^6$ GeV the cross section can be parametrized as

$$\sigma^{\text{inel}}_{\text{p-air}} \cong \sigma_0 \left[1 + \delta \ln \frac{E(\text{GeV})}{100} \right],$$

with $\sigma_0 \cong 258$ to 269 mb and correspondingly $\delta \cong 0.07$ to 0.05. The
inelastic cross section of pions in air is about 75% of $\sigma^{\text{inel}}_{\text{p-air}}$. Using Equation 3
to relate σ to λ, we find the surviving flux:

$$N_s(E, y) = N(E, 0) e^{-y/\lambda_0} \left(\frac{E}{100} \right)^{-y\delta/\lambda_0}.$$

To the extent that the energy variation is slow, we can also estimate the flux
of all hadrons by replacing $\lambda_\pi \to \lambda_\pi(E)$ and $\lambda_N \to \lambda_N(E)$ in the attenuation
lengths Λ^π and Λ_N and using Equations 7 and 8. The measured flux of

hadrons as determined by C-jets and Pb-jets at Mt. Chacaltaya is 10 per $m^2 \cdot sr \cdot year$ with measured $\sum E_\gamma > 10\,TeV$. Taking $E_0 \approx 40\,TeV$, and using the primary proton flux from Table 2, we estimate the bounds as 0.7–1.0 below and 41–50 per $m^2 \cdot s \cdot sr$ above, which brackets the data nicely.

An important point is that an increasing cross section leads to a spectrum of hadrons in the atmosphere that is steeper than the primary spectrum, as observed in the data.

It is possible to account roughly for the observed fluxes of C-jets and Pb-jets at Chacaltaya by assuming that this class of events consists of all hadrons that do not interact in the last 100 g cm^{-2} above the detector. Conversely A-jets can be interpreted roughly as arising from nucleons that survive to this depth and then interact within the last 100 g cm^{-2}. Similar considerations have been applied to a large body of data at various depths, including the Concorde experiment at 100 g cm^{-2} (Capdevielle et al 1979) as well as the three large emulsion chamber experiments and the hadron calorimeter experiments referred to above. These experiments appear to be mutually consistent and, within the broad limits of these crude estimates, consistent with calculations based on hadronic scaling in the fragmentation region, an increasing cross section, and the composition and energy spectra of Table 2.

In view of the complications, the correct method to calculate fluxes of C-jets and A-jets is a full simulation that incorporates experimental selection and resolution. Such a calculation has been carried out by Kasahara et al (1979). They find that hadronic scaling can be consistent with the observed fluxes of A-jets, but only if the cross section continues to rise to at least 10^6 GeV and if the composition is mixed with an increasing fraction of iron that becomes dominant around 10^{14} eV (Akashi et al 1979b). This is similar to the conclusion of Astafiev & Mukchamedshin (1979).

5.2 Summary of the Data

5.2.1 SCANNING AND SELECTION   The x-ray film from the lower chamber is scanned for dark spots of appropriate configuration for interactions in the carbon layer, i.e. C-jets. Energy measurements of individual photons (γ-jets) within the C-jet is by track counting in the nuclear emulsion under a microscope. Threshold for detection of individual photons is 100–200 GeV (Bellandi Filho et al 1979). Each C-jet is scanned out to a radius of 2.5 mm. Because the interaction could have occurred anywhere in the thick carbon target (or in the bottom of the upper Γ block) there is a 5% uncertainty in the angles of the photons. For data analysis, events with more than four photons are selected. The photons come predominantly from decay of neutral pions produced in the interaction of an incident hadron in the target layers. This has been confirmed by a peak in the invariant mass distribution of all possible photon pairings at the pion mass (Lattes et al 1974). Since the incident hadron, and indeed the interaction itself, is not seen, the angles of the photons must be defined relative to the energy-weighted center of the event. Events are grouped in bins of total measured electromagnetic energy ($\sum E_\gamma$), starting at 5 TeV to be clear of the scanning threshold.

All of these selection criteria, which are necessary to obtain well-defined classes of events, lead to biases that are exacerbated by the steep energy spectrum of the interacting hadrons. Of particular importance is the fact that only the electromagnetic component of the energy can be measured—the interaction energy of each event is unknown. The effects of secondary interactions in the thick target must also be considered.

5.2.2 DISTRIBUTIONS OF MOMENTA AND ANGLES   So far a total of 350 C-jets with $\sum E_\gamma > 5$ TeV have been analyzed. The data are shown in Figures 8a and 8b as integral distributions of γ-ray energies for four groups of $\sum E_\gamma$.

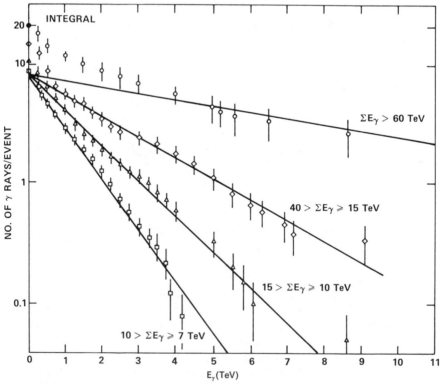

Figure 8a   Energy spectra of γ rays for different energy groups of C-jets from the data of the Brazil-Japan group (Lattes et al 1974) on interactions in the producer layer.

Figure 8b shows the data plotted vs $f = E_\gamma/\sum E_\gamma$ to remove the gross energy dependence. (Since the total energy of the event is not measured, the normal scaling variable cannot be used here.) Among the data are 57 events with $\sum E_\gamma > 20$ TeV and mean visible energy of 40.5 TeV, corresponding to a mean interaction energy somewhat above 100 TeV.

5.3   Interpretation and Implications

Because of the selection effects and biases enumerated above, a simulation including these effects is very helpful in interpreting the experimental

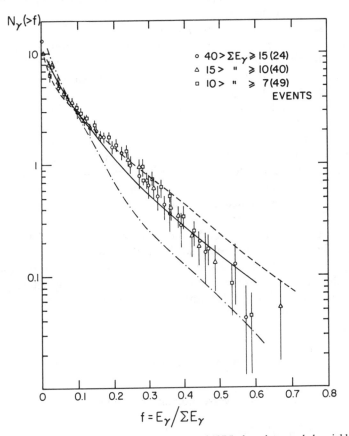

Figure 8b   Data of the Brazil-Japan group (Lattes et al 1974) plotted vs a scaled variable; the fractional energy of each γ ray in a C-jet, $f = E_\gamma/\sum E_\gamma$. The lines are from simulations of Ellsworth et al (1979a). The upper (*dashed*) line corresponds to an interaction model scaled from accelerator data using fits of the form $F_{P\pi}(x) \approx e^{-ax}$. The middle (*solid*) line corresponds to $a \to 1.5a$. The lowest (*dash-dot*) line results from cutting off all secondary mesons with $x > 0.1$ in interactions of more than 50 TeV in the lab.

results. Such a simulation has been performed by Ellsworth et al (1979a) using a simple interaction model without correlations and with momentum distributions extrapolated from accelerator data using scaling. It was found, for example, that the distribution of the ratio of measured electromagnetic energy to true (but unobservable) interaction energy was very broad (standard deviation of 0.17) with a mean of 0.37, as compared to 1/6, which is the mean fraction of energy that went into π^0's in an unbiased sample of events. This difference is a consequence of the large fluctuation in the fraction of interaction energy going to neutral pions together with the steep spectrum, the selection of events with a minimum of four photons, and the fixed threshold for $\sum E_\gamma$.

These same biases prevent the measurement of secondary photons from being an effective test of scaling, even though the center-of-mass interaction energies involved are about an order of magnitude above the highest accelerator range. Although scaling in x implies scaling in $f = E_\gamma / \sum E_\gamma$ (apart from threshold effect), a breakdown of x scaling does not necessarily show up in f. This is shown in Figure 8b, where only the rather drastic effect of cutting off all secondaries with more than 10% of the incident momentum gives a significant deviation from the data. The simulation result for angular distribution reflects these same effects: in the model that goes into the simulation the pseudo-rapidity density is two. This becomes four in the simulated events, as compared to the experimental value of six.

A feature of the Chacaltaya data that is often emphasized is the M_γ distribution. This is a histogram of events classified by the invariant mass of γ rays within a certain angle of the energy-weighted center of the event. The cone is defined (event by event) so that if the γ rays come from isotropic decay of a cluster of particles and if selection effects can be ignored, then M_γ is the cluster mass times the fraction of π^0's in the cluster. Specifically, M_γ is obtained by the following algorithm: Define the invariant mass of γ's inside a cone θ as $M_\gamma(\theta)$.

Then

$$M_\gamma(\theta) = \left[\left(\sum_\theta p_\mu^\gamma \right)^2 \right]^{1/2} \cong \left| \sum_\theta E_\gamma \sum_\theta E_\gamma \theta_\gamma^2 \right|^{1/2}.$$

Also define the total transverse momentum inside θ:

$$P_T(\theta) = \int^\theta P_T \, d\theta = \sum E_\gamma \theta_\gamma.$$

For small θ we find $M_\gamma(\theta) < (4/\pi) P_T(\theta)$, and for large θ we find $M_\gamma(\theta) > (4/\pi) P_T(\theta)$. Define $M_\gamma \equiv M_\gamma(\theta_0)$, where θ_0 is the solution of $M_\gamma(\theta_0) = (4/\pi) P_T(\theta_0)$. In practice this involves an interpolation.

Nearly half the events among the 57 have $M_\gamma > 3$ GeV, a much larger fraction than found in the simulation of Ellsworth et al (1979a). Events with large M_γ are also seen among the lower energy data, but with a lower frequency. Moreover, the events with large M_γ also have high multiplicity. On the basis of these features of the data the Japan-Brazil group argue that there is a threshold (around 50 TeV) for production of events with large clusters or fireballs.

It is clear from the definition of M_γ that there is a correlation between transverse momentum and M_γ. Recently Ellsworth et al (1980) included a fraction of events in the simulation with production of high transverse momentum, as expected in constituent models of hadrons. With a sufficiently large hard-scattering component it is possible to get a third or so of the events with $M_\gamma > 3$ GeV. The results are also sensitive to the correlation between longitudinal and transverse momentum, to rise of the multiplicity of particles with small center-of-mass momenta, and to nuclear target effects, which can also contribute to an increased multiplicity in high energy events for which the photon energy threshold (200 GeV) is in the central region of the interaction.

T. Shibata (1979) has argued that the transverse momentum distributions require a component of high transverse momentum in agreement with the standard quark-parton model of hard collisions (Feynman et al 1978, Cutler & Sivers 1978, Combridge et al 1977). Shibata extended the hard-scattering calculations, which are normally done for near 90° in the center of mass, to the forward fragmentation region, which is relevant for the cosmic ray calculation in which only fast secondaries are visible. A test of this result will be to look for asymmetries corresponding to jet production (Gaisser & Sidhu 1977). It should also be noted that Shibata took the selection effects into account only in an average way.

We believe that the C-jet data require both increasing multiplicity and large transverse momenta in the 100-TeV range. At least some part of this may involve what are now considered conventional ideas, particularly if hard-scattering processes are to be associated with the C-jets of large M_γ. Halzen (1975) first proposed this interpretation of these events. If it is correct it would suggest that hard-scattering processes involve large associated multiplicities. Finally, we note that the cosmic ray experiments are probing the hard-scattering picture in a different kinematical regime from that usually emphasized in accelerator studies.

6   EVENTS AROUND 500 TeV

6.1   Event Selection and Analysis

To study interactions in this energy range with the large emulsion chambers it is necessary to accept interactions in the overlying atmosphere. This is the study of γ families or A-jets. Events are selected with more than a minimum number of γ-jets incident on the chamber ($n_\gamma \geq 3$ or 4) to eliminate single π° events. Requirements for minimum and maximum spread of the families are also imposed to avoid overlap of separate photons on the one hand and to avoid excessive atmospheric cascading on the other. These criteria lead to selection of events in which the most important interaction typically occurred one to several kilometers above the detector, depending on the specific scanning radius chosen (this corresponds to 1–2 interaction lengths in air).

All of the ambiguities in the C-jet analysis arising from selection effects are also present for A-jets. Cascading in the atmosphere further obscures matters. Simulations show that roughly half the observed electromagnetic energy comes from the main interaction in a family (Dunaevskii & Slavatinsky 1977, M. Shibata 1979). It can be shown, however, that the cascade equations (Equations 1 and 2) preserve scaling under both hadronic and electromagnetic cascading, provided the data are treated in a way that eliminates the effects of the energy thresholds.[4] This follows essentially from dimensional analysis (Zhdanov et al 1975, Dunaevskii & Slavatinsky 1979). Thus a violation of scaling in the γ-family data would imply a violation of hadronic scaling, but consistency of the data with scaling is a rather insensitive test of the underlying strong interactions.

It is very difficult at present to draw any firm conclusion from the systematics of the data on photons in A-jets. A considerable number of events with $\sum E_\gamma > 100$ TeV have been measured; some of these data were obtained with the large-area thin chambers, but there is disagreement between the Fuji and the Pamir groups on the data itself. The Pamir group reported a large scaling violation in the average f' spectra[4] corresponding to a decrease in the fraction of high energy secondaries (and an increase in the multiplicity) as the visible energy increases (Bayburina et al 1977). The Fuji group, on the other hand, found a distribution that scales for $\sum' E_\gamma$ from 80 to ≥ 250 TeV (Akashi et al 1978). They suggest that the effect observed by the Pamir group is due to a systematic underestimate of E_γ that increases as E_γ increases. Krys et al (1979), however, have estimated this bias and believe it cannot be so large.

The experimental situation is currently unresolved, but we can ask what inferences can be drawn if the Fuji group is right and there is no observed scaling violation in the data. Simulations by some of the Lodz members of the Pamir collaboration (Wrotniak 1977, Krys et al 1979) as well as by M. Shibata of the Fuji group give similar results for the f' distributions that are

---

[4] This is done by a process known as "rejuvenation" in which one defines $\sum' E_\gamma$ to be the sum of the energies of the n fastest γ rays in the family, where n is defined by $E_n/\sum_{i=1}^{n} E_i \cong f_{th}$. Here f_{th} is an arbitrary fixed number, usually chosen to be 0.04 in the Pamir analysis. The resulting scaling variable is $f \equiv E_\gamma/\sum' E_\gamma$.

expected if hadronic scaling is valid up to 1000 TeV. Dunaevskii et al (1979) and Akashi et al (1979b) show that these distributions are also nearly independent of composition. These expectations are generally in good agreement with the Mt. Fuji data. On the other hand, extreme models with no fast secondaries ($x > 0.1$) would be exluded. The simulations also show that distributions in $E_\gamma r_\gamma$ should be somewhat sensitive to models. According to Fujimoto's (1979) rapporteur talk, the lateral distributions require either high p_T (average of 650 MeV/c for pions) or drastic breakdown of scaling. We believe, however, that effects of heavy primaries have not yet been considered in this context. Akashi et al (1979c) find scaling with a mixed composition to be consistent with these data.

6.2 Hadron-to-Photon Ratios

We expect that much more information can be obtained from those events in which hadrons as well as photons can be detected. This expectation is borne out by simulations of Dunaevskii et al (1979), who show that the hadron distributions are quite sensitive to the primary mass, whereas electromagnetic quantities alone are insensitive to it. The exposure factors of thick chambers, which can detect hadrons as well as photons in A-jets, are summarized in Table 4 of the previous section. The number of events with $\sum E_\gamma > 100$ TeV measured by each group is also listed. Here $\sum E_\gamma$ includes all visible energy, whether from photons incident on the chamber or from hadrons that happen to interact.

6.2.1 CENTAUROS   This work was largely stimulated by the discovery of the original Centauro[5] event by the Brazil-Japan group (Lattes et al 1973, Tamada 1977). This event was found as a family of 43 jets in the lower chamber containing ~ 200 TeV of visible energy. The corresponding family in the upper chamber contained only eight jets, of which six were Pb-jets (i.e. hadrons). The energy deposited in the upper chamber was only ~ 30 TeV. The primary interaction responsible for the family happened to be close enough to the chamber so that its height could be estimated by triangulation as 50 ± 15 m. Because of the lack of photons incident from the atmosphere, it appears that at most one π° was produced in the atmospheric interaction, which produced at least 49 hadrons. Correcting for hadrons not interacting in the chamber, one estimates 75 hadrons were produced in the original interaction. Keeping in mind that only the electromagnetic portion of hadronic interactions in the chamber is seen, we

---

[5] It is not clear whether the name originates from the fact that the spot in the x-ray film of the lower chamber resembles the constellation Centaurus or because the body of the event in the lower chamber looks like a different beast from its head in the upper chamber.

infer that the interaction energy was probably somewhat greater than 500 TeV.

Four other events of the Centauro type have been found and three of these have been analyzed in detail to determine the amount of cascading and thus to see whether the events are indeed unusual in the sense that the π^0 component is virtually absent in the primary interaction. This analysis, carried out by the Brazil-Japan group (Bellandi Filho et al 1979 and references therein) and by Yodh (1977), depends on an indirect determination of the interaction height, which cannot be determined by triangulation as for the first Centauro. Instead, it is estimated by assuming that the transverse momentum distribution of the hadrons is the same as that observed in Centauro I. With interaction heights established in this manner, it is concluded that all four events are consistent with no π^0 production in the original interaction. All observed photons are consistent with an origin in secondary interactions in the atmosphere of other hadrons produced in the original interaction.

The identity of the hadrons produced in Centauro interactions is unknown. It is usually assumed that they are nucleons and antinucleons for purposes of the analyses described above, but there is no way actually to make such an identification. In Centauro I the distribution of depths of hadron interactions in the chamber is consistent with the nucleon interaction length, but it is also consistent with an interaction length up to several times λ_N. The produced hadrons might therefore have cross sections considerably smaller than nucleons and could as well be charged pions, quarks, or other particles with lifetimes long enough to reach the detector. Taking 100 meters as the minimum production height and 3 TeV as a typical energy, we can estimate $\tau_0 > M(\text{GeV}) \times 10^{-10}$ s for this minimum lifetime, where M is the mass chosen for the produced particle.

Another characteristic feature of the Centauro events is an apparently large value of the mean transverse momentum of the secondaries. Figure 9 shows the integral distribution of the quantity $E_h^{(\gamma)} \cdot r/h$ for Centauro I. Here $E_h^{(\gamma)}$ is the electromagnetic energy deposited by the secondary hadron as it interacts in the chamber, and r is the distance of this jet from the energy-weighted center of the family. The quantity h is the height of the interaction, which is only determined directly for Centauro I. If the hadron deposited all its energy in visible form in the chamber, then Figure 9 would be the integral transverse momentum distribution of secondaries produced in Centauro interactions. If the visible energy is from produced neutral pions, however, then the distribution in this figure is obtained by folding the transverse momentum distribution of Centauro secondaries with the k_γ distribution, where k_γ is the fraction of energy given to neutral pions by the hadrons that interact in the chamber. The Brazil-Japan group used $k_\gamma = 0.2$ to estimate $\langle p_T \rangle = 1.7$ GeV from $\langle E_h^{(\gamma)} \cdot r/h \rangle \approx 0.3$—0.4 in Figure 9.

It is important also to take into account the threshold constraint $E_h^{(\gamma)} > \varepsilon \cong 2$ TeV. If we define $x = \sum E_h^{(\gamma)} \cdot r/h = k_\gamma p_T$ then

$$\frac{dn}{dx} \cong \int_x \frac{dp_T}{p_T} \int_{\varepsilon p_T/x} dE \frac{dn}{dE\,dp_T} \left(\frac{dP}{dk_\gamma} \right)_{k_\gamma = x/p_T}. \qquad 11.$$

Here $dn/dE\,dp_T$ is the energy-p_T distribution of secondaries of the Centauro interaction and dP/dk_γ is the normalized distribution of energy given to the electromagnetic component when a secondary hadron interacts in the chamber. The energy distribution dn/dE can be estimated from the observed distribution of $E_h^{(\gamma)}$, given an assumption about dP/dk_γ. To get an indication of the size of the bias effect we took for dP/dk_γ a distribution with a mean of 0.17 and a median of 0.10 to characterize pion production in nucleon interactions. Using a p_T distribution with a mean of 1.2 GeV we found $\langle x \rangle \cong 0.44$. For a distribution with a mean p_T of 0.9 GeV, we found $\langle x \rangle \cong 0.34$ GeV. This suggests that the mean p_T of hadrons produced in Centauro interactions could be closer to 1 GeV than to 1.7 GeV.

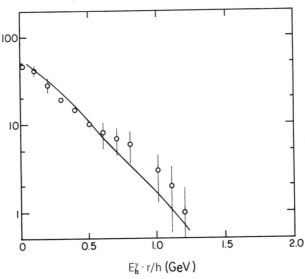

Integral number of C-jets & Pb-jets

$E_h^\gamma \cdot r/h$ (GeV)

Figure 9 The integral distribution of $k_\gamma P_T$ for secondaries produced in Centauro I. The abscissa is $E_h^\gamma \cdot r/h$, which is a measure of $k_\gamma \cdot p_T$. The solid line results from a calculation (see text) that folds the energy threshold of the detector with the k_γ distribution. The true mean P_T required to get agreement with the observed lateral distribution is ~1 GeV/c.

6.2.2 OTHER EXPERIMENTS AND SIMULATIONS   Having seen the special nature of these few events, we now ask, how do they compare with other events and with simulations? The Chacaltaya and Mt. Fuji events with $\sum E_\gamma > 100$ TeV are shown in Figure 10 (Fujimoto 1979) together with a portion of the events in this category from the Pamir experiment. The first diplot compares observed numbers of γ-jets and hadronic jets; the second compares the visible energies in the γ-jets with the hadronic jets. The feature that is most evident is that $n_\gamma > n_h$ for most of the events. The unusual Centauro events are those five on the extreme right with $n_h \approx 50$ and $n_\gamma < h_h$. Also note that there is only one event with $n_\gamma > 100$ and $n_h > 70$.

A Monte Carlo simulation was carried out by Acharya et al (1979a) to determine the distribution of events due to air shower cores. A high energy interaction model with hadronic scaling and increasing cross section was used. All hadrons were followed and all γ rays were propagated using approximation A, as γ-ray energies are much greater than critical energy in air. Experimental cuts were imposed and characteristics of events were

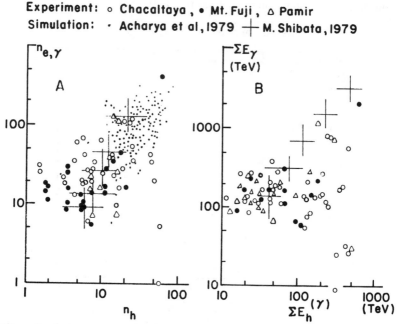

Figure 10   Scatter plots of a portion of the events with total visible energy greater than 100 TeV. The left plot shows number of γ rays and hadrons for each event. The right plot shows the total visible energy in γ rays and hadrons for each event. Also shown are the results of two simulations. The five events in (*B*) with $\sum E_\gamma < 50$ TeV and $\sum E_h^{(\gamma)} > 200$ TeV are beyond the range of any conventional explanation so far proposed.

determined. Events were generated using primary protons with energy greater than 5×10^{15} eV. A total of 190 events were generated and they are shown in Figure 10a. Here n_γ and n_h refer to the jets seen within a 5-cm radius. The jets outside this radius are not included in the jet counts. Several points are to be noted about this simulation. First, the Monte Carlo events populate a region beyond the data points, indicating possibly that relevant energies are somewhat lower than 5×10^{15} eV. Second, two events in the simulation look quite similar to the upper Centauro points, including the lateral distributions. The fraction of simulated events that look like Centauros is much smaller than the observed fraction, however. Acharya et al point out that the simulated rate for this type of event estimated from the primary spectrum is not inconsistent with the observed rate for these two Centauros. Thus, there seems to be an excess of high multiplicity events in the simulation (presumably as a result of cascading, since the interaction model only has $n \propto \log E$). There may be more high multiplicity events not yet analyzed due to efforts, particularly by the Chacaltaya group, to select events that occurred near the chamber. Another detailed point suggested by the simulation is that there must exist a sizeable γ-jet density beyond the 5-cm radius that should be observable in the experiment.

The simulations of M. Shibata (1979, shown as crosses in Figure 10) cover the full range of energies and multiplicities of the data. Only the averages of Shibata's simulation are available, but it is clear that the Centauro events lie well outside the range of the simulations.

The Japan-Brazil group had also identified a class of events called mini-Centauro, characterized by n_h values of 5–20 but with a large ratio of hadronic to electromagnetic energies. There are ten such events reported (Bellandi Filho et al 1979) with $\sum E_\gamma > 100$ TeV. These appear normal in the $n_h - n_\gamma$ diplot but tend to lie outside the simulation results in the $\sum E_h^{(\gamma)} - \sum E_\gamma$ plot. There are events in this region from all three experiments.

The simulations mentioned above were performed for proton primaries. Because the Centauro events are characterized by an excess of the hadronic component, it is necessary to consider the possibility of heavy primaries, which could contribute many energetic nucleons through their breakup in the atmosphere. The original Centauro event clearly could not be due to breakup of a heavy nucleus near the chamber because (a) the interaction was too deep in the atmosphere (more than 30 interaction lengths for iron) not to be accompanied by significant atmospheric cascading from previous interactions of the nucleus, and (b) the transverse momentum was much higher than the Fermi momentum for an ordinary nucleus. There may, however, be some contribution of heavy primaries to some of the other Centauro events (Dunaevskii et al 1979, Fujimoto 1979).

Energies of the individual γ rays involved are too low ($E_\gamma \lesssim 10$ TeV) for the Landau-Pomeranchuk-Migdal (LPM) effect to be important. This is a density effect (Landau & Pomeranchuk 1953, Migdal 1956) that suppresses cascading at high energies and could in principle lead to a misidentification of γ rays as Pb-jets. For $E_\gamma \lesssim 10$ TeV the effect, however, is negligible (Konishi et al 1978, Ellsworth et al 1979b).

6.2.3 POSSIBLE EXPLANATIONS   We can therefore summarize the situation by saying that there is a class of events with total visible energy between 200 and 500 TeV that lies beyond the range of any conventional explanation so far proposed. This group comprises five events out of a total of 95 events with visible energy above 100 TeV found in scans of three experiments. The anomalous group includes two Centauro events (the original and one other), the two most energetic mini-Centauros (Bellandhi Filho et al 1979), and one event from the Pamir experiment. Thus at least 5% of the events around 1000 TeV appear to be anomalous. The fraction could be much higher, since Centauro interactions high in the atmosphere would probably be obscured by subsequent atmospheric cascading (Tamada 1979).

Two classes of explanations for Centauro events can be imagined: (a) those involving a new kind of interaction of ordinary hadrons beyond some threshold energy and (b) those involving exotic components of the primary beam. An example of (a) proposed by Sutherland (1979) involves liberation of quarks in nucleon-nucleon collisions of sufficiently high energies. Such an idea is only marginally consistent with existing quark searches (Jones 1977). The alternative class is exemplified by the suggestion of Bjorken & McLerran (1979) that Centauros may be produced by metastable globs of compressed nuclear matter perhaps condensed around a free quark.

Centauro type events have also been mentioned by Chin & Kerman (1979) and by Mann & Primakoff (1980) in association with production of possible stable or metastable multiquark states. Quantitative estimates of multiplicity and production cross sections have not been attempted by these authors.

6.3   Multiple Cores

There exists a subclass of events among the emulsion chamber data on A-jets with distinct subcores (Y. Fujimoto, private communication, Akashi et al 1979a). Experimental data on the double-core events among these are discussed systematically by Bellandi Filho et al (1979). There is one event in which the interaction apparently occurred in the roof of the shed (3 m above the chamber) over the emulsion chamber based on observed divergence of the cores. The cores had respectively 11 and 8 TeV visible energy and a separation of 0.16 cm, corresponding to a minimum relative p_T of 10 GeV. If this event was due to hard scattering of constituents, one visible jet would consist of fragments of the projectile and the other would be a jet scattered into the forward center-of-mass hemisphere. (Target fragments and the backward jet would be low energy in the lab and so below threshold.) It could also be due to decay of a massive object produced in the interactions, as suggested by the Brazil-Japan group.

Whether the other events also involve very large p_T is more difficult to decide because the heights of the interactions responsible for the cores are not directly determined. Primaries with $A > 1$ could lead to multiple cores (McCusker 1975); so could nearby interactions superimposed on the hadronic cascade. Simulations are required to clarify these issues.

Multiple cores in air showers have also been the subject of study for many years (McCusker 1975, Sakata et al 1979, Hazen et al 1979). They, too, contain fragmentary evidence for processes with high p_T. However, the difficulties of interpretation mentioned above also affect these experiments, which are sensitive primarily to the numerous low energy particles in the electromagnetic cascades of the showers. Preliminary results of simulations (Green et al 1979) suggest that at least some subcores can be accounted for by fluctuations, both statistical and in shower development.

7. EXTENSIVE AIR SHOWERS: INTERACTIONS UP TO MILLIONS OF TeV

So far we have considered cosmic ray experiments in the energy range up to $\sim 10^{15}$ eV. The only experiments that probe energies significantly beyond this are done with extensive air shower (EAS) arrays, which overcome the very low flux with very large areas. Showers with primary energies up to $\sim 10^{20}$ eV have been observed (see Edge et al 1978 and Lloyd-Evans et al 1979 for accounts of the largest showers and their astrophysical implications). Table 1a indicates the range of acceptance areas of air shower arrays. (The fraction of the area of an array that is actually instrumented to detect particles is very small. A typical setup would consist of some 10–100 detectors each with an area of ~ 1–10 m^2.) In view of new machines now under construction, which will study hadronic interactions up to the equivalent of $\sim 10^{15}$ eV (see Table 1b), an important role of small air shower experiments in the future will be to study primary composition. We expect large air shower experiments to remain the only source of information on interactions as well as composition above $s^{1/2} = 10^4$ GeV for an indefinite period.

The basic techniques of air showers are admirably described in the classic reviews of Greisen (1956, 1960, see also LaPointe 1970), who also describes the morphology of showers: energy balance, energy spectra, and lateral distributions of the various shower components. There are three components: the hadronic core, the muons, and the electromagnetic cascade, which originates primarily from π^0's produced in hadronic interactions along the core. Muons originate from decay of charged mesons in the core.

The subject of air showers and particle physics above 10^{15} eV has been reviewed recently (Gaisser et al 1978). In this section we therefore emphasize some new experiments and include only a brief update of the implications of existing experiments for particle physics. (See Watson 1979, for a survey of existing air shower facilities and techniques.)

7.1   Longitudinal Profiles

The longitudinal profiles of some simulated showers are shown in Figure 11 (Gaisser 1979). The shapes are dominated by the electromagnetic component. The electrons (and positrons) are the most numerous particles in the shower (by an order of magnitude) because of the rapid multiplication in pair production and bremsstrahlung initiated by high energy photons. This process is characterized by the radiation length $t_R \cong 36$ g cm^{-2} in air, which is about half (or less) of the interaction length of hadrons in air at high energies. In contrast, the smaller muon component grows to a maximum and stays fairly constant because of the long interaction length of muons.

Standard arrays of particle detectors sample any shower at only one depth. A basic goal of some current experiments is to measure a signal that traces the entire development of individual showers directly, and so to achieve instead the kind of information typical of a calorimeter experiment, including direct determination of primary energy. There are two examples of this approach now being pursued, both of which require clear, moonless nights.

7.1.1 ATMOSPHERIC CHERENKOV TECHNIQUE   Detection of Cherenkov light produced in the atmosphere by relativistic particles in the shower has a long history, dating back to the original experiment of Jelley & Galbraith (1953). It is now used in several experiments (see Khristiansen 1979 for a review). Both the Moscow State University (MSU)/Yakutsk group (Kalmykov et al 1979, Khristiansen 1979) and the Durham group (Orford & Turver 1976, Hammond et al 1978) have developed techniques to study longitudinal profiles of individual showers by analyzing the time structure of the detected Cherenkov pulses, as proposed by Fomin & Khristiansen (1971).

Hammond et al (1977, 1978) demonstrated the feasibility of a self-contained technique with a particularly simple geometrical interpretation (Orford & Turver 1976). The Durham array contains eight phototubes with nanosecond timing, separated by distances of several hundred meters. Demonstration of this Cherenkov imaging method consists of showing that the arrival times of the 10%, 50%, and 90% points on the rising and falling edge of the signal in each of several detectors determine a set of spheres, the centers of which fall on a single line (the trajectory of the shower core) with the highest point first, and so on. Amplitudes of the signals can be related to shower size at each depth with the help of simulations (Protheroe & Turver 1979).

The calibration experiment (Hammond et al 1977, 1978) was done at the Haverah Park air shower array. It confirms that the Cherenkov imaging technique can locate shower axes and determine primary energies well. Because of the leverage afforded by the large separation between centers of

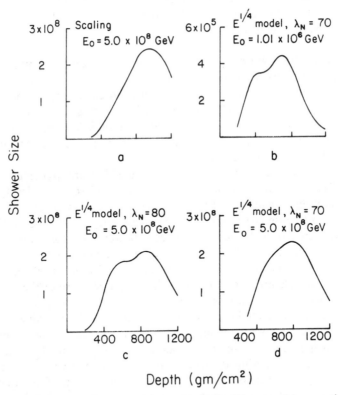

Depth (gm/cm^2)

Figure 11 Four showers with the same core histories for different model assumptions and energies (see text) initiated by protons. The ordinate gives the number of shower particles.

successive spheres, the potential accuracy of definition of the shower axis is significantly better than with conventional methods. The technique is most naturally suited to showers in the range 10^{17}–10^{18} eV, though smaller showers can also be studied (Thornton & Clay 1979).

The full Durham experiment was conducted over a three-year period at Dugway Proving Ground in Utah and completed in the Spring of 1980; data are now being analyzed. Some preliminary results from the MSU/Yakutsk experiment, showing profiles of three individual showers, were presented by Khristiansen (1979) in his rapporteur talk at Kyoto. Two full cascades from the Durham calibration experiment, also showing the rising edge of the showers, were shown by Orford & Turver (1976). Final results from these experiments should be of great value in understanding composition and the nature of hadronic interactions around 10^{17}–10^{18} eV.

The use of fast timing techniques to map shower development may also be applicable in modified form to other shower components, such as muons (Aguirre et al 1979a, T. J. L. McComb and K. E. Turver, private communication, Watson 1979).

7.1.2 ATMOSPHERIC SCINTILLATION TECHNIQUE   The idea of detecting the atmospheric scintillation light produced by nitrogen molecules due to excitation by a passing shower front was discussed by Suga (1962), Chudakov et al (1962), and Greisen (1965). The signal is proportional to the number of ionizing particles so that the effective collection area of a detector increases with shower energy. This allows detection of very rare, very high energy showers at large distances.

The idea was recently put into practice by the Utah group (Bergeson et al 1977, Cassiday et al 1979) with the use of modern microelectronics. Unlike Cherenkov light, emission of the scintillation light is isotropic. The idea therefore is to observe showers from the side as they pass across the sky at some distance from the detector. Indeed showers that come toward the compact detector array (Fly's Eye) are contaminated with Cherenkov light and are more difficult to interpret. The shower trajectory is defined by the timing of signals arriving in different facets (phototubes) of the detector which, together, cover a large portion of the sky. The amplitudes allow determination of the full shower profile. Because of the side view, signals from different portions of the shower do not overlap and the shower profile is in principle simpler to interpret than in the case of Cherenkov light.

A calibration experiment carried out in conjunction with the conventional array at Volcano Ranch (Bergeson et al 1977) confirms that the technique can determine shower sizes correctly. An array of 48 mirror units with 12–14 phototubes each has been built on a hill at Dugway Proving Ground. Its field of view includes the Durham Cherenkov array, so intercalibration of the two techniques is possible. Preliminary data taken at various stages of construction have been reported at conferences. Shower profiles clearly show the rising edge of the shower, indicating that calorimetric measurements will be possible. Khristiansen (1979) has compared average shower profiles from the MSU/Yakutsk Cherenkov array (10 showers) with the average of 19 showers from Fly's Eye. The agreement in shape is good, with both techniques displaying the rising portion as well as the region of shower maximum.

A second and smaller Fly's Eye, to be located about 7 km from the first, is planned. This will allow experimental determination of atmospheric attenuation of the signal as well as give the geometrical advantage of a stereoscopic view. Arrays of detectors of muons, electrons, and possibly other shower components to be deployed near Fly's Eye are also being considered (see, for example, Linsley 1979, Jones 1979b, Bowen, Jones & Jones 1979). These would complement Fly's Eye by providing information on individual shower components and on lateral distributions. The new Akeno array in Japan (Kamata 1979) includes detectors of atmospheric scintillation light among a variety of particle and Cherenkov light detectors.

Design studies (Cassiday et al 1979, 1977) indicate that Fly's Eye eventually should be able to see showers with $E \approx 10^{21}$ eV at distances up to 50 km if the primary spectrum is not cut off. Much more frequent will be showers in the 10^{17}–10^{18}-eV range, where statistics should be sufficient for studies of cross sections and composition. We discuss this somewhat further in Section 8.

7.1.3 SENSITIVITY TO MODELS   To see how an air shower cascade reflects the physics of the particle interactions and the primary composition, recall the simple model of an electromagnetic cascade described in Heitler's (1944) textbook. The number of particles at depth x (g cm^{-2}) is estimated as $n(x) \approx 2^{x/t_R}$ for $E \gg E_{critical} = \varepsilon$. A typical electron energy at depth x is $E(x) = E_0(\frac{1}{2})^{x/t_R}$, so that

$$x_{max} \approx \frac{t_R}{\ln 2} \times \ln \left(\frac{E_0}{\varepsilon} \right)$$

and

$$n_{max} \approx \frac{E_0}{\varepsilon}.$$

The corresponding relations for air showers are similar, especially if hadronic scaling holds so that the bremsstrahlung analogy is good. In this case the energy E_0 of the initial photon is replaced by an effective energy, which is an injection energy for photons that initiate electromagnetic cascades along the shower core. This effective energy reflects the way in which the interaction energy is distributed among secondaries in hadronic interactions. One still has n_{max} proportional to total primary energy E_0, but one finds (Hillas 1979a, Linsley 1977) that

$$x_{max} \approx t_R(1-\alpha) \ln \left(\frac{E_0}{A'}\right) + \frac{(\lambda_\pi + \lambda_N)}{K}.$$

12.

Here α is a parameter that reflects the energy dependence of the multiplicity of fast secondaries (not the total multiplicity). If hadronic scaling holds in the fragmentation region then $\alpha = 0$ even if the total multiplicity increases faster than $\ln E$. K is related to inelasticity in meson- and nucleon-initiated collisions, and A' is the effective primary mass.

Showers with energies below 10^{18} eV are past maximum at most arrays, so x_{max} cannot be measured directly with conventional techniques. An exception is the Chacaltaya array (Aguirre et al 1979b). Linsley (1977) has shown how to infer the rate of change of depth of maximum with energy indirectly from the dependence on zenith angle and shower energy of various parameters measurable at the surface (for example, shapes of lateral distributions of various shower components). A recent summary (Gaisser et al 1979a) suggests values of "elongation rate" d x_{max}/d ln $E \approx 40$ g cm^{-2} from several experiments over a range $\sim 10^{17}$–10^{18} eV, close to the value one would expect in models with hadronic scaling in which $\alpha = 0$. Shower maxima are intrinsically broad so the uncertainties are correspondingly large, but extreme models with essentially no fast secondaries would appear to be ruled out. An exception is the very much smaller result of Kalmykov et al (1979), who report an elongation rate of about one half the radiation length.

It is evident from Equation 12 that the depth of maximum and its rate of change also depend on energy dependence of interaction lengths, inelasticity, and composition. If the cross section increases at the rate expected from conventional extrapolations (see Figure 12 below) then one would expect a negative contribution to the elongation rate of 10–15 g cm^{-2}. This tends to strengthen the conclusion about fragmentation scaling.

Changes in composition also affect d x_{max}/d ln E. Thornton & Clay (1979) find the surprising result, based on measurements of full width and half maximum (FWHM) of the atmospheric Cherenkov signal, that the elongation rate is about 60 g cm^{-2} between about 10^{15} and about 5×10^{16} eV. Since other energy-dependent factors in Equation 12 tend to make the elongation rate less than the radiation length (~ 36 g cm^{-2}) they conclude that the composition changes from relatively heavy to relatively light in this range. However, their analysis has been criticized recently by Orford &

Turver (1980). They point out an inconsistency in the model used to relate FWHM to depth of maximum, as well as a systematic change in the sensitivity of the technique that can be misleading.

Monte Carlo simulations based on various assumed interaction models and compositions have been compared to a wide range of air shower data by many authors. In view of the complexity of the problem, it is perhaps not surprising that a variety of conclusions have been reached. (Compare the reviews of Gaisser et al 1978, Grieder 1977, and Khristiansen 1979.) There is general agreement that if nearly all showers are due to protons, then many observed features of showers require changes in properties of hadronic interactions at high energies in order to achieve rapid development of the showers. From Equation 12 it is clear that this can be achieved by subdividing the interaction energy among many particles at high energy ($\alpha > 0$) and/or by increasing the inelasticity. It can also be achieved by increasing A'. Note also that an appreciable fraction of Centauro-type interactions could effectively increase α. The rather large values of elongation rate would appear to point to a large value of A' as the source of the observed early development of showers [but not if Kalmykov et al (1979) are right that the elongation rate is really small].

Ouldridge & Hillas (1978, Hillas 1979a,c) used a mixed composition (intermediate between the low energy composition and that described in Table 2) and an interaction model based on radial scaling of accelerator data. They find agreement with many features of air showers, such as the μ/e ratio, which had previously been difficult to reconcile with hadronic scaling. A similar conclusion had been reached by Gaisser et al (1978) by comparing simulated proton and Fe showers with data in the context of a somewhat more simplified calculation (for example, treating kaons as pions). Hillas emphasizes (1979a,c) that the continued increase of the cross section to 10^{17} eV is essential for obtaining agreement with the data.

7.1.4 FLUCTUATIONS   Fluctuations in various measurable parameters that reflect fluctuations in shower development are an important potential source of information on composition and models. Fluctuations are expected to be small in showers generated by primary heavies because they are a superposition of smaller showers. However, fluctuations due to a mixture of primaries as well as fluctuations in development of proton and helium showers should be more prominent. Elbert et al (1976) concluded that observed fluctuations in muon number for showers of fixed size at sea level were too large to be accounted for either by pure protons or pure Fe primaries. On the other hand they found measured fluctuations to be smaller than expectations based on the low energy composition. This refers to the energy range 10^{15}–10^{17} eV.

Reported fluctuations in depth of maximum of showers in the range 10^{17}–10^{19} eV vary widely. Linsley (1977) reports $\sigma_x \cong 225$ g cm^{-2}, inferred from measurements of fluctuations in the shape of lateral distributions of charged particles. Preliminary results from fluctuations in the shape of the lateral distributions measured at Haverah Park suggest values in the range of 100 g cm^{-2} (Craig et al 1979, England et al 1979). Fluctuations in rise time of the signal in the water Cherenkov detectors at Haverah Park give $\sigma_x = 88 \pm 4$ g cm^{-2} (Lapikens et al 1979). Kalmykov et al (1979) find a still smaller result, $\sigma_x = 64 \pm 7$ g cm^{-2}, after correcting for the instrumental contribution to the experimentally determined variance. This result is obtained from analysis of pulse shapes of atmospheric Cherenkov light from about 100 showers in the energy range 2×10^{16} to 4×10^{17} eV. (Recall that this experiment also gives a very small value of elongation rate.) For comparison, expected values of σ_x are respectively 20, 70, and 81 g cm^{-2} for Fe, proton, and a half-and-half mixture of Fe and proton primaries. [These estimates, quoted by Craig et al, are based on the calculations of Gaisser et al (1978) with scaling and increasing cross section.] In each case mentioned above the fluctuation in depth of maximum is inferred indirectly from an accessible shower parameter. Further experiment and analysis are needed to clarify the situation.

Figure 11 illustrates how fluctuations in shape of showers could also be interesting. Unlike the situation in dense calorimeters, most high energy showers in the atmosphere have no appreciable longitudinal structure. This is because the radiation and interaction lengths are comparable in air. Some structure may develop, however, in rare cases when there is a large gap in energy deposition. Figure 11 shows four simulated showers with identical nuclear interaction histories. In this example (one out of 100), half the energy was deposited before one λ_N and the rest around $5\lambda_N$. No structure at all is visible in the case of the scaling model. The structure is visible in the "$E^{1/4}$" model. (This is a crude version of a model with $\alpha = \frac{1}{4}$ in Equation 12.) The difference is that in the first case most of the energy deposited in the first interaction is subdivided among a very few π^0's so that each γ cascade has relatively high energy and long penetration. In the second case, the energy is subdivided into many short, low energy showers, so that the fluctuation in the core remains visible provided λ_N is not too short. Figure 11d illustrates how the gap is filled in as λ_N decreases relative to t_R. Figure 11 also illustrates how a gap that is visible in a low energy shower (11b) is filled in in the same shower at higher energy (11d) because energies of individual photons are higher and their showers more penetrating.

Observation of significant bumpiness in 10^{17}–10^{18} eV showers would indicate significantly new particle physics. Attenuation will, of course, have to be understood very well to ensure that an observed dip is not due to locally increased absorption of the signal.

7.2   High Energy Muons

Rates of multiple high energy muons from the same primary are especially
sensitive to primary composition and to early development of the shower,
i.e. to properties of the highest energy interactions in the shower (Elbert
1978). The standard way to select very high energy muons ($\gtrsim 1$ TeV) is to
place a detector deep underground. The most extensive information to date
on multiple high energy muons comes from the Utah experiment (Lowe et
al 1975, 1976). The detector had a fiducial area of 80 m^2 (defined normal to
the muon direction). Events with as many as 50 visible muons were
observed. Showers from different directions had to penetrate different
thicknesses of rock in the overlying mountain. The range covered
corresponded to muons with energies between ~ 1 and ~ 10 TeV, with the
best statistics at the lower end of this range. In one experiment (Lowe et al
1976) outrigger detectors were used to extend the range of observable
lateral separations between pairs of muons.

The Utah group carried out an extensive series of simulations (Bergeson
et al 1975 and references therein) to compare their observations with a
model based on hadronic scaling in the fragmentation region and with
rising cross sections, as shown in Figures 12a and 12b. The transverse
momentum distribution of the parent mesons was taken to have an x-
dependent mean (sea gull effect) in agreement with accelerator data;

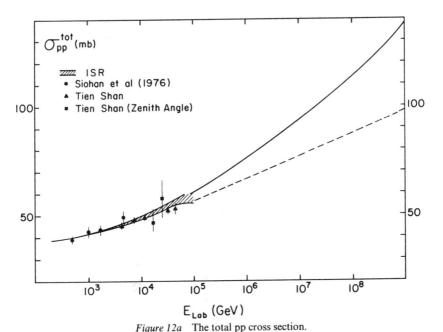

Figure 12a   The total pp cross section.

however, inclusive cross sections were scaled from ∼20-GeV/c hadron-nucleus data rather than from higher energy hadron-hadron data.

The most important result of the comparison was the conclusion that the p_T distribution of the parent mesons had an effective mean of about 600 MeV/c rather than 400 MeV/c as at accelerator energies. According to the simulations, typical interactions producing the observed muons had energies of only 10–20 TeV per nucleon (independent of primary mass), with members of pairs usually coming from different interactions in the cascade. A parametrization of p_T that was motivated by hard-scattering ideas (Halzen & Luthe 1974, Halzen 1975) was also used in the Utah simulation. The version with a p_T^{-4} component (proportional to A rather than $A^{2/3}$) superimposed on a low energy p_T distribution gave results close to but still not as broad as the observed lateral distribution. We note, moreover, that the parametrization is based on data near 90° in the center of mass, so its use in this context neglects possible x dependence of the hard-scattering component.

Rates of multiples to singles depend on assumed composition as well as on the lateral spread. It was found that either the overall intensity or the fraction of heavies in the primary beam had to be increased between 10^{14} and 10^{16} eV, as compared to an extrapolation of low energy spectra with all

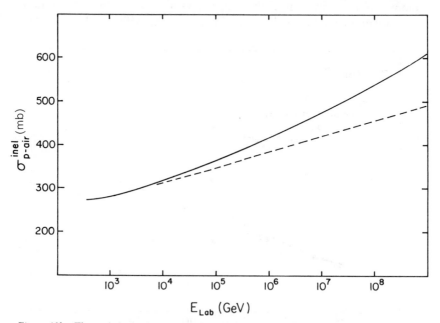

Figure 12b   The p-air inelastic cross section. Data and extrapolations are discussed in the text.

slopes equal to 2.7. The spectrum of Table 2 satisfies this requirement because of the increasing fraction of heavy nuclei. It would be interesting to see in detail whether or not this composition can account for the rates of multiples in this and other deep experiments, particularly the 200-m^2 detector at Homestake. It has a threshold for vertical muons of ~ 3 TeV and thus should be able to extend the range of the Utah detector somewhat. DUMAND could also explore the composition in this way, as discussed by Elbert (1978).

Increased sensitivity, particularly to composition around 10^{14}–10^{15} eV, could be obtained with a surface EAS array above a deep underground detector to tag the primary energy of the shower accompanying the muons. This has been done by the KGF group, but with a fairly small underground detector of 4-m^2 total area. They conclude from an analysis of their data that the lateral distribution of muons of energy greater than 220 GeV in 10^{14} eV showers cannot be understood by either pure iron or pure proton primaries (Acharya et al 1979b). The possibility of a surface array over the much larger and somewhat deeper detector at Homestake is being studied and appears feasible (Deakyne et al 1979).

The very deep detectors can also be sensitive to prompt muons, for example from production of heavy flavors (Gaisser et al 1979b). Observation of zenith-angle dependence at the Utah detector shows, however, that the contribution of prompt muons is less than 10% for $E_\mu \approx 1.5$ TeV, which corresponds roughly to $\sigma \cdot B_\mu < 0.05$ mb at 15 TeV. The Homestake detector, because of its greater depth, should be more sensitive to prompt muons.

8   ENERGY DEPENDENCE OF CROSS SECTIONS

8.1   Cross Section to 50 TeV

The basic cosmic ray technique for determination of cross section in the energy range up to 50 TeV is to measure the flux, j_u, of unaccompanied charged hadrons at an atmospheric depth x (typically a mountain altitude). Because of the contamination with unaccompanied charged secondaries, this flux is an upper bound to the flux, j_s, of surviving primary protons at depth x. The latter is related to the primary flux of protons, $j_p(E)$, and to the cross section by

$$j_u(E, x, A, E') \geqq j_s = e^{-x/\lambda(E)} j_p(E), \qquad\qquad 13.$$

where E is the measured hadron energy, A is the area of the anticoincidence array, and E' is the minimum energy needed to trigger it. The energy-dependent interaction length is related to the inelastic p-air cross section by Equation 3. Thus Equation 13 translates into a lower bound on $\sigma_{\text{p-air}}^{\text{inel}}$.

As the energy increases the efficiency of the anticoincidence array increases. It can be estimated experimentally from the measured flux of unaccompanied charged pions and protons (Siohan et al 1978) and directly by a numerical calculation of accompaniment (Gaisser & Yodh 1973). It is estimated that heavy primaries are unimportant in the energy range of interest because of the high probability of accompaniment and their small contribution to the all-nucleon spectrum of primary cosmic rays below 10^5 GeV per nucleon (MacKeown 1970). These corrections can be applied to the measured flux to improve the original lower bound (Yodh et al 1972) and obtain an estimate of $\sigma_{\text{p-air}}^{\text{inel}}$ (Gaisser & Yodh 1974, Gaisser et al 1977). Because of the logarithmic dependence of $\sigma_{\text{p-air}}^{\text{inel}}$ on j_p/j_s the results are rather insensitive to uncertainties in the primary and surviving fluxes. A 50% uncertainty in j_p/j_s gives rise to less than 10% uncertainty in the cross section for j_p/j_s of about 1000.

Glauber theory (Glauber 1970, Glauber & Matthiae 1970) has been used to derive σ_{pp}^{tot} from $\sigma_{\text{p-air}}^{\text{inel}}$. The results are shown in Figure 12a for the University of Maryland (Siohan et al 1978) and Tien Shan (Nam et al 1975) cosmic ray experiments. Also shown is the ISR measurement of σ_{pp}^{tot} up to 2000 TeV (solid line—Amaldi et al 1978) and the dispersion theory extrapolation to 100 TeV (shown as cross-hatched region). The latter is based on the measurement of the ratio of the real to imaginary of the forward amplitude (Amaldi et al 1977) and use of dispersion relations. The accelerator results confirm the early conclusion from cosmic ray data that σ_{pp}^{tot} continues to increase to 50 TeV, and they improve the precision of the measurements. It is therefore possible to invert the process described above and to infer the primary proton flux from the surviving flux. This exercise has been done by Gaisser et al (1977), who find a result consistent with the primary proton spectrum as stated in Table 2.

Also shown in Figure 12a are two extrapolations of σ_{pp}^{tot} based on fits (Bartel & Diddens 1974) to the ISR data. The solid line shows an extrapolation of the form $a + b(\ln E)^2$ and the dashed line a form $a + b(\ln E)$. (Here a also contains terms that vanish like inverse powers of energy for large E.)

8.2   Relation Between σ_{pp} and $\sigma_{\text{p-air}}$

In using the Glauber theory to relate $\sigma_{\text{p-air}}$ to σ_{pp}, it is important to take into account correctly the nearly elastic processes that contribute to

$$\sigma_{\text{p-air}}^{\text{abs}} = \sigma_{\text{p-air}}^{\text{tot}} - \sigma_{\text{p-air}}^{\text{el}}$$

but do not contribute to the $\sigma_{\text{p-air}}^{\text{inel}}$ that is measured in the cosmic ray experiments. There are two such processes: (a) quasi-elastic scattering, in which the target nucleus is excited but no pions are produced, and (b) diffraction dissociation, in which a target nucleon in the nucleus fragments but the leading nucleon retains nearly all the incident energy. In addition a small inelastic screening correction to the Glauber calculation must be included to account for the contribution of diffractive processes to elastic nucleon-nucleus scattering. These effects have been discussed by Gaisser et al (1975) and are included in the calculations of $\sigma_{p\text{-air}}^{inel}$ shown in Figure 12b. These corrections make $\sigma_{p\text{-air}}^{inel}$ smaller than $\sigma_{p\text{-air}}^{abs}$.

The relation between σ_{pp} and $\sigma_{p\text{-air}}$ depends both on the nuclear shape and on the profile of the nucleon, as well as on σ_{pp}. The nucleon profile is usually characterized by the slope parameter $B \equiv d/dt \, (\ln d\sigma/dt)|_{t=0}$, where t is the invariant momentum transfer in pp elastic scattering. In specific models (e.g. Cheng et al 1973, Ball & Zachariasen 1975) there is a self-consistent relation between the energy dependence of σ_{pp}^{tot} and that of B. In deriving the results shown in Figure 12b we have assumed the phenomenological relation

$$B/\sigma = \text{const} \cong 0.296 \text{ GeV}^{-2} \text{ mb}^{-1},$$

observed to hold for protons over the accelerator energy region (Barger 1974). We have taken the effective rms nuclear radius of air to be 2.6 fm.

8.3   Cross Section at Air Shower Energies

Existing determinations of the cross section in the air shower energy range fall into two categories: (a) an attempt to reconstruct the distribution of starting points of air showers, and hence $\lambda(E)$, from the zenith-angle dependence of measured μ/e ratio in showers at ground level, and (b) indirect determinations from comparison of Monte Carlo calculations of showers to many features of the data. The μ/e method depends on the assumption that the muon number determines the primary energy and that the electron number determines the depth of shower initiation with negligible fluctuations. If these assumptions were valid then the interaction length λ could be determined from

$$R(N_\mu, N_e, \theta) = R(N_\mu, N_e, 0) \exp\left[-\frac{y(\sec\theta - 1)}{\lambda}\right],$$

where R is the rate of showers of fixed N_μ and N_e and θ is the zenith angle. In fact, however, there are large θ-dependent fluctuations in the determination of N_μ and in the relation between N_μ and E_0. Barger et al (1974) concluded that these fluctuations make attempts (Ganguli & Subramanian 1974) to determine $\sigma_{p\text{-air}}$ from existing data (Fukui et al 1960) very unreliable. Tonwar (1979) has reviewed the situation with more recent data and reached the same conclusion.

In the indirect approach, the energy dependence of the assumed cross section is only one of several features of the input model of hadronic interactions to which the calculated properties of air showers are sensitive. The general conclusion is, however, that if hadronic scaling remains valid to $\sim 10^{17}$ eV then the cross section must continue to increase at least as fast as shown in Figure 12 in order to prevent calculated showers from penetrating further into the atmosphere than is observed (Ouldridge & Hillas 1978). This conclusion also depends on the primary composition, as discussed in Section 7 above, but since a considerable fraction of heavy nuclei is also required the conclusion stands.

Hadronic interactions could be considerably different at 10^{17}–10^{18} eV than the extrapolation of accelerator data based on hadronic scaling. A direct determination of the cross section in this energy region ($s^{1/2} \approx 30{,}000$ GeV) is therefore of great interest. It is one of the fundamental goals of the Fly's Eye experiment. Cassiday et al (1977, 1979) have described the technique for reconstructing the distribution of starting points of showers from the data, taking into account effects of heavy primaries and the very early shower development when scintillation light reaches a detectable level.

The basic method consists of fitting a distribution of starting points of the form

$$\sum_i w_i \frac{e^{-x/\lambda_i}}{\lambda_i}$$

to the observed distribution. Here the sum extends over the several primary components (as in Table 2), and w_i and λ_i are respectively the relative weight and interaction length of the ith nuclear group. A similar analysis could presumably also be applied to the air Cherenkov experiment (Orford & Turver 1976). Cassiday et al estimate that with Fly's Eye it will be possible to determine λ_P to $\pm 10\%$ for $w_P > 0.2$. This estimate was, however, obtained by applying the fitting procedure to sets of showers with $\lambda_P = 75$ g cm^{-2}, corresponding to $\sigma_{\text{p-air}}^{\text{inel}} = 320$ mb. If the cross section continues to increase to the Fly's Eye energy range ($\gtrsim 10^{17}$ eV) as shown in Figure 12, $\lambda_{\text{p-air}}$ will be ~ 45 g cm^{-2}, corresponding $\sigma_{\text{p-air}}^{\text{inel}} = 530$ mb. This is comparable to the low energy value for $\lambda_{\alpha\text{-air}}$, and may make the problem of disentangling cross section and primary composition considerably more difficult.

It is an unfortunate fact (Barger et al 1974) that, because of nuclear shadowing effects, $\sigma_{\text{p-air}}^{\text{inel}}$ changes less rapidly than the underlying σ_{PP}. This is apparent in Figure 12, which shows that a three-fold increase in $\sigma_{\text{PP}}^{\text{tot}}$ between 10^3 and 3×10^8 GeV results in a doubling of $\sigma_{\text{p-air}}^{\text{inel}}$.

Based on the discussion above, we conclude that the Fly's Eye experiment will most probably be able to distinguish between a cross section that remains constant at its value around 10^5 GeV ($\lambda_p \approx 65 \, \mathrm{g \, cm^{-2}}$) and an increasing cross section which leads to $\lambda_p \approx 40$–$50 \, \mathrm{g \, cm^{-2}}$ at 3×10^8 GeV. Distinguishing among different models of increasing cross section, however, requires a sensitivity of better than $\pm 10\%$ around 3×10^8 GeV (Barger et al 1974, see Figure 12b). This will present a significant challenge to the new air shower experiments.

9 CONCLUSION AND OUTLOOK

From the study of cosmic ray components in the atmosphere and of individual hadronic interactions in balloon-borne and ground-based detectors, we can infer several properties for hadronic collisions between 10 and about 200 TeV in the lab or between 140 and 650 GeV in the center of mass.

1. Hadron-nuclear cross sections continue to increase with energy up to 1000 TeV and possibly to the highest air shower energies. The increase of proton-proton cross section that is required is such that its values reach 70–80 mb at 10^3 TeV and probably greater than 100 mb by 10^6 TeV.
2. No clear evidence exists for violation of scaling in the fragmentation region ($x \gtrsim 0.05$) at least up to about 500 TeV.
3. Multiplicity, however, must increase more rapidly than for Feynman scaling to describe the increase in rapidity density of C-jets observed at Chacaltaya.
4. The high energy events, C-jets, require the increase of the large transverse momentum component due to hard collisions in order to account for the $E_\gamma \cdot r$ and M_γ distributions. The increase needed is consistent with current models of hard collisions.
5. Cross sections for production of charm (and possibly other heavy flavors) in cosmic ray interactions in the 10 to 100 TeV range must be of the order of one millibarn.

These inferences are relatively insensitive to elemental composition of primary cosmic rays, being more directly related to the incoming all-nucleon flux (except for air showers). The analysis depends, however, on details of particle production in hadron-nucleus collisions in the large x region, and it could be placed on much firmer basis if accurate data from accelerator experiments with nuclear targets were available.

The only information about weak interactions comes from deep underground studies. The anomalous cascades, multitrack events, and double-core events seen cannot be explained in terms of conventional weak interaction ideas and may imply existence of new particles and/or new processes. Experiments now underway at Baksan and Homestake, the proton decay detectors under construction, and new KGF experiments should provide additional observations with better and bigger detectors. Understanding these events certainly should be given very high priority.

At still higher energies, above 500 TeV, the Centauro events of the Japan-Brazil group, which are characterized by the deficiency of electromagnetic jets, defy conventional explanations. They may indicate new hadronic states of matter, either produced in high energy collisions or remaining as relics of the big bang. If they are produced in high energy interactions with sufficient cross sections at collider energies, one can expect to see an exciting new regime of elementary particle interactions in the near future. In any case, completing the task of comparing and understanding the Pamir, Chacaltaya, and Fuji data on A-jets with energies above 100 TeV is of the utmost importance.

New techniques in studying longitudinal development of giant air showers, as well as highly instrumented air shower arrays, promise a detailed look at the early development and particle distributions in individual events. This has the potential of simultaneously unraveling the nature of the shower initiator and behavior of highest energy interactions. The quality of the understanding will depend on the ability to simulate such high energy cascades. This in turn will depend on knowledge of particle production in the large x region at the highest energy colliders on which to base extrapolation to higher energies. Information on σ_{pp}^{tot} at $s^{1/2} = 540$ GeV should be forthcoming soon from the CERN $\bar{p}p$ collider, but detailed information on the fragmentation region may have to wait until the pp collider, ISABELLE, comes into operation. It would be very desirable to instrument the $\bar{p}p$ colliders to explore this region also.

If Fe is a dominant component at 10^6 GeV, as discussed in this report, then we will have an opportunity to study nucleus-nucleus collisions at 1 TeV per nucleon both directly with balloon-borne detectors and through observations of EAS at still higher energies.

Direct studies of hadronic interactions at energies equivalent to $E_{Lab} \approx 10^6$ GeV at the new colliders can potentially contribute to our understanding of the primary composition in this energy range, which is accessible only to indirect observation through atmospheric cascades. This is of fundamental importance to cosmic ray astrophysics because it is a region where there is known structure in the spectrum, and it is near the high energy limit of typical pulsar acceleration mechanisms. Here again it will be desirable to have data from the fragmentation region in addition to the central region from the new colliders, as well as detailed understanding of particle production on nuclear targets from the existing fixed-target machines. Studies of fragmentation and pion production in nucleus-

nucleus collisions will also be required for interpretation of air showers in view of the importance of heavy nuclei among the primaries. These studies will have to come from emulsions exposed to the cosmic ray beam and from high energy, heavy ion accelerators. This requires high statistics at energies well above meson production threshold.

ACKNOWLEDGMENTS

We would like to thank all of our cosmic ray colleagues for discussions over the past several years, without which we would have been unable to prepare this review. A complete list of those who have contributed to our understanding of the subject would include a large fraction of the reference list. In the actual preparation of the manuscript we have benefited from the critical advice of R. W. Ellsworth, K. E. Turver, Y. Fujimoto, T. Stanev, J. W. Elbert, A. A. Watson, and J. Linsley about various aspects of interpreting the data discussed here.

Literature Cited[a]

Acharya, B. S., et al. 1979a. *Proc. 16th ICRC, Kyoto, 1979* 6:289
Acharya, B. S., et al. 1979b. *Proc. 16th ICRC, Kyoto, 1979* 8:304, 272
Adair, R. K., et al. 1974. *Phys. Rev. Lett.* 33:115
Adair, R. K., et al. 1977. *Phys. Rev. Lett.* 39:112
Aguirre, C., et al. 1979a. *Bartol Conf.*, 1978, p. 13
Aguirre, C., et al. 1979b. *J. Phys. G* 5:139
Akashi, M., et al. 1964. *Suppl. Prog. Theor. Phys.* 32:1
Akashi, M., et al. 1978. *J. Phys. G* 4:L41
Akashi, M., et al. (Mt. Fuji, E.C. Group). 1979a. *Bartol Conf.*, 1978, p. 334
Akashi, M., et al. 1979b. *Proc. 16th ICRC, Kyoto, 1979* 13:98
Akashi, M., et al. 1979c. *Proc. 16th ICRC, Kyoto, 1979* 13:87
Albini, E., et al. 1976. *Nuovo Cimento A* 32:101
Alexeyev, E. N., et al. 1979. *Proc. 16th ICRC, Kyoto, 1979* 10:12
Allkofer, O. C. 1979. *Proc. 16th ICRC, Kyoto 1979* 14:385
Allkofer, O. C., et al. 1979. *Proc. 16th ICRC, Kyoto, 1979* 10:411
Amaldi, U., et al. 1977. *Phys. Lett. B* 66:390

Amaldi, U., et al. 1978. *Nucl. Phys. B* 145:367
Amineva, T. P., et al. 1973. *Proc. 13th ICRC, Denver, 1973* 3:1788
Anisovich, V. V., Shekhter, V. M. 1978. *Sov. J. Nucl. Phys.* 28:554
Aseikin, V. S., et al. 1975. *Proc. 14th ICRC, Munich* 7:2462
Astafiev, V. A., Mukchamedshin, R. A. 1979. *Proc. 16th ICRC, Kyoto 1979* 7:204
Balasubrahmanyan, V. K., Ormes, J. F. 1973. *Astrophys. J.* 186:109
Ball, J. S., Zachariasen, F. 1975. *Nucl. Phys. B* 85:317
Barger, V. 1974. *Proc. 17th Int. Conf. High Energy Physics, London, 1974* I:193
Barger, V., et al. 1974. *Phys. Rev. Lett.* 33:1051
Bartel, W., Diddens, A. N. 1974. As quoted by A. N. Diddens, *Proc. 17th Int. Conf. High Energy Physics, London, 1974* I:41
Bayburina, S. G., et al. 1977. *Proc. 15th ICRC, Plovdiv, 1977* 11:459
Bayburina, S. G., et al. (Pamir collaboration) 1979a. *Proc. 16th ICRC, Kyoto, 1979* 7:75
Bayburina, S. G., et al. 1979b. *Proc. 16th ICRC, Kyoto, 1979* 7:241
Bellandi Filho, J., et al. (Brazil-Japan collaboration) 1979. *Bartol. Conf., 1978*, pp. 94, 145, 317

[a] We have adopted the following abbreviations for several frequently cited references: "ICRC" means International Cosmic Ray Conference; "Bartol Conf." refers to *Cosmic Rays and Particle Physics—1978* (Bartol Conference), ed. T. K. Gaisser, New York: AIP, published 1979; "Utah Workshop" refers to *Proceedings of the Air Shower Workshop, University of Utah, May 1979*, ed. T. K. Gaisser, Newark, Del: Bartol Res. Found., published 1979.

Benecke, J., et al. 1969. *Phys. Rev.* 188:2159

Benvenuti, A., et al. 1975. *Phys. Rev. Lett.* 35:1486

Bergeson, H. E., et al. 1975. *Phys. Rev. Lett.* 35:1681

Bergeson, H. E., et al. 1977. *Phys. Rev. Lett.* 39:847

Bhat, P. N., et al. 1979. *Proc. 16th ICRC, Kyoto, 1979* 13:8

Bjorken, J. D., McLerran, L. D. 1979. *Phys. Rev. D* 20:2353

Bowen, T., Jones, J. J., Jones, W. V. 1979. *Proc. Utah Workshop, 1979*, p. 95

Budilov, V. K., et al. (Pamir Emulsion Chamber Collaboration). 1977. *Proc. Pamir Collaboration Meet., Lodz, 1976.* In *Zesz. Nauk. Uniw. Lodz. Ser. 2* 60:7 (In English)

Caldwell, J. H. 1977. *Astrophys. J.* 218:269

Capdevielle, J. N., et al. 1979. *Proc. 16th ICRC, Kyoto, 1979* 6:324

Cassiday, G. L., et al. 1977. *Proc. 15th ICRC, Plovdiv, 1979* 8:270

Cassiday, G. L., et al. 1979. *Bartol Conf., 1978*, p. 417

Chatterjee, B. K., et al. 1965. *Proc. 9th ICRC, London, 1965* 2:808

Chaudhary, B. S., Malhotra, P. K. 1975. *Nucl. Phy. B* 86:360

Cheng, H., et al. 1973. *Phys. Lett. B* 44:97

Chin, S. A., Kerman, A. K. 1979. *Phys. Rev. Lett.* 43:1292

Chudakov, A. E., et al. 1962. *Proc. 5th Interamerican Seminar on Cosmic Rays, La Paz, Bolivia, 1962* 2:XLIX–1

Cline, D. B. 1979a. *Proc. Bartol Conf., 1978*, p. 451

Cline, D. B. 1979b. *Proc. 16th ICRC, Kyoto, 1979* 14:271

Cline, D. B., Rubbia, C. 1979. *Proc. 16th ICRC, Kyoto, 1979* 13:119

Combridge, B. L., et al. 1977. *Phys. Lett. B* 70:234

Craig, M. A. B., et al. 1979. *Proc. 16th ICRC, Kyoto, 1979* 8:180

Crouch, M. F., et al. 1978. *Phys. Rev. D* 18:2239

Cutler, R., Sivers, D. 1978. *Phys. Rev. D* 17:196

Cutts, D., et al. 1978. *Phys. Rev. Lett.* 40:141

Damgard, G., et al. 1965. *Phys. Lett.* 17:152

Das, K. P., Hwa, R. 1977. *Phys. Lett. B* 68:459

Deakyne, M., et al. 1978. *Proc. Neutrino—'78 Conf., Purdue, 1978*, p. 887

Deakyne, M., et al. 1979. *Proc. Bartol Conf., 1978*, p. 409

Duke, D. W., Taylor, F. E. 1978. *Phys. Rev. D* 17:1788

Dunaevskii, A. M., Slavatinsky, S. A. 1977. *Proc. 15th ICRC, Plovdiv, 1977* 7:349

Dunaevskii, A. M., Slavatinsky, S. A. 1979. *Proc. 16th ICRC, Kyoto, 1979* 7:87

Dunaevskii, A. M., et al. 1979. *Proc. 16th ICRC, Kyoto, 1979* 7:154

Edge, D. M., et al. 1978. *J. Phys. G* 4:133

Eisenberg, Y., et al. 1978. *Nucl. Phys. B* 135:189

Elbert, J. W., et al. 1976. *J. Phys. G* 2:971

Elbert, J. W. 1978. *Proc. DUMAND Summer Workshop, La Jolla, 1978*, ed. A. Roberts. 2:101

Ellsworth, R. W., et al. 1977. *Astrophys. Space Sci.* 52:415

Ellsworth, R. W. 1979. *Proc. 16th ICRC, Kyoto, 1979* 7:333

Ellsworth, R. W., et al. 1979a. *Proc. Bartol. Conf., 1978*, p. 111

Ellsworth, R. W., et al. 1979b. *Proc. 16th ICRC, Kyoto, 1979* 7:55

Ellsworth, R. W., et al. 1980. In *Proc. XX Int. Conf. High Energy Phys., Madison, 1980*

England, C. D., et al. 1979. *Proc. 16th ICRC, Kyoto, 1979* 8:88

Faissner, H., et al. 1976. *Phys. Lett. B* 60:401

Fermi, E. 1951. *Phys. Rev.* 81:683

Feynman, R. P. 1969. *Phys. Rev. Lett.* 23:1415

Feynman, R. P., et al. 1978. *Phys. Rev. D* 18:3320

Fomin, Yu. A., Khristiansen, G. B. 1971. *Sov. J. Nucl. Phys.* 14:360 (642).

Fuchi, H., et al. 1979a. *Proc. Bartol Conf., 1978*, p. 49

Fuchi, H., et al. 1979b. *Proc. 16th ICRC, Kyoto, 1979* 6:235

Fuchi, H., et al. 1979c. *Proc. 16th ICRC, Kyoto, 1979* 6:112

Fuchi, H., et al. 1979d. *Nuovo Cimento A* 51:18; *Phys. Lett. B* 85:135

Fujimoto, Y. 1979. *Proc. 16th ICRC, Kyoto, 1979* 14:308

Fukui, S., et al. 1960. *Suppl. Prog. Theor. Phys.* 16:1

Fumuro, F., et al. 1977. *Proc. 15th ICRC, Plovdiv, 1977* 7:59

Gaisser, T. K., Yodh, G. B. 1973. *Proc. 13th ICRC, Denver, 1973* 3:2140

Gaisser, T. K., Yodh, G. B. 1974. *Nucl. Phys. B* 76:182

Gaisser, T. K., et al. 1974. *Phys. Lett. B* 51:83

Gaisser, T. K., et al. 1975. *Proc. 14th ICRC, Munich, 1975* 7:2161

Gaisser, T. K., Halzen, F. 1976. *Phys. Rev. D* 14:3153

Gaisser, T. K. 1977. *Proc. 15th ICRC, Plovdiv, 1977* 10:267

Gaisser, T. K., et al. 1977. *J. Phys. G* 3:L241

Gaisser, T. K., Sidhu, D. P. 1977. *Proc. 15th ICRC, Plovdiv, 1977* 7:18

Gaisser, T. K., et al. 1978. *Rev. Mod. Phys.* 50:859

Gaisser, T. K. 1979. *Proc. Utah Workshop, 1979*, p. 57

Gaisser, T. K., et al. 1979a. *Proc. 16th ICRC, Kyoto, 1979* 9:275

Gaisser, T. K., et al. 1979b. *Proc. Workshop on the Production of New Particles in Super-High Energy Collisions, Madison, 1979*, ed. V. Barger, F. Halzen. Madison: Univ. Wisconsin

Ganguli, S. N., Subramanian, A. 1974. *Nuovo Cimento Lett.* 10:235

Giacomelli, G. 1978. *Proc. 19th Int. Conf. High Energy Physics, Tokyo, 1978*, ed. S. Homma, M. Kawaguchi, H. Miyazawa. Phys. Soc. Jpn.

Glashow, S. L., et al. 1970. *Phys. Rev. D* 2:1285

Glauber, R. J. 1970. *High Energy Physics and Nuclear Structure*, ed. S. Devons. New York: Plenum

Glauber, R. J., Matthiae, G. 1970. *Nucl. Phys. B* 21:135

Goodman, J. A., et al. 1979a. *Phys. Rev. D* 19:2752; also in *Proc. Bartol. Conf., 1978*, p. 207

Goodman, J. A., et al. 1979b. *Phys. Rev. Lett.* 42:854

Goodman, J. A., et al. 1979c. *Proc. Utah Workshop, 1979*, p. 29

Green, B. R., et al. 1979. *Proc. 16th ICRC, Kyoto* 13:211

Greisen, K. 1956. *Prog. Cosmic Ray Phys.* 3:3

Greisen, K. 1960. *Ann. Rev. Nucl. Sci.* 10:63

Greisen, K. 1965. *Proc. 9th ICRC, London* 2:609; see also *Proc. 5th Interamerican Seminar on Cosmic Rays, La Paz, Bolivia, 1962* 1:I–5

Grieder, P. K. 1977. *Rev. Nuovo Cimento* 7:1

Grigorov, N. L., et al. 1971. *Proc. 12th ICRC, Hobart, 1971* 5:1746

Hahn, H., et al. 1977. *Rev. Mod. Phys.* 49:625

Halzen, F. H., Luthe, J. 1974. *Phys. Lett. B* 48:440

Halzen, F. 1975. *Nucl. Phys. B* 92:404

Halzen, F. 1977. *Phys. Rev. D* 15:1929

Hammond, R. T., et al. 1977. *Proc. 15th ICRC, Plovdiv, 1977* 8:287

Hammond, R. T., et al. 1978. *Nuovo Cimento C* 1:315

Hayakawa, S. 1969. *Cosmic Ray Physics*, p. 313 ff. New York: Wiley-Interscience

Hayashi, T., et al. 1972. *Prog. Theor. Phys.* 47:280, 1988

Hazen, W. E., et al. 1979. *Proc. Bartol Conf., 1978*, p. 31

Heitler, W. 1944. *The Quantum Theory of Radiation*, pp. 232–40, London: Oxford Univ. Press. 272 pp. 2nd ed.

Heitler, W., Janossy, L. 1949. *Proc. Phys. Soc. London Sect. A* 62:374, 669

Hillas, A. M. 1975. *Phys. Rep.* 20:59

Hillas, A. M. 1979a. *Proc. Bartol. Conf., 1978*, p. 373; *Proc. 16th ICRC, Kyoto, 1979* 8:7

Hillas, A. M. 1979b. *Proc. 16th ICRC, Kyoto, 1979* 6:13

Hillas, A. M. 1979c. *Proc. 16th ICRC, Kyoto* 9:13

Hoffman, H. J. 1975. *Phys. Rev. D* 12:82

Hoshino, K., et al. 1975. *Proc. 14th ICRC, Munich, 1975* 7:2442

Isgur, N., Wolfram, S. 1979. *Phys. Rev. D* 19:234

Jelley, J. V., Galbraith, W. 1953. *Philos. Mag.* 44:619

Johnson, J. R., et al. 1978. *Phys. Rev. D* 17:1978

Jokisch, H., et al. 1979. *Phys. Rev. D* 19:1368

Jones, L. W. 1977. *Rev. Mod. Phys.* 49:717

Jones, L. W. 1979a. *Proc. Utah Workshop, 1979*, p. 23

Jones, L. W. 1979b. *Proc. Utah Workshop, 1979*, p. 85

Juliusson, E. 1974. *Astrophys. J.* 191:331

Kalmykov, N. N., et al. 1979. *Proc. 16th ICRC, Kyoto* 9:73

Kamata, K. 1979. *Proc. Bartol Conf., 1978*, p. 443

Kasahara, K., et al. 1979. *Proc. 16th ICRC, Kyoto, 1979* 13:70, 76

Khristiansen, G. B. 1979. *Proc. 16th ICRC, Kyoto, 1979* 14:360

Kirkby, J. 1979. *Proc. Int. Symp. on Lepton and Photon Interactions at High Energies*, Fermilab, Batavia, Aug. 1979

Konishi, E., et al. 1978. *Nuovo Cimento A* 44:509

Krishnaswamy, M. R., et al. 1971. *Proc. R. Soc. London Ser. A* 323:489

Krishnaswamy, M. R., et al. 1976. *Proc. Int. Neutrino Conf., Aachen, 1976*, ed. H. Faissner et al., p. 197

Khrishnaswamy, M. R., et al. 1977. *Pramana* 5:59

Krishnaswamy, M. R., et al. 1979a. *Proc. 16th ICRC, Kyoto, 1979* 13:24

Krishnaswamy, M. R., et al. 1979b. *Proc. 16th ICRC, Kyoto, 1979* 13:14

Krishnaswamy, M. R., et al. 1979c. *Proc. 16th ICRC, Kyoto, 1979* 6:128

Krys, A., et al. 1979. *Proc. 16th ICRC, Kyoto, 1979* 7:182, 188; *Zesz. Nauk. Uniw. Lodz., Ser. 2* 32:5

Landau, L. D. 1953. *Izv. Akad. Nauk SSSR* 17:51 (Transl. in *Collected Papers of L. D. Landau*, ed. D. Ter Haar, 1965. New York: Gordon & Breach)

Landau, L. D., Pomeranchuk, I. 1953. *Dokl. Akad. Nauk SSSR* 92:535, 735.

Lapikens, J., et al. 1979. *Proc. 16th ICRC, Kyoto, 1979* 8:95

LaPointe, M. 1970. *Introduction to Experimental Techniques of High Energy Astrophysics*, ed. H. Ögelman, J. R. Wayland, p. 177. Wash. DC: *NASA SP-243*

Lattes, C. M. G., et al. (Brazil-Japan Collaboration). 1971. *Suppl. Prog. Theor. Phys.* 47:1

Lattes, C. M. G., et al. (Japan-Brazil Collaboration). 1973. *Proc. 13th ICRC, Denver 1973* 3:2227; 4:2671

Lattes, C. M. G., et al. (Brazil-Japan Collaboration). 1974. *CKJ Rep.—13, Cosmic Ray Lab., Univ. Tokyo,* June 5, 1974. (Unpublished)

Li, L.-F., Paschos, E. A. 1971. *Phys. Rev. D* 3:1178

Linsley, J. 1977. *Proc. 15th ICRC, Plovdiv, 1977* 12:89

Linsley, J. 1979. *Proc. Utah Workshop, 1979,* p. 81

Lloyd-Evans, J., et al. 1979. *Proc. 16th ICRC, Kyoto* 13:130

Lowe, G. H., et al. 1975. *Phys. Rev. D* 12:651

Lowe, G. H., et al. 1976. *Phys. Rev. D* 13:2925

McCusker, C. B. A. 1975. *Phys. Rep.* 20C:229

MacKeown, P. K. 1970. *Proc. 6th Interamerican Seminar on Cosmic Rays, La Paz, Bolivia* 3:684

Mann, A. K., Primakoff, H. 1980. *Phys. Rev.* In press

Migdal, A. B. 1956. *Phys. Rev.* 103:1811

Muraki, Y., et al. 1979. *Phys. Rev. Lett.* 43:974

Nam, R., et al. 1975. *Proc. 14th ICRC, Munich, 1975* 7:2258

Nishimura, J. 1964. *Suppl. Prog. Theor. Phys.* 32:72

Nishimura, J. 1967, *Handbuch der Physik,* ed. S. Flügge, Vol. 66/2:1–114. Berlin, Heidelberg, New York: Springer

Niu, K. 1979. *Proc. Bartol. Conf., 1978,* p. 181

Niu, K., et al. 1971. *Prog. Theor. Phys.* 46:1644

Ochs, W. 1977. *Nucl. Phys. B* 118:397

Ogata, T., et al. 1979. *Proc. 16th ICRC, Kyoto, 1979* 6:100

Ohta, I. 1971. *Suppl. Prog. Theor. Phys.* 47:271

Orford, K., Turver, K. E. 1976. *Nature* 264:727

Orford, K., Turver, K. E. 1980. *Phys. Rev. Lett.* 44:959

Ormes, J. F., Freier, P. 1978. *Astrophys. J.* 222:471

Orth, D. D., et al. 1978. *Astrophys. J.* 226:1147

Ouldridge, M., Hillas, A. M. 1978. *J. Phys. G* 4:L35

Peters, B. 1961. *Nuovo Cimento* 22:800

Prentice, J. 1979. *Proc. Int. Symp. on Lepton and Photon Interactions at High Energies,* Fermilab, Batavia, Aug. 1979

Protheroe, R. J., Turver, K. E. 1979. *Nuovo Cimento A* 51:277

Quigg, C. 1977. *Revs. Mod. Phys.* 49:297

Ramana Murthy, P. V. 1977. *Rep. No. CRL 51-77-10, Cosmic Ray Lab., Univ. Tokyo.* Unpublished

Ren, J. R., et al. 1979. *Proc. 16th ICRC, Kyoto, 1979* 7:273

Ringland, G., Wachsmuth, H. 1980. *Univ. Wisc. Preprint C00-088-116*

Rosner, J. 1979. *Proc. Bartol. Conf., 1978,* p. 297

Rossi, B., Greisen, K. 1941. *Rev. Mod. Phys.* 13:240

Ryan, M. J., et al. 1972. *Phys. Rev. Lett.* 28:985, E1497

Sakata, M., et al. 1979. *Proc. Bartol. Conf., 1979,* pp. 18, 23, 29

Sarma, K. V. L., Wolfenstein, L. 1976. *Phys. Lett. B* 61:77

Sato, Y., Sugimoto, H., Saito, T. 1976. *Phys. Soc. Jpn.* 41:1821

Sawayanagi, K. 1979. *Phys. Rev. D* 20:1037

Schmidt, W. K. H., et al. 1976. *Astron. Astrophys.* 46:49

Shibata, M. 1979. *Proc. 16th ICRC, Kyoto, 1979* 7:176

Shibata, T. 1979. *Proc. 16th ICRC, Kyoto, 1979* 7:346

Simon, M., et al. 1980. *Astrophys J.* In press

Siohan, F., et al. 1978. *J. Phys. G* 4:1169

Suga, K. 1962. *Proc. 5th Interamerican Seminar on Cosmic Rays, La Paz, Bolivia, 1962* 2:XLIX–1

Sugimoto, H., Sato, Y., Saito, T. 1975. *Prog. Theor. Phys.* 53:1541

Sutherland, D. 1979. *Proc. Bartol Conf., 1978,* p. 503

Tamada, M. 1977. *Nuovo Cimento B* 41:245

Tamada, M. 1979. *Proc. 16th ICRC, Kyoto, 1979* 6:295

Taylor, F. E., et al. 1976. *Phys. Rev. D* 14:1217

Thomé, W. et al. 1977. *Nucl. Phys. B* 129:365

Thornton, G., Clay, R. 1979. *Phys. Rev. Lett.* 43:1622

Tonwar, S. C. 1979. *J. Phys. G.* 5:L193

Van Hove, L., Pokorski, S. 1975. *Nucl. Phys. B* 86:243

Watson, A. A. 1979. *Proc. Utah Workshop, 1979,* p. 1

Wdowczyk, W., Wolfendale, A. W. 1979. *Proc. 16th ICRC, Kyoto, 1979* 6:8

White, D. H. 1979. *Proc. Bartol Conf., 1978,* p. 486

Wrotniak, J. A. 1977. *Zesz. Nauk. Uniw. Lodz* 60:165

Yakovlev, V. I., et al. 1979. *Proc. 16th ICRC, Kyoto, 1979* 6:59

Yen, E. 1974. *Phys. Rev. D* 10:836

Yodh, G. B. 1977. *Proc. Conf. on Prospects for Strong Interaction Physics at ISABELLE,* Brookhaven, Apr. 1977, *BNL* 50701:109

Yodh, G. B., Pal, Y., Trefil, J. S. 1972. *Phys. Rev. Lett.* 28:1005

Zhdanov, G. B., et al. 1975. *Preprint No. 163.* Moscow: FIAN (In Russian)

Ann. Rev. Nucl. Part. Sci. 1980. 30: 543–81

POSITRONIUM

※5624

Stephan Berko and Hugh N. Pendleton
Department of Physics, Brandeis University, Waltham, Massachusetts 02254

CONTENTS

1 INTRODUCTION

Positronium is the quasi-stable bound system of an electron and its antiparticle, the positron. Since its discovery by Deutsch (1951), positronium—now denoted Ps—has been the subject of many experimental and theoretical investigations. In 1954 DeBenedetti & Corben reviewed the fundamental aspects of the formation, structure, and annihi-

0163-8998/80/1201-0543$01.00

lation of Ps in one of the first volumes of this review series. Since then our knowledge of Ps physics and of slow positron (e^+) physics in general has broadened and deepened, leading to a host of applications in atomic physics, in chemistry, and in solid state physics. The most fundamental aspect of Ps physics remains, however, unchanged: as the simplest purely leptonic electromagnetically bound state, Ps, together with muonium, forms an ideal system to study the bound-state aspects of quantum electrodynamics (QED). During the last few years interest in such QED tests has increased because of new developments in experimental technique as well as in quantum field theory.

The Ps parameters of interest for QED tests are the energy levels and the annihilation rates of the low-lying states. As pioneered by Deutsch, the classical Ps production technique is the use of a gas that serves as an e^+ energy moderator as well as a source of electrons for Ps formation. The energy separation between the spin singlet (para-) and spin triplet (ortho-) levels of the Ps ground state ($n = 1$) has been measured to ever increasing precision using Ps formation in a gas (Mills & Bearman 1975, Egan et al 1975). Owing to the more recent development of slow positron beams of variable energy (see the review by Griffith & Heyland 1978), a new method of forming Ps in vacuum was discovered at Brandeis University (Canter et al 1974). Through the use of this technique the Lyman-α photons of Ps were directly observed (Canter et al 1975), thus ending a twenty-five-year search for the $n = 2$ states of Ps. The same method led to a fine-structure measurement within the $n = 2$ Ps states (Mills et al 1975), paralleling the Lamb shift measurement in hydrogen some thirty years earlier. Using these new techniques researchers at the University of Michigan (Gidley et al 1976a,b) obtained high precision ortho-Ps ($n = 1$) annihilation rate values, which then led to a theoretical reevaluation of this rate (Caswell et al 1977).

New developments in the theory of the bound-state problem in QED have been inspired by the challenge of the problem itself, as well as by the resurgence of interest in relativistic quantum field theory, signaled by the emergence of the Weinberg-Salam (WS) theory unifying the electromagnetic and weak interactions and by the development of quantum chromodynamics (QCD) for strong interactions. (For a review of both WS and QCD, see Appelquist et al 1978). The discovery of charmonium (presumably the bound state of a charmed and anticharmed quark) and of the upsilon particles (perhaps bound states of a quark even heavier than the charmed quark, and its antiquark) present theorists with the possibility of "quark analogs" of positronium.

In their 1954 review DeBenedetti & Corben wrote, "In these days when the attention of the physicist is concentrated on the study of queer particles and of interactions which escape our comprehension, it is somewhat refreshing to turn our minds toward a subject which does not lack fundamental importance and about which we can claim a certain amount of understanding." It is interesting to note that developments in the study of the new "queer particles" (quarks and gluons) and of unanticipated interactions (WS and QCD) may ultimately lead to a close connection with the subject of "fundamental importance" we treat again in this review. In 1980 it is impossible to present a brief review of all aspects of Ps physics, as was done in 1954; in our review we shall concentrate on the recent developments mentioned in this introduction, with emphasis on the fundamental QED aspects of Ps physics. For other details the reader is referred to the several earlier Ps reviews such as those of Deutsch & Berko (1965), Hughes (1972), Stroscio (1975), and Berko et al (1979), and to the recent general reviews of QED by Lautrup et al (1972), Kinoshita (1978a), and Drell (1979).

2   POSITRONIUM THEORY

2.1   *Early Development: Annihilation Selection Rule*

The first theoretical discussion of positronium (Ps) is found in the work of Pirenne (1944, 1946, 1947) and Wheeler (1946), who set the stage for the many subsequent Ps studies regarding structure, means of formation, and modes of decay. Fundamental to the understanding of Ps physics is the selection rule governing the e^+e^- annihilation process (Yang 1949, Wolfenstein & Ravenhall 1952). Conservation of energy and momentum forbids the single-photon (1γ) annihilation of free Ps. The general selection rule for the annihilation of Ps from a state of orbital angular momentum l and total spin s into n photons is given by

$$(-1)^{l+s} = (-1)^n.$$

This follows from the charge conjugation properties of Ps and of n-photon states: each photon contributes a factor of (-1), whereas in Ps the electron and positron are interchanged, yielding a factor of $(-1)^{l+s}$, since they have opposite intrinsic parity. One concludes that spin singlet S states ($l = 0$, $s = 0$) annihilate into an even number of photons only, while spin triplet S states ($l = 0$, $s = 1$) annihilate into an odd number greater than one. We should also mention the unique perpendicular linear polarization correlation of the 2γ annihilation radiation discussed by Wheeler as early as 1946 and first measured by Bleuler & Bradt (1948) and Hanna (1948) using Compton polarimeters. For a modern high precision measurement and a discussion of its relevance to the Einstein-Podolsky-Rosen paradox, hidden variables, the Bell inequality, and so forth, see Kasday et al (1975). The polarization correlation in the 3γ decay of 1^3S_1 has also been studied (see Faraci & Pennisi 1980, and references therein).

2.2   Energy Levels: Positronium vs Hydrogen

The spectroscopic differences between positronium and hydrogen are due to the particle-antiparticle nature of Ps, which assures the equality of the positron and electron masses and magnitudes of magnetic moments and the possibility of self-annihilation. The Schrödinger levels are rescaled by the reduced mass ratio, so $E_n(\text{Ps}) \approx \frac{1}{2}E_n(\text{H})$; thus $E_n(\text{Ps}) = chR_\infty/2n^2$ $= m_e c^2 \alpha^2/4n^2$, where R_∞ is the Rydberg constant and α the fine-structure constant. The binding energy of Ps is ~ 6.8 eV compared to ~ 13.6 eV for H, and the Lyman-α emission from Ps is expected at 243.0 nm, whereas that of H is at 121.5 nm. Figure 1 gives a schematic comparison of the $n = 1$ and $n = 2$ energy levels of H and of Ps. We follow the accepted notation for Ps states (borrowed from He): $n^{2s+1}l_j$, where $\mathbf{J} = \mathbf{l} + \mathbf{s}$; the spectrum divides into $s = 0$ singlet (para-) and $s = 1$ triplet (ortho-) states. In our subsequent discussions we use for the unit of energy $E_{\text{RYD}} = hf_{\text{RYD}} = hcR_\infty$, since the Rydberg constant R_∞ has recently been measured to a remarkable accuracy (Goldsmith et al 1978):

$$R_\infty c = 3\ 289\ 841\ 941.8\ (7.9)\ \text{MHz}.$$

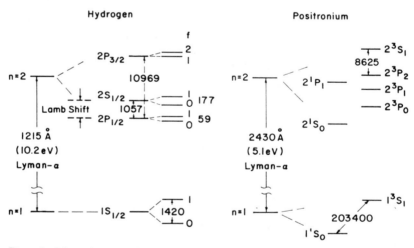

Figure 1   Schematic comparison of the $n = 1$ and the $n = 2$ energy levels of hydrogen and positronium. Energy level differences are given in MHz to the nearest MHz; for the exact experimental and theoretical values see Drell (1979) for hydrogen, and consult the text for positronium.

The splitting of the Schrödinger levels of Ps is different from that of H not only in scale, but also in general structure. The large magnetic moment of the positron ($\mu_{e^+} = \mu_{e^-} \simeq 657\mu_p$) makes the magnetic spin-spin interaction in Ps comparable to the spin-orbit interaction, eliminating the clear distinction between the fine structure and the hyperfine structure observed in H. The electron-positron annihilation mechanism, acting virtually, causes spin-dependent fine-structure shifts of the same order as those caused by magnetic spin-spin and spin-orbit interactions.

In hydrogen the large proton-to-electron-mass ratio allows one to obtain a good first approximation to the spectrum by solving the Dirac equation for an electron moving in a fixed Coulomb field; recoil effects and proton magnetic moment effects introduce a small (hyperfine) perturbation. Without these small effects one finds the celebrated $2P_{1/2} - 2S_{1/2}$ degeneracy, which is then removed principally by radiative QED corrections to the electron-photon vertex, causing the Lamb shift (see Figure 1).

In Ps one is forced to go beyond the one-particle Dirac equation to a relativistic two-particle equation, even in a first approximation of the fine structure. The Breit equation (Breit 1929), augmented by the lowest-order virtual annihilation interaction (one-photon virtual annihilation of the triplet S states), yields a satisfactory account of the spectrum to order $\alpha^2 E_{RYD}$ (Bethe & Salpeter 1957), as calculated by Pirenne (1946, 1947), Berestetski (1949), Berestetski & Landau (1949), and Ferrell (1951). This spectrum shows no analog of the $2P_{1/2} - 2S_{1/2}$ degeneracy of H; the "Lamb shift" in Ps will make no qualitative change in the Ps spectrum. To this lowest order the expression for the "fine-structure" splitting $E(1^3S_1) - E(1^1S_0) = \Delta_1$ is given by $\Delta_1 = E_{RYD}\alpha^2 7/6$, where 4/6 comes from the spin-spin interaction (Fermi 1930) and 3/6 from the one-photon virtual annihilation of the 1^3S_1 ortho-Ps state. The expression for the same splitting in the $n = 2$ states is simply $\Delta_2 = \Delta_1/8$.

2.3 First-Order Annihilation and Decay Rates

The leading term in $\Gamma_{2\gamma}(n^1S_0)$, the 2γ annihilation rate of the n^1S_0 para-Ps states, was computed by Pirenne (1944, 1946) and Wheeler (1946) based on the Dirac (1930) cross section $\sigma_{2\gamma}$ for the annihilation of free positrons with electrons. At low relative e^+e^- velocity v, we find $\sigma_{2\gamma}^1 = 4\pi r_0^2 c/v$ for spin singlet collisions, where $r_0 = \alpha\hbar/m_e c$. One obtains in lowest order

$$\Gamma_{2\gamma}^{(0)}(n^1S_0) = \sigma_{2\gamma}^1 v |\psi_n(0)|^2 = \frac{1}{2}\frac{m_e c^2}{\hbar}\frac{\alpha^5}{n^3} = 2\pi f_{RYD}\frac{\alpha^3}{n^3}, \qquad 1.$$

where $\psi_n(0)$ is the n^1S_0 Ps wavefunction evaluated at the origin. For $n = 1$ the 2γ annihilation rate is approximately 8×10^9 s^{-1}.

The lowest-order decay rate of the ortho-Ps n^3S_1 states was obtained by Ore & Powell (1949):

$$\Gamma^{(0)}_{3\gamma}(n^3S_1) = \frac{2}{9\pi}(\pi^2 - 9)\frac{m_e c^2}{\hbar}\frac{\alpha^6}{n^3} = \frac{4}{9\pi}(\pi^2 - 9)\alpha\Gamma^{(0)}_{2\gamma}(n^1S_0). \qquad 2.$$

For the $n = 1$ triplet state, the 3γ annihilation rate is $\sim 7 \times 10^6\,\text{s}^{-1}$.

Since the electron and positron do not overlap in the P states of Ps, the 2γ and 3γ annihilation have much lower expected rates. Alekseev (1958, 1959) calculates 2γ rates for P states of the order of $f_{RYD}\alpha^5$ and 3γ rates of the order of $f_{RYD}\alpha^6 \ln \alpha$. Since the rate for the electric dipole transition to S states is of order $f_{RYD}\alpha^3$, the branching ratio for annihilation is of order α^2 (or worse), which implies that such processes would be extraordinarily difficult to detect. The Lyman-α rate from 2P to 1S states (Bethe & Salpeter 1957) is given by

$$\Gamma^{(0)}_{LY\alpha} = 2\pi f_{RYD}\frac{256}{6561}\alpha^3, \qquad 3.$$

corresponding to a 3.19×10^{-9} s 2P lifetime against Lyman-α decay. The decay of the 1^1S_0 state into four photons is $\Gamma_{4\gamma}(1^1S_0) \approx 2\pi f_{RYD}(0.192)\alpha^5/3\pi^2$ (McCoyd 1965).

2.4   Radiative Corrections: Early Phase

To obtain radiative corrections theorists have found it convenient to employ the relativistically covariant two-body equation of Schwinger (1951), Gell-Mann & Low (1951), and Salpeter & Bethe (1951) since that formalism enables one to disentangle unobservable infinite renormalization effects from observable finite radiative effects. (In this review we call all higher-order corrections "radiative," although strictly speaking some of the higher-order terms are kinematic rather than radiative.) The first application of this formalism to positronium was given by Karplus & Klein (1952), who found that the triplet-singlet splitting

$$\Delta_1 = E_{RYD}\left\{\frac{7}{6}\alpha^2 - \frac{1}{\pi}\left(\frac{16}{9} + \ln 2\right)\alpha^3 + \cdots\right\}, \qquad 4.$$

and that to this order Δ_2 is still $\Delta_1/8$. Shortly thereafter the shifts for all the $n = 2$ levels were computed by Fulton & Martin (1954). They find that

$$E(2^1S_0) = E_{RYD}\left\{-\frac{1}{8} - \frac{53}{512}\alpha^2 - \frac{3}{16\pi}\alpha^3 \ln \alpha \right.$$
$$\left. + \frac{1}{\pi}\left[\frac{931}{2880} + \frac{7}{48}\ln 2 - \frac{1}{6}\ln R(2,0)\right]\alpha^3 + \cdots\right\}, \qquad 5.$$

where $R(2, l)$ is a dimensionless version of Bethe's logarithmic average of hydrogenic excitation energies. The $R(2, l)$ are independent of α and can be computed efficiently to any desired precision (Bethe et al 1950, Harriman 1956). For $l = 0$ and 1 we have $\ln R(2, 0) = 2.8118\ldots$, $\ln R(2, 1) = -0.0300164\ldots$. The corresponding value of $E(1^1S_0)$ does not seem to have been published, presumably because it is only recently that the possibility of measuring it via two-photon Doppler-free spectroscopy has emerged (see Hänsch 1977 for the hydrogen measurements).

In addition, Fulton & Martin find that

$$E(2^3P_0) = E_{RYD}\left\{ -\frac{1}{8} - \frac{95}{1536}\,\alpha^2 \right.$$
$$\left. -\frac{1}{\pi}\left[\frac{25}{576} + \frac{1}{6}\ln R(2, 1)\right]\alpha^3 + \cdots \right\}. \qquad 6.$$

Letting $\Delta(J) = E(2^3P_J) - E(2^3P_0)$, and $\Delta' = E(2^1P_1) - E(2^3P_0)$, we have

$$\Delta(2) = E_{RYD}\left\{ \frac{9}{160}\,\alpha^2 + \frac{13}{320\pi}\,\alpha^3 + \cdots \right\},$$

$$\Delta' = E_{RYD}\left\{ \frac{1}{24}\,\alpha^2 + \frac{1}{32\pi}\,\alpha^3 + \cdots \right\}, \qquad 7.$$

$$\Delta(1) = E_{RYD}\left\{ \frac{5}{160}\,\alpha^2 + \frac{5}{192\pi}\,\alpha^3 + \cdots \right\}.$$

Shortly thereafter Harris & Brown (1957) computed the first radiative correction to the singlet Ps annihilation rate $\Gamma_{2\gamma}(1^1S_0)$, obtaining

$$\Gamma_{2\gamma}(1^1S_0) = \Gamma^{(0)}_{2\gamma}(1^1S_0)\left[1 - \frac{1}{\pi}\left(5 - \frac{1}{4}\pi^2\right)\alpha + \cdots \right]. \qquad 8.$$

At this point the early phase ended because further corrections presented far more difficulty than those already computed.

2.5 Radiative Corrections: Recent Phase

There have been further radiative corrections computed in the last decade: the $\alpha^4 \ln \alpha$ term of Δ_1, some of the α^4 terms of Δ_1, and the α^5 and $\alpha^6 \ln \alpha$ terms of $\Gamma_{3\gamma}(1^3S_1)$. The difficulty of these calculations is associated with the fact that there cannot be a straightforward perturbation expansion in α since one cannot obtain bound states by perturbing free electron-positron states (the $\ln \alpha$ in Equation 5 bears witness to that). One must start with an approximate bound-state wavefunction (which cannot be a polynomial in α) and hope to improve it systematically. In work based on the Bethe-

Salpeter equation it is desirable to have a convenient analytic expression for that starting wavefunction, and also to have it be a solution of a covariant, if truncated, version of the Bethe-Salpeter equation. The calculations of the early phase did not satisfy this desideratum and so a double expansion was necessary: the starting wavefunction was kinematically improved so that it would more closely approximate a solution of a truncated covariant equation; and, simultaneously, interactions omitted in the truncation were allowed to improve the wavefunction dynamically. A further difficulty is associated with sums over intermediate electron-positron states during the course of such dynamical improvement: if those states are taken to be free it happens that infinitely many diagrams must be summed to get results of a given order. Systematic use of the Coulomb Green's function (Schwinger 1964) takes care of that infinite summation.

During the 1970s a series of computations of contributions to Δ_1 were published by Fulton et al (1971), Barbieri et al (1973), Owen (1973, 1977), Cung et al (1976), Barbieri & Remiddi (1976, 1978), Lepage (1977), Bodwin & Yennie (1978), Caswell & Lepage (1978, 1979). The final result of these computations was the addition of an $\alpha^4 \ln \alpha$ term to Δ_1:

$$\Delta_1 = E_{\text{RYD}}\left\{\frac{7}{6}\alpha^2 - \frac{1}{\pi}\left(\frac{16}{9} + \ln 2\right)\alpha^3 - \frac{5}{12}\alpha^4 \ln \alpha + \cdots\right\}. \qquad 9.$$

Some of the contributions of order α^4 have also been computed (Barbieri et al 1973, Samuel 1974, Douglas 1975, Cung et al 1976, 1978, 1979), but much remains to be done. The partial results obtained so far contribute with a magnitude of the order of 10 MHz to the frequency Δ_1/h (see table in Kinoshita 1978a).

Recent work on this problem has made use of a starting wavefunction that satisfies the criterion mentioned above: all the corrections can be viewed as the result of systematically taking into account the difference between the exact interaction kernel and a truncated interaction kernel for a suitably transformed Bethe-Salpeter equation (Gross 1969, Lepage 1977, Caswell & Lepage 1978, Barbieri & Remiddi 1978, Buchmuller & Remiddi 1980). Perhaps the most powerful transformations are those of Caswell & Lepage and Barbieri & Remiddi, which map the Bethe-Salpeter problem onto a perturbed nonrelativistic reduced-mass Schrödinger problem with the usual Coulomb interaction—that is, the starting problem becomes the familiar hydrogen atom of elementary quantum mechanics.

Stroscio & Holt (1974) attempted to calculate α^5 corrections to $\Gamma_{3\gamma}(1^3S_1)$ without satisfying the starting wavefunction desideratum; their results were corrected by Caswell et al (1977), and improved by Caswell & Lepage (1979). One coefficient in the result is currently obtainable only by multidimensional numerical integration: that result is

$$\Gamma_{3\gamma}(1^3S_1) = \Gamma_{3\gamma}^{(0)}(1^3S_1)\left[1 - (10.266 \pm 0.011)\frac{\alpha}{\pi} + \frac{1}{3}\alpha^2 \ln \alpha + \cdots\right]. \qquad 10.$$

3 POSITRONIUM DETECTION AND FORMATION METHODS

3.1 Positronium Detection

The principles of most of the techniques used to detect Ps were developed by Deutsch and his students during the years 1949 to 1952. A detailed personal account of these early experiments can be found in a collection of papers on the discovery of Ps (Deutsch 1975). In a gas or in condensed matter, fast positrons injected from a radioactive source will slow down to a few eV prior to annihilation, since the cross section for inelastic processes such as ionization and excitation is orders of magnitude larger than the annihilation cross section. At low energies positrons that do not form Ps, or some other bound chemical complex, annihilate with atomic electrons during collisions; the small 3γ to 2γ annihilation cross-section ratio $\sigma_{3\gamma}/\sigma_{2\gamma} \approx 10^{-3}$ results in the small 3γ to 2γ yield ratio $Y_{3\gamma}/Y_{2\gamma} = 3(\sigma_{3\gamma}/\sigma_{2\gamma}) \approx 1/378$ for spin-averaged collisions. On the other hand, when Ps is formed in a low density gas $Y_{3\gamma}/Y_{2\gamma}$ approaches $3/1$ with decreasing gas density, reflecting the statistical ratio of the triplet and singlet states. In this case the small $\sigma_{3\gamma}/\sigma_{2\gamma}$ ratio produces the large difference between the singlet and triplet lifetimes $\tau(1^1S_0) \approx 1.25 \times 10^{-10}$ s; $\tau(1^3S_1) \approx 1.42 \times 10^{-7}$ s. One can thus detect Ps formation by measuring the 3γ to 2γ yield ratio with a multiple-coincidence γ detector system or by measuring the positron lifetime. The lifetime is usually obtained by using ^{22}Na as a positron source and measuring the time-delayed coincidence spectrum between the prompt ($< 10^{-11}$ s) nuclear 1.28-MeV gamma ray (that signals the emission of the positron) and one of the annihilation quanta. The observation of a long-lived component that was nearly independent of gas pressure led Deutsch to the discovery of Ps. In a low density gas this long-lifetime component approaches the characteristic 3γ decay of ortho-Ps. At higher densities or in condensed matter the interaction of Ps with atomic electrons leads to a decrease in the 3γ decay mode associated with a shortening of the long lifetime. The so-called "pick-off" process in which a positron in ortho-Ps annihilates during a collision with an atomic electron of the appropriate spin into two quanta is principally responsible for this decrease.

Another technique of observing the increased 3γ yield signaling the formation of Ps is to measure the energy spectrum of one of the annihilation photons. The energy distribution of a photon from 3γ annihilation exhibits a characteristic three-body decay spectrum ranging from 0 to $m_e c^2$, whereas the photon from 2γ decay is nearly monoenergetic ($E \simeq m_e c^2$). The $m_e c^2$ energy (0.511 MeV) is Doppler-broadened by a few keV, which reflects the momentum distribution of the center of mass of the $e^+ e^-$ pair. It is particularly simple to measure the "peak-to-valley" ratio of the annihilation spectrum in a NaI detector and to calibrate this ratio against the nearly pure 2γ spectrum obtained from positrons annihilating in a metal where no Ps forms (see Hughes et al 1955).

In the 2γ annihilation mode the momentum distribution of the $e^+ e^-$ pairs also creates, by momentum conservation, a small angle correlation between the photons (mrads), i.e. a deviation from antiparallel (180°) emission. This angular correlation of annihilation radiation (ACAR) is measured by a multiple detector coincidence apparatus to obtain data on electronic momentum densities in matter (West 1973, Berko 1977, 1979a, and Mijnarends 1979). In the ACAR curves Ps formation is signaled by a narrow peak around zero momentum, reflecting the low momentum of the nearly thermal Ps, compared to the broad distribution by e^+ annihilation with atomic electrons.

Perhaps the most conclusive way of detecting Ps is to study the decrease in the 3γ yield due to the Zeeman mixing of Ps states when a magnetic field is applied, a method to be discussed in a later section.

3.2 Positronium Formation in Gases

Until recently the only way of forming and studying "free" Ps was in low density gases. Fast positrons from a long-lived e^+ source such as ^{22}Na, ^{58}Co, or ^{64}Cu are injected into a gas where they slow down to a few eV prior to Ps formation or to direct annihilation from scattering states. Ps formation takes place by radiationless electron capture as long as the kinetic energy T of the e^+ is greater than $E_i - E_{Ps}$, where E_i is the ionization energy of the gas atom or molecule, and E_{Ps} is the Ps binding energy, 6.8 eV. For $T > E_i$ the kinetic energy of the Ps atom exceeds E_{Ps} and Ps can dissociate in subsequent gas collisions instead of annihilating; for $T > E_{exc}$, where E_{exc} is the lowest excitation energy of the gas, inelastic collisions compete and can dominate Ps formation. The narrow energy region $E_{exc} \gtrsim T \gtrsim E_i - E_{Ps}$, or $E_i \gtrsim T \gtrsim E_i - E_{Ps}$, is called the Ore gap (Ore 1949). Using these two limits for T and assuming that the kinetic energy distribution after the last ionizing collision is roughly uniform, one estimates for the fraction f of positrons forming Ps

$$E_{exc} - (E_i - E_{Ps})/E_{exc} \lesssim f \lesssim E_i - (E_i - E_{Ps})/E_i = E_{Ps}/E_i. \qquad 11.$$

In rare gases other than xenon one obtains experimental f values from 10 to 40%, in reasonable agreement with the Ore gap predictions. In molecular gases the situation is more complex since in most cases there is no Ore gap $(E_i - E_{Ps} > E_{exc})$; however, Ps formation can still be abundant since the cross section for excitation of low-lying molecular levels by the positron can be small. Once formed the fate of Ps depends on the details of the Ps interaction with the gas. After Deutsch showed that gases such as NO can effectively quench ortho-Ps, many experiments were performed to study atomic and molecular collisions with Ps as well as the formation of various Ps-molecule and e^+-molecule complexes. The study of f values and of lifetimes, including experiments with external fields, has led to a reasonable understanding of e^+ and Ps behavior in gases (see for example the review by Massey 1973). In simple gases such as He it was possible to obtain information about the energy dependence of the e^+-atom elastic collision cross section by carefully analyzing the time dependence of the annihilation yields as the positrons are moderated to kinetic energies below the Ore gap. The importance of such cross sections in atomic physics has been emphasized by Massey (1976).

A comparison between e^--atom and e^+-atom scattering cross sections can serve as an important test of scattering theory, since the difference stems only from the opposite sign of the Coulomb interaction and the lack of Pauli exclusion effects; the onset of the Ps formation channel is, however, a complicating factor. These indirect measurements of e^+-atom cross sections using moderating gases have recently been supplanted by direct cross-section experiments using the newly developed, variable energy, slow e^+ beams—see the review by Griffith & Heyland (1978).

3.3 Positronium Formation in Condensed Matter

Positronium can also be formed in many liquids and in some insulating solids. An essential requirement is the short time for positron thermalization in condensed matter ($\sim 10^{-12}$ s).

Ps formation in liquids and solids does not have to occur via the Ore mechanism; instead Ps can be formed by the capture of an electron in the ionization trail or "spur" created by the slowing down of the positron (Mogensen 1974). By now the many Ps formation and annihilation studies in liquids have led to the development of Ps chemistry, an acknowledged branch of radiation chemistry. For a detailed description of the "physical chemistry" of positrons and of Ps see Goldanskii & Firsov (1971), and for an up-to-date review of Ps chemistry see Ache (1979).

In metals, thermalized positrons annihilate from delocalized Bloch states without forming Ps. In its lowest energy state the e^+ wavefunction is periodic, with minima at the positive ions of the lattice and maxima at the interstitial sites. Annihilation then proceeds mainly via conduction electrons, by the 2γ channel. The ACAR technique measures the density in momentum space of the electrons sampled by the e^+. In metals, discontinuities in these distributions reflect the size and topology of the Fermi surface; in simple systems one can also obtain information about the e^+e^- many-body correlation effects. Positron annihilation, particularly the ACAR technique, has become a valuable tool in testing electronic band theory in solids (Berko 1977, 1979a) particularly in pure metals and alloys (Berko 1979b, Mijnarends 1979).

Experiments in solids also indicate that positrons have a strong affinity to low density regions of the material and annihilate preferentially in such regions. Thus in organic materials "free-volume" considerations are important, in ionic crystals various color centers trap positrons, and in metals atomic vacancies and dislocations can capture positrons prior to annihilation. This trapping effect has led to the recent development of e^+ annihilation as a tool in metallurgy for various defect studies (for reviews see West 1973 and Doyama 1979).

Ps can be formed in insulators, as indicated by the appearance of the

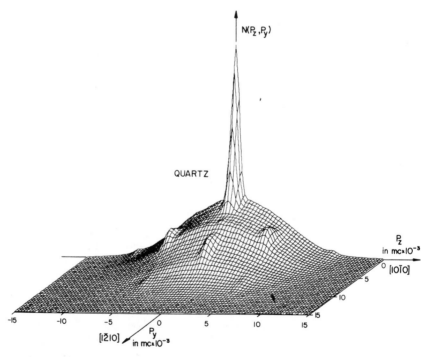

Figure 2 Two-dimensional angular correlation of annihilation radiation (ACAR) from quartz oriented with the c axis along p_x, exhibiting characteristic delocalized positronium peaks. Momenta are in units of $mc \times 10^{-3}$ (angles in mrad). From Berko et al (1977).

characteristic zero momentum peak in the 2γ angular correlation curves. In some oriented crystals such as quartz, satellite peaks have been discovered at momenta determined by the reciprocal lattice of quartz (Brandt et al 1969); these peaks were shown by a Zeeman effect experiment to be due to Ps with an itinerant Bloch wavefunction analogous to that of a delocalized exciton (Greenberger et al 1970). In Figure 2 we show these features in a recent measurement of the two-dimensional ACAR surface from oriented quartz (Berko et al 1977). We conclude that in solids Ps, like low energy positrons, can be found in delocalized quantum states that are repelled by the atomic cores of the host material. It is therefore reasonable that when slow positrons or Ps atoms formed in the bulk diffuse to a surface they can be ejected from the solid into the vacuum with positive energies, i.e. they have a negative work function. Investigation of this mechanism led to the development of slow positron beams of controllable energy.

The earliest observation of Ps escape from a solid was made by Paulin & Ambrosino (1968), who discovered that in fine oxide powders (~ 100 Å diameter) a large fraction of positrons form Ps, which annihilates almost as ortho-Ps does in pure vacuum. The effect was described theoretically as Ps formation in the bulk followed by Ps diffusion into the spaces between the powder particles, with subsequent annihilation (Brandt & Paulin 1968). A recent review by Paulin (1979) presents relevant information about the diffusion of Ps as well as e^+ in various solids.

4 SLOW POSITRON BEAMS

4.1 *Production of Slow Positrons*

During the last decade groups at several laboratories have developed slow positron beams with controllable energy and of sufficient intensity to serve as a source for further experiments. This development is based on the discovery that when fast positrons from a β^+ source are moderated in solids a fraction is ejected from the surface. The history of the early development of the technique has been recently reviewed in detail by Griffith & Heyland (1978). It is interesting to note that a very early attempt by Madansky & Rasetti (1950) to search for slow positrons diffusing out of thin ^{64}Cu activated foils failed; no slow positrons were observed, even though one slow e^+ per 10^3 fast β^+ was expected. They suggested reasons for this lack of slow positrons which, in retrospect, sound prophetic: "(*a*) positrons do not diffuse through a crystal lattice, but are trapped in potential minima and annihilate before moving appreciable distances; (*b*) positrons diffuse through the lattice but are trapped at the surface, and (*c*) Ps is formed, and even if it should diffuse out of the material, it would not be detected in these experiments." All three suggestions have been proven correct in recent experiments; their effects reduce, but do not completely stop the expected slow positron yield.

The first report of slow e^+ emission was made by Cherry (1958). The slow positrons were emitted from a Cr surface, with a conversion efficiency of approximately one slow e^+ per 10^7 incident fast positrons. The first slow e^+ beam of sufficient yield to be used in experiments was obtained by Costello et al (1971), using fast positrons pair-produced by a 55-MeV linear accelerator bremsstrahlung beam and a thin gold-plated mica foil as a fast-slow converter. Since then the efficiency ε (slow e^+ yield divided by total β^+ yield from a radioactive source) of converters has been greatly improved. MgO-coated gold foils in a "venetian blind" geometry with a ^{58}Co source behind the converter yield an ε of up to 3×10^{-5} (Canter et al 1972) and an energy width $\Delta E \sim 2$ eV. Stein et al (1978) report for a carbonized boron converter $\varepsilon \approx 10^{-6}$ with $\Delta E \cong 0.15$ eV, and a carbonized gold converter yields $\varepsilon \approx 10^{-4}$ and $\Delta E < 0.2$ eV (Pendyala & McGowan 1980). An efficient converter can be produced using vacuum-annealed tungsten ribbons ($\varepsilon \approx 10^{-4}$), as reported by Dale et al (1980).

The most efficient generation of slow positrons to date has been described by Mills (1979a): using a carefully annealed single crystal of Cu activated by a one-third monolayer of S as a converter and a ^{58}Co β^+ source in the backscattering mode, one can achieve a conversion efficiency of 10^{-3} and a nearly thermal energy distribution. Figure 3 shows the simple geometry of this converter, which requires ultra-high vacuum conditions (10^{-10} torr)

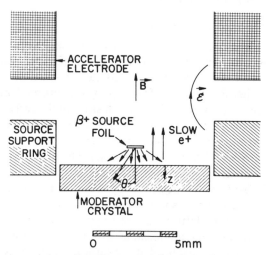

Figure 3 Geometry of the radioactive source foil and the slow positron copper crystal moderator used in the high efficiency "fast-slow" positron converter by Mills (1979a). Pressure in the vacuum chamber $\sim 10^{-10}$ torr.

for operation. An intensity of over 10^6 slow positrons per second has been achieved using a 200-mCi ^{58}Co source with the converter cooled to liquid N_2 temperature. Using the nearly exponential implantation profile of the fast positrons from radioactive sources (Brandt & Paulin 1977, Mourino et al 1979), the probability for thermalized positrons to diffuse to the surface (Mills & Murray 1980), and the probability of emission once the e^+ has reached the surface (~ 0.5), Mills obtains a theoretical estimate for the efficiency of the Cu converter of $\varepsilon = (6 \pm 3) \times 10^{-3}$. In insulators, such as MgO-coated converters, the e^+ emission is more complex and less well understood. The possibility that slow positrons from MgO are produced by field ionization of Ps has been discussed by Griffith et al (1978).

4.2 Experiments with Slow Positron Beams

4.2.1 POSITRONIUM PRODUCTION WITH SLOW POSITRONS   The first experiments designed to study the interaction of slow positrons with surfaces yielded the unexpected result that such positrons (~ 20 eV) are reemitted into the vacuum as Ps with a probability that can exceed 50% (Canter et al 1974). Figure 4 shows the slow beam apparatus used at Brandeis for these experiments and for subsequent Ps Lyman-α detection studies. The slow positrons from a MgO converter were guided by a magnetic field from the ^{58}Co source into a target region; the energy of the positrons was controlled by biasing screens. Ps formation was observed by the "peak-to-valley"

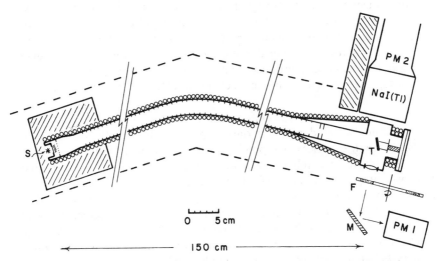

Figure 4   Slow positron beam apparatus used to study Ps formation by slow positrons incident on solid targets, and for the observation of Ps Lyman-α radiation (Canter et al 1974, 1975). S indicates the ^{58}Co source behind MgO-coated gold converter; T indicates the target and heater stage; F is the optical filter wheel; M is the front-surface aluminized mirror.

energy analysis as well as by the total 3γ yield measurement. It was found that the Ps production yield increases with increasing temperature of the target, an increase attributed at that time to a decrease in surface contaminants (mainly water) in the rather poor vacuum conditions of the experiment (4×10^{-8} Torr). This discovery of the possibility of producing Ps in vacuum efficiently and far removed from the background radiation of the radioactive source led to a major new impetus in Ps research.

The combination of the slow e^+ beam technique with ultra-high vacuum (UHV) technique ($\sim 10^{-10}$ torr) has recently resulted in the first studies of slow e^+ interactions with well-characterized solid surfaces. It was found that a temperature-dependent Ps formation effect is also characteristic of atomically clean metal surfaces (Mills 1978, Lynn 1979, Rosenberg et al 1979, 1980a). A successful model used to explain Ps formation involves e^+ surface states, already postulated in order to explain experiments on the capture of bulk positrons by voids produced by radiation damage (Hodges & Stott 1973, Nieminen & Hodges 1976). This model (Mills 1979b, Lynn & Welch 1980) involves e^+ implantation, diffusion to the surface, capture into surface states produced by the image potential, and Ps formation from the trapped state by thermal activation. It successfully explains the observed yields and temperature dependence. At high temperatures the results also indicate competition from vacancy trapping in the bulk and possibly at the surface. From these results the binding energy into the surface trap was deduced using a Born-Haber cycle, and was found to range from 2 to 3 eV, consistent with the theoretical estimates. Using a pulsed e^+ beam, one finds the average energy of Ps emitted at high temperatures (790°C) to be ~ 0.1 eV with a non-Maxwellian distribution; at lower temperature a higher energy Ps beam is also produced, with an average energy of ~ 3 eV, which thus suggests two Ps formation mechanisms (Mills & Pfeiffer 1979).

4.2.2 OTHER POSITRON-SURFACE INTERACTION STUDIES   The positrons that are not captured into the surface trap are emitted into the vacuum with characteristic energies corresponding to a negative work function. Mills et al (1978) have been able to measure the energy spectrum of the reemitted slow positrons by a retarding field analyzer and to obtain the work function for e^+ in various metals; their values range from (-0.2 ± 0.1) eV for Al to (-1.7 ± 0.2) eV for Cr. These values are in reasonable agreement with theoretical estimates (Nieminen & Hodges 1976, Hodges & Stott 1973), which are themselves rather uncertain since the work function stems from a delicate balance between the large repulsive effect of the positive ions and the large attractive correlation effect of the conduction electrons. For a review of the early phase of e^+-surface studies, see Brandt (1976).

The first observation of low energy positron diffraction (LEPD) from a solid surface, Cu(111), has recently been reported by Rosenberg et al (1980b). They used a tungsten converter, an electrostatic monochromater, and a channeltron for detecting the angular distribution of elastically scattered slow positrons (20–400 eV). It has been known for some time that positrons retain part of the spin polarizations they acquire in β^+ decay when thermalizing in bulk metals (Berko 1967). Recently a polarization experiment on a slow e^+ beam indicated that the surface emission effect from a MgO-coated converter does not depolarize the slow positrons (Zitzewitz et al 1979); LEPD experiments with polarized e^+ beams are thus possible in the future.

5   POSITRONIUM ENERGY LEVEL MEASUREMENTS

5.1   The Fine-Structure Splitting of the $n = 1$ State

The fine structure interval $\Delta_1 = E(1^3S_1) - E(1^1S_0)$ could be detected, in principle, by inducing radiofrequency (rf) transitions and observing a resonant decrease in the 3γ yield from ortho-Ps. The high frequency corresponding to Δ_1 ($\sim 2 \times 10^5$ MHz) and the low power available at these frequencies render such an experiment hard to perform. Instead, the measurement of Δ_1, pioneered by Deutsch, is based on the Zeeman mixing of the $n = 1$ Ps levels in the presence of a magnetic field.

5.1.1 ZEEMAN EFFECT IN THE POSITRONIUM GROUND STATE   Because the expectation value of the total magnetic moment of Ps vanishes in the triplet as well as in the singlet spin states, no linear Zeeman effect can exist. In higher order, the magnetic field mixes the $M = 0$ triplet state with the singlet, while leaving the $M = \pm 1$ triplet states undisturbed.

A simple nonrelativistic computation of the effect of magnetic field coupling yields for the energy eigenvalues $E_{J,M}$ in the presence of the field B, $E_{1,\pm 1} = E(1^3S_1)$ and

$$E_{1/2 \pm 1/2, 0} = \tfrac{1}{2}\Sigma_1 \pm \tfrac{1}{2}\Delta_1(1 + x^2)^{1/2}, \qquad\qquad 12.$$

where $\Sigma_1 = E(1^3S_1) + E(1^1S_0)$ and $x = 2\mu_B g B(\Delta_1)^{-1} \sim B(\text{kG})/36$. Also, μ_B is the Bohr magneton and $g = g_e$ is the dimensionless magnetic moment of the free electron or positron. Higher-order relativistic computations of the Zeeman effect (Grotch & Kashuba 1973, Lewis & Hughes 1973) show that one obtains Expression 12, but with $g = g_e(1 - 5\alpha^2/24) \approx g_e(1 - 11 \times 10^{-6})$. We sketch in Figure 5 the Zeeman levels as a function of B. The $n = 1$ spin eigenstates $\psi_{J,M}$ with B present are

$$\psi_{1,0} = [1 + y^2]^{-1/2}[\chi_{1,0} + y\chi_{0,0}]; \; \psi_{0,0} = [1 + y^2]^{-1/2}[\chi_{0,0} - y\chi_{1,0}] \qquad 13.$$

and

$$\psi_{1,\pm 1} = \chi_{1,\pm 1},$$

with

$$y = x[1 + (1 + x^2)^{1/2}]^{-1};$$

the χ's are the eigenstates without the field.

The total decay rates $\Gamma_{J,M}$ of the "mixed" states $\psi_{J,M}$ are

$$\Gamma_{1,0} = [1 + y^2]^{-1}[\Gamma_{3\gamma} + y^2\Gamma_{2\gamma}]$$

and

$$\Gamma_{0,0} = [1 + y^2]^{-1}[\Gamma_{2\gamma} + y^2\Gamma_{3\gamma}],$$

where the $\Gamma_{3\gamma}$ and $\Gamma_{2\gamma}$ are the zero field annihilation rates for $n = 1$ Ps (Eqs. 8 and 10). Thus the pseudo-"triplet" state $\psi_{1,0}$ is "quenched" by the field, annihilating into the 2γ channel at a partial rate of $y^2(1+y^2)^{-1}\Gamma_{2\gamma}$.

The above Zeeman splitting was obtained without taking into account the annihilation channels, whereas the rates were obtained by the golden rule. For a justification of this treatment of decaying states, see Grisaru et al (1973). A substantial improvement in the measurement of Δ_1 will require the complete solution of the full QED problem in the presence of the B field.

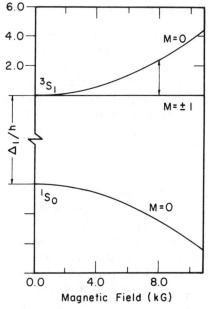

Figure 5   Zeeman splitting of the $n = 1$ levels of positronium. Frequency scale is in GHz.

5.1.2 MEASUREMENTS OF TRIPLET-SINGLET SPLITTING IN THE GROUND STATE   The first value of Δ_1 was obtained by measuring the 3γ quenching of the $\psi_{1,0}$ state as a function of B (Deutsch & Dulit 1951) by forming Ps in a gas; Δ_1 was obtained with a precision of $\pm 15\%$, sufficient to show the existence of the lowest-order virtual annihilation term (see Section 2.2). In more precise quenching experiments one has to take into account the angular distribution of the 3γ decays with respect to B (Drisko 1956).

The precision in Δ_1 was greatly improved by the introduction of the rf resonance method (Deutsch & Brown 1952). An rf field induces transitions between the undisturbed $\psi_{1,\pm 1}$ states and the "mixed" $\psi_{1,0}$ state at a given field B. A resonant decrease in the 3γ decay is observed at the proper frequency $f_{01} = [(1+x^2)^{1/2} - 1]\Delta_1/2h$ obtained from Equation 12.

During the last 25 years a sequence of Δ_1 measurements with increasing precision has been performed by Hughes and collaborators using the rf technique (see Egan et al 1977, and references therein). Another recent high precision Δ_1 measurement was performed by Mills & Bearman (1975). Instead of sweeping the rf frequency, these experiments are performed at a fixed frequency, with the value of B swept. Δ_1 is obtained at various gas densities and its value extrapolated to zero density. To obtain the required precision the modern Δ_1 experiments use a multidetector system. Figure 6 shows the microwave cavity and the detector system used by Mills & Bearman (1975); they measure simultaneously the 3γ as well as the 2γ yield. The derivative ("first difference") signals of the resonant increase (decrease) of the 2γ (3γ) yields are shown in Figure 7 as a function of B measured in terms of the frequency of an NMR probe. Since the natural linewidth (full width at half-maximum, FWHM) of the Zeeman transition is given approximately by $(\Delta B)/B = \Gamma_{2\gamma}/(4\pi\Delta_1) \cong 0.3\%$ (for small B), the line center has to be measured to about 10^{-3} of the linewidth for a precision of 5 ppm in Δ_1. Thus a very large amount of data must be collected ($\approx 10^{10}$ total counts). For a detailed discussion of the experimental pressure shifts and the extrapolation to zero density, see Egan et al (1977); for a theoretical evaluation of the pressure shifts, see Bearman & Mills (1976). A list of the Δ_1 measurements can be found in Berko et al (1979). We let ν_1 denote the frequency Δ_1/h. To date the best experimental values are

$$\nu_1(\text{exp}) = 203\,387.0\ (1.6)\ \text{MHz (Mills & Bearman 1975)} \qquad 14.$$

$$\nu_1(\text{exp}) = 203\,384.9\ (1.2)\ \text{MHz (Egan et al 1975).} \qquad 15.$$

These values are to be compared with the numerical value of Equation 9. In evaluating this expression, we use $\alpha^{-1} = 137.035987$ (29), i.e. the Josephson effect α (see Olsen & Williams 1976). The same value of α will be used in all subsequent evaluations of theoretical formulas. We present the numerical value of v_1 in terms of the values of the α^2, α^3, and $\alpha^4 \ln \alpha$ terms of Equation 9. We append an order of magnitude estimate of the lowest-order term not yet fully computed, of order $f_{RYD}\alpha^4$, preceded by a $\pm$ sign. In our last expression the number in parentheses represents the effect of the uncertainty in α alone.

$$v_1(\text{th}) = (204\,386.666 - 1\,005.497 + 19.125)\ \text{MHz} \pm 10\ \text{MHz}$$

$$= 203\,400.294\ (86)\ \text{MHz} \pm 10\ \text{MHz}. \qquad 16.$$

Since the $f_{RYD}\,\alpha^4$ terms, which have not yet been evaluated completely, can contribute a magnitude of ~ 10 MHz to v_1, experiment and theory are in reasonable agreement. Future improvements in the experimental value might involve the use of line-narrowing techniques as well as the use of slow e^+ beams to produce Ps in vacuum.

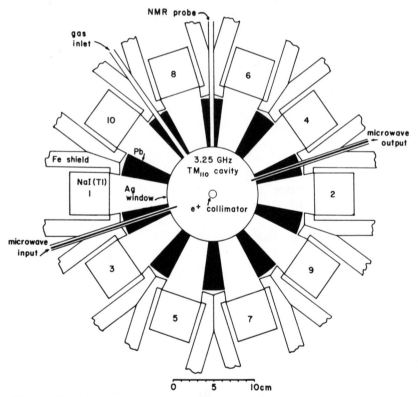

Figure 6   Multiple detector system used to observe the Zeeman resonance for Ps formed in a buffer gas in a resonant cavity. The regions shaded in black correspond to Pb collimators. The magnetic field (perpendicular to the plane of the drawing) is measured by the removable NMR probe (from Mills & Bearman 1975).

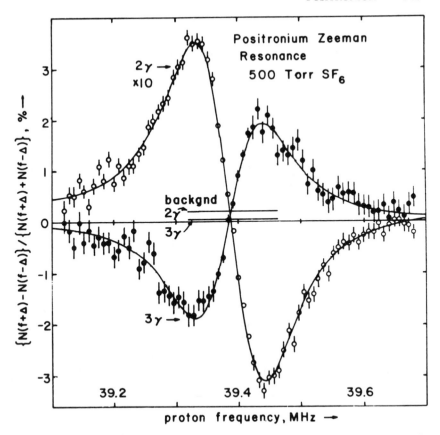

Figure 7   First difference resonance signals indicating the 2γ (3γ) resonant increase (decrease)
from Ps formed in a 500-torr SF$_6$ buffer gas (from Mills & Bearman 1975).

5.2   *Lyman-α Radiation from the n = 2 Positronium State*

Since the first attempt by Kendall and Deutsch (Kendall 1954) to form Ps
$n = 2$ excited states in a gas by optically pumping the $n = 1$ states with an
arc lamp, numerous searches have been made to obtain Ps excited states,
either by direct formation in a gas, or by optical excitation from the ground
state (for references see Canter et al 1975 and Berko et al 1979). It is not
surprising that the attempts to observe Lyman-α radiation from naturally
formed $n = 2$ Ps in a moderating gas were unsuccessful: the Ore gap
argument (Equation 11) predicts a considerably smaller fraction of $n = 2$ Ps
than $n = 1$ Ps production in most gases. Also, the slowing-down process of
fast positrons produces many excited atoms which undergo optical decay,
contributing substantially to the photon background. The $n = 1$ to $n = 2$

excitation technique utilizes a magnetic field in order to observe a change in the 2γ yield when 2P states are excited, thereby exploiting the differential Zeeman mixing of the $n = 2$ states. For theoretical computations of the $n = 2$ Zeeman and Stark shifts see Kendall (1954), Lewis & Hughes (1973), and Curry (1973). A small, but perhaps significant, signal in the 2γ coincidence rate was reported using the excitation technique by Varghese et al (1974). The large mixing and quenching of the $n = 2$ Ps states in the presence of the magnetic field as well as the large motional Stark effect render this technique relatively ineffective for $n = 2$ Ps spectroscopy studies. Curry & Schawlow also searched for $n = 2$ Ps with their technique of using a thin film of MgO powder as β^+ moderator as well as Ps producer (Curry & Schawlow 1971), but they were unable to observe Lyman-α radiation (Curry 1972).

The first clear observation of Lyman-α radiation was made with the slow beam technique developed at Brandeis. Using the same slow beam that showed a large $n = 1$ Ps formation cross section (see Figure 4), we observed $n = 2$ Ps formation by detecting directly the Lyman-α photons in coincidence with the subsequent annihilation γ's of the $n = 1$ states. Using a UV phototube and an interference filter wheel, we observed a positive signal at the proper frequency (2430 Å) when the Lyman-α photon was measured followed by the time-delayed annihilation γ's from the 3γ decay of the 1^3S_1 state. In Figure 8 we show this effect, signaling the formation of excited $n = 2$ Ps states, when a 25-eV e^+ beam interacts with a Ge target.

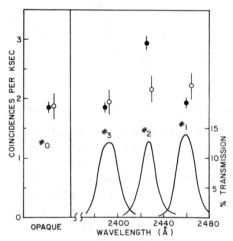

Figure 8 UV photon-annihilation γ coincidence rate using different interference filters in front of the photon detector (see Figure 3). The transmission characteristics of the three filters are also shown (~ 20 Å FWHM). The solid points correspond to 25-eV incident e^+ energy and the open circles to 400-eV incident e^+ energy (Canter et al 1975).

The effect disappears when the energy of the incident e^+ beam is increased to 400 eV; the implantation of the positrons in the target is deeper, and $n = 2$ Ps formation at the surface decreases substantially. The intensity of the Lyman-α signal at 25-eV e^+ energy indicates that the ratio of the yield of $n = 2$ Ps states to the ground-state yield is $\sim 1:10^3$. Experiments are being prepared to study excited-state Ps formation from metal surfaces that are atomically well-characterized in UHV conditions.

5.3   Measurement of the $(2^3S_1 - 2^3P_2)$ Fine-Structure Interval

Following the observation of the Lyman-α signal from Ps, an experiment to induce 2S to 2P rf transitions was performed (Mills et al 1975). The 3γ lifetime of the 2^3S_1 state is $8\tau(1^3S_1) \approx 1.1 \times 10^{-6}$ s (Equation 2); the two-photon decay to the ground state is unobservably rare. On the other hand, the 2P states are metastable against annihilation, but decay with a 3.19×10^{-9} s lifetime (Equation 3) to the ground state (triplet to triplet, singlet to singlet).

Of the three fine-structure intervals predicted to lie in common microwave bands, the $2^3S_1 - 2^3P_2$ splitting δ_2 was selected for the experiment, since it is predicted to be in the X band (Fulton & Martin 1954). The experiment was designed to search for a resonant increase in Lyman-α decay when $2^3S_1 \rightarrow 2^3P_2$ transitions are induced at the proper rf frequency. The apparatus was a modification of the slow beam setup used to measure

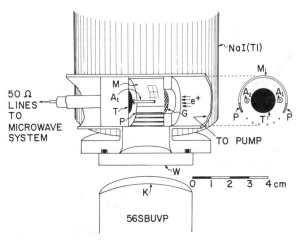

Figure 9   Positron target chamber and microwave cavity used to measure the Ps fine-structure splitting $E(2^2S_1) - E(2^3P_2)$ (Mills et al 1975). G is the grid; T is the copper target; M is the aluminized Suprasil quartz mirror; W is the Suprasil quartz window; K is the CsTe photocathode; P are the support posts; A_1 is the input antenna; A_2 is the output antenna; NaI(Tl) is the annihilation detector.

the Lyman-α signal, The experiment was performed with a ∼30-eV e^+ beam at a low magnetic guiding field of ∼50 G, thus circumventing the problems of the large Zeeman mixing and motional Stark shifts of the $n = 2$ Ps states. The cut-away sketch in Figure 9 shows the target area consisting of a microwave cavity with wire-mesh windows for introducing the e^+ beam and for observing the Lyman-α photons. The filters were omitted in front of the UV photomultiplier in order to increase the intensity of the Lyman-α signal. The slow positrons collide with the back wall of the rf copper cavity and form Ps in the $n = 1$ as well as in the $n = 2$ states. At the proper rf frequency f, the 2^3S_1 state is pumped into the 2^3P_2 state, increasing the Lyman-α intensity.

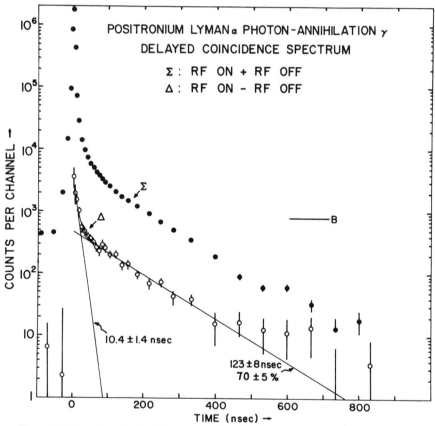

Figure 10 Time-delayed coincidence spectra between the (Lyman-α) photon converter start pulse and the (Ps ground-state) annihilation detector stop pulse, obtained by Mills et al (1975). The constant background due to accidentals is indicated by solid line B, and has been subtracted from the raw data.

In order to reduce the large background the time delay spectrum $n(t)$ between the Lyman-α photons and the γ's from the subsequent annihilation of the ground state, observed with NaI(Tl) detectors, was measured. Thus $2^1P \rightarrow 1^1S \rightarrow 2\gamma$ or $2^3P \rightarrow 1^3S \rightarrow 3\gamma$ events were registered. Such a lifetime spectrum, shown in Figure 10, reveals a long-lived component characteristic of the $\sim 1.4 \times 10^{-7}$ s lifetime of the 3γ decay of the 1^3S_1 state. The slight shortening of the lifetime (to $\sim 1.2 \times 10^{-7}$ s) was attributed to collisions with the walls. Subsequent coating of the cavity with MgO indeed increased the lifetime. The appearance of an intermediate lifetime (~ 10 ns) is not fully explained, but could result from a high velocity $n = 2$ Ps component.

With microwaves on at the appropriate frequency the intensity of the delayed coincidence spectrum increases, as indicated in Figure 10, by the finite difference signal $n(t)_{rf-on} - n(t)_{rf-off}$. The signal used to study the resonance is the integral N under the time-delayed component of the spectrum $n(t)$ (Figure 10). In Figure 11 the signal $S(f) = [N_{on}(f) - N_{off}]/N_{off}$ is plotted against the X-band frequency f. In separate runs a "derivative signal"

$$S'(f) = \frac{[N(f + \Delta) - N(f - \Delta)]}{[N(f + \Delta) + N(f - \Delta)]}$$

was measured with $\Delta = 30$ MHz: it is also shown in Figure 11. A Lorentzian shape $S(f) = \frac{1}{4}A\delta^2[(f - f_0)^2 + \frac{1}{4}\delta^2]^{-1}$ was fitted to these data resulting in $A = (11.4 \pm 0.6)\%$, $f_0 = (8628.4 \pm 2.8)$ MHz, and $\delta = (102$

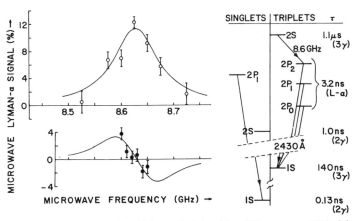

Figure 11   Observed Lyman-α signal S (open circles) and first difference signal S' (solid circles) as a function of microwave frequency (see text). The inset is the schematic term diagram for the $n = 2$ and $n = 1$ Ps states indicating the relevant transitions and the lifetime for each level (Mills et al 1975).

$\pm\ 12$) MHz, with $\chi^2/\nu = 12/10$. That the width δ of the resonance is twice the natural linewidth of the 2^3P_2 state is attributed to power broadening.

The theoretical value for the fine-structure splitting can be obtained from appropriate combinations of Equations 5, 6, and 7, yielding

$$v_2(\text{th}) = [E(2^3S_1) - E(2^3P_2)]/h$$

$$= f_{\text{RYD}}\left[\frac{23}{480}\alpha^2 - \frac{3}{16\pi}\alpha^3 \ln\alpha - (0.113\ 1197)\alpha^3 \pm O(\alpha^4 \ln \alpha)\right]$$

$$= (8\ 394.4523 + 375.4129 - 144.6137)\,\text{MHz} \pm 10\ \text{MHz}$$

$$= 8\ 625.2515(36)\ \text{MHz} \pm 10\ \text{MHz}. \qquad\qquad 17.$$

The 0.0036 MHz error is that associated with the uncertainty in α alone. Prior to comparison with experiment this value has to be corrected for the small Zeeman and motional Stark shifts due to the 54 G rms magnetic field in the cavity. Using a Ps kinetic energy between 0 and 1 eV, we find that the corrected theoretical value of v_2 ranges between 8620 and 8624 MHz.. The experimental error of ± 2.8 MHz quoted above is that of counting statistics only; uncertainty in the constancy of the rf power as a function of f, as well as uncertainty in the velocity of the Ps atoms, adds an estimated uncertainty of ± 6 MHz. We note that the radiative corrections to δ_2, i.e. the terms of order $f_{\text{RYD}}\ \alpha^3 \ln \alpha$ and $f_{\text{RYD}}\ \alpha^3$, add 231 MHz, so the present experiment confirms these contributions to an accuracy of approximately 5%. Since the next uncomputed term, $f_{\text{RYD}}\ \alpha^4 \ln \alpha$, can be of order 10 MHz, experiment and theory are presently in good agreement.

A subsequent rf measurement, using a nonresonant circular waveguide with a wave propagating along the direction of the Ps emission, detected a Doppler shift corresponding to a Ps beam of ~ 1 eV kinetic energy (Mills et al 1975). Experiments are presently in progress using a traveling X-band waveguide in a geometry that cancels the first-order Doppler effect. Improved precision in δ_2 will necessitate the computation of the $f_{\text{RYD}}\ \alpha^4 \ln \alpha$ terms for detailed interpretation.

6   ANNIHILATION RATE EXPERIMENTS

6.1   *Annihilation Rate of the 1^1S_0 State*

The 2γ lifetime $\tau_{2\gamma} = [\Gamma_{2\gamma}(1^1S_0)]^{-1}$ of the $n = 1$ para-Ps state is too short ($\sim 1.25 \times 10^{-10}$ s) to be directly measured electronically with high precision. The rate $\Gamma_{2\gamma}(1^1S_0)$ has been determined indirectly by Theriot et al (1967) by studying the rf resonance linewidth as a function of power in a Zeeman experiment (see Section 5.1.2) and extrapolating to zero power. Their result is $\Gamma_{2\gamma}(1^1S_0, \exp) = 7.99(11) \times 10^9\ \text{s}^{-1}$. The value of the theoretical expression (Equation 8) is

$$\Gamma_{2\gamma}(1^1S_0, \text{theor}) = 2\pi f_{\text{RYD}}\, \alpha^3 [1 - (0.8061512)\alpha \pm O(\alpha^2 \ln \alpha)]$$

$$= (8\,032.5050 - 47.2534) \times 10^6\ \text{s}^{-1} \pm 2 \times 10^6\ \text{s}^{-1}$$

$$= 7\,985.2516\,(51) \times 10^6\ \text{s}^{-1} \pm 2 \times 10^6\ \text{s}^{-1}. \qquad 18.$$

Thus the experiment agrees with theory within the experimental uncertainty, which is unfortunately too large to check the first radiative correction term of order $f_{\text{RYD}}\, \alpha^4$.

An experiment presently being performed at the University of Michigan (D. W. Gidley and A. Rich, private communication), is designed to study the 2γ lifetime of the Zeeman "mixed" state $\psi_{1,0}$ in the presence of a magnetic field (Section 5.1.1). Using the experimental value of Δ_1 and the equations of Section 5.1.1, one then obtains $\Gamma_{2\gamma}(1^1S_0)$.

6.2   Annihilation Rate of the 1^3S_1 State

The predicted 3γ lifetime of the $n = 1$ ortho-Ps state ($\sim 1.4 \times 10^{-7}$ s) is amenable to direct electronic measurement. The usual experiments involve the e^+ lifetime measurement in a moderating gas (see Section 3.2) and the linear extrapolation of the ortho-Ps decay rate to zero density. It is usually assumed that $\Gamma(\text{gas}) = \Gamma_{3\gamma}(1^3S_1) + q\rho$, where q is the quenching coefficient (mostly due to the pick-off process) for Ps collisions with the gas molecules, and ρ is the gas density.

The first two experiments on $\Gamma_{3\gamma}(1^3S_1)$ were by Beers & Hughes (1968) (see also Hughes 1973), who obtained $7.275\ (15) \times 10^6\ \text{s}^{-1}$; and by Coleman & Griffith (1973), who obtained a value of $7.262\ (15) \times 10^6\ \text{s}^{-1}$. These values were in good agreement with the theoretical value of $7.242 \times 10^6\ \text{s}^{-1}$ obtained by Stroscio & Holt (1974) (see also Stroscio 1975) by incorporating radiative corrections (Section 2.5).

Gidley et al (1976a) reported a high precision measurement of the ortho-Ps lifetime with Ps formed in loose SiO_2 powder of average grain radii between 35 and 70 Å. The annihilation rates were obtained as a function of $\rho^* = \rho/(\rho_{\text{solid}} - \rho)$, a dimensionless variable proportional to the free volume of the powder. The powder density ρ was varied from 0.03 to 0.26 g cm^{-3} by compression, with ρ_{solid} being 2.2 g cm^{-3}. Good linearity with ρ^* was found; the annihilation rates extrapolated to $\Gamma_{3\gamma}(1^3S_1, \text{exp}) = 7.104\,(6) \times 10^6\ \text{s}^{-1}$, a value $(1.9 \pm 0.1)\%$ below the theoretical result. In a theoretical study Ford et al (1976) concluded that they could not attribute this low rate (long lifetime) to the influence of the powder environment since most interactions result in shortening rather than lengthening of the lifetime.

The powder experiment was followed by a direct measurement of the vacuum decay rate of ortho-Ps by Gidley et al (1976b). The experiment used the technique of Ps formation in vacuum by slow positrons developed at Brandeis (see Section 4.2.1). Slow positrons (400 eV) from a converter are focused by an electrostatic system onto the MgO-coated entrance cone of a channel electron multiplier (CEM). Ps is formed on this surface and is emitted into the vacuum, while a timing signal is obtained from the CEM that counts the secondary electrons produced by the incident positron—see Figure 12 for the sketch of the target region. The Ps atoms leaving the CEM cone were confined to annihilate within a MgO-coated cavity ("copper can"). The maximum Ps kinetic energy was measured directly to yield (0.8 ± 0.2) eV in qualitative agreement with the estimate of the kinetic energy of the $n = 2$ Ps states from the Doppler measurement using the rf technique (see Section 5.3). A time spectrum between the $t = 0$ CEM signal and the 3γ signals from NaI detectors mounted close to the interaction region was obtained with a time-to-amplitude converter. The decay rate of ortho-Ps was measured to be $\Gamma_{3\gamma}(1^3S_1, \exp) = 7.09\,(2) \times 10^6\ s^{-1}$.

The possibility that the low annihilation rate was due to a small "contamination" of 2^3S_1 Ps annihilations was studied by incorporating an rf cavity with a CEM detector (Canter et al 1978); no $n = 2$ effect was observed at the 0.1% level of statistical accuracy.

The disagreement between experiment and theory was resolved in 1977 with the reevaluation of the theoretical QED rate by Caswell et al (1977), who found that the previous theoretical expression contained a sign error in one of the terms contributing to the radiative correction to $\Gamma_{3\gamma}(1^3S_1)$. The

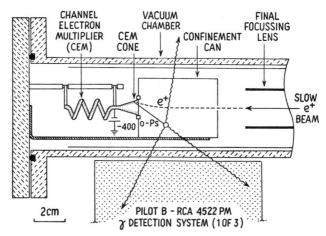

Figure 12   Interaction region of apparatus used to measure $\Gamma_{3\gamma}(1^3S_1)$ in vacuum by Gidley et al (1976a,b) and Gidley & Zitzewitz (1978).

most recent theoretical expression of $\Gamma_{3\gamma}(1^3S_1)$ by Caswell & Lepage (1979), Equation 10, yields the value

$$\Gamma_{3\gamma}(1^3S_1, \text{th}) = \Gamma_{3\gamma}^{(0)}(1^3S_1)\{1 - [3.2678\,(35)]\alpha + \tfrac{1}{3}\alpha^2 \ln \alpha + O(\alpha^2)\}$$

$$= (7.21117 - 0.17196 - 0.00063) \times 10^6 \text{ s}^{-1}$$

$$\pm 0.00038 \times 10^6 \text{ s}^{-1}$$

$$= 7.03858\,(18) \times 10^6 \text{ s}^{-1} \pm 0.00038 \times 10^6 \text{ s}^{-1}. \qquad 19.$$

Here the uncertainty in parentheses stems from the uncertainty in the numerical evaluation of the radiative correction term in Equation 10. The uncertainty associated with the value of α is only $\pm 6 \text{ s}^{-1}$. More recently, Gidley & Zitzewitz (1978) have remeasured the vacuum decay rate using the slow e^+ beam technique with various dimensions for the confinement can as well as the entrance aperture. They find a linear dependence on A/V, the ratio of the entrance hole area to the cavity volume (see Figure 13); an extrapolation to zero A/V yields the value $\Gamma_{3\gamma}(1^3S_1, \text{exp}) = 7.050\,(13) \times 10^6 \text{ s}^{-1}$, in reasonable agreement with the new theoretical value (Equation 19).

Recently two high precision 3γ rate experiments have been reported using the standard Ps gas formation technique. Gidley et al (1978) measure the 3γ rate by forming Ps in a mixture that is 80% Freon-12 and 20% isobutane gas. They obtain the timing signal by letting the fast β^+ from a ^{68}Ge source pass through a scintillator prior to entering the gas (see Figure 14). Linear extrapolation to zero gas density yields $\Gamma_{3\gamma}(1^3S_1, \text{exp})$

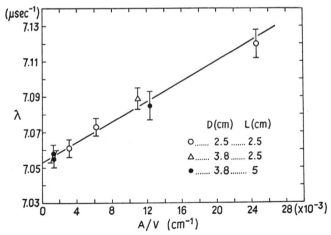

Figure 13 Decay rate versus the ratio of the entrance hole area to the cavity volume for three different cavities (see text) measured by Gidley & Zitzewitz (1978).

$= 7.056\,(7) \times 10^6$ s^{-1}. They also report a reevaluation of their 3γ rate measurement in SiO$_2$ powders and obtain $7.067\,(21) \times 10^6$ s^{-1}, in agreement with the gas data.

Griffith et al (1978) determined $\Gamma_{3\gamma}$ in various gases; they obtain by extrapolation $\Gamma_{3\gamma}(1^3S_1, \text{exp}) = 7.045\,(6) \times 10^6$ s^{-1}. Both high precision gas experiments showed excellent linearity of $\Gamma_{3\gamma}$ with gas pressure.

The most recent high precision experimental values of $\Gamma_{3\gamma}(1^3S_1)$ are thus in good agreement with each other, but are somewhat higher (by 0.2%) than the latest theoretical value. Whether this is due to some systematic experimental error or the radiative correction terms yet to be evaluated remains to be seen. The value of $\Gamma_{3\gamma}(1^3S_1)$ is important in QED since it represents the only available experimental test of a fundamental decay rate.

In closing our review of Ps experiments designed to test QED calculations, we collect the results in Table 1. As we have seen, four fundamental quantities have been measured so far and the agreement with theory is reasonable to good. Except for $\Gamma_{2\gamma}(1^1S_0)$, further improvement in the experimental values will require further computation of radiative corrections in order to make sensitive comparisons: the size of the lowest-order uncomputed term for each theoretical value is also indicated in Table 1.

In addition to these four values, the Lyman-α photon energy has also been measured—see Section 5.2. With increased slow e$^+$ intensities it will perhaps become feasible to produce sufficiently many Ps atoms in the ground state to perform a two-photon Doppler-free laser absorption experiment, thus measuring the QED shifts between the $n = 1$ and $n = 2$ Ps states—see, for example, the theoretical considerations of Letokhov &

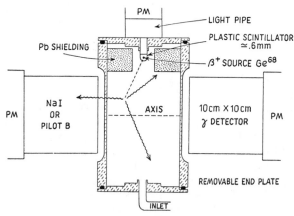

Figure 14   Gas chamber and detector arrangement used by Gidley et al (1978) to measure the decay rate of ortho-Ps formed in a gas mixture of 80% Freon-12 and 20% isobutane.

Table 1 Positronium fine structure and decay rates[a]

	$v_1 = [E(1\,^3S_1) - E(1\,^1S_0)]/h$ in MHz	$v_2 = [E(2\,^3S_1) - E(2\,^3P_2)]/h$ in MHz
Theory	$203\,400.294(86) \pm 10^{a,b}$	$8625.2515(36) \pm 10^e$
Experiment	$203\,387.0(1.6)^c$ $203\,384.9(1.2)^d$	$8628(6)^{f,\dagger}$

	$\Gamma_{2\gamma}(1\,^1S_0)$ in 10^9 s^{-1}	$\Gamma_{3\gamma}(1\,^3S_1)$ in 10^6 s^{-1}
Theory	$7.985\,2516(51) \pm 0.002^g$	$7.038\,58(18) \pm 0.000\,38^i$
Experiment	$7.99(11)^h$	$7.056(7)^j$ $7.045(6)^k$

[a] As in the text, for the theoretical values the numbers with $\pm$ indicate the estimate of the next correction term, not yet computed. The numbers in parentheses indicate the uncertainty introduced by the error in α, except for $\Gamma_{3\gamma}$ (see text). We use $\alpha^{-1} = 137.035987$ (29) (Olsen & Williams 1976). The most recent value is $\alpha^{-1} = 137.035963$ (15) (Williams, E. R., Olsen, P. T. 1979. *Phys. Rev. Lett.* 42:1575–79).

[b] Caswell & Lepage 1979.
[c] Mills & Bearman 1975.
[d] Egan et al 1975.
[e] Fulton & Martin 1954.
[f] Mills et al 1975.
[g] Harris & Brown 1957.
[h] Theriot et al 1967.
[i] Caswell & Lepage 1979.
[j] Gidley et al 1978.
[k] Griffith et al 1978.

[†] Experiment performed in 54G B field; for comparison with theory add (1–5) MHz for Zeeman and motional Stark shifts (see text).

Minogin (1976). However, one is far from being able to realize experimentally the coherent emission schemes using annihilation radiation proposed by Letokhov (1974) and Bertolotti & Sibilia (1979).

7 OTHER TOPICS ASSOCIATED WITH POSITRONIUM

We finish our review by mentioning briefly several important topics associated with fundamental Ps physics.

7.1 Tests of Charge Conjugation Invariance

In Section 2.1 we discussed the $e^+ e^-$ annihilation selection rules. In particular the $1\,^1S_0$ state is strictly forbidden to decay into three photons by charge conjugation (C) invariance. Although C invariance is believed to be obeyed in pure QED interactions, experiments have been performed to set a limit on possible C-violating terms. Mills & Berko (1967) searched for the C-forbidden $^1S_0 \rightarrow 3\gamma$ decay by using the theoretically predicted difference in the 3γ angular correlation between the C-allowed $^3S_1 \rightarrow 3\gamma$ decay and the C-forbidden $^1S_0 \rightarrow 3\gamma$ decay. For a limit of the annihilation rate branching ratio, they obtain $\Gamma_{3\gamma}(1^1S_0)/\Gamma_{2\gamma}(1^1S_0) < 2.8 \times 10^{-6}$. C invariance also forbids the $(1^3S_1) \rightarrow 4\gamma$ decay; Marko & Rich (1974) obtained an experimental limit for this C-forbidden rate $\Gamma_{4\gamma}(1^3S_1)/\Gamma_{3\gamma}(1^3S_1) < 8 \times 10^{-6}$. For theoretical discussions see Berends (1965), Weisberg (1966), and Mani & Rich (1971).

7.2   Positronium Negative Ion

After the earliest computation by Wheeler (1946), the stability and binding energy of the Ps negative ion (Ps^-), i.e. the ($e^+e^-e^-$) three-particle system, was discussed theoretically in several papers using various approximation methods. The system is bound; perhaps the most precise value of the ground-state energy is given by Kolos et al (1960) with a 50-term Hylleraas expansion, and by Frost et al (1964) with a Pekeris series solution of equivalent length, resulting in $E_{Ps^-} = -7.129$ eV, i.e. a binding energy ($E_{Ps} - E_{Ps^-}$) of 0.327 eV. The e^+e^- annihilation rate and the 2γ angular correlation from Ps^- were theoretically investigated by Ferrante (1968). More recently Ho (1979) studied the autoionization states of Ps^- (resonances in e^--Ps scattering).

To date no experiment verifies the existence of Ps^-. It is interesting to note that high precision e^+e^- annihilation rate experiments in alkali metals (Weisberg & Berko 1967) indicate an asymptotic limit for this rate with decreasing conduction electron density; the limit should be the annihilation rate of Ps^-. With the advent of the technique for forming Ps in vacuum using slow e^+ beams, perhaps Ps^- and even e^--Ps scattering experiments are becoming feasible.

7.3   Positron Spin Polarimeter

Since the discovery of parity violation in beta decay the technique of measuring the spin polarization of electrons and positrons has been well developed—see, for example, the reviews of Grodzins (1959) and Page (1962). It was Page who first developed e^+ polarimeters using the Zeeman mixing of Ps (Page & Heinberg 1957, Page 1962). The polarimeters are based on the effect that when positrons spin-aligned parallel (antiparallel) to an external field B form Ps, "singlet" ("triplet") Ps formation is favored. Briefly, one expands the initial Ps spin states ($\uparrow\downarrow$) and ($\downarrow\uparrow$) (the first arrow indicates the spin of the e^+), in terms of the Zeeman mixed pseudo-singlet state $\psi_{0,0}$ and the pseudo-triplet state $\psi_{1,0}$ (Section 5.1.1). The (time-dependent) mixing amplitudes depend on the direction of the initial e^+ spin with respect to B (on the y parameter of Section 5.1.1). The ratio of singlet to triplet states changes when the magnetic field is reversed. For more complete algebra, see Bisi et al (1962) and Gilleland & Rich (1972). Thus any experiment that can separate the singlet from the triplet state annihilation can be used as a polarimeter.

In their original paper Page & Heinberg (1957) used the 2γ angular correlation from a moderating gas in the presence of a field B to observe the singlet-triplet separation. Greenberger et al (1970) used 2γ ACAR measurements with a field B to prove the delocalized nature of Ps in quartz. Telegdi suggested using a lifetime technique to separate singlet from triplet annihilations (see Grodzins 1959). Based on this technique Dick et al (1963) developed an efficient e^+ polarimeter: a change in the intensity of the slow lifetime component of Ps annihilation in a plastic scintillator upon magnetic field reversal was used to measure the polarization component of the positrons along the field direction. A similar polarimeter was used in the $g-2$ experiments of the free e^+ by Gilleland & Rich (1972). More recently a variation of the "timing-polarimeter" using Ps formation in MgO powder was developed by Gerber et al (1977) and used to measure the e^+ polarization from a ^{68}Ga source. In the use of these polarimeters for absolute polarization measurements one has to include the positron depolarization in the e^+ thermalizer (Braicovich et al 1967) as well as the various interactions of Ps with atomic electrons; for a detailed study of the influence of these interactions on the magnetic field effect, see the density matrix treatment by Mills (1974).

7.4   Magnetic Moment of the Positron

Although it is not directly related to Ps physics we describe briefly the status of e^+ magnetic moment measurements. At present the best experimental value of the e^+ magnetic moment anomaly $a_{e^+} = (g-2)/2$ is a_{e^+} = $11602(11) \times 10^{-7}$ (Gilleland & Rich 1972). The accuracy of the corresponding electron anomaly has been greatly improved by Dehmelt and collaborators using their single-electron resonance technique in a Penning trap; their most recent value is $a_{e^-} = 1\,159\,652\,200(40) \times 10^{-12}$ (Van Dyck et al 1977, 1979). For the excellent agreement of the QED calculation with the 1977 value, see Kinoshita (1978b). Computations are in progress (T. Kinoshita, private communication) to evaluate the next order terms (over 700 Feynman diagrams!); these computations, combined with the new 1979 value of a_{e^-} will perhaps necessitate a new value of α.

The University of Washington group has recently succeeded in trapping positrons in a Penning trap (Schwinberg et al 1979). We thus can expect a new high precision value for a_{e^+}. The value of a_{e^+} tests the CPT theorem, which requires $a_{e^+} = a_{e^-}$. The best test of this equality to date is by Serednyakov et al (1977), who compare directly a_{e^+} with a_{e^-} by observing the simultaneous beam resonance depolarization of e^+ and e^- beams trapped in the same storage ring; they obtain $|a_{e^-} - a_{e^+}|/a_e \leq 1.0 \times 10^{-5}$. For future a_{e^+} experiments planned at the University of Michigan, see Newman et al (1978).

7.5   Astrophysical Sources of Annihilation Radiation

During the last several years various observations of astrophysical γ lines interpreted as red-shifted 0.511-MeV 2γ annihilation radiation were reported (see Leventhal et al 1978 and references cited therein). In particular, Leventhal et al (1978) reported the observation by a balloon-borne germanium γ-ray telescope of a γ line at (510.7 ± 0.5) keV, with a FWHM of less than 3.2 keV, emitted from the direction of the galactic center. They also report some evidence for a γ continuum, interpreted as being due to 3γ decay of Ps, Ps being formed when positrons stop in the interstellar gas. Recently Bussard et al (1979) made a detailed theoretical study of the interaction of positrons with galactic hydrogen, including the Ps formation mechanism. They find that for gas temperatures less than 10 K, Ps formation is the dominant channel. The width of the 0.5-MeV line observed by Leventhal et al (1978) suggests a temperature of less than 10 K and a degree of ionization larger than 5%.

Perhaps the most spectacular γ event observed so far was that reported at the 1980 APS Washington Meeting (Evans 1980, Ramaty 1980). A γ-ray transient burst (March 5, 1979) was registered simultaneously by nine different spacecraft; by long baseline wave front triangulation they identify the source of the burst as being inside a supernova remnant in the Large Magellanic Cloud. They propose as a source a vibrating neutron star. A γ-line energy observed to be 25% lower than 0.511 MeV is assigned to e^+e^- 2γ annihilation, gravitationally red-shifted by the neutron star. Computations show that the superposition of synchrotron radiation and annihilation radiation of an e^-e^+ plasma can account for the observed spectrum.

In the large magnetic fields believed to be characteristic of neutron stars $(10^{12} \text{ G} \leq B \leq 10^{13} \text{ G})$ Ps could annihilate by emission of a single photon, the field providing the necessary extra momentum. A theoretical computation of the 1γ vs 2γ annihilation yields of Ps in such strong fields by Carr & Sutherland (1978) indicates that the 1γ channel starts competing only when B exceeds 10^{13} G; they conclude that the 1.022-MeV line is not sufficiently intense to be observed in neutron stars. For a review of astrophysical γ ray spectroscopy see Lingfelter & Ramaty (1978).

8   CONCLUSIONS

We have seen that the physics of positronium and of slow positrons, rather than being esoteric and of narrow scope, has fundamental implications ranging from the tests of QED to astrophysics, with application from chemistry to metallurgy. Even in medical physics the use of positron annihilation has become important as "positron emission tomography" in metabolic studies (see the review by Correia et al 1979). These wide-ranging applications will surely grow as new high intensity sources and better detection techniques are developed.

We indicated that Ps is an ideal pure leptonic system to study the relativistic bound-state problem in QED. The other such system, muonium, is also being used for the same purpose; however, in muonium only the ground-state splitting Δ_1 has been measured, to about the same precision as in Ps (see for example Kinoshita 1978a). In hydrogen the corresponding (hyperfine) splitting of the ground state is one of the most precisely measured quantities in physics (to 2 parts in 10^{12}), but the proton's size and internal structure complicate the theory at the parts per million level; so the truly remarkable experimental precision cannot be fully exploited to test pure QED. Of course strong interactions enter even positronium calculations via their effect on vacuum polarization, and weak interactions enter via single Z^0 exchange and double $W^{\pm}$ exchange, but such contaminations have less effect on positronium than on any other experimentally accessible system. In positronium they do not reach levels detectable by current experimental technique.

It is hoped that the improved theoretical technology of bound states will have implications for other relativistic two-body systems. It is interesting to note that the level structure of Ps resembles that of the charmonium levels (Appelquist et al 1978) more than it does that of hydrogen (Figure 1).

In the future we can expect new high precision measurements of Δ_1 and δ_2 and of the annihilation rates. Unfortunately, because of the decay channels, the accuracy of the Δ_1 and δ_2 measurements will be limited by the natural linewidth of the resonance lines used to measure these splittings. It is unlikely that Δ_1 can be improved by much more than another factor of ten (i.e. to ± 0.1 MHz). Perhaps the $n = 2$ splitting δ_2 can be measured to the same accuracy, since the $n = 2$ line has about the same width as the Zeeman resonance line. However, δ_2 is more sensitive to higher order correction terms than Δ_1—compare Equations 17 and 16. Most important for such measurements will be the development of more efficient ways of producing Ps in excited states, perhaps by resonant photon absorption.

Improved measurements of these fundamental frequencies should inspire the computation of higher-order terms, perhaps to order $\alpha^5 f_{RYD}$, an arduous task indeed. From Table 1 we see that even at such a level we do not expect the comparison with experiments to yield a new determination of α. The annihilation rate measurements could be improved with existing technique by perhaps a factor of ten; given the nonresonant nature of such measurements, this is probably an upper limit.

With the increased intensity of slow positron and positronium sources we can look forward to other experiments of fundamental nature. The direct weighing of antielectrons remains to be performed (see for example Fairbank et al 1974), although strong indirect evidence exists that most forms of antimatter satisfy Einstein's principle of equivalence (Schiff 1959, Good 1961).

Finally, we can expect an increased activity in e^+ and Ps astrophysics. The creation and study in the laboratory of an e^+e^- plasma and of other ionized gas environments of astrophysical interest should become feasible, leading to tests of models proposed to explain the astronomical observations of the annihilation line.

ACKNOWLEDGMENTS

The authors are indebted to Larry Abbott, David Schoepf, and Frank Sinclair for reading the manuscript. We would like to express our appreciation to K. F. Canter, H. G. Dehmelt, R. Drachman, D. Gidley, T. Kinoshita, G. P. Lepage, A. P. Mills, Jr., and A. Rich for illuminating discussions regarding various theoretical and experimental aspects of the material reviewed above.

This work was partially supported by a grant from the National Science Foundation.

Literature Cited

Ache, H. J. 1979. *Proc. 5th Int. Conf. on Positron Annihilation*, pp. 31–47. Sendai, Japan: The Japan Institute of Metals
Alekseev, A. I. 1958. *Sov. Phys. JETP* 7:826–30
Alekseev, A. I. 1959. *Sov. Phys. JETP* 9:1312–15
Appelquist, T., Barnett, R. M., Lane, K. 1978. *Ann. Rev. Nucl. Part. Sci.* 28:387–499
Barbieri, R., Christillin, P., Remiddi, E. 1973. *Phys. Rev. A* 8:2266–71
Barbieri, R., Remiddi, E. 1976. *Phys. Lett. B* 65:258–62
Barbieri, R., Remiddi, E. 1978. *Nuc. Phys. B* 141:413–22.
Bearman, G. H., Mills, A. P. Jr. 1976. *J. Chem. Phys.* 65:1841–50
Beers, R. H., Hughes, V. W. 1968. *Bull. Am. Phys. Soc.* 13:633
Berends, F. A. 1965. *Phys. Lett.* 16:178–81
Berestetski, V. B. 1949. *J. Exp. Theor. Phys. USSR* 19:1130.
Berestetski, V. B., Landau, L. 1949. *J. Exp. Theor. Phys. USSR* 19:673
Berko, S. 1967. *Positron Annihilation*, pp. 61–79. New York: Academic
Berko, S. 1977. *Compton Scattering*, pp. 273–322. London: McGraw-Hill

Berko, S., Haghgooie, M., Mader, J. J. 1977. *Phys. Lett. A* 63:335–38
Berko, S., Canter, K. F., Mills, A. P. Jr. 1979. *Progress in Atomic Spectroscopy, Part B*, pp. 1427–1452. New York: Plenum
Berko, S. 1979a. *Proc. 5th Int. Conf. on Positron Annihilation*, pp. 65–87. Sendai, Japan: The Japan Institute of Metals
Berko, S. 1979b. *Electrons in Disordered Metals and at Metallic Surfaces*, pp. 239–91. New York: Plenum
Bertolotti, M., Sibilia, C. 1979. *Appl. Phys.* 19:127–30.
Bethe, H. A., Brown, L. M., Stehn, J. R. 1950. *Phys. Rev.* 77:370–74
Bethe, H. A., Salpeter, E. E. 1957. *Quantum Mechanics of One- and Two-Electron Atoms*, New York: Academic
Bisi, A., Fiorentini, A., Gatti, E., Zappa, L. 1962. *Phys. Rev.* 128:2195–99.
Bleuler, E., Bradt, H. L. 1948. *Phys. Rev.* 73:1398
Bodwin, G. T., Yennie, D. R. 1978. *Phys. Rep.* 43:267–303
Braicovich, L., De Michelis, B., Fasana, A. 1967. *Phys. Rev.* 164:1360–66
Brandt, W., Paulin, R. 1968. *Phys. Rev. Lett.* 21:193–95

Brandt, W., Coussot, G., Paulin, R. 1969. *Phys. Rev. Lett.* 23:522–24

Brandt, W. 1976. *Advances in Chemistry Series, No. 158; Radiation Effects on Solid Surfaces*

Brandt, W., Paulin, R. 1977. *Phys. Rev. B* 15:2511–18

Breit, G. 1929. *Phys. Rev.* 34:553–73

Buchmuller, W., Remiddi, E. 1980. *Nucl. Phys. B* 162:250–70

Bussard, R. W., Ramaty, R., Drachman, R. J. 1979. *Astrophys. J.* 228:928–34

Canter, K. F., Coleman, P. G., Griffith, T. C., Heyland, G. R. 1972. *J. Phys. B* 5:L167

Canter, K. F., Mills, A. P., Berko, S. 1974. *Phys. Rev. Lett.* 33:7–10

Canter, K. F., Mills, A. P., Berko, S. 1975. *Phys. Rev. Lett.* 34:177–80

Canter, K. F., Clark, B. O., Rosenberg, I. J. 1978. *Phys. Lett. A* 65:301–3

Carr, S., Sutherland, P. 1978. *Astrophys. Space Sci.* 58:83–88

Caswell, W. E., Lepage, G. P., Sapirstein, J. 1977. *Phys. Rev. Lett.* 38:488–91

Caswell, W. E., Lepage, G. P. 1978. *Phys. Rev. A* 18:810–19

Caswell, W. E., Lepage, G. P. 1979. *Phys. Rev. A* 20:36–43

Cherry, W. 1958. PhD thesis, Princeton University, N. J.

Coleman, P. G., Griffith, T. C. 1973. *J. Phys. B: 6:2155–61*

Correia, J. H., Burnham, C. A., Chester, D. A., Elmalch, D. R., Alpert, N. M., Brownell, G. L. 1979. *Proc. 5th Int. Conf. on Positron Annihilation*, pp. 391–401. Sendai, Japan: The Japan Institute of Metals

Costello, D. G., Groce, D. E., Herring, D. F., McGowan, J. W. 1971. *Phys. Rev. B* 5:1433–36

Cung, V. K., Fulton, T., Repko, W. W., Schnitzler, D., 1976. *Ann. Phys. NY* 96:261–85

Cung, V. K., Devoto, A., Fulton, T., Repko, W. W. 1978. *Nuovo Cimento* 43A:643–57

Cung, V. K., Devoto, A., Fulton, T., Repko, W. W. 1979. *Phys. Rev. A* 19:1886–92

Curry, S. M., Schawlow, A. L. 1971. *Phys. Lett. A* 37:5–6

Curry, S. M. 1972. PhD thesis, Stanford University, California

Curry, S. M. 1973. *Phys. Rev. A* 7:447–50

Dale, J. M., Hulett, L. D., Pendyala, S. 1980. *Surface and Interface Analysis*. To be published

De Benedetti, S., Corben, H. C. 1954. *Ann. Rev. Nucl. Sci.* 4:191–218

Deutsch, M. 1951. *Phys. Rev.* 82:455–56; 83:866–67

Deutsch, M., Dulit, E. 1951. *Phys. Rev.* 84:601–2

Deutsch, M., Brown, S. C. 1952. *Phys. Rev.* 85:1047–48

Deutsch, M., Berko, S. 1965. $\alpha\beta\gamma$-*Ray Spectroscopy*, Vol. 2, pp. 1583–98. Amsterdam: North-Holland

Deutsch, M. 1975. *Adventures in Experimental Physics*, Vol. 4, pp. 64–81. Princeton, NJ: World Science Education

Dick, L., Feuvrais, L., Madansky, L., Telegdi, V. L. 1963. *Phys. Lett.* 3:326–29

Dirac, P. A. M. 1930. *Proc. Cambridge Philos. Soc.* 26:361–75

Douglas, M. 1975. *Phys. Rev. A* 11:1527–38

Doyama, M. 1979. *Proc. 5th Int. Conf. on Positron Annihilation*, pp. 13–30. Sendai, Japan: The Japan Institute of Metals

Drell, S. 1979. *Themes in Contemporary Physics*, pp. 3–16. Amsterdam: North-Holland

Drisko, R. M. 1956. *Phys. Rev.* 102:1542–44

Egan, P. O., Frieze, W. E., Hughes, V. W., Yam, M. H. 1975. *Phys. Lett. A* 54:412–14

Egan, P. O., Hughes, V. W., Yam, M. H. 1977. *Phys. Rev. A* 15:251–60

Evans, W. D. 1980. *Bull. Am. Phys. Soc.* 25:535

Fairbank, W. M., Witteborn, F. C., Madey, J. M. J., Lockhart, J. M. 1974. *Gravitazione Sperimentale*, pp. 310–30. New York: Academic

Faraci, G., Pennisi, A. R. 1980. *Nuovo Cimento* 55B:257–63

Fermi, E. 1930. *Z. Phys.* 60:320–33

Ferrante, G. 1968. *Phys. Rev.* 170:76–80.

Ferrell, R. A. 1951. *Phys. Rev.* 84:858–859

Ford, G. W., Sander, L. M., Witten, T. A. 1976. *Phys. Rev. Lett.* 36:1269–72

Frost, A. A., Inokuti, M., Lowe, J. P. 1964. *J. Chem. Phys.* 41:482–89

Fulton, T., Martin, P. C. 1954. *Phys. Rev.* 95:811–22

Fulton, T., Owen, D. A., Repko, W. W. 1971. *Phys. Rev. A* 4:1802–11.

Gell-Mann, M., Low, F. 1951. *Phys. Rev.* 84:350–54

Gerber, G., Newman, D., Rich, A., Sweetman, E. 1977. *Phys. Rev. D* 15:1189–93

Gidley, D. W., Marko, K. A., Rich, A. 1976a. *Phys. Rev. Lett.* 36:395–97

Gidley, D. W., Zitzewitz, P. W., Marko, K. A., Rich, A. 1976b. *Phys. Rev. Lett.* 37:729–32.

Gidley, D. W., Rich, A., Zitzewitz, P. W., Paul, D. A. L. 1978. *Phys. Rev. Lett.* 40:737–40

Gidley, D. W., Zitzewitz, P. W. 1978. *Phys. Lett. A* 69:97–99.

Gilleland, J. R., Rich, A. 1972. *Phys. Rev. A* 5:38–49

Goldanskii, V. I., Firsov, V. G. 1971. *Ann. Rev. Phys. Chem.* 22:209–58

Goldsmith, J. E. M., Weber, E. W., Hänsch, T. W. 1978. *Phys. Rev. Lett.* 41:1525–28

Good, M. L. 1961. *Phys. Rev.* 121:311–13.

Greenberger, A., Mills, A. P. Jr., Thompson, A., Berko, S. 1970. *Phys. Lett. A* 32:72–73.
Griffith, T. C., Heyland, G. R. 1978. *Phys. Rep. C* 39:171–77
Griffith, T. C., Heyland, G. R., Lines, K. S., Twomey, T. R. 1978. *Phys. Lett. A* 69:169–71
Grisaru, M. T., Pendleton, H. N., Petrasso, R. 1973. *Ann. Phys. NY* 79:518–41.
Grodzins, L. 1959. *Prog. Nucl. Phys.* 7:165–241
Gross, F. 1969. *Phys. Rev.* 186:1448–62
Grotch, H., Kashuba, R. 1973. *Phys. Rev. A* 7:78–84
Hanna, R. C. 1948. *Nature* 162:332
Hänsch, T. W. 1977. *Phys. Today* (May):34–43
Harriman, J. M. 1956. *Phys. Rev.* 101:594–98
Harris, I., Brown, L. M. 1957. *Phys. Rev.* 105:1656–61
Ho, Y. K. 1979. *Phys. Rev. A* 19:2347–52
Hodges, C. H., Stott, M. J. 1973. *Phys. Rev. B* 7:73–79
Hughes, V. W., Marder, S., Wu, C. S. 1955. *Phys. Rev.* 98:1840–48
Hughes, V. W. 1972. *Atomic Physics*, Vol. 3. New York: Plenum
Hughes, V. W. 1973. *Physics 1973, Plenar v. Physikertag*, Vol. 37, pp. 123–35. Weinheim: Physik Verlag
Karplus, R., Klein, A. 1952. *Phys. Rev.* 87:848–58
Kasday, L. R., Ullman, J. D., Wu, C. S. 1975. *Nuovo Cimento B* 25:633–60
Kendall, H. W. 1954. *The first excited state of positronium*, PhD thesis, MIT, Cambridge, Mass.
Kinoshita, T. 1978a. *Proc. 19th Int. Conf. High Energy Physics*, pp. 571–77. Tokyo, Japan: Physical Society of Japan
Kinoshita, T. 1978b. *New Frontiers in High Energy Physics*, pp. 127–43. New York: Plenum
Kolos, W., Roothaan, C. C. J., Sack, R. A. 1960. *Rev. Mod. Phys.* 32:178–79
Lautrup, B. E., Peterman, A., de Rafael, E. 1972. *Phys. Rep.* 3:193–260
Lepage, G. P. 1977. *Phys. Rev. A* 16:863–76
Letokhov, V. S. 1974. *Phys. Lett. A* 49:275–76
Letokhov, V. S., Minogin, V. G. 1976. *Soviet Phys. JETP* 44:70–77
Leventhal, M., MacCallum, C. J., Stang, P. D. 1978. *Astrophys. J.* 225:L11–L14
Lewis, M. L., Hughes, V. W. 1973. *Phys. Rev. A* 8:625–39
Lingfelter, R. E., Ramaty, R. 1978. *Phys. Today* 31:40–47
Lynn, K. G. 1979. *Phys. Rev. Lett.* 43:391–94; E43:803
Lynn, K. G., Welch, D. O. 1980. *Phys. Rev. July 1980*
Madansky, L. Rasetti, F. 1950. *Phys. Rev.* 79:397–98

Mani, H. S., Rich, A. 1971. *Phys. Rev. D* 4:122–27
Marko, K., Rich, A. 1974. *Phys. Rev. Lett.* 33:980–83
Massey, H. S. W. 1973. *Fundamental Interactions in Physics*, pp. 189–212. New York: Plenum
Massey, H. S. W. 1976. *Phys. Today* 29 (March):42–51
McCoyd, G. C. 1965. PhD thesis. St. John's University, New York
Mijnarends, P. E. 1979. *Positrons in Solids*, pp. 25–81. Berlin-Heidelberg: Springer
Mills, A. P. Jr., Berko, S. 1967. *Phys. Rev. Lett.* 18:420–25
Mills, A. P. Jr. 1974. *J. Chem. Phys.* 62:2646–59
Mills, A. P. Jr., Bearman, G. H. 1975. *Phys. Rev. Lett.* 34:246–50
Mills, A. P. Jr., Berko, S., Canter, K. F. 1975. *Phys. Rev. Lett.* 34:1541–44.
Mills, A. P. Jr. 1978. *Phys. Rev. Lett.* 41:1828–30
Mills, A. P. Jr., Platzman, P. M., Brown, B. L. 1978. *Phys. Rev. Lett.* 41:1076–79
Mills, A. P. Jr. 1979a. *Appl. Phys. Lett.* 35:427–29
Mills, A. P. Jr. 1979b. *Sol. State Commun.* 31:623–26
Mills, A. P. Jr., Pfeiffer, L. 1979. *Phys. Rev. Lett.* 43:1961–64
Mills, A. P. Jr., Murray, C. A. 1980. *Appl. Phys.* 21:323–25
Mogensen, O. E. 1974. *J. Chem. Phys.* 60:998–1004.
Mourino, M., Lobl, H., Paulin, R. 1979. *Phys. Lett. A* 71:106–9
Newman, D., Rich, A., Sweetman, E. 1978. *New Frontiers in High Energy Physics*, pp. 199–220. New York: Plenum
Nieminen, R. M., Hodges, C. H. 1976. *Solid State Commun.* 18:1115–18
Olsen, P. T., Williams, E. R. 1976. *Proc. 5th Int. Conf. on Atomic Masses and Fundamental Constants.* New York: Plenum
Ore, A. 1949. Universitet: Bergen Arbok, Naturvitenskapelig rekke No. 9
Ore, A., Powell, J. L. 1949. *Phys. Rev.* 75:1696–99
Owen, D. A. 1973. *Phys. Rev. Lett.* 30:887–88
Owen, D. A. 1977. *Phys. Rev. A* 16:452–56
Page, L. A., Heinberg, M. 1957. *Phys. Rev.* 106:1220–24.
Page, L. A. 1962. *Ann. Rev. Nucl. Sci.* 12:43–79
Paulin, R., Ambrosino, G. 1968. *J. Phys.* 29:263–70
Paulin, R. 1979. *Proc. 5th Int. Conf. on Positron Annihilation*, pp. 601–12. Sendai, Japan: The Japan Institute of Metals
Pendyala, S., McGowan, J. W. 1980. *J. Electron Spectrosc.* To be published

Pirenne, J. 1944. PhD thesis, University of Paris

Pirenne, J. 1946. *Arch. Sci. Phys. Nat.* 28:233

Pirenne, J. 1947. *Arch. Sci. Phys. Nat.* 29:121; 29:207; 29:265

Ramaty, R. 1980. *Bull. Am. Phys. Soc.* 25:535

Rosenberg, I. J., Weiss, A. H., Canter, K. F. 1979. *Proc. 5th Int. Conf. on Positron Annihilation.* Sendai, Japan: The Japan Institute of Metals

Rosenberg, I. J., Weiss, A. H., Canter, K. F. 1980a. *J. Vac. Sci. Technol.* 17:253–55

Rosenberg, I. J., Weiss, A. H., Canter, K. F. 1980b. *Phys. Rev. Lett.* 44:1139–42

Salpeter, E. E., Bethe, H. A. 1951. *Phys. Rev.* 84:1232–42

Samuel, M. A. 1974. *Phys. Rev. A* 10:1450–51.

Schiff, L. I. 1959. *Proc. Natl. Acad. Sci. USA* 45:69–80

Schwinberg, P. B., Van Dyck, R. S. Jr., Dehmelt, H. G. 1979. *Bull. Am. Phys. Soc.* 24:758.

Schwinger, J. 1951. *Proc. Natl. Acad. Sci. USA* 37:452–59

Schwinger, J. 1964. *J. Math. Phys.* 5:1606–8

Serednyakov, S. I., Sidorov, V. A., Skrinsky, A. N., Tumaikin, G. M., Shatunov, J. M. 1977. *Phys. Lett. B* 66:102–4

Stein, T. S., Kauppilla, W. E., Pol, V., Smart, J. H., Jesion, G. 1978. *Phys. Rev. A* 17:1600–8

Stroscio, M. A., Holt, J. M. 1974. *Phys. A* 10:749–55

Stroscio, M. A. 1975. *Phys. Rep.* 22C:215–77

Theriot, E. D. Jr., Beers, R. H., Hughes, V. W. 1967. *Phys. Rev. Lett.* 18:767–69

Van Dyck, R. S. Jr., Schwinberg, P. B., Dehmelt, H. G. 1977. *Phys. Rev. Lett.* 38:310–14

Van Dyck, R. S. Jr., Schwinberg, P. B., Dehmelt, H. G. 1979. *Bull. Am. Phys. Soc.* 24:758.

Varghese, S. L., Ensberg, E. S., Hughes, V. W., Lindgren, I. 1974. *Phys. Lett. A* 49:415–17

Weisberg, H. L. 1966. *Univ. Calif. Radiat. Lab. Rep. UCRL-16801*, pp. 1–13

Weisberg, H., Berko, S. 1967. *Phys. Rev.* 154:249–57

West, R. N. 1973. *Adv. Phys.* 22:263–383

Wheeler, J. A. 1946. *Ann. NY Acad. Sci.* 48:219–38

Wolfenstein, L., Ravenhall, D. G. 1952. *Phys. Rev.* 88:279–82

Yang, C. N. 1949. *Phys. Rev.* 77:242–45

Zitzewitz. P. W., Van House, J. C., Rich, A., Gidley, D. W. 1979. *Phys. Rev. Lett.* 43:1281–84

AUTHOR INDEX

(Names appearing in capital letters indicate authors of chapters in this volume)

583

CUMULATIVE INDEXES

CONTRIBUTING AUTHORS VOLUMES 21-30

596

CHAPTER TITLES, VOLUMES 21–30

PARTICLE INTERACTIONS AT HIGH ENERGIES

PARTICLE SPECTROSCOPY

The η and η^1 Particles in the Pseudoscalar Nonet	G. De Franceschi, A. Reale, G. Salvini	21:1–54
Kaonic and Other Exotic Atoms	R. Seki, C. E. Wiegand	25:241–81
The Particle Data Group: Growth and Operations—Eighteen Years of Particle Physics	A. H. Rosenfeld	25:555–98
The Physics of e^+e^- Collisions	R. F. Schwitters, K. Strauch	26:89–149
Psionic Matter	W. Chinowsky	27:393–464
Quarks	O. W. Greenberg	28:327–86
Charm and Beyond	T. Appelquist, R. M. Barnett, K. Lane	28:387–499
Light Hadronic Spectroscopy: Experimental and Quark Model Interpretations	S. D. Protopopescu, N. P. Samios	29:339–93
The Tau Lepton	M. L. Perl	30:299–335
Charmed Mesons Produced in e^+e^- Annihilation	G. Goldhaber, J. E. Wiss	30:337–81

PARTICLE THEORY

Evidence for Regge Poles and Hadron Collision Phenomena at High Energies	C. B. Chiu	22:255–316
Production Mechanisms of Two-to-Two Scattering Processes at Intermediate Energies	G. C. Fox, C. Quigg	23:219–313
Proton-Nucleus Scattering at Medium Energies	J. Saudinos, C. Wilkin	24:341–77
Gauge Theories of Weak Interactions (Circa 1973–74 C. E.)	M. A. B. Beg, A. Sirlin	24:379–449
The Parton Model	T.-M. Yan	26:199–238
Diffraction of Hadronic Waves	U. Amaldi, M. Jacob, G. Matthiae	26:385–456
Quarks	O. W. Greenberg	28:327–86
Charm and Beyond	T. Appelquist, R. M. Barnett, K. Lane	28:387–499
Light Hadronic Spectroscopy: Experimental and Quark Model Interpretations	S. D. Protopopescu, N. P. Samios	29:339–93

RADIATION EFFECTS

Large Scale Radiation-Induced Chemical Processing	V. T. Stannett, E. P. Stahel	21:397–416
Radiation Damage Mechanisms as Revealed through Electron Spin Resonance Spectroscopy	H. C. Box	22:355–82
Nuclear Applications in Art and Archeology	I. Perlman, F. Asaro, H. V. Michel	22:383–426
K-Shell Ionization in Heavy-Ion Collisions	W. E. Meyerhof, K. Taulbjerg	27:279–331
Synchrotron Radiation Research	H. Winick, A. Bienenstock	28:33–133

WEAK AND ELECTROMAGNETIC INTERACTIONS

Electromagnetic Transitions and Moments in Nuclei	S. Yoshida, L. Zamick	22:121–64
Perturbation of Nuclear Decay Rates	G. T. Emery	22:165–202
Atomic Structure Effects in Nuclear Events	M. S. Freedman	24:209–47
Gauge Theories of Weak Interactions (Circa 1973–74 C. E.)	M. A. B. Bég, A. Sirlin	24:379–449
CP Violation and K^0 Decays	K. Kleinknecht	26:1–50
The Weak Neutral Current and Its Effects in Stellar Collapse	D. Z. Freedman, D. N. Schramm, D. L. Tubbs	27:167–207
Neutrino Scattering and New-Particle Production	D. Cline, W. F. Fry	27:209–78
Hypernuclei	B. Povh	28:1–32